Perspectives in Neural Co

Singapore
Tokyo

Also in this series:

Adrian Shepherd
Second-Order Methods for Neural Networks
3-540-76100-4

Dimitris C. Dracopoulos
Evolutionary Learning Algorithms for Neural Adaptive Control
3-540-76161-6

John A. Bullinaria, David W. Glasspool and George Houghton (Eds)
4th Neural Computation and Psychology Workshop, London, 9–11 April 1997: Connectionist Representations
3-540-76208-6

Maria Marinaro and Roberto Tagliaferri (Eds)
Neural Nets - WIRN VIETRI-97
3-540-76157-8

Gustavo Deco and Dragan Obradovic
An Information-Theoretic Approach to Neural Computing
0-387-94666-7

Thomas Lindblad and Jason M. Kinser
Image Processing using Pulse-Coupled Neural Networks
3-540-76264-7

L. Niklasson, M. Bodén and T. Ziemke (Eds)
ICANN 98
3-540-76263-9

Maria Marinaro and Roberto Tagliaferri (Eds)
Neural Nets - WIRN VIETRI-98
1-85233-051-1

Amanda J.C. Sharkey (Ed.)
Combining Artificial Neural Nets
1-85233-004-X

Dirk Husmeier
Neural Networks for Conditional Probability Estimation
1-85233-095-3

Achilleas Zapranis and Apostolos-Paul Refenes
Principles of Neural Model Identification, Selection and Adequacy
1-85233-139-9

Mark Girolami
Self-Organising Neural Networks
1-85233-066-X

Walter J. Freeman

Neurodynamics: An Exploration in Mesoscopic Brain Dynamics

Springer

Professor Walter J. Freeman, MD
Department of Molecular and Cell Biology, University of California
Berkeley, CA 94720-3200 USA
http://sulcus.berkeley.edu/

Series Editor
J.G. Taylor, BA, BSc, MA, PhD, FlnstP
Centre for Neural Networks, Department of Mathematics, King's College, Strand, London WC2R 2LS, UK

ISBN 1-85233-616-1 Springer-Verlag London Berlin Heidelberg

British Library Cataloguing in Publication Data
Freeman, Walter J.
Neurodynamics : an exploration in mesoscopic brain dynamics. – (Perspectives in neural computing)
1.Neural networks (Neurobiology) 2.Brain
I.Title
612.8'2

ISBN 1852336161

Library of Congress Cataloging-in-Publication Data
Freeman, Walter J.
Neurodynamics : an exploration in mesoscopic brain dynamics / Walter J. Freeman.
p. cm. -- (Perspectives in neural computing)
Includes bibliographical references and index.
ISBN 1-85233-616-1 (acid-free paper)
1. Brain--Mathematical models. 2. Neural networks (Neurobiology) 3. Electroencephalography. 4. Cognition--Research--Methodology. I. Title. II. Series.

QP376 .F673 2000
573.8'619--dc21 99-058323

Typesetting: Ian Kingston Editorial Services, Nottingham
Printed and bound at The Cromwell Press, Trowbridge, Wiltshire
34/3830-543210 Printed on acid-free paper SPIN 10727044

Dedication

To my family: without which, nothing.

Acknowledgments

The research on which these reports are based was supported for nearly 40 years by the National Institute of Mental Health Grant MH06686 and for 10 years by the Office of Naval Research Grant N00014-90-J-4054. I am grateful to Mark D. Lenhart for his dedication in the arduous work of scanning and editing the reports from their original formats, and to Helen Cademartori for proof-reading so many equations and technical terms.

The cover shows a portion of an AM pattern as a contour plot prepared by John M. Barrie.

Contents

Prolog

The Need for Mesoscopic Models in Nonlinear Brain Dynamics

Brain systems operate on many levels of organization, each with its scales of time and space. Dynamics, the modeling of change, is applicable to every level, from the atomic to the molecular, and from macromolecular organelles to the neurons into which they are incorporated. In turn the neurons form populations, these form the subassemblies of brains, and so on up to embodied brains interacting purposively with the material, interpersonal, and politico-social environments. Each level is **macroscopic** to the one below it and **microscopic** to the one above it. Among the most difficult tasks that scientists face are those of conceiving and describing the exchanges between levels, seeing that the measures of time and distance are incommensurate, and that causal inference is far more ambiguous between levels than it is within levels, especially when the gap between levels is wide. In that case, which holds for the relation between neuron and brain, the best recourse is to conceive, identify and model intervening levels, which physicists call **mesoscopic** (Ingber 1992; Namba *et al.* 1992; Imry 1997). It is the task of this book to present data, models, and experimental techniques comprising a mesoscopic approach to brain dynamics.

The principal level of understanding brains at present is that of the neuron and its networks. Owing to rapid and continuous advances during the past century, the structural and functional properties of neurons and their local networks are well known in all their complexity (Koch and Laurent 1999). Sophisticated mathematical tools have been devised to model the dynamics of single neurons (Johnston and Wu 1995), their axons (Hodgkin 1964) and dendrites (Rall 1995), and of nerve nets in great variety (Traub and Miles 1991; Churchland and Sejnowski 1992), including feedforward and feedback topologies ranging in number from 2 to in excess of 10 000 (Bower and Beeman 1995). The biochemical machinery that supports these functions is equally well understood, including identification of a variety of molecular neurotransmitters and neuromodulators, the membrane organelles that release them and to which they attach in acting, the enzyme systems that make, degrade and recycle them, the ways in which neuromodulators can bring about long-term changes in the functions and structures of neurons, and the oxidative metabolism that provides their energy. Seeing that neurons are studied with microscopes and microelectrodes, for my purposes I will call this the microscopic level.

As for the macroscopic level, recent improvements in non-invasive functional brain imaging using PET, SPECT and fMRI to observe the spatial patterns of metabolism and blood flow, which are secondary to neuronal activity, in conjunction with various intentional behaviors (Pfurtscheller and Lopes da Silva 1988; Hayman 1992; Roland 1993) have brought remarkable advances in detection in the brains of awake volunteers of areas showing high levels of metabolic activity in conjunction with a variety of cognitive states and tasks. The resulting localizations provide a substantial advance over phrenology, but they give little insight into the neurodynamics that occasions the metabolic demands and supports the correlated behaviors. Brain imaging using spatiotemporal maps of the electric and magnetic fields of brains (electro- and magnetoencephalography – EEG and MEG) provide data that are much closer than metabolic studies to the time-scales of brain dynamics. Substantial advances have been made in modeling the global brain dynamics (Nunez 1981, 1995; Kelso 1995; Wright and Liley 1996; Tononi and Edelman 1998; Uhl 1999; Robinson *et al.* 1998), especially as manifested in the electromagnetic fields observed at the scalp, with emphasis on alpha (Liley *et al.* 1999), beta (von Stein *et al.* 1999) and gamma rhythms (Llinás and Ribary 1993; Joliot *et al.* 1994).

These models show the outlines of the forms that are likely to be taken by field theories describing the global cooperative interactions governing the functions of entire cerebral hemispheres through the formation of standing and traveling waves of large-scale neural activity. However, the theory that is required for interpreting the observed and simulated global patterns in terms of lower levels is incomplete, because the gulf between the local actions of neurons and the global construction of brain states cannot be crossed in one giant leap, but stepwise with empirical models and experimental data at an intervening mesoscopic level. Our situation is a bit similar to that of early 19th century physicists, when Newton's equations provided a grand theory of everything, but some local relations such as Oersted's and Faraday's laws were needed before Maxwell could synthesize his overarching equations. Those laws are examples of mesoscopic bridges.

Two kinds of approach are currently being undertaken from opposite sides of the gulf to be bridged in neuron-to-brain modeling. One abutment rests in data from single neurons, which are conceived to coordinate and even synchronize their firing patterns, so as to constitute sparsely connected neural nets and nerve cell assemblies. Examples include the demonstrations by Abeles (Abeles 1991; Vaadia *et al.* 1995) of precisely timed "synfire chains" and of repeated transitions between quasi-stationary states (Abeles *et al.* 1998; Vaadia *et al.* 1995); by Singer and Gray (1995) and Eckhorn (1991) of synchronized firing of visual cortical neurons indicative of "binding" of "feature detector" neurons to form percepts; by Georgeopolis *et al.* (1986) of multineuronal firing in motor cortex constituting "vector codes" for movement; by Miyashita (1993, 1995) and Amit (1995, 1998) of the maintenance of representations in local neuronal netlets by "reverberations" to form short term memories; and by Grossberg (1988, 1990) of "adaptive resonance" in multilayer neural networks that adjudicates the feedback comparison of images deriving from sensory input in a lower layer (F1) with those from prior inputs stored in a higher layer (F2).

The other bridge abutment is directed toward understanding the formation of neural ensembles with state variables representing pulse and wave densities that are continuous in time and space. Concepts to define and analyze these quantities are largely taken from theoretical physics. Examples include the concept of circular causality in the "enslaving" of microscopic particles (atoms or neurons) by the "order parameters" that the particles create (Haken 1983, 1999); local mean fields created by sparse and weak but widespread synaptic interactions (Schuster and Wagner 1990; Cooper 1995; Bear and Cooper 1998); description of background neural firing patterns using statistical mechanics (Wilson and Cowan 1973; Ingber 1995b, 1997) to account for the emergence of oscillations in the gamma range of the EEG in local domains that are variously described as "activity density functions" (Freeman 1975), "dissipative structures" (Prigogine 1980), "patches" (Calvin 1996), "bubbles" (Zhang 1991; Amari 1977; Taylor 1997), the dendritic "bath" (Bulsara, Maren and Schmera 1993), and "collective chaos" (Shibata and Kaneko 1998), which appears to differ from noise-driven "stochastic chaos" (Kaneko 1997) only in that in collective chaos the elements are governed by chaotic attractors, whereas in stochastic chaos the elements (neurons) are governed by point attractors (Skarda and Freeman 1987; Freeman 1988b; Chapter 14). Heavy reliance is placed on the EEG (Lopes da Silva 1991) measured simultaneously at multiple sites on the subdural cortical surface to provide evidence for cooperative activity on a millimeter scale (Bullock et al. 1995; Wright and Liley 1994) as well as over distances of centimeters (Freeman and van Dijk 1987; Bressler *et al.* 1993; Bressler 1995), and on models of oscillations particularly in the alpha (Liley *et al.* 1999; Wright, Sergejev and Liley 1994; Gevins *et al.* 1997) and gamma (Ingber 1995a; Kruse and Eckhorn 1996; Traub *et al.* 1996; Roelfsma *et al.* 1997; Basar 1980, 1998; Basar and Bullock 1989) ranges of the EEG spectrum. It is these concepts, however they are named and formalized, that make feasible the project of constructing local dynamic models as mesoscopic bridges, putting the theory into practice first to guide the design of experiments and then to analyze the newly acquired data.

Selection and Definition of Dynamic State Variables

For centuries brains have been described as dynamic systems, beginning with Descartes, who conceived the pineal as a kind of valve used by the soul to regulate the pumping of spiritual fluid into the muscles, an idea that was quickly disproved by Dutch and Italian physiologists using plethysmographs (Freeman 1997). This was followed by metaphors of clocks, telegraph and telephone systems, thermodynamics (which underlay Darwinian "nerve energy" and the Freudian "Id"), digital computers, and holographs (Pribram 1971). Yet brains are not "like" any artificial machine. If anything, they are "like" natural self-organizing processes such as stars and hurricanes. With the guidance and constraint of genes, they create and maintain their structures while exchanging matter and energy with their surrounds. They are unique in the way they move themselves through their personal spaces and incorporate facets of their environments for their own purposes, to flourish and reproduce their kind. It is these intrinsic processes of self-realization that must be described mathematically, which means to construct, solve, and evaluate sets of descriptive differential equations or their equivalent.

The inadequacy of machine metaphors stems from the difference between two distinctive uses of differential equations, one to build classical control systems, and the other to describe semi-autonomous systems. As control systems seen from the outside, brains take inputs in the form of stimuli and give outputs in the form of logically coherent responses. Neurobiologists usually begin not with whole brains but with the smallest part that suffices. For many purposes this is the neuron. They measure its input, such as a volley of afferent action potentials, and its output, such as its postsynaptic potential (PSP) and its action potentials that are controlled by its PSP. They use the ratio of output to input to specify the transform designated by an input–output (I–O) pair. They repeat this test under varying conditions, until there are no more surprises. The inductive collection of I–O pairs constitutes an experimental database, from which to generalize to a model. It is usually expressed in a differential equation that, when solved for the input and the initial conditions at the start of the input, yields a curve that can be fitted to the observed output. This equation describes an operator that transforms the input into the output. For example, a nerve impulse is transformed by an axodendritic synapse into an exponentially decaying PSP, and a model consisting of either a first-order linear differential equation or the cable equation suffices to describe this operation. From this simple beginning, which forms the foundation of modern cellular neurophysiology, they extend the method to distributed, time-varying and nonlinear models, such as the Hodgkin–Huxley equations with time- and voltage-dependent coefficients, and more recently to the field of neural networks, in which they elaborate to arrays of interconnected neurons with trains of impulses in an unlimited variety of network configurations.

This classical input–output approach has been enormously productive in studies of the dynamics of single neurons, when they are isolated by use of tissue cultures, slices, and anesthetics to suppress synaptic interactions. It is inadequate in studies of normal neural tissues, because it fails to explain how brains create spatiotemporal patterns of activity that precede the stimuli that animals have learned to expect. Modeling such systems with self-organizing capabilities and semi-autonomous background activity requires nonlinear equations with multiple feedback pathways. These sets of equations seldom have the possibility of analytic solutions, though solutions can be estimated by numerical integration with digital computers and by piecewise linear analytic approximations that are implemented with describing functions and root locus techniques.

Herein lies the necessity for collecting a different kind of I–O pair during recording and measurement from brains in animals that are active in the pursuit of their own goals. The electroencephalogram (EEG) is especially important, but in ways that have not been widely foreseen. A Task Force of the American Psychiatric Association some years ago compiled a list and assessment of then current uses of quantitative EEG ("qEEG") in psychiatry (Luchins 1990). The list included temporal spectral analysis, cross-correlation, multiple discriminant analysis, and visual imaging of color-coded brain activity (maps of averaged evoked potentials and EEGs recorded at the scalp). These are all linear techniques. The Task Force failed to perceive the uses of qEEG as databases for modeling nonlinear brain dynamics, which includes guiding and testing the construction of unstable dynamic models operating far from equilibrium and subject to rapid and repeated state transitions.

The first step in modeling a neuron or a mesoscopic brain system is to define its state in terms of those quantities that can be measured directly or indirectly, such as the membrane potentials and firing rates of selected neurons, or the field potentials and pulse densities of populations of neurons. These quantities are represented by the variables in the differential equations that are used to predict the forms of change in the system, so that a relation is established at the outset between the model and the dynamics that it represents. The next step is to formulate the I–O relation in terms of the state variables. Here "I–O" refers both to the "Input–Output" for classical systems, and also to the "Inner–Outer" for self-organizing systems. In classical dynamics, where the initial conditions and inputs of a system are under the control of the experimenter, the system can be modeled by cascaded feedforward equations that express causal chains. The relations of inputs and outputs are well defined by **linear causality**. But brains continually interact with their surrounding environments, and each brain state is at once an input for a motor action with its sensory consequences and an output consequent to a sensory stimulus. In other words, a self-organized brain state is an **operator** that produces a behavioral act that mediates the interaction of the body with its environment, by which the goals of the animal are achieved. Therefore a set of Inner–Outer pairs need not, and generally does not, reflect a linear causal relation between the two kinds of observation and measurement, but it does provide an assay of the nature of a putative relation between a brain and its world and the reliability of that relation as it is formulated in a model. Such feedback models, even at the primitive level of physiological homeostasis, are representations in the relation of **circular causality** (Freeman 1995).

The most important instance of circular causality in neurodynamics is the relation between a collection of neurons and the population that they form by virtue of their synaptic interactions. The microscopic contributions of the individuals support the formation of the mesoscopic state of the population, which Haken (1983) calls an "order parameter", because it shapes, regulates and constrains or "enslaves" the activity of the individuals into an ordered pattern with the degrees of freedom reduced far below the magnitude given by the number of otherwise autonomous members. Therefore, a neurodynamic model of brain activity must have three kinds of state variables: microscopic, representing the activities of the neurons as individual members of a population; mesoscopic, representing the activities of the neural ensembles comprising the parts of the brains; and macroscopic, representing the cognitive actions of the brains under study. The microscopic state variables are most commonly used in linear causal chains to represent the relations between stimulus intensities and the firing rates of sensory neurons, and between the firing rates of motor neurons and the strengths of muscle contractions. The definition of mesoscopic and macroscopic state variables requires consideration of circular causality, because the local mean fields that govern these states are created by synaptic interactions, in which each neuron transmits (in round numbers) to 10 000 others and receives from 10 000 others. The selections of the 10 000 from the million others within the radius of its dendritic tree, however, have not been shown to be invariant in sequential observations. Furthermore, the driving of sensory systems by stimuli and the measurement of muscle twitches, as in reflex responses, can be modeled by microscopic state variables, but goal-directed behaviors cannot be modeled without reference first to mesoscopic populations and then to the interactive self-organization of the assemblies into global, macroscopic order, because the intentional states that direct and shape the behaviors emerge from the interactive dynamics of numerous semi-autonomous populations of neurons.

Thus the immediate task of neurodynamics is to build models with sets of differential equations having mesoscopic state variables. These models can be used to explain how local masses of neurons of the perceptual systems can organize their activity when they are destabilized by microscopic sensory inputs. One of the key findings in support of this approach to neurodynamics is the value of the EEG as a means for estimating the magnitudes of the mesoscopic state variable of neural populations. Because the dendritic current manifested in the EEG is formed by summation in the volume conductor of the local areas of cortex, it is the best available assay of the local mean field intensities of cortical populations. Moreover, owing to the divergent–convergent topology of corticocortical connections, as distinct from the topographic mapping that characterizes cortical relations with sensory and motor systems, the surface EEG is by far the better measure of the output of a cortical population, whereas the activity of its individual neurons is the better measure of the cortical response to its input.

Spatial Dimensions for Models of Prototypical Brain Subsystems

Spatial patterns of the EEG were first observed at the pial surface of the cortex by John Lilly (1954), using 25 electrodes and techniques for "toposcopy" that were developed for displaying spatial patterns of alpha waves at the surface of the scalp by Walter (1953) and Livanov (1977) with 100 electrodes. The most remarkable results in this direction were achieved by DeMott (1970) using 400 silver electrodes spaced 0.25 mm apart in 16×25 arrays giving 4×7 mm windows onto the pial surface of the visual cortex of monkeys performing conditioned responses to visual stimuli. He used 400 amplifiers to drive 400 neon tubes, which he photographed at a frame rate of 270/sec (3.7 msec interval) to display graphs of the spatial patterns with 10 steps of intensity. He documented interesting changes in the spatial patterns with learning (DeMott 1970). Though he did not isolate and work explicitly with gamma band activity, he led the way in acquiring the data needed to pursue the cortical neurodynamics by which the spatial patterns are generated.

Verification of spatial EEG patterns requires spatial spectral analysis (Freeman 1978b; Freeman *et al.* 2000) in order to establish the spacing between electrodes (equivalent to using temporal spectral analysis to optimize the digitizing step in time series analysis) and prevent undersampling and aliasing (Gonzalez and Wintz 1977), but this procedure has not been used with the EEG as widely as it should be. Numerous investigators have reported finding gamma activity in human subdural and scalp EEG recordings (reviewed by Sheer 1989; Schippers 1990), but verification has been difficult owing to contamination with electromyographic (EMG) potentials having similar temporal spectral properties. Substantial improvements in signal identification techniques applied to EEG and MEG derivations have recently provided strong evidence for the correlation of brief bursts of gamma EEG in widely separated cortical areas during cognitive processing in perception and decision making by human volunteers (Tallon-Baudry *et al.* 1998; Müller *et al.* 1996; Rodriguez *et al.* 1999; Miltner *et al.* 1999), as well as correlations of activity at lower frequencies (Gevins *et al.* 1997; von Stein *et al.* 1999). The significance of the multichannel synchrony in the gamma range is thought to lie in manifesting neural operations by which "binding" of neural activity occurs between networks of neurons separated by relatively great distances in the forebrain in the cerebral cortex of either or both hemispheres (Milner 1974; von der Malsburg 1983; Engel *et al.* 1991; König and Schillen 1991; Singer and Gray 1995). Multiple channels for recording of impulse firing from groups of neurons are also used in local domains (Georgeopolis *et al.* 1986; Miyashita 1995; Wilson and McNaughton 1993). However, spatial sampling at multiple sites of recording followed by pairwise correlation and spectral coherence do not yield the data that are necessary for discerning spatial patterns, either at the millimeter scale of mesoscopic dynamics or at the centimeter scale of macroscopic dynamics. The distinction between spatial sampling and spatial mapping becomes most clear in pattern classification, when the assignment of membership depends on high-dimensional relations of the intensity at every point relative to the intensities at all other points. This is precisely the problem that must be addressed and solved by the sensory neocortices and by the olfactory bulb and cortex (Chapters 11 and 13).

My experiments were focused on deriving mesoscopic spatial patterns of EEG in the olfactory system in small laboratory animals. This is the simplest and phylogenetically oldest sensory system, and it is the preeminent sense for most mammals. Studies in comparative neuroanatomy (Herrick 1948) have indicated that olfaction was the most advanced in phylogenetic development of early vertebrates, and that it may have provided the prototype for cortical processing in all other sensory systems. Studies of behavioral dynamics leading to the formation of multisensory percepts called Gestalts (Koffka 1935; Köhler 1940) gave evidence that at some stage in the sequential transformations of sensory stimuli, all sensory-perceptual systems must yield activity patterns that can be combined. In other words, they must share the same "code", which indicates that an understanding of visual and auditory perception may well emerge from studies of olfaction (Freeman 1991c).

The crucial experiments were simple in conception, though complex in execution. Rabbits were implanted with an array of 64 electrodes in an 8×8 grid placed onto the surface of the olfactory bulb (Viana Di Prisco and Freeman 1985; Freeman and Davis 1990). They were trained to respond by licking to one odorant, such as amyl acetate accompanied by a reward (water), and merely to sniff in response to another odorant (such as butyric acid) as an unrewarded stimulus. On each trial a set of 64 EEG traces was recorded for 6 sec, which included a control period and a test period. Brief EEG segments lasting about 100 msec were selected during the times of inhalation of the background control air or either of the two odorants presented on randomly interspersed trials. Sets of several hundred of these EEG segments were analyzed for each animal. The membership of each of segment in the three odorant classes was known, as the Outer member of each I–O pair. The crucial question was, what aspect or aspects of the EEG segments, as the Inner member, would enable us to classify the segments correctly in respect to the antecedent odorant conditions?

The hypotheses were that, between the time of inhalation and the performance of a correct response, odorant information existed in the bulb as a pattern of neural activity, on the basis of which the animal made the discrimination, and that this information would be detectable in some as yet to be determined properties of the EEGs. The results showed that the information sought was indeed manifested in the EEGs. It was identified as a spatial pattern of amplitude modulation of the EEG oscillation in the high-frequency range of 20–80 Hz. We named this the "gamma" range, in analogy to the high end of the X-ray spectrum (Bressler and Freeman 1980). A common waveform was found in each segment, and the spatial pattern of its amplitude tended to have a reproducible shape with each inhalation of the background or a test odorant. In principle this form of information is like a frame in a black-and-white movie, in which the carrier wave is the light, and the 2D pattern is formed by the highs and lows of the amplitude modulation (**AM pattern**) of the light (the carrier wave).

The finding that the bulbar "code" for olfaction is spatial is not surprising. It was predicted by Adrian (1950) on the basis of his pioneering studies in the hedgehog. After all, the role of the receptors forming a sheet of neurons in the nasal cavity is to transform an incident chemical species into a spatial pattern of action potentials, much as a retinal pattern of light is transformed to a pattern of ganglion cell activity. This spatial pattern of receptor activity is transmitted to the bulb by unbranched axons that have some degree of topographic order, so that a new spatial pattern is injected into the bulb for each new odorant. Evidence for its existence has been presented by several groups using metabolic labeling techniques (e.g. Lancet *et al.* 1982) and genetic studies (Ressler *et al.* 1994; Wang *et al.* 1998). Nor is it surprising that the information should exist in an induced burst of activity accompanying each inhalation, which replaces the background activity during inhalation, because it is well known that detection of odorants happens with inhalation. But there are five aspects that are very surprising.

First, the information that serves to classify EEG segments with respect to odorants is distributed over the entire olfactory bulb with uniform density for every odorant. Unlike the single-cell activity that is driven by sensory input, it is not localized to particular mitral cells with dendrites in particular glomeruli. By inference every neuron participates in every discrimination, even if it does not fire, because a spatial pattern requires both dark and light. This fact is established by repeating the classification test many times while deleting randomly selected groups of channels. No channel is any more or less important than any other channel in effecting correct classifications (Figure IX in comments and Figure 13 in Chapter 13). Furthermore, numerous attempts to demonstrate odorant specificity in the discharge rates of action potentials from single neurons have failed. In view of the facts that the minimal number of channels for correct classification of EEG segments is 8 to 16, and that each segment reflects the activity of many thousands of neurons, it is likely that the percepts of odors from odorants are carried by populations of neurons, and they are not demonstrable in the activity of the handful of neurons that can be simultaneously accessed by multiple microelectrode recording. In brief, the reception of an odorant requires a sparse network of neurons, while the perception of an odor requires all the neurons in the bulb. EEG potentials are formed by the passage across the extraneuronal tissue resistance of the same dendritic current that determines the firing rates and therefore the outputs of the projection neurons (Freeman 1975, 1992a; Freeman and Baird 1987; Eeckman and Freeman 1991). That is, the recording of an EEG trace provides the best available measure of the output of a local neighborhood of the bulb, while the recording of cortical action potentials provides the best available measure of bulbar responses to sensory inputs.

Second, the bulbar AM patterns do not relate to the stimulus directly but instead to the meaning of the stimulus, as shown in several ways. The simplest way is to switch a reward between two odorants, so that the previously rewarded stimulus is no longer reinforced and vice versa. Both AM patterns change, though the two odorants do not. So also does the spatial pattern for the control segments in background air. So also do all pre-existing AM patterns change when a new odorant is added to the repertoire. The AM patterns for all odorants that can be discriminated by a subject change whenever there is a change in the odorant environment. Furthermore, when rabbits are trained serially to odorants A, B, C, D, and then back to A, the original pattern does not recur, but a new one appears (Freeman and Schneider 1982; Freeman 1991a). This property is to be expected in a true associative memory, in which every new entry is related to all previous entries by learning, because the context and significance are subject to continual and unpredictable change.

The point here is that brains do not process "information" in the commonly used sense of the word. They process meaning. When we scan a photograph or an abstract, we take in its import, not its number of pixels or bits. The regularities that we should seek and find in patterns of central neural activity have no immediate or direct relations to the patterns of sensory stimuli that induce the cortical activity but instead to the perceptions and goals of human and animal subjects. Far from being fed into our cortices in the way that we program and fill our computers, everything that we know and can remember has been constructed by the global self-organizing dynamics of activity within our brains.

Third, the carrier wave is aperiodic. The EEG does not show oscillations at single frequencies, but instead has waveforms that are erratic and unpredictable in each segment, nor is it ever the same in any two segments, irrespective of odorant condition. Yet the same waveform appears on all the channels, although of course at different amplitudes and phases. Moreover, during exhalation and prior to each inhalation there is persisting background activity, which is highly irregular, with its energy broadly distributed over the gamma range of the spectrum. Odorant segments in many instances show no increase and often a decrease in mean bulbar EEG amplitude from this basal state, yet they always display a remarkable similarity of waveform across channels. There is no possibility that this seemingly random waveform could occur by chance almost simultaneously on all channels. The possibility that the common activity might be imposed in this gamma spectral range onto the whole bulb by receptors or by centrifugal pathways from the forebrain and brainstem was ruled out by three findings. First, the temporal dispersions in the primary olfactory nerve and in the medial olfactory tract acted as low-pass filters that would prevent driving in the gamma range (Freeman 1974a). Second, the spatial patterns of phase of the oscillatory evoked potential on electrical stimulation of the primary olfactory nerve conformed to the direction and velocity of the axons, but the spatial patterns of phase in the EEG oscillations were compatible with the self-organizing dynamics within the bulb and not with external driving (Freeman *et al.* 1995). Third, the same aperiodic basal activity persists after the bulb, nucleus and prepyriform cortex have been surgically isolated as a unit from the receptors and the rest of the brain, though in that circumstance without evidence of modification with respect to changes induced by learning, hunger, thirst, and other conditions relating to intentional behavior (Becker and Freeman 1968).

These findings led to the conclusion that the mesoscopic activity of the olfactory system is chaotic (Freeman and Skarda 1985; Skarda and Freeman 1987). More generally, many semi-autonomous subsystems of brains are capable of generating controlled but locally unpredictable activity that looks like noise but is not. This is significant in two respects. The lesser is that it directs search for carrier waveforms that are aperiodic and not for oscillations at specific frequencies such as the alpha or the "40 Hz". The greater significance of chaos attaches to the property of chaotic systems that gives them an ability to jump suddenly and completely from one global activity pattern to another with minimal duration of transient "ringing", just as, for example, we jump from one word to the next in the rapid flow of speech and from one gait to another in walking, jogging or running. The EEG shows sequences of patterns, each carried by an aperiodic waveform manifesting a chaotic attractor, and not by a wave at a single frequency governed by a limit cycle attractor perturbed by noise, as previous EEG interpretations had suggested. Dynamicists realize that the positive Lyapunov exponents in high-dimensional chaotic systems manifest the capacity to create information as well as to destroy it. These opposing capabilities are at play in the creation of each new perceptual class in the process of learning by induction, including deletion of irrelevant details whereby membership is assigned to examples of the class in the process of generalization. Chaotic dynamics endows brains with the capacity to create new kinds of trials in the course of learning by trial and error, by which to generalize across inputs from equivalent sensory receptors.

Fourth, surgical, pharmacological, or cryogenic inactivation of the main pathway from the bulb to the prepyriform cortex causes the cortical activity to go silent both in respect to action potentials and to EEGs, and the bulbar activity to display nearly periodic oscillation with each inhalation (Chapter 5; Freeman 1975; Gray and Skinner 1988). On recovery from the interruption of the pathway, the normal action potentials and EEGs return. These findings demonstrate that the background unit and EEG activity is the semi-autonomous global property of the entire olfactory system. The mechanism includes interaction between the bulb, anterior olfactory nucleus (AON) and prepyriform cortex. Action potentials are transmitted from the bulb to the nucleus and cortex by surface paths and in the reverse direction by deep paths.

The findings mean that the chaotic activity is produced by the olfactory bulb, nucleus and cortex interacting. It is not the local sum of the noise of idle neurons, nor is it the action of any one part driving the others. It is the controlled and directed product of the whole system. Further, the background state is not a steady state or equilibrium that is perturbed by noise. It is a global chaotic state that keeps the semi-autonomous and locally self-regulating olfactory system in constant readiness to jump to any desired perceptual state at any time, as well as to respond to limbic inputs that determine in advance what that desired state may be (Kay and Freeman 1998). Strong evidence supporting the state transition hypothesis is provided by the pattern of EEG phase at the common instantaneous frequency in successive bursts of gamma activity. Independently of the AM pattern of amplitude each burst has a two-dimensional phase gradient in the form of a cone (Figure I). The radially symmetric pattern and circular isophase contours of the "phase cone" are centered on a point of extreme phase, the location of which varies randomly from each burst to the next. The sign of the extreme (maximum or minimum lag) also varies randomly, showing that the pattern is not due to a pacemaker in the bulb or some other part of the brain such as the thalamus. The phase cone can be attributed to a state transition, because in a distributed medium the transition does not occur everywhere at the same instant, but it begins at a site of nucleation and spreads radially at the group velocity of the medium. In the case of the neuropil the limiting velocity is provided by axons extending parallel to the pial surface. Their small size and low conduction velocity determines the phase gradient in m/sec, though it varies inversely with the peak frequency of the gamma bursts in radians/mm (Freeman and Baird 1987). This hypothesis accounts for the random variation in location and sign of the apex of the phase cone (Freeman 1990). The interdependent function of the bulb and prepyriform is often not demonstrable by covariance of their time series, showing that linear techniques of correlation and spectral analyses are deficient. Thereby, the olfactory system can be viewed as a prototype of the dynamics of other sensory systems. The determination of what kind of patterns the olfactory system makes, and how it makes them, is a necessary prelude to analyzing the higher order processes by which they are assembled into Gestalts in the larger perceptual apparatus of the forebrain. This led to the successful search for similar AM patterns (Barrie *et al.* 1996 (Ch. 13 in this volume); Ohl *et al.* 1999) and phase cones (Freeman *et al.* 1995) in the visual, auditory, and somatosensory cortices.

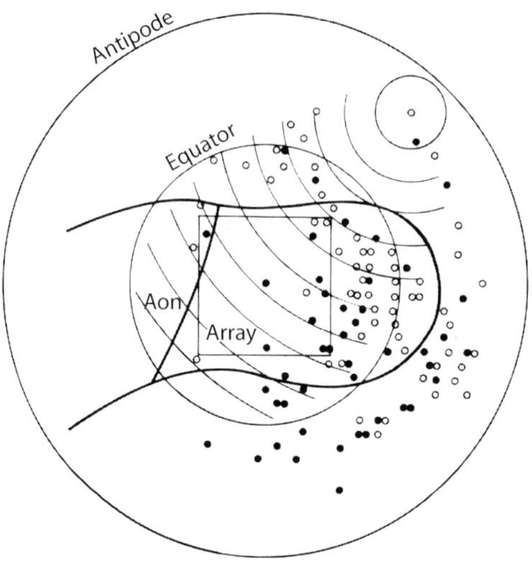

Figure I Phase patterns measured at the surface of the olfactory bulb were optimally fitted with a cone that incorporated on average >65% of the variance of the 64 measurements of phase with respect to the phase of the spatial ensemble average. Example from 60 bursts of the locations of phase maxima (solid dots) and minima (open dots) have been projected into the flattened surface of the bulb of the rabbit. The 64 phase values from each burst after low-pass spatial filtering of the real and imaginary parts from the amplitude and phase values of the dominant component, were fitted with a cone in spherical coordinates. The sketch is a projection of the outline of the bulb as it would appear on looking through the left bulb onto the array an the lateral surface of the bulb. A representative set of isophase contours is shown at intervals of 0.25 radians/mm. The locations of the apices of the cone of the sphere (2.5 mm in radius) at the center of the array to the antipodes, which takes the form of the largest circle of the display. The square is the outline of the 8×8 electrode array. Many of the phase values could be fitted almost as well with a plane, the direction of which varied from burst to burst, which indicated that the degree of confidence concerning the location of the conic apex diminished with distance from the array. The standard error of each point was estimated to be twice the radius of the dots. The apex was rarely extrapolated outside the bulb for 3/60 bursts; those values could be interpreted as replaceable by their antipodes. The absence of extrapolations into the anterior olfactory nucleus (AON) is expected, because the bulbar surface is interrupted there by the connections to and from the bulb (the "punctured sphere" of Rall and Shepherd 1968). The same patterns of phase cones have been found (Freeman *et al.* 1995) in the neocortical EEG described in Chapter 13. From Freeman and Baird (1987, Figure 10, p. 403).

The significance of the conic phase gradient is that it can provide a spatial boundary condition for cooperative cortical states. The bulb has a histological boundary, outside of which its local interactive activity cannot extend. Although cytoarchitectonic differences exemplified in Brodmann's areas serve to distinguish cortices, the neuropil of the neocortex is continuous across and between the areas, as shown by the phenomenon of spreading depression. Simultaneous EEG recordings from the several sensory areas show that they do not have the same waveforms, as the recordings from within each area do, so there must be boundaries between them. The aperiodic bursts result from cooperative interactions among millions of neurons, but cooperativity must weaken with phase dispersion. The radial phase gradient may attenuate action at a distance, so the half-power radius ($\pm$ cosine 45°) can serve to mark the boundary of a cooperative domain in the neocortex. The diameter can be calculated from the peak frequency of the gamma burst and the gradient. Modal diameter values of 5 to 10 mm (Freeman *et al.* 1995) indicate that AM patterns are far larger than the sizes of cortical columns of various types, which are in the submillimeter range, corresponding to the glomeruli in the olfactory bulb averaging 125 microns in diameter (Freeman 1975), so that a typical AM pattern would involve at least 10^3 columns, each with at least 10^4 neurons, giving minimally 10^7 neurons involved in a cortical mesoscopic event in the gamma range.

Fifth, AM pattern formation is controlled by neurohormones, and the process can be modified or blocked by chemical manipulations. Modification in the spatial EEG patterns in the olfactory bulb with conditioning requires the participation of other parts of the brain by means of centrifugal pathways (Gray *et al.* 1986). Rabbits with an array implanted onto one bulb to measure EEG patterns were injected with a beta blocker, propanolol, through a cannula in each bulb so as to prevent the action of norepinephrine in the bulb, which is normally released into the bulb by the locus coeruleus following an unconditioned stimulus that accompanies an odorant. Normally a new AM pattern of EEG activity appears as an animal acquires a conditioned response to the odorant. But when the bulbs were both perfused with propanolol, the spatial patterns of the EEG did not change during the training, nor did the rabbits acquire a conditioned response to the odorant. When the rabbits were given an odorant without reinforcement after injection of norepinephrine directly into the bulb, there were dramatic changes in spatial pattern. Such changes did not occur when no reinforcement is presented, or when it was presented in conjunction with a visual or auditory cue but not with an odorant, at least in the first several hundred milliseconds after the onset of the cues.

These findings give important clues to the roles of the several centrifugal pathways into the olfactory system. In keeping with the conception that local brain subsystems regulate themselves semi-autonomously through homeostatic feedback under the most general modulating bias controls of set points, it appears likely that the modulating pathways do not carry detailed information in the form of specific and structured activity patterns. Instead, they carry global commands, such as “turn on”, “turn off”, “imprint”, “habituate”, “bifurcate”, and so forth. Clearly the connections to and from the forebrain into the olfactory system are essential for the operations by which olfaction is integrated into the genesis and regulation of intentional behavior, but that they do not carry site-specific information. The modulations are implemented by well-known transmitters and neuromodulators, including the monoamines, acetylcholine, histamine, and various neuropeptides operating on the olfactory system in common with the rest of the forebrain.

Dynamic Models of Semi-autonomous Neural Populations

Perhaps the most critical question that has been addressed to the mechanisms of gamma oscillations is whether to model the dynamics at the microscopic or mesoscopic level. It is amply documented that single neurons not depending on feedback can and do generate aperiodic as well as periodic oscillations (Aihara *et al.* 1990; Hayashi and Ishizuka 1995; Fox 1997; Dogaru and Chua 1998; Ghose and Freeman 1997), often in the gamma range, and cellular properties underlying the induction of and recovery from action potentials have been invoked to explain gamma activity, particularly in thalamocortical circuits (Llinás 1990; Vonkrosigk *et al.* 1993; Pedroarena and Llinás 1997). However, it does not necessarily follow that single neurons drive gamma activity. The following distinction is clear. Either single neurons rely on calcium- or potassium-dependent hyperpolarizing afterpotentials to serve as pacemakers to drive excitatory populations through spatially divergent pathways, or the gamma oscillations arise by mass action between excitatory and inhibitory neural populations. The microscopic models have been used explain the spatial patterns of coherence in gamma activity over broad spectral domains, in particular the most challenging recent finding of "zero lag" correlation between the gamma oscillations of neurons separated by millimeter distances (Roelfsema *et al.* 1997), between which the axonal transmission times are expected to cause phase shifts between oscillations. Usher *et al.* (1993) and Schillen and König (1994) have modeled this phenomenon by assuming that the feedback delay within each target matches the transmission delay between targets in an excitatory feedback network. Traub *et al.* (1996) have overcome some of the rigidity of that model by invoking doublet firing of single neurons, which they show is enhanced in states of high amplitude gamma. The tendency of neurons embedded in excitatory feedback networks to fire twice on impulse input is well known (Freeman 1968b, 1975; Nicoll 1971), and the increase in gain of the excitatory loop with learning has been shown to increase gamma amplitude substantially (Freeman 1979b, 1987b), so these models do not rule out the possibility that gamma oscillations are mesoscopic, nor do they explain the otherwise puzzling quarter-cycle phase lag of the inhibitory neural firing with respect to the excitatory firing, which is essential for a mesoscopic explanation (Freeman 1975; Eeckman and Freeman 1990). As already noted, the random variations in the sign and the spatial location of the apex of the phase cone are not consistent with external driving. The persistence of conic phase gradients through the ensuing bursts appears to manifest the sparseness of cortical connectivity, which, despite the immense number of synapses and the distances of connection, is not sufficiently high in density to converge the oscillations to zero time lag. However, the phase differences in the conic patterns are predominantly less than a quarter cycle, so there is a strong basis for finding correlations between traces from pairs of sites. When the mean phase difference between neurons sampled at points in areas sustaining AM patterns with common oscillation is computed over multiple trials in time, as is commonly done in processing single unit data, the correlation value must often be significantly positive, and when it is it must be maximal at zero time lag within the limits of experimental error, even though the lag of maximal correlation is not zero on any single trial.

To the extent that cortical transmissions undergo spatial integral transforms, the correlated variance in the gamma activity constitutes the significant cortical output. This feature gives reason to suppose that a high degree of gamma coherence can be expected at zero lag in macroscopic activity over large brain areas, perhaps an entire hemisphere. Recent reports (Müller *et al.* 1996; Tallon-Baudry 1998; Rodriguez *et al.* 1999; Miltner *et al.* 1999) have shown that patterns of gamma activity in scalp recordings are closely related to visual attention and cognition. The findings suggest that the distance from the cortex to the recording sites outside the head, which is imposed by the intervening scalp and skull, may work to improve access by observers to the spatially coherent activity of global AM patterns, because the spatial summation that blurs the local activity reveals the shared components of the global activity. Everyone knows that if you get too close to a newsprint picture, you see only meaningless dots, and that an appropriate distance reveals the image. However, the global macroscopic dynamics that would coordinate collections of mesoscopic domains remains obscure.

Two criteria are needed to test and evaluate mesoscopic models of sensory neurodynamics. One criterion is a measure of how well the state variables of a model conform to the spatiotemporal patterns of EEGs and pulse densities during background and induced activity, as well as activity that is evoked by direct electrical stimulation of the brain (impulse responses). The other criterion measures the capacity of a model for classification of inputs into categories that have been instilled through reinforcement learning, because that is primarily what the sensory systems do. The main tool by which to build, adapt and test a model for the quality of its waveforms is to construct digital adaptive filters and to use nonlinear regression to fit their impulse responses to evoked potentials. The best tool by which to evaluate classification performance is multivariate statistics, in which each spatial pattern is expressed by a 1×64 column vector given by the 64 amplitudes of the common carrier wave that specifies a point in 64-space. Each class of responses to stimuli has the form of a cluster of points that is measured by a center of gravity (centroid) and radius (standard deviation). The probability of membership in a class for each sample is given by a *t*-value in the number of standard deviations to the nearest centroid.

The unexpected EEG results listed above supported several key decisions about the forms of proposed dynamical models to be described in the following chapters of this book. First, the proper element for a mesoscopic model is the local population and not networks of neurons. This provides an enormous simplification, because superposition studies of cortical evoked potentials (impulse responses) in small-signal, near-linear, physiological ranges (Biedenbach and Freeman 1965) have shown that the masses of dendrites of neurons in a local neighborhood can be modeled as a linear integrator. It is the operation of axons that is spectacularly nonlinear, particularly at the trigger zones, where in populations the density of dendritic currents is transformed to the density of pulses. That relationship is linear but time-varying for single neurons (Freeman 1975), whereas for populations it has the form of a static nonlinear curve now called a sigmoid that expresses the bilateral saturation of pulse density output. At the inhibitory extreme the axonal thresholds limit the expression of neuronal output as inhibition increases, and at the excitatory extreme the relative refractory periods limit the firing of neurons in the population as excitation increases. The sigmoid is a static nonlinearity, because at rates <1 pulse/sec most cortical neurons spend $>99.9\%$ of their lifetimes just below threshold, and, when they do fire a pulse, it is largely at random with respect to the firing times of their neighbors in their population. Except when they are driven by stimulation, cortical neurons give very low correlation coefficients (often ~0.01) of their firing times with those of their neighbors (Abeles 1991; Kreiter *et al.* 1989; Martignon *et al.* 1995), indicating that the fraction of the variance of the activity of the single neuron that is covariant with the neighborhood is also very low (1 part in 10 000) (Freeman 1975; Wilson and McNaughton 1993). The time-varying refractory periods that follow action potentials are smoothed by spatial ensemble averaging. Most of the complexities of the single neuron therefore remain below at the hierarchical level of the neuron.

Second, measurements of the spectra of olfactory EEGs show broad peaks in the gamma range, which serve to predict the waveforms of impulse responses, that is, cortical evoked potentials and post-stimulus time histograms to which curves are fitted that are solutions to the equations. Families of impulse responses at various stimulus intensities are fitted by varying the parameters in the models in piecewise linearization. A key result is that the time constants in the models representing the passive membrane integration and the cable and conduction delays in populations are shown not to change when the waveforms are changed by increasing stimulus intensity, so the time parameters need not be modified. The parameters that are changed represent the synaptic coupling strengths in dendrites and the sensitivities of trigger zones in axons. This step, too, provides enormous simplification, because once the time constants are determined in the open loop state, they need not be modified and adapted during curve fitting, which halves the number of free parameters.

Third, oscillations in the gamma range and the steady state background activity that supports them with an excitatory bias are properties of neuron populations, not of the single neurons acting as pacemakers. Biases arise when excitatory populations excite each other in positive feedback and generate background noise (Chapter 8), and oscillations arise when an excitatory population is coupled with an inhibitory population to form a negative feedback loop. The oscillations are periodic in a model of the bulb standing alone to represent the surgically isolated bulb. Chaos in the model arises when the bulbar populations are interconnected with the AON and prepyriform cortex, so that each excites the other. Each has its own characteristic frequency, but they differ and cannot agree. Neither can escape the other, so that stable aperiodic oscillation persists. If the three parts are disconnected, the chaos in the model disappears, as it does in experimental animals, when the bulb, AON, and cortex are surgically disconnected. Within linear domains the cortical models can be solved by use of the Laplace transform for families of measurements of impulse responses, and the state-dependent variations in the coupling coefficients can be represented by means of root locus techniques with supplementary describing functions relating the changes in coefficients to behavioral and pharmacological manipulations.

Fourth, the simplicity of the central olfactory cytoarchitecture and the results of spatial pattern analysis of EEGs led to modeling the dynamics in two spatial dimensions. The lack of geometric forms in the spatial patterns of amplitude modulation (AM patterns) led away from partial differential equations and instead to integrodifferential equations, which in digital embodiments could be represented by arrays of coupled ordinary difference equations. The bulb is simulated by constructing an array of local oscillators that are interconnected between the excitatory elements by simulated synapses, so that they excite each other. These connections couple the oscillators into a layer, and they ensure that the whole array oscillates with a common waveform. The inhibitory populations are likewise interconnected by simulated synapses to give mutual inhibition, which facilitates the emergence of amplitude differences in the common waveform, that is, spatial amplitude modulation giving AM patterns. There is one pair of input and output connections for each local oscillator. The mutually excitatory synapses between the oscillators are selectively strengthened during learning under simulated reinforcement by presentation of examples of a class of stimuli, leading to the formation of a Hebbian nerve cell assembly (Hebb 1949) that consists of groups of local oscillators that have been co-excited by the stimuli. A global command is required in the model to "form an assembly", which simulates the effect of the release of norepinephrine into the bulb by the locus coeruleus in response to reinforcement by an unconditioned stimulus. Two or more classes of stimuli are formed in sequential "training" sessions, leading to formation of multiple "nerve cell assemblies" that are separated by simulated "habituation".

The resulting model called "KIII" is an array of 64 coupled oscillators that interact with lumped oscillators representing the AON and prepyriform cortex through distributed delay lines. The array simulating the bulb generates a reproducible spatial pattern of amplitude of a common chaotic waveform, whenever an example of a learned class of stimuli is presented to the model. That pattern serves by Euclidean distance measures in 64-space to identify and classify the stimulus presented. The olfactory system and its model have certain preferred patterns of activity, any one of which it naturally falls into, when given the opportunity. That opportunity is provided by the presence of any stimulus in a class that it has learned to respond to, whether or not the example was included in the original training set. A stimulus of this class provides the input and starting conditions that are needed to place the system into the basin of attraction, so named in analogy to a bowl in which a ball will roll to the bottom and stay there. There is a learned chaotic attractor and its attendant basin of attraction for each class of odorant that an animal can discriminate. Its basin defines the range of generalization for identification of a stimulus of that class.

The KIII dynamical model classifies industrial objects such as bolts and screws (Freeman *et al.* 1988; Yao *et al.* 1991) and Japanese vowel sounds (Shimoide and Freeman 1995). When presented with a few examples of each class of object that it is to identify and is trained by increasing the synaptic weights between pairs of local oscillators that are co-activated by the inputs, the KIII model establishes the equivalent of nerve cell assemblies. Thereafter, with each test input the system jumps from a basal chaotic state to the basin of an appropriate chaotic attractor (Freeman 1990), and the output is expressed as a spatial pattern of amplitude of a common chaotic waveform, comparable to the way that the olfactory bulb jumps to a distinctive spatial pattern of chaotic activity when the nasal receptors are presented with a familiar odorant.

Implications of Mesoscopic Neurodynamics for Brain Function

These experimental data and descriptive dynamic models have profound consequences for understanding how sensory cortices work at the interface between the brain and the outside world. The crucial point is made by tracing the course of a stimulus into the receptor layer, where it is transduced into a pattern of action potentials, and then into the cerebral cortex, through the thalamus for other systems but directly for olfactory input. What happens in the bulb is a sudden destabilization of the bulbar system, so that it makes an explosive jump from a preexisting state, expressed in a spatial AM pattern of activity, to a new state that is expressed in a different spatial pattern. The AM pattern is selected by the stimulus, and, if reinforcement is provided, there is further modification of the AM pattern, but in the main the pattern is determined by prior experience with this class of stimulus. The AM pattern expresses the nature of the class and the meaning for the subject, not the identity of the particular stimulus. The information that specifies the locations and types of receptors that are activated is irrelevant and is not retained, because the activated receptors belong in a class of equivalent receptors. The output does not express the identity of a chemical material but a collection of experiences that the individual has had with the material. The sensory input serves to trigger the formation of the AM pattern that is transmitted to other parts of the forebrain.

The need for this process can be comprehended by noting that the olfactory environment is indefinitely rich in odorant substances, only a small portion of which ever come to the attention of a subject or form the basis for action. That portion is different for every individual, because it depends in part on genetic determinants of the system but mostly on previous experience with selected odorants and the contexts of reinforcement. In a word, the rate of flow of information from the environment is infinite. Any system for information processing must reduce such a flow to a finite rate, lest it be overwhelmed. Man-made systems do this by means of filters, which are designed by observers to accept the portions that are desired by the observers. Brains have no homunculi to specify their goals and desired inputs, and they rely instead on self-organized chaotic processes to generate activity patterns that are finite in dimension. Perception and recall are essentially operations of the same dynamical process by which meaning is created.

Because each chaotic pattern is created from within and not imposed from outside, raw sense data are not incorporated and stored in the cortex as whole patterns or episodic representations. There is only the modification of synaptic weights among populations of neurons, such that after some experience the brain creates appropriate patterns of neural activity and of behavior that precede and follow the presentation, by stimulation or recall, of an expected example of a learned class of stimuli. Perception is a creative process rather than a look-up table of imprinted data. How is the product transmitted, and what happens with the sense data that triggered the construction? The discovery of meaningful AM patterns in the EEG is not sufficient, because two modes of spatially patterned neural activity coexist in the bulb concurrently with each inhalation. One mode consists of the mesoscopic AM pattern of cooperative bulbar activity, which expresses the meaning of a stimulus. The other mode is microscopic stimulus-evoked activity that consists of spatial patterns of microscopic pulse firings in sparse networks of interconnected neurons, which expresses the properties of the stimulus. Both patterns can be inferred to exist over the entire bulb, but they can be observed and measured only in part, and then only after averaging. The mesoscopic pattern is extracted by spatial ensemble averaging, which is done by recording a collection of waveforms simultaneously, measuring them as a set of time series, and calculating the average. The microscopic pattern is obtained by recording evoked potentials or action potentials in a collection of serial trials and combining them by time ensemble averaging (Freeman 1987a, 1992a). Both of these patterns are transmitted simultaneously by the bulb. How can we know which of them is accepted and acted upon by the prepyriform cortex to which the bulb transmits?

It is immediately clear that the global pattern that is transmitted from the bulb is accepted by the cortex, because the EEG waveforms of the prepyriform cortex are often highly correlated with those of the bulb in the gamma range of frequencies at which the bulb is driving the cortex (Bressler 1988), and the phase gradient of the prepyriform waves corresponds closely to the directions and velocities of the lateral olfactory tract and its terminal branches (Boudreau and Freeman 1963; Freeman 1978a). But is the stimulus-evoked activity pattern accepted by the cortex after it has been transmitted from the bulb? The answer is clear from the structure of the transmission pathway and the functional properties of the cortical neurons that receive the bulbar output. Unlike the primary olfactory nerve that carries receptor input to the bulb, which has a significant degree of topographic order, the lateral olfactory tract from the bulb to the prepyriform cortex is a divergent–convergent projection. Each bulbar axon branches and distributes its output broadly over the cortex. Conversely each cortical neuron receives input from many thousands of widely distributed bulbar neurons, and its dendrites sum that input continually over time. The only portion of the input from the bulb to the cortex that survives this operation of space–time integration is the common waveform with the spatially coherent instantaneous frequency, which is the product of the mesoscopic cooperativity between the bulb and cortex. The stimulus-evoked microscopic pattern is localized spatially, and it is poorly coordinated temporally, so that even though it is transmitted to the cortex, as time ensemble averages of recordings from bulbar neurons have demonstrated, it is expunged by the smoothing process of spatiotemporal integration in the transmission path receiving neurons. The integration embodies the processes of generalization and induction.

This process must occur during many forms of corticocortical transmission as the means for enhancement of signal (the product of self-organizing dynamics) and reduction of noise (the residue of the particular circumstances of the receipt and initial preprocessing of a unique stimulus in one or another of the differing sensory ports of the brain), so that it deserves the epithet “brain laundry” (Freeman 1991a, 1991c). The spatial integration by divergent–convergent pathways is qualitatively the same process by which summation of extracellular dendritic currents gives the

EEG, which is the main reason why the EEG, when it is recorded from arrays on the cortical surface, is the optimal available electrophysiological measure of cortical output. In contrast, brains have no neural machinery to perform temporal ensemble averaging, and the information that observers extract by using that technique is precisely what brains remove by laundering. Calculation of averaged evoked potentials and post stimulus time histograms gives estimates of cortical responses to inputs, so the techniques of time ensemble averaging provide measures of what brains discard, that is, brain trash. In brief, mesoscopic techniques retrieve signals that brains actually use.

Tools for Mesoscopic Linear and Nonlinear Neurodynamics

There are two main extrapolations from the above epistemology to be dealt with. First, neurodynamics is largely based on the study of the self-organization of spatiotemporal brain activity patterns along an evolutionary trajectory, and, second, changes take place by rapid and repeated jumps from each pattern to the next. Physicists call discontinuities in the states of matter “phase transitions”, while engineers call similar switches domains of solutions of systems and their models “state transitions”, because they use the term “phase” to denote a time delay between two oscillations instead of a change in the physical condition of a substance. So far, state transitions have been modeled only in the first two stages of the olfactory system, and then only for one frame at a time in what is clearly a life-long sequence of patterns. These models are a beginning for enormous tasks ahead. They provide a framework in which to pursue the construction of population neurodynamics in the visual, auditory and somesthetic sensory systems, and to model the dynamics of gamma activity in relation to higher processes, by which the global patterns of an entire cerebral hemisphere emerge (Baars 1988; Serafetinides 1993; Taylor 1997; Lumer *et al.* 1998; Pettigrew and Miller 1998; Uhl 1999) that can be expected to involve the limbic and motor systems directly in the shaping of intentional behavior. These fully macroscopic aspects of brain dynamics go beyond the scope of this book.

Most neurobiologists see nervous systems from a perspective dominated by empiricism, linear causality, and the stimulus-determined operation of sparse neural networks. In contrast, I urge the primacy of intentionality as the mechanism by which goal-directed behavior emerges, bringing the requirement for circular causality to describe the flows between the microscopic and mesoscopic levels, and the necessity of hierarchical organization of neurons in populations prefatory to attempting to explain macroscopic behavior. The focus of this book is on detailed descriptions of the tools by which a theory of self-organizing neurodynamics can be constructed and tested with experimental data, consisting of measurements of brain activity that are guided by the growing body of mesoscopic theory. These tools are listed here and described in the following reprints of the articles in which the details were originally published. They are applicable to both neocortical and paleocortical dynamics in descriptions of homeostatically self-regulating brain subsystems that are locally autonomous while contributing to behavior by cooperating among each other without seriously deflecting their ongoing operations.

Tools in Mesoscopic Neurodynamics

Chapter 1. Spatial mapping of brain potentials. This can be assisted by principle component analysis and Karhunen–Loève decomposition of spatial basis functions ("modes" and "AM patterns" of "carrier waves") to characterize overlapping monopole ("closed"), dipole, and tripole fields of extracellular electrical potential, and to distinguish standing from traveling waves.

Chapter 2. (a) The Dirac delta function to model evoked potentials. The single shock electrical stimulus has been the mainstay for physiological perturbation in brains for over a century. Its model, the delta function, serves as the input to linearized differential equations that are solved to predict impulse responses.

(b) Nonlinear regression to fit the sums of basis functions to digitized measurements of brain potentials. The algorithm was invented by Gauss in the mid-19th century, but it became useful only after the development of high-speed digital computers in mid-20th century. It is optimal for evaluating parameters in descriptive equations by fitting solutions to data when the curves cannot be transformed to straight lines. The equations are digital adaptive filters.

(c) Superposition and paired shock testing in search of linear domains. In order to use linear basis functions (damped cosines, exponentials, steps, ramps) to fit cortical evoked potentials, it is necessary to determine a near-linear range of function of the system under study by superposition of pairs of inputs, the classical physiological technique of paired shock testing, and to show that the range is physiological. The basis functions decompose the evoked potential into additive desired signal, clutter, and white noise.

Chapter 3. Rational approximations for Laplace transforms of transcendental linear operators such as $\exp(-sT)$ and $K_2 \exp(-\sqrt{sT_2})$for axonal, synaptic and cable feedback delays. These are essential for solving single and multiple loop feedback models of linearized differential equations by partial fraction expansion to obtain the inverse transforms that are to be fitted to experimental and artificial data.

Chapter 4. (a) Root locus techniques. These are tailor-made to organize and model the data obtained by induction of changes using behavioral, pharmacological, and parametric manipulations, yielding families of evoked potentials that offer valuable overviews of the stability properties of neural populations.

(b) Piecewise linearization. This gives families of root loci in the complex plane that serve to parameterize the sets of gains in multiple feedback loops and relate them to changes in synaptic strengths induced by various manipulations.

(c) Subsidiary describing functions for root loci. An example is the equation for what is now known as the "sigmoid" curve, which describes the dependence of a feedback gain on the intensity of activating input, and which is a static nonlinearity that expresses bilateral saturation.

Chapter 5. The K_2–T_2 operator to model undefined multiple loop distributed synaptic feedback in cortex. This generalized operator serves to linearize the background covariance of frequency and decay rate of damped cosines of cortical impulse responses. It facilitates principal component analysis of collections of simultaneous measurements of behavior and brain activity, and helps to validate the heuristic forms of digital adaptive filters for measuring brain activity.

Chapter 6. Analog computers to complement digital embodiments that are subject to numerical instabilities. They are used to simulate bilateral saturation in order to show how increases in stimulus intensity change the waveforms of evoked potentials due to bilateral saturation at trigger zones. They also serve to model the changes in evoked potential frequencies and decay rates with discriminative training and habituation, showing that changes in evoked potentials are due to changes in the gains at specified synapses that are subject to modification, not to changes in time delays in membranes, synapses, dendritic cables and axon transmissions by propagation of action potentials. The fixation of time constants exemplifies a basic technique in approaching optimization of thousands of parameters in a large system, which is to group them at the outset into those that remain fixed and those that are to be adapted.

Chapter 7. Parametric, surgical and pharmacological manipulations of synapses and trigger zones in neural populations by which to estimate the gain coefficients in heuristic models. The most important is application of local and general anesthetics and surgical transection to induce open loop states, in which to measure the time and space constants of populations.

Chapter 8. Stability analyses of neural populations based on piecewise linearization. These lead to understanding the role played by positive feedback (mutual excitation) in determining the set points (bias levels) of homeostatic (negative feedback) loops, which are often neglected in studies of homeostasis. The analyses also support the distinction between point, limit cycle, and strange attractors in neural populations.

Chapter 9. Simultaneous 64-channel recording at high multiplexing rates for adequate temporal and spatial resolution of cortical activity patterns. The variations in latencies, frequencies, and temporal waveforms of cortical self-organized patterns preclude serial sampling and time ensemble averaging, and enable us to recognize the importance of spatial ensemble averaging for extracting the common carrier waveform.

Chapter 10. (a) Conditional probability of pulse occurrence depending on time lags, frequency ranges, and wave amplitudes using simultaneous measurements of pulse trains from single neurons and the local field potential potentials and EEGs. This technique was essential in the discovery in 1966 of the sigmoid function that implemented bilateral saturation in neural populations.

(b) The asymmetric nonlinear gain sigmoid curve, which is necessary in models to simulate the bistability of cortical populations and the induction of state transitions by excitatory sensory inputs. The equation is derived from the Hodgkin–Huxley system and has a single adaptive parameter that relates bias offset, maximal asymptote, and maximal gain to the level of behavioral arousal, and that suffices to fit the same curve to the normalized conditional pulse probability density dependent on EEG amplitude from a broad variety of neurons in numerous cortical areas and the brains of several species.

Chapter 11. (a) Euclidean distance in 64-space. This is used to measure similarity and difference of spatial patterns of amplitude modulation (AM) and phase modulation (PM) of the carrier waves generated by cooperative and competitive interactions in cortical populations. The technique also serves to calculate centroids and radii of clusters of AM patterns, and to classify examples on individual trials.

(b) Spatial spectral analysis, spatial filtering and signal identification techniques applied to multiple aperiodic EEG time series (Freeman 1978b; Freeman *et al.* 2000). The tools are used to specify the design, size, and spacing between electrodes in arrays, to estimate the point spread functions and the depths of cortical dendritic generators, and to provide confidence intervals on measurements of amplitude and phase of EEG oscillations.

Chapter 12. Modeling chaotic olfactory dynamics with the criterion of simulating the waveforms, spatial patterns, and $1/f^{\alpha}$ spectra of the bulbar and prepyriform EEGs, under the constraints of the known anatomical and physiological properties of the neurons and their connections.

Chapter 13. Tuning curves, using the degree of behavioral correlation and classification as the criterion ("touchstone") for temporal segmentation of EEGs, and for parameter optimization in adaptive filter design. The aim of temporal and spatial filtering is to extract neural signals. The only way to designate a spatiotemporal pattern of brain activity as a "signal" is to correlate it with aspects of normal behavior at known or inferred times of behavioral state transitions.

Chapter 14. Stochastic differential equations and random number generators to handle problems of numerical instabilities in digital simulation of predicted signals embedded in noise for testing models, and to distinguish between deterministic and stochastic chaos. A related technique involves generating an artificial data set with a predicted signal that is embedded in clutter and noise, preprocessed, and filtered in the same way as unit and EEG data. This step is essential for testing complex software and making sure that it does what is intended, for example, distinguishing aperiodic time series with $1/f^{\alpha}$ spectra from quasiperiodic limit cycles embedded in noise.

These techniques are not ends in themselves but are devised and adapted to meet the needs of theory and the exigent data from experimental measurements. They are in principle suited for use with a variety of sources for mapping the spatial patterns of activity in brain populations, including EEG, MEG, multiple unit recording from microelectrode arrays, impregnation of neural tissues with voltage-sensitive optical dyes (Villringer and Dirnagl 1997), and measures such as PET, SPECT and fMRI of the spatial patterns of metabolic rates and levels of blood flow, that are secondary to neuronal activity in conjunction with various intentional behaviors. The techniques will find increasing utility as improvements occur in the spatial and temporal resolution and the size of the windows of observation, and as the requisite measurements of concomitant behaviors keep suitable pace with the measurements of gamma activity.

Part I

The Dynamics of Neural Interaction and Transmission

1. *Spatial Mapping of Evoked Brain Potentials and EEGs to Define Population State Variables*

In the 1950s the brain was conceived as a collection of "centers" for the storage and release of stereotypic behaviors, such as feeding, respiration, shivering, sleep, rage, fear, and sexual activity. The experimental strategy for localization of the center for a specific behavior was three-fold: focal ablation to abolish it; focal stimulation to reproduce it; and electrical recording with a depth electrode to observe the neural activity that drove it. This phrenological conception still governs a large proportion of neurobehavioral research. To a large extent it is the main principle used to guide the use of new techniques for brain imaging by fMRI, SPECT, PET, and related methods for measuring metabolic activity and blood flow in brains of subjects engaged in specific behaviors. The aim is to identify complex cognitive functions with the locations colored spots on 2D projections of brain images. Owing to its simplicity this conception has widespread appeal, but it poorly serves thinking about brain organization and activity, and it led me into a series of failed predictions (Freeman 1961a).

My first research project was a study of the hypothalamic "centers" for temperature regulation in cats (Freeman and Davis 1959). Prior studies had shown that lesions of the anterior hypothalamus impaired resistance to body overheating, whereas posterior lesions prevented shivering and other means for defense against overcooling, so that a "heat center" and a "cold center" were thought to operate in tandem. I placed a thermode in the anterior hypothalamus and showed that both local heating and local cooling elicited appropriate behavioral responses in waking cats, whereas the same stimuli given to the posterior hypothalamus caused reversed responses. I concluded that a thermosensory "center" was located anteriorly, and that a motor "center" was located posteriorly.

I then undertook recording of the electrical activity of the anterior hypothalamus in order to detect the activity of a neural "thermostat". I started by recording trains of action potentials from single neurons ("units") in the lateral hypothalamus and the fields of Forel, for which the firing rates increased with the onset and intensity of shivering (Freeman and Hemingway 1958), but found no units that changed rates in direct response to changes in blood temperature. The cells in the anterior hypothalamus are very small, and their action potentials are difficult to isolate, so I decided to look for oscillations induced by thermal stimulation of the hypothalamus in the electroencephalogram (the EEG), which in cortex had recently been identified with dendritic currents (Chang 1951). The idea prevailed then that an electrode that was *in* a nucleus would detect the activity *of* that nucleus. I used an array of four wires insulated to the tips 1 mm apart in a line, flanking a thermistor in the center. I used the outer pair to apply heat with radio frequency diathermy current and the inner pair to measure the local electrical activity, the "hypothalamogram" as it was called. I heated the tissue first gently, then strongly, and found no change in the EEG even up to 70 °C, although all unit activity was terminated by heating more than 42 °C. Obviously the EEG currents were coming from sites outside the hypothalamus.

By using linear arrays of electrodes and tetrodes forming a pyramid to measure the voltage gradients through the hypothalamus, I tracked down three sources, finding components from the thalamus dorsally, the hippocampus laterally, and by far the largest from the prepyriform cortex (paleocortex located in the frontal lobe anterior to the pyriform lobe in the temporal lobe). It became obvious that the EEG could be recorded at a distance from cortex as easily from inside the brain as from outside on the scalp, and that the commonly used term "local field potentials" ("LFP") can be misleading, because most components of extracellular recordings ("open fields") are distributed throughout the brain, and they are so powerful that they obscure local contributions ("closed fields").

Rather than return to the study of unit activity, which would only tell me what I already knew or suspected in the further experiments I had proposed, I shifted my attention to the prepyriform generator. Its activity, as shown in Figures 1–4 in Chapter 1, had been seen by several researchers before me, most notably by Henry Lesse (1957), who with depth electrodes detected its 40 Hz oscillations in (not of) the amygdala of human subjects experiencing emotional arousal and inferred he had found a "center" for emotion. However, no one had mapped the field in detail or asked how the generator worked. I dropped the idea of "centers" and instead explored this strange entity that spread its signal broadly through the forebrain, and that appeared in the form of an oscillating "dipole field", as seen in Figure 7 in Chapter 1.

The dipole field of the prepyriform cortex offered an excellent preparation for analysis of the behavioral correlates of its evoked potentials and EEGs, because I could place a stimulating pair of electrodes stereotaxically in the olfactory bulb to excite the cortex with impulses applied to the connecting path, the lateral olfactory tract, and then move a pair of electrodes through the brain while recording so as to "home in" on the cortex, as if following a radio beacon. With one electrode in the superficial pole of the dipole and the other one in the deep pole, I could use bipolar recording with a differential amplifier to retain the cortical signal and reject the EEGs overlapping from other cortical generators as common mode, thus obtaining a "clean" prepyriform EEG. The study also showed (Figure 6 in Chapter 1) that the synaptic electromotive forces (emf) of the EEG currents for the surface negative wave peaks (manifesting excitation of pyramidal cells) and the surface positive wave peaks (representing inhibition of those same cells) both lay in the superficial side of the dipole in the cortical dendritic layer.

The most important result of this study was access to the mesoscopic output of a neural population that was not obscured or contaminated by outputs of other populations. This access gave me the basis for defining a population state variable in explanatory differential equations. The raw EEG is a mix of activities from many sources, and it cannot be used until it has been decomposed into its elements. An optimal way is by placing electrodes so as to enhance certain components and diminish others. The key finding is shown in Figure 9 in Chapter 1, which compares the spatial distributions of EEGs and evoked potentials, by properly placed stimulating electrodes the generator of this component of the EEG can be activated in a controlled manner. This kind of analysis has now been done for the various parts of the olfactory system and the hippocampus, but it needs to be done for the major parts of the neocortex. Even today, no one knows which neural populations in the six layers of neocortex generate the synaptic currents that flow across the tissue resistance and reveal the alpha, beta and gamma oscillations so widely reported in the human EEG. Which populations generate the various components of cortical EEGs, and which do not? What spatial forms do their fields of potential take – open or closed? Where should electrodes be placed to record them optimally? Where are the synaptic endings that release the dendritic currents? Which afferent tracts can be electrically stimulated to activate those endings and re-create the patterns of the EEG? Can the same fields be activated with controlled inputs such as sensory or magnetic stimuli? Are the currents from excitatory or inhibitory neurons? Which polarities manifest excitation or inhibition of the neurons? What are the characteristic frequency ranges of their oscillations? Is the feedback intracellular or extracellular? Answers to these questions are essential for defining the state variables in models of brain dynamics in such a way that EEG data can be used to support or refute the models.

Distribution in Time and Space of Prepyriform Electrical Activity

Introduction

In a previous study (9) evidence was found that a large proportion of the spontaneous electrical activity recorded from the basal forebrain of the anesthetized cat was generated bilaterally by the prepyriform cortex, not by the subcortical structures in which the recording electrodes were located. This finding raised the question of how the oscillating potentials of the brain are transmitted, i.e. to what extent the spread of "EEG activity is due (i) to continuous spread of membrane depolarization along axonal or dendritic fibers; (ii) to the formation of intracellular-extracellular current loops such as occur on a shorter time scale in peripheral nerve, with instantaneous field distribution of the current through the brain; and (iii) to successive activation of contiguous cells. To approach this question a study was made of the nature of the various waveforms generated by the prepyriform cortex and of their rates, directions, and distances of spread in and around the cortex.

The prepyriform cortex was found to be well suited to this type of analysis, since it generates a very high amplitude evoked potential (28), which at the same time is relatively simple in form (4, 10, 13). It has many of the electrical attributes of sensory neocortex, including both surface-positive and surface-negative responses, but with the major exception that a transcommissural evoked potential is absent. The dimensions of the cortex are sufficiently large with respect to the size of recording electrodes so that damage resulting from placement usually does not noticeably alter the records of cortical electrical activity. Its boundaries are clearly defined, structurally, histologically, and electrically. Knowledge of its histological structure, which is simpler than that of sensory neocortex, is adequate for the cat (3, 23), and some excellent studies are available from the mouse (21, 25).

In the anesthetized cat the cortex can be exposed surgically by resection of the eye and the orbital plate, and it can be reached stereotaxically from above without disturbing the molecular layer or its relations to the circulatory systems or the underlying skull. A major afferent system, the lateral olfactory tract, lies directly on the surface (cf. 10, 22), so that there can be separation of afferent and efferent pathways at the surface and the base of the cortex. On the other hand, some of these features limit the applicability of the information so derived, so that the results and interpretations cannot be applied without reservation to the understanding of other cortical structures.

The method of analysis consisted of four steps. (i) The spontaneous prepyriform waveforms were examined and classified. (ii) The prepyriform evoked potentials were examined and classified, while varying the site, frequency, and intensity of electrical stimulation of the prepyriform cortex and its afferent pathways. (iii) The distributions of a representative evoked potential in space and time were mapped. (iv) The distributions of this evoked potential were compared with those of other evoked potentials and with those of the various spontaneous waveforms. In this way a clear picture was obtained of the patterns of distribution of the spontaneous electrical activity, which would be exceedingly difficult to derive from a study of the spontaneous activity alone.

Acute experiments were carried out on adult cats anesthetized with pentobarbital. Surgical exposure of the prepyriform area required removal of the eye and orbital plate, with retention of the supraorbital ridge. The recording electrodes were either silver balls resting on the cortex, stainless steel needles sharpened and insulated to a tip diameter of 10–50y, or coaxial electrodes of #26 nichrome wire in #23 needle tubing. The latter types were inserted stereotaxically normal to the surface either vertically from above or horizontally from below. Electrical stimulation was applied either by paired silver balls on the cortex or the olfactory bulb, or by coaxial electrodes inserted into deeper structures. Sites of deep stimulation were determined histologically by means of electrolytic deposits. Records were made with an ink-writing recorder with a flat frequency response from 0.5 to 50 c./sec., or with a preamplifier and oscilloscope with 3 db. fall-off points at 0.2 and 40,000 c./sec. and an input impedance of 10 M[EQ]. High- and low-frequency filters were sometimes used to examine appropriate frequency components selectively.

Electrodes for chronic implantation consisted of insulated #26 nichrome wires, which were inserted in pairs cemented together with one exposed tip 1.5 mm. directly above the other. One of the tips was placed in the molecular layer of the cortex and the other at the "base," i.e. in the fusiform cell layer or immediate, subjacent white matter. This was done by placing a stimulating electrode in the olfactory bulb and lowering the paired electrode to the vicinity of the prepyriform cortex by means of stereotaxic coordinates. Final positioning was obtained by evoking activity in the field with the stimulating electrode and lowering the recording electrode until the evoked potential from the lower tip was a mirror image of that from the upper tip. From one to four pairs were implanted in each side in the prepyriform cortex of 15 cats. Histological confirmation has been obtained in all of them. The wires were attached to the skull with dental cement and soldered to the terminals of one or two plugs. These in turn were cemented to the skull and to steel screws set in the skull. Stimulation and recording were done over periods up to 12 months.

Results

1. Forms of Spontaneous Electrical Activity

a. Characteristics of Dominant Waveform

In the normally active cat the dominant waveform of the prepyriform potentials was a sinusoidal wave with a frequency of approximately 34–38/sec. (Fig. 1, A–B). This waveform came in bursts lasting up to 1 sec. and at amplitudes ranging up to 2 mV. peak-to-peak. In addition, there were sinusoidal waves at approximately the same amplitude but at half the dominant frequency (Fig. 1, C–D). The maximum amplitude of the former was found in the anterior prepyriform cortex (A–B), whereas that of the latter occurred in the posterior prepyriform cortex predominantly in the temporal lobe (C–D). At both frequencies the spontaneous activity at the surface was an approximate mirror image of the activity at the base of the cortex. Deviations from the dominant frequency were common, and usually took the form of modulation in single bursts, e.g. a continuous shift in frequency from 56/sec. to 38/sec. in a burst lasting up to 1 sec. (A-B, at left); more stable variations of 5–10% either upwards or less commonly downwards also occurred. Oscilloscopic records showed the presence of sinusoidal waves at higher frequencies (predominantly 95–110/sec., but only rarely at double the dominant frequency) at amplitudes up to 300–500 μV.

b. Respiratory Wave

In the low-frequency range there was a repetitive waveform, the amplitude of which ranged up to 300 μV., and the frequency of which was related to that of respiration (Fig. 2, AB). It consisted of a blunt, surface-negative peak following inspiration by

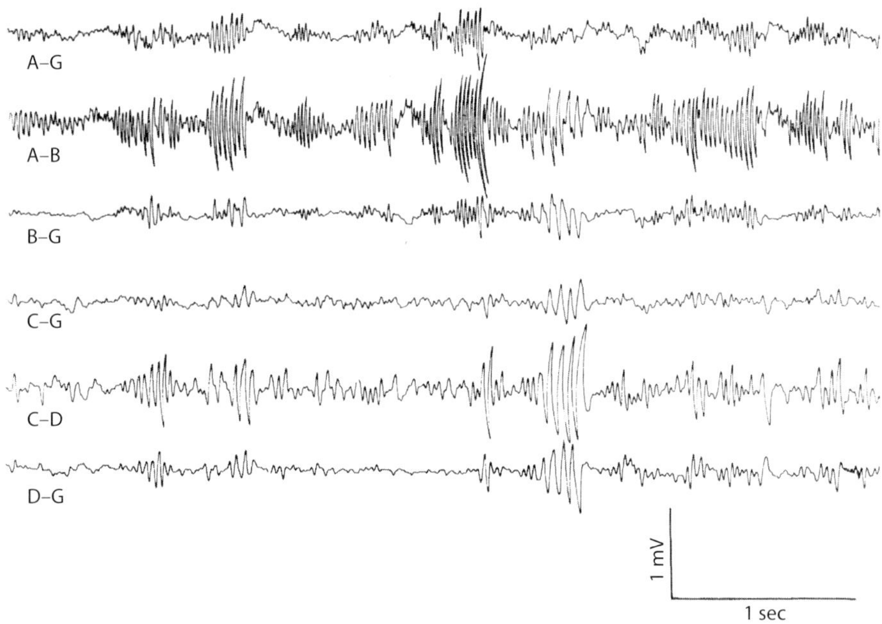

Fig. 1. Fast wave activity of left prepyriform cortex with low frequencies filtered out. Lines A–G and C–G are monopolar from the surface of the frontal and temporal prepyriform cortex respectively. Lines B–G and D–G are monopolar from base 1.5 mm. deep to A–G and C–G. Lines A–B and C–D are corresponding bipolar records at same gain as monopolar records. Note that antiphasic monopolar signals are accentuated in bipolar records.

150–250 msec. and lasting about 200 msec., and after a short delay a surface-positive peak of approximately equal amplitude and duration. During rapid respiration (e.g. sniffing) the two peaks coalesced into a continuous waveform. The bursts at the dominant frequency were closely related to this slow wave, occurring predominantly from the peak of the positive wave to the beginning of the negative wave (Fig. 2, A′B′).

According to Liberson (15), respiratory waves in the basal forebrain electrical activity were reported in 1953 by Novikova and Khvoles. Those observed here were regarded as intrinsic electrical changes rather than movement artifacts for several reasons. The surface-negative and surface-positive peaks were each accompanied by mirror-image peaks at the base of the cortex. Similar and approximately synchronous respiratory waves were seen in both the left and right prepyriform areas (Fig. 3). The high-frequency bursts had a precise temporal relation to the slow wave, and the ratio of their amplitudes was fairly constant at different points of the cortex. The waves were stable in form from week to week in each animal and were very similar although not identical in all of six cats with appropriately placed electrodes. The peak amplitudes were found in the olfactory bulbs, and in the anterior frontal prepyriform cortex; had these waves been due to respiratory movements of the brain within the skull, such a gradient would be difficult to account for. Moreover, the waves at anterior points slightly preceded those at posterior points on each cortex. The same waves were found despite frequent replacement of the recording

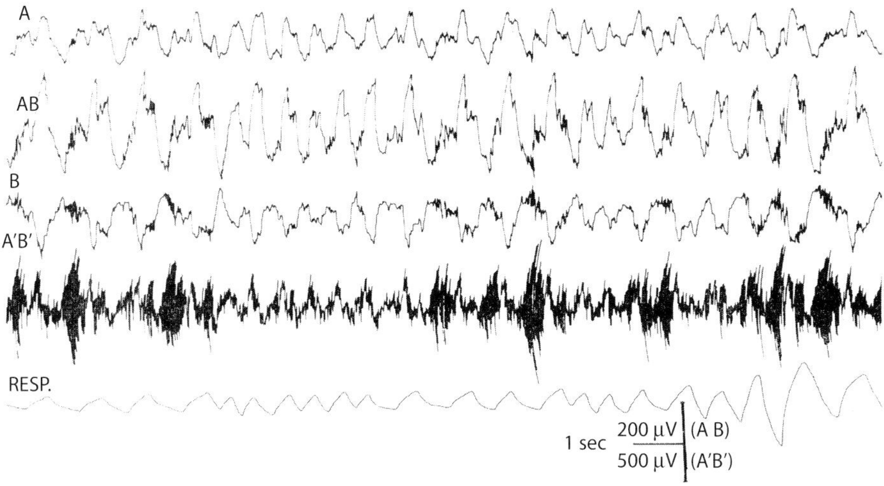

Fig. 2. Low-frequency (lines A, AB, B) and high-frequency (line A′B′) activity of frontal prepyriform cortex recorded near site of maximum amplitude of respiratory wave is compared with respiration (obtained with pneumograph around chest attached to strain gauge – inspiration upwards). Acceleration in respiration occurred during sniffing. Records A and B are monopolar from surface and base of cortex respectively; records AB and A′B′ are bipolar from same electrodes.

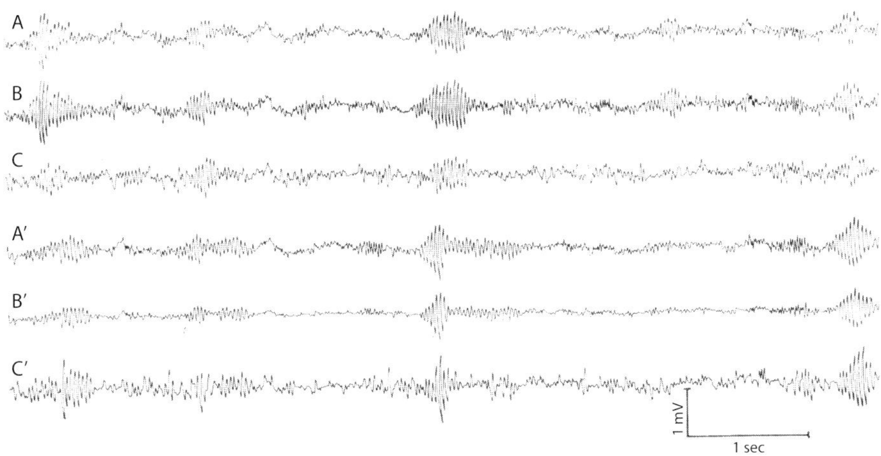

Fig. 3. Simultaneous bipolar records without filters from prepyriform cortex of waking cat with implanted electrodes, three on left side (A, B, C) and three from right side (A′, B′, C′). Site A is about 3 mm. posterior to left olfactory bulb, site B is 3 mm. behind A, and site C is 3 mm. behind B, all near left lateral olfactory tract. Sites A′, B′, and C′ are symmetrically located on right side. Placement is representative of most of the cats. Monopolar records from electrodes showed that at each site electrical activity at base was antiphasic with respect to activity at surface (cf. Figs. 1, 2, 4).

cable; they occurred with or without perceptible movement of the head or cable and were independent of posture. When movement artifacts did occur, they were clearly discerned because of the association with movement, their irregular distribution, and the lack of a turnover. The source of these waves did not lie in the nasal mucosa, since the instillation of 2–4 ml. of procaine HCl (1%) in both nostrils did not alter their relationship to respiration. Tetanization of the lateral olfactory tract at 500–1000/sec. at intensities several times threshold (for single shock) did not alter either the respiratory or fast waves, unless sniffing occurred.

c. Comparison of Tracings from Multiple Prepyriform Sites

Similar activity was present on both sides, usually at slightly different frequencies (Fig. 3) and unequal in amplitude. In 10 of 15 cats with implanted electrodes the amplitude was up to five times greater on one side than on the other. In two of the 10 it was higher on the right and in the other eight cats on the left. In Fig. 3 there are three bipolar records from the left cortex and three from the right, each from a point on the surface with respect to a point at the base 1.5 mm. distant. The bursts tend to occur approximately synchronously on the two sides. In the center set of bursts the frequency at all points on the left is 36/sec. and on the right 34/sec. The waveforms recorded anteriorly on both sides lead those recorded posteriorly by about 1/8 cycle, i.e. about 3 msec. Since the distance from A to C is about 6 mm., the rate of spread indicated by phase shift is about 2m./sec. This phase lag is barely detectable in Fig. 3. Alternatively, measures were obtained by placing the signal from A on the horizontal deflection plates of an oscilloscope and the signal from C on the vertical plates. The resulting Lissajous ellipse was then photographed and the amount of phase shift was computed from the ratio of the lengths of the major and minor axes. This method gave approximately the same results.

d. Sleep and Anesthesia

The presence of bursts of fast waves and respiratory waves was associated with the presence of an active state of behavior (Fig. 4, right). During sleep there were suppression of the electrical activity at the dominant and higher frequencies and the appearance of irregular slow waves, which were aperiodic and not related to respiration (Fig. 4, left). Many of the irregular waves at the surface had mirror-image waves at the base. Spindle bursts were frequently seen in monopolar recordings from the surface and the base (Fig. 4, first and second arrows), but they were in-phase at the surface and base of the cortex and were not present in bipolar recordings on the surface with respect to the base of the cortex (Fig. 4, second line). An additional state was found to accompany behavioral arousal and the alerting response of the neocortical EEG, in which the irregular waves and spindle bursts disappeared, but the bursts at the dominant frequency were not present (Fig. 4, beginning at third arrow). The respiratory waves might or might not be present in this state, but continuous low-amplitude electrical activity at the dominant and higher frequencies was readily detectable. This state was associated with a lack of overt behavioral activity.

The spontaneous activity of the prepyriform cortex in cats under pentobarbital was lower in amplitude and more variable in appearance, although no new waveforms appeared. Irregular waves were commonly present, but regular slow waves synchronous with respiration were not observed. Fluctuations at 19/sec. and 38/sec. were rarely present, nor could bursts at this frequency be distinguished; there was no perceptible relation between the lower and higher frequencies, and there was much less synchronization of frequency and amplitude at different sites within each

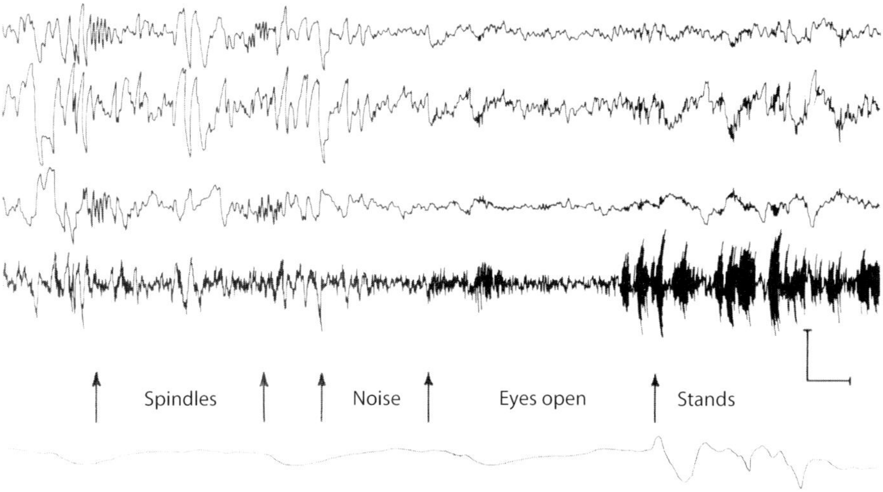

Fig. 4. Same recording arrangement as in Fig. 2 was used in sleeping cat. At third arrow brief noise was made by striking two pieces of metal together. At fourth arrow eyes opened, and at fifth arrow cat stood up. Calibration: 200 μV (top three lines), 500 μV. (4th line), 1 sec.

cortex. There were in many cats sinusoidal waves at intermediate frequencies of 8–15/sec., which did not have the form of spindle bursts. Spindle bursts were frequently noted in recordings from the neocortex and in monopolar recordings from the prepyriform cortex, but as with sleep spindles they were in phase at the surface and base of the cortex, i.e. a "turnover" for them could not be located in or below the cortex. In general, the electrical activity under pentobarbital resembled that in the sleeping cat but at lower amplitude.

2. Forms of Evoked Potential

a. Stimulation of Cortical Surface and Olfactory Tract

Direct, single-shock, bipolar stimulation of the exposed surface of the cortex of the anesthetized cat produced a polyphasic change in potential, which consisted of a dicrotic surface-negative peak (cf. Fig. 5) lasting 15–25 msec. and followed by a surface-positive wave and then a small second negative wave, each lasting 30–40 msec. The magnitude of the potential as well as to some extent its form was dependent on both the intensity and the duration of stimulation. At high stimulus intensities (6–10×threshold) the duration of the positive wave increased and the second negative wave disappeared. The chronaxie for both peaks of the initial dicrotic negative wave was 0.085–0.095 msec. The minimal start-latency was 0.75 msec. The waveform spread concentrically from the site of stimulation with decremental amplitude to the approximate histological limits of the prepyriform cortex at a rate of 1.7–2.2 m/sec.

Several sites were found outside the cortex, the electrical stimulation of which evoked prepyriform potentials with the same basic form as that evoked by direct

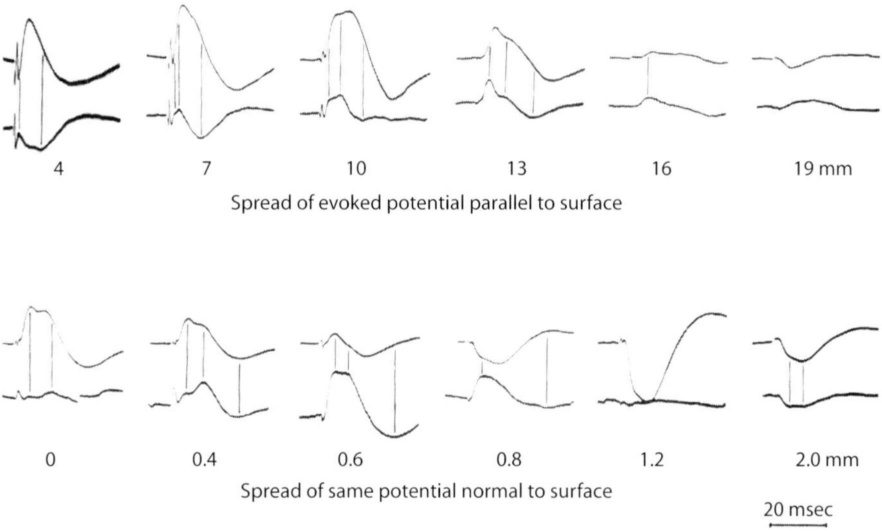

Fig. 5. Comparison is made between monopolar records (upper tracing of each pair) and bipolar records (lower tracing, at twice gain of monopolar record) of prepyriform evoked potential. Distances in parallel series are posteriorly from base of olfactory bulb; distances in normal series are toward base from surface. Retouched.

stimulation but differing in significant ways. Stimulation with a single shock at the base of the olfactory bulb caused the potential to move as a wave from anterior to posterior tangentially to the surface of the cortex at the same rate (Fig. 5, top line). Its amplitude was initially incremental (4–7 mm.) and showed a plateau in the central portion of the cortex along the lateral olfactory tract (7–10 mm.). Near the borders the amplitude diminished (16 mm.) and the wave length increased; at the borders the amplitude decreased rapidly and an initial positive wave appeared (19 mm.). The second peak of the dicrotic first negative wave was most prominent in the posterior frontal prepyriform cortex (10 mm.). Repetitive stimulation at rates above 3/sec. caused augmentation of the first peak and suppression of the second peak; there was post-tetanic potentiation of the first peak (and not the second) lasting up to 3–4 min. following tetanization for 20 sec. of the ipsilateral olfactory bulb.

b. Stimulation of Other Sites

Stimulation of the periamygdaloid cortex in most cats evoked a potential of the same form, polarity, and velocity, which spread tangentially over the cortex from posterior to anterior. The amplitude at any one point of the cortex was constant during repetitive stimulation of the lateral periamygdaloid cortex at rates up to 10/sec., but it decreased to zero during the first 10–20 stimuli delivered to the medial periamygdaloid cortex. The latency of the potential increased with increasing distance of the site of stimulation from the temporal prepyriform cortex. No prepyriform potential was evoked by stimulation of the entorhinal cortex. In some cats the polarity of the wave was reversed (Fig. 6, PA), so that the initial wave was positive at the surface and negative at the base of the cortex. In all other respects the waves were similar. No explanation of this reversal is available. A similarly inverted waveform of long latency (6–8 msec.) was sometimes found during repetitive stimulation of the anterior commissure. This was an augmenting response, which became maximal at stimulus rates of 10–15/sec. after the first 1 or 2 sec. of stimulation. In addition, a highly variable and complex response was found following stimulation in the vicinity of the anterior olfactory nucleus, which it is thought might represent the result of mixed stimulation of the medial and lateral olfactory tracts.

Three types of potential could not be evoked: an antidromic potential, e.g. by stimulation of the medial forebrain bundle (cf. 14); a propagated orthodromic potential, e.g. in the globus pallidus (23) or claustrum (3) following direct stimulation of the cortex [for an opposing view see Berry *et al.* (4)]; or a transcommissural potential following stimulation at symmetrical points on the opposite cortex.

Stimulation of the midline nuclei of the thalamus in some cats evoked an initially surface-negative prepyriform potential. During repetitive stimulation the second peak of the dicrotic first wave was accentuated, whereas the first part was diminished. Following tetanization of the thalamus the second peak (and not the first) was potentiated, whether the subsequent evoking stimulus was applied to the thalamus or the lateral olfactory tract. No recruiting response was found in the cortex, which could not be attributed to current spread from the adjacent neocortex.

The prepyriform potentials evoked by direct stimulation or by stimulation of the lateral olfactory tract or periamygdaloid cortex in cats with implanted electrodes were similar in form to those evoked in cats under pentobarbital, differing only in the greater prominence of the second surface-negative wave in the former group. Light doses of pentobarbital (25–30 mg./kg.) brought about an increase in amplitude in the second peak of the dicrotic negative wave and a decrease in amplitude of

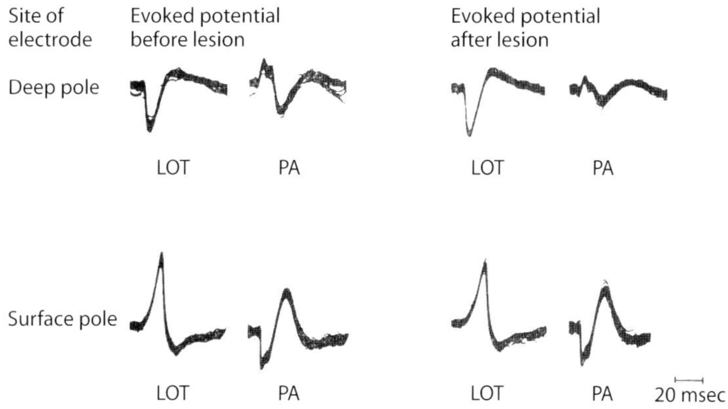

Fig. 6. Each tracing consists of 100 superimposed evoked potentials following 1/sec. stimulation of either lateral olfactory tract (LOT) or periamygdaloid cortex (PA). Recording microelectrode was placed in deep pole by insertion from above through caudate nucleus. Tracings were recorded before and after electrolytic lesion was made at site of recording with recording electrode (50 μA. for 15–30 sec.). Electrode was then moved across zero isopotential to surface pole. Tracings were again made before and after second lesion was made. Results show that lesions caused diminution of both phases of initially negative potentials but not of initially positive potentials.

the second negative wave. Doses in excess of 40 mg./kg. suppressed both the second peak and the second negative wave.

Since there is as yet no way to join these variations into a comprehensive analysis, the observations may be summarized by pointing out that the important changes in waveform were in the site of initial negativity and in the relative amplitudes of the two peaks of the initial dicrotic wave. Apart from this the waveform was remarkable for its simplicity and universality. Furthermore, despite changes in form and amplitude due to changes in the site, frequency, or intensity of stimulation, all these surface potentials were accompanied by mirror-image potentials at the base of the cortex. Both surface and deep potentials were abolished simultaneously by 4–5 min. of anoxia, by the intravenous administration of massive doses of curare (24), or by continuous electrical stimulation at rates in excess of 200/sec.

3. Spread of Evoked Potential in Time

a. Surface Spread and Triphasic Action Potential

It has already been noted that the tangential spread of the evoked potential over the surface of the cortex occurred with a definite latency and without a "turnover." Preceding the potential evoked by stimulation of the lateral olfactory tract and having the same threshold was a triphasic action potential (initially positive), which was continuously "decremental," and the velocity of which was the same as the start velocity of the evoked potential (Fig. 5, top lines 4–10 mm.). Lesions of the tract or the application to the surface of the cortex of small squares of filter paper soaked in procaine HCl (1%) blocked the further progression of the action potential as well as the evoked potential beyond the affected region, but did not alter the potentials proximal to the block. The triphasic potential did not have a mirror image potential at the base of the cortex (Fig. 5, bottom lines). It was resistant to anoxia and massive intravenous doses of curare, and did not diminish in amplitude at stimulus rates of up to 500/sec. Its maximal amplitude occurred over the lateral olfactory tract. At the margins of the cortex it became diphasic but remained initially positive. Similar diphasic waves were frequently observed on the surface of the cortex immediately preceding the potentials evoked by stimulation of the thalamus and periamygdaloid cortex, but only if the wave was initially surface-negative. The same phenomenon was noted preceding the wave evoked by direct stimulation, but only if great care was taken to eliminate the shock artifact.

b. Spread to Base and Zero Isopotential

The spread of the evoked potential from the surface to the base of the cortex was a process with different characteristics. (A steel microelectrode tapered to 5 microns, insulated, and transected to leave a flat recording surface 15 microns in diameter was used to obtain accurate measurements. A coaxial electrode was used to obtain the tracings in Fig. 5.) As the exploring electrode was inserted through the outer layers in a direction normal to the surface, decrement in amplitude became prominent at a depth of about 300 microns and continued rapidly until a small polyphasic evoked potential remained (Fig. 5, bottom lines, 0.6 mm.); at times no change in potential could be detected. Electrolytic deposits examined histologically showed that this region of zero isopotentiality lay in the vicinity of the junction of the molecular and pyramidal cell layers. It was also found that unit potentials could be recorded in large numbers at and below the zero isopotential, whereas they were virtually absent on the superficial side of the zero isopotential. Still deeper the approximate mirror-image potential characteristic of the base appeared, became maximal at depths of 1.0–1.5 mm. and then gradually decreased at greater depths (2.0 mm.). No triphasic action potential was present. It was not possible to demonstrate crest latency in the negative peak of the molecular layer nor in the positive peak of the deep layers in directions normal to the surface; furthermore, the two peaks recorded along a line normal to the cortical surface occurred at approximately the same time.

c. Effects of Lesions in Superficial and Deep Poles

Electrolytic lesions made in the molecular layer (i.e. in the surface pole), resulted in diminution of the amplitude of both the surface-negative and surface-positive waves, when the evoked potential was subsequently recorded in the lesion with the same electrode (Fig. 6, LOT, before lesion). Similar lesions made in the polymorphic cell layers (i.e. in the deep pole) did not decrease the deep-positive and deep-negative potentials similarly recorded, and in fact they often increased slightly (Fig. 6, LOT, after lesion).

d. Comparison of Monopolar and Bipolar Recordings

A significant difference between the patterns of spread tangential to the cortex and normal to the cortex was also shown by comparison of concomitant monopolar and bipolar recordings (Fig. 5). Since the bipolar record is a measure of a difference in potential, close approximation of two electrodes with respect to the wave length of an electrical disturbance recorded monopolarly from either one permits treatment of the bipolar record as an increment. A precise definition of the relationship between such records requires the use of partial derivatives, but an adequate understanding of the two patterns found in the prepyriform cortex can be gained by visualizing two waves of potential in a volume-conductor – one moving through the volume with a constant form, velocity and amplitude, and the other remaining stationary in the volume and changing in amplitude.

In the first case the bipolar record from two electrodes placed on the line of movement gives the slope of the moving wave, so that the peaks of the bipolar record occur concomitantly in time with the maximum slopes of the monopolar records. This is the case of peripheral nerve, in which the bipolar record of the action potential tends to approach the first derivative with respect to time of the monopolar record. The prepyriform potential is more complex, since it is variable in amplitude, variable in form, and composed of a dicrotic initial negative peak. Despite these complexities the negative and positive peaks of the bipolar record from one electrode with respect to a second electrode 2 mm. distal to it on the lateral olfactory tract tend to occur at the times of the maximum negative and positive slopes of the wave recorded monopolarly from the first electrode (Fig. 5). This is evidence that the prepyriform evoked potential moves as a wave across the surface.

In the second case (the stationary wave) bipolar records give the gradient of potential in space. The gradient changes at the same instant as the entire wave, so that the peaks of the monopolar and bipolar records are concomitant in time; but the minimum amplitudes of the bipolar record are found at locations of the electrode in space which give the maximum amplitudes of the monopolar record, and vice versa. Records from the prepyriform cortex taken from one electrode with respect to another 300 microns deep to the first are complicated by the occurrence of the dicrotic negative peak and by a delayed deep-negative wave without a mirror-image surface wave, which may accompany and partly obscure the surface-positive phase at moderately high stimulus intensities (cf. Fig. 5, monopolar record at 0.6 mm.). They still show clearly that the minimum amplitudes of the bipolar record coincide in space with the maximum amplitudes of the monopolar records, and that the maximum amplitude of the bipolar record occurs across the zero isopotential of the monopolar record. On the other hand, the peak amplitudes tend to coincide in time. The bipolar records therefore approach the first derivative of the monopolar records with respect to distance and not to time. This is evidence that the prepyriform evoked potential does not move as a wave from the surface to the base of the cortex.

e. Reversal of Poles and of Effects of Lesions

Stimulation of the anterior commissure or the periamygdaloid cortex in some cats resulted in a reversed wave, which was initially deep-negative and surface-positive (Fig. 6, PA, before lesion). The loci of maximal amplitude of these responses were lateral and posterior in the prepyriform cortex to those following stimulation of the lateral olfactory tract. In these doubly inverted potentials microlesions in the deep pole (i.e. the site of initial negativity) resulted in diminution of both the positive and negative waves, whereas microlesions in the surface pole did not (Fig. 6, PA, after lesion). In four cats in which arrangements were made to stimulate the anterior commissure and the lateral olfactory tract alternately, it was found that the location of the zero isopotential was the same for both types of evoked potential.

4. Distribution of Evoked Potential in Space

a. Methods and Conditions of Mapping

Comparison of these data makes it apparent that, irrespective of the site of stimulation, during the establishment of the evoked potential two fundamentally different processes are at work (i) there is distribution in time of the evoked potential over the surface of the cortex; (ii) there is instantaneous distribution of the potential in space through the base of the cortex. A reduction in the complexity of the data can be attained by making a series of measurements of potential in space and time, and from them making a series of maps at each of a series of instants of the spatial distribution of potential. A still further reduction can be attained by mapping the spatial distribution of peak amplitudes without regard to their time of occurrence. Both methods were tried, and the second was found to be superior for three reasons: it was simpler; it provided in a single series of maps the loci of peak amplitudes in the field of the cortex; and it provided a more accurate picture of the spatial distribution of the spontaneous electrical activity of the cortex. The first method simply showed the development of the peaks of activity from anterior to posterior along the cortex. The second method involved some distortion, e.g. the overlapping of isopotentials at the margin of the cortex, but this was minimized by placing the planes of the maps in the coronal plane, which lies at an angle of about 60° from the major direction of movement of the evoked potential along the surface of the cortex. Still further reduction in distortion could have been attained by taking the plane of mapping 90° to the cortical surface, i.e. perpendicular to the line of movement of the potential, but this would have involved presentation of the maps at an unusual plane of histological section of the brain.

Among the more important conditions of the mapping procedure were the use of a monitoring electrode at a moderately distant point in the field to control spontaneous variations in field amplitude and the use of a steel microelectrode just large enough to exclude unit potentials and to have a stable impedance. The duration of stimulation (of the root of the lateral olfactory tract) was kept at chronaxie, the intensity at four times threshold, and the frequency at 4/sec., a rate at which fatigue and potentiation were not apparent, and at which the second peak of the initial surface-negative dicrotic peak was always suppressed below the first. The electrode was moved through the tissue along each of a series of tracks until the evoked potential had reached predetermined positive or negative amplitudes. It was then photographed and the coordinates were recorded. Transfer of the resulting maps to the histological sections required compensation for the distortion of the tissues resulting during mounting. Measurements from point to point are not possible in such distorted planes, but measurements from known lines are considerably more reliable. The vertical positions of points were determined along the electrode tracks (most of which could be seen) and the horizontal positions were measured along these lines from the zero isopotential. The latter was placed on the pyramidal cell layer of the cortex, which required making changes in its position but not in its curvature. The procedure was done on seven cats.

b. Characteristics of Dipole Field

The resulting maps (Fig. 7) showed that the potentials fell into a series of continuous distributions compatible with an electrical event in a relatively homogeneous conductor (save for the underlying skull). There were two major poles of opposing sign during the first (surface-negative) phase, which were approximately reversed during the second (surface-positive) phase. The superficial pole lay in the molecular layer in the vicinity of the lateral olfactory tract. The deep pole lay in the middle of the concavity of the gyrus, somewhat lateral and posterior to the superficial pole. The boundaries of the poles were not sharply defined except at the zero isopotential. The peak amplitude in the superficial pole exceeded that in the deep pole by a factor of 1.5–2.5. On the other hand, planimeter measurements of the areas enclosed by the 2 mV. and 4 mV. contours (allowing 1.0 and 0.5 mm. respectively for the distance of the negative contours from the surface of the cortex) indicated that the deep areas exceeded the surface areas by about the same factor. Assuming a direct proportionality between area and volume, the volume of the deep pole appeared to exceed that

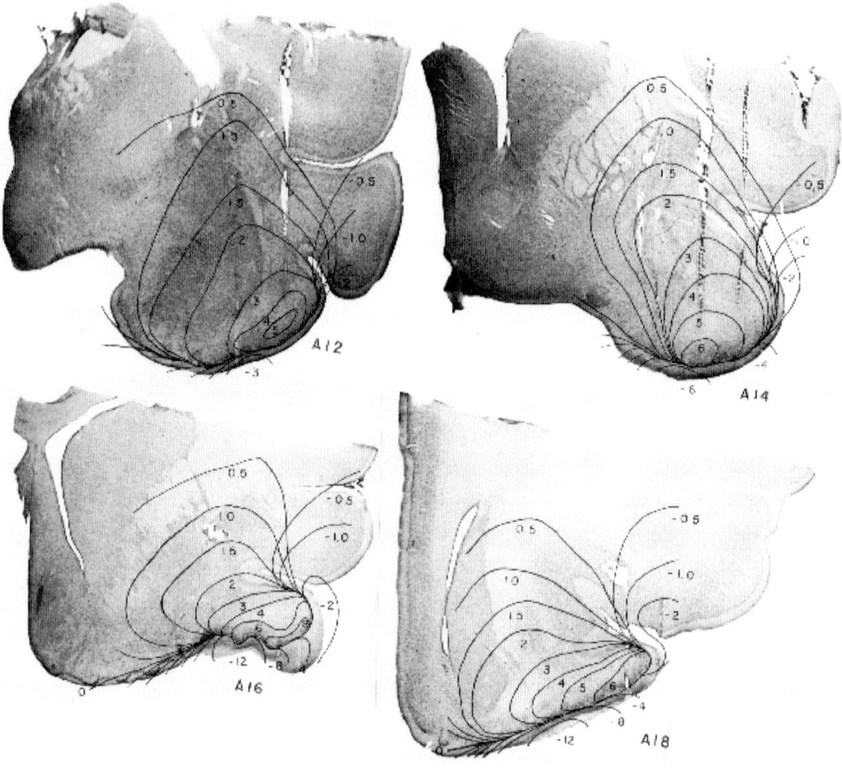

Fig. 7. Coronal histological sections of cat brain at designated stereotaxic levels. Superimposed on these are isopotentials in millivolts of prepyriform evoked field at peak amplitudes independently of time of occurrence.

of the superficial pole by approximately the same amount that the potential of the superficial pole exceeded that of the deep pole. Since potential in a volume conductor is proportional to charge density, the products of potential times volume indicate that the total charges of the two poles at the peak surface-negative amplitude were approximately equal, but opposite in sign. Hence the field is an approximate dipole field. Since the field exists in a volume conductor it must be inferred that during the first phase current flows from the base of the cortex to the surface and during the second phase in the opposite direction. This current is of maximal extracellular density across that part of the zero isopotential surface located between the poles, but it appears to spread through adjacent tissues with little relation to histological boundaries. Since the evoked potential is the IR product of this field of current, it is clear how such distant factors as the curvature of the cortex and the impedance of the skull can in part determine the magnitude of the evoked potential recorded at any point in the field. Bishop and O'Leary (5) observed a similar field in the superior colliculus of the cat, giving it a slightly different interpretation.

c. Oscillating Injury Potentials

During the procedure of mapping the dipole the microelectrodes frequently struck the inner surface of the skull and were bent into the shape of a hook. (Planes of measurement in which this occurred were discarded.) During withdrawal of such an electrode across the zero isopotential (and occasionally during insertion of a normal microelectrode) there occurred oscillations in potential at the same frequency as the spontaneous electrical activity but at 5–20 times greater amplitude. When a steel microelectrode was deliberately bent into a 50 micron hook and insulated to leave an exposed surface 20 microns in diameter, such oscillations (Fig. 8, A) could regularly be elicited at many points in and below the cellular layers of the cortex but not in the outer two-thirds of the molecular layer. They were seen most often and at greatest amplitude in the vicinity of the zero isopotential. These oscillations resembled the "injury potentials" seen during microelectrode recordings from single units in several respects. The fields were of small size with extremely steep gradients. The discharges were transient, lasting from a few seconds up to several minutes, and were repetitive and regular but usually intermittent. The amplitude was variable and tended to diminish rapidly with time. The discharges were elicited only by movement of the electrode with respect to the tissue. There were two major differences: (I) the frequency of the injury potential is variable, whereas the frequency of these oscillations was not; and (ii) the duration and form of a single injury potential are roughly the same as those of a unit potential, whereas the duration and form of a single oscillation were very similar to those of the spontaneous waves of the prepyriform cortex. If during one of these bursts the lateral olfactory tract was stimulated at 4/sec., it was found that the evoked potential could often be observed with roughly the same increase in amplitude as that of the spontaneous oscillation (Fig. 8, B, a, b). This was the only circumstance in which highly localized, high-amplitude evoked potentials could be recorded with microelectrodes (down to 10 microns in diameter). Oscillating injury potentials as often did not contain the evoked potential (Fig. 8, B, c, d), even when they were found at the zero isopotential of the cortex.

5. Comparison of Distribution of Spontaneous and Evoked Potentials

Precise mapping of the amplitude distribution of spontaneous potentials was not feasible, due to the mixture of prepyriform potentials with potentials generated in neighboring structures, as well as the continuous and uncontrolled variation in amplitude of the spontaneous potentials. This was particularly true of anesthetized cats with low-amplitude spontaneous activity. The method used to map such potentials was to place two electrodes in the field, one stationary and the other exploratory, and to compare the relative amplitudes of spontaneous and evoked potentials at different points in the field with these monopolar records (Fig. 9).

a. Distribution Normal to Surface

Along lines normal to the cortex there was no significant phase shift in the peak amplitudes of the sinusoidal waves at the dominant frequency other than the inversion of the waveform, i.e. a 180° phase shift (Fig. 9, a). The ratios of differences in

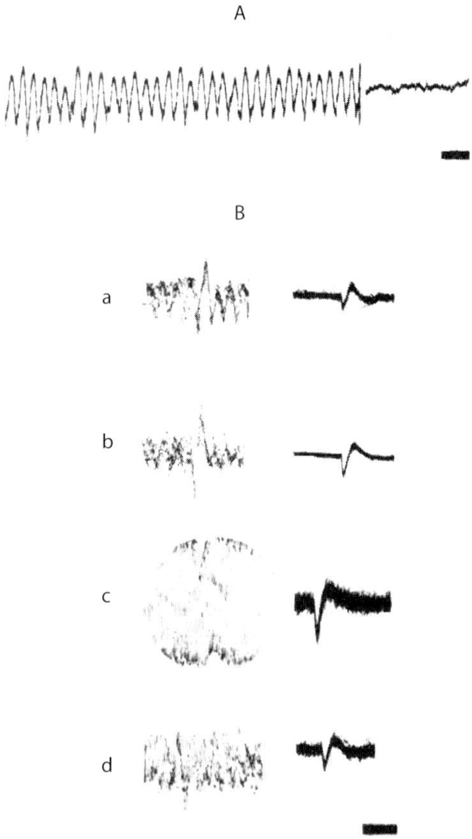

Fig. 8. A: oscilloscopic tracing of oscillating injury potential and of subsequent level of spontaneous activity at same site and same gain. Time, 100 msec. B: by use of up to 30 superimposed sweeps evoked potential in some injury oscillations was found to be increased in amplitude in comparison to similar records made after abatement of oscillations (a, b), but in other injury oscillations evoked potential was superimposed on oscillations without alteration in its magnitude (c, d). Time, 100 msec.

amplitude of the evoked and spontaneous potentials at different points in the field along normal lines were identical; in particular the positions of the maximum amplitudes and of the zero isopotential (Fig. 9, f) for the two types were identical. When the amplitude of the evoked potential was altered by anoxia (6), locally applied procaine, or intravenously administered curare (24), simultaneous changes of the same order of magnitude occurred in the spontaneous activity. The mirror-image quality of the activity at a point on the surface and an underlying point also persisted, there being no circumstance in which dissociation could be observed other than the previously noted injury potential.

b. Distribution Tangential to Surface

Along lines tangential to the cortical surface the spontaneous activity showed phase shifts that were commensurate in amount with the rate of spread of evoked potentials. The phase shifts of the peak amplitudes of the spontaneous and evoked

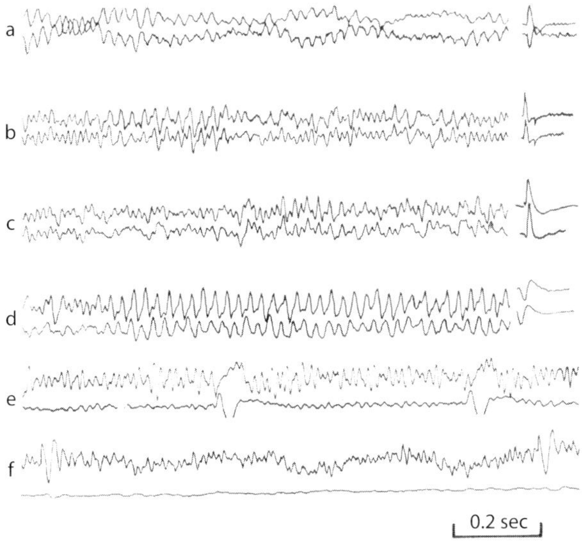

Fig. 9. A series of comparisons of spontaneous electrical activity of prepyriform cortex and prepyriform potential evoked by stimulation of lateral olfactory tract. Points in (a–d) are located in or near the two poles of field. The upper tracing is from surface of olfactory bulb and lower (at decreased gain) is from surface of prepyriform cortex. In (f) upper tracing is from surface of cortex and lower (at equal gain) is from zero isopotential just beneath site of upper tracing. Stimulus interval in (e–f): 1 sec.

potentials often coincided (Fig. 9, b–d). However, comparison of records from different points on the surface of the cortex showed that there was no clear correlation between the amplitudes of evoked and spontaneous activity. It has already been noted that the poles of the fields of potential evoked by stimulation of various extracortical structures did not coincide with one another, and it may further be recalled that although there was a coincidence of major changes in amplitude of spontaneous activity and a coincidence of bursts, the amplitude at any point on the cortical surface was not nearly as closely correlated with the amplitude at neighboring points on the surface as with points on lines normal to the surface (Fig. 3).

Discussion

In the system of analysis of potentials in a volume conductor developed by Wilson *et al.* (29), an initially positive diphasic wave is ascribed to depolarization advancing toward the exploring electrode, whereas an initially negative diphasic wave is ascribed to receding depolarization. The assumption is required that each cell in the system under study be continuous in the direction of movement of the potential. This system was loosely adapted to the study of the cortex by Marshall *et al.* (19), and it is now widely accepted that a surface-positive wave represents an ascending depolarization, whereas a surface-negative wave represents depolarization moving from the surface to the base. While this analysis is useful for describing the electrical activity of many parts of the nervous system, e.g. peripheral nerve (16) and some central nuclei (17), it does not appear to be valid for the prepyriform evoked potential. This potential is distinct from those systems in that it moves tangentially over the surface without an initially positive wave, until the borders of the cortex are reached. The total distance of spread (up to 14 mm. or more) is greater than the horizontal extension of the indigenous cells. Since a triphasic action potential can usually be recorded prior to the evoked potential, and since the latter follows the former by an interval compatible with monosynaptic transmission, it seems clear that the superficial spread of the evoked potential is due to successive activation of contiguous cells by the propagated action potential. In one sense the surface spread of activity can be ascribed to cells with continuous extension which produce an initially positive moving potential, but this description applied only to the exogenous fibers of the lateral olfactory tract and their triphasic action potential.

The spread of activity from the surface to the base of the cortex seems at first glance to follow the basic pattern described by Wilson *et al.* (29), since there is an inversion of the waveform. However, evidence was found excluding the possibility that the surface wave moves to the base (cf. 22). First, there was no demonstrable latency of either potential in directions normal to the surface, whereas movement was readily measured in directions parallel to the surface. Second, there was little and often no change in potential recorded monopolarly at the turnover, which is incompatible with the passage across the turnover of a continuously moving e.m.f. Third, bipolar records from electrodes in a line normal to the cortex had waveforms which approached the first derivative of the concomitantly recorded monopolar forms with respect to distance. Bipolar records taken from electrodes in a line parallel to the surface had forms which approached the first derivative of the monopolar forms with respect to time. The latter case is compatible with a moving e.m.f., as in peripheral nerve (16), the former only with a stationary e.m.f. Fourth, small lesions made in the superficial layers of the cortex resulted in diminution of both the negative and positive phases of the evoked potentials subsequently recorded in the lesions, whereas identical lesions in the deep layers did not cause diminution of either phase so recorded. This finding implies that the spread of current from the surface to the base occurs in the same fashion as the spread of action currents in peripheral nerve into the region of a nerve block (16), i.e. the e.m.f. of the evoked potential does not leave the superficial layer of the cortex. These findings do not exclude the possibility that there is a delay in the movement of the evoked potential from the surface to the zero isopotential; although no evidence was found to support this possibility, the precision of the measurements was not sufficient to permit an inference to be made one way or the other.

In any case, the deep cortical potential must be explained on some other basis than the propagation of an e.m.f. from the surface. An explanation can be derived from the finding that the evoked potential is the product of an approximate dipole field, which is spatially distorted by the high impedance of the underlying skull. This finding implies that during the surface-negative phase a net positive charge is moving into cells in the surface of the cortex, while approximately the same net positive charge is moving out of cells in the base of the cortex. Whether the "extracellular" tissue space is inside or between the glia is for the moment not important (18). Either there are two sets of cells, one at the surface undergoing "depolarization" and one at the base undergoing concomitant "hyperpolarization," or else a single set of cells is present with an intracellular bridge extending across the zero isopotential. Since only superficial e.m.f. could be demonstrated for the evoked field of potential, it is inferred that a single set of cells is present, with parts lying on both sides of the pyramidal cell layer. [O'Leary and Bishop (22) observed what appears to have been the same phenomenon in the avian optic lobe; they concluded that it represented evocation of successive responses by fast fibers.] Since the pyramidal cells rather than the glia have been shown to have this orientation and extension, and since the dendrites ramify through the superficial pole, it is inferred that the e.m.f. of the dipole field are associated with the dendrites (cf. 6, 7, 24). The development of the surface-negative evoked potential may therefore be seen to occur in five stages: (i) the activation of the tips of the dendrites by the action spikes of fibers of the lateral olfactory tract (21, 25); (ii) the spread of this change to the vicinity of the cell body at an indeterminate rate; (iii) the activation of e.m.f. causing the movement of a net positive charge into the cell; (iv) the departure of this charge across the axon membrane; and (v) the migration of this charge through the extraneuronal tissue space back to the surface of the cortex.

The validity of this interpretation postulated by Gesell (11) depends on the core conductor property of the neuron. The ability of peripheral nerve axon to sustain a longitudinal current has been known since the discovery of the demarcation current. Accurate measurements of this property in both peripheral nerve axons and the cell bodies and dendrites of brainstem nuclei have been made by Lorente de Nó (16, 17) and by Shanes (26). The theory implies the existence of intraneuronal currents of far greater density than those in the extraneuronal medium. This is reflected in peripheral nerve in the high current density at the nodes of Ranvier (27). Such localized peaks of potential could not be discerned in the extraneuronal prepyriform field; intracellular recordings were not attempted. However, some indirect evidence of the existence of intraneuronal currents of high density was found in the form of an injury potential of very high amplitude with the waveforms of both evoked and spontaneous activity. Such potentials were found in the vicinity of the main dendritic and axonal shafts of the cortical pyramids with barbed microelectrodes most likely to transect those shafts.

Comparison of the prepyriform evoked field with the antidromically or orthodromically evoked field of spinal motoneurons (8) shows that the former is predominantly dipolar whereas the latter is predominantly monopolar and negative. Hence the term "dendritic depolarization" encompasses two processes, one in which a net positive charge moves into the cell and transiently remains there in association with a change in membrane potential, and the other in which it immediately moves across and out of the cell in association with a change in dendritic (not necessarily membrane) e.m.f. Since the dendrite constitutes an electric double layer (16) and since intracellular recordings are not available, it cannot yet be known whether the e.m.f. lies in the membrane, in the mitochondria or other internal structure, or even in the satellite cells adjacent to the dendrite. The observed excess of negative potential of the superficial pole over the positive potential of the deep pole may perhaps in part be explained on the basis of the coexistence of these processes involving the prepyriform dendrites during the evoked potential.

The occurrence in some cats of evoked potentials with an initially deep negative wave and e.m.f. localized by microlesions to the deep layers indicates that the same explanations hold for the surface-positive responses as for the surface-negative responses. It is not clear whether these two types of response are due to the presence in the prepyriform cortex of large numbers of bipolar pyramidal cells (21) or to two functionally distinct sets of cells, each with processes extending across the pyramidal cell layer. In the four cats examined the zero isopotential for the two types of response was the same, but the poles of the two types of response did not coincide in position in the cortex.

Comparison of the evoked and spontaneous potentials showed that along lines normal to the surface the distribution of amplitudes and phase shifts of the two types were identical, whereas along lines tangential to the surface there was not such a close correlation of amplitudes. These results are consistent with the anatomical orientation of the prepyriform pyramidal cells normal to the surface. Presumably activity at one end of a cell will determine the electrical activity at the other, but activity in one cell will not necessarily determine the activity of its neighbor, although each may reflect the activity of a common input. In view of the many other similarities between the two types of electrical activity which confirm previously established evidence derived from studies of the neocortex (6, 7, 24), it is concluded that the e.m.f. for the two types are the same. In answer to the question raised in the introduction the spread of "EEG waves" from the prepyriform cortex involves all three of the processes listed. There is spread of activation by propagation or electrotonus along the dendrites of each pyramidal cell. Ions (or electrons) move through the interior of the cell forming pulsating or alternating currents of high density, which cross the axon membrane in amounts presumably determined by neuronal, glial, or vascular impedances around the axon. There is movement of this current to and from the surface of the cortex in the form of an extraneuronal field causing an IR drop in adjacent tissues. Finally there is successive activation of adjacent cells, which gives the illusion of a potential continuously propagated tangentially over the surface.

Some tentative conclusions suggested by these answers are (i) that the site of origin of the prepyriform dipole field is in or adjacent to the dendrites; (ii) that this is not the only type of field generated by the prepyriform cortex; (iii) that the most probable site of action of the prepyriform dipole field is in the immediate vicinity of the axons, whether in or beneath the cortex; (iv) that the recording of potentials with the type of electrodes used in this study is a measure of the "line voltage" of the return current path; and (v) that the magnitude of the line voltage may be irrelevant to the local activity of many of the structures the current traverses.

Summary

The surface-negative potential evoked in the prepyriform cortex of the anesthetized cat by stimulation of the lateral olfactory tract spreads tangentially over the surface of the cortex with a measurable latency and without inversion or a preceding positive wave; lesions of the surface block the spread of the potential beyond the lesions. The surface potential spreads to the base without demonstrable latency and with inversion; small lesions in the superficial layers diminish the amplitude of the evoked potential recorded in the lesion, whereas similar lesions in the base do not diminish the evoked potential recorded there. An electrical map of the evoked potential shows that it is the product of a dipole field of current. One pole lies in the molecular layer and contains the electromotive forces (e.m.f.) of the dipole; the other pole lies in and below the cellular layers of the cortex and is passive. Evidence was found that extraneuronal and intraneuronal currents of highest density occur across that part of the zero isopotential lying between the poles. It is concluded that the evoked potential spreads tangentially over the surface because of successive activation of the e.m.f. in or around the dendrites of the pyramidal cells. This e.m.f. drives current into the interior of the cells; the deep potential is produced by the passive emergence of the current from the axons. In several cats an evoked field was also found with e.m.f. in the basal layers of the cortex; this field had its zero isopotential at the same location, but the active and passive poles were reversed, and the potential was initially surface-positive. Evidence is presented that the dominant waveform of the spontaneous prepyriform electrical activity is generated by the same e.m.f. and has substantially the same distribution in time and space as the prepyriform evoked potential.

Acknowledgments

Reprinted with permission from *J. Neurophysiol* (1958), **22**, 644–665.

This investigation was supported by a Postdoctoral Fellowship (BF-6317-C) from the National Institutes of Health, United States Public Health Service, and currently by a Fellowship from the Foundations' Fund for Research in Psychiatry.

References

1. Adrian, E. D. Olfactory reactions in the brain of the hedgehog. *J. Physiol.*, 1942, **100**: 459–473.
2. Allen, W. F. Distribution of cortical potentials resulting from insufflation of vapors into the nostrils and from stimulation of the olfactory bulbs and pyriform lobe. *Amer. J. Physiol.*, 1943, **139**: 553–555.
3. Berlucchi, C. Ricerche di fine anatomia sul claustrum e sull' insula del gato. *Riv. sper. Freniat.*, 1927, **51**: 125–157.
4. Berry, C. M., Hagamen, W. D., and Hinsey, J. C. Distribution of potentials following stimulation of olfactory bulb in cat. *J. Neurophysiol.*, 1952, **15**: 139–148.
5. Bishop, G. H. and O'Leary, J. L. The polarity of potentials recorded from the superior colliculus. *J. cell. camp. Physiol.*, 1942, **19**: 289–300.
6. Chang, H.-T. Dendritic potential of cortical neurons produced by direct electrical stimulation of the cerebral cortex. *J. Neurophysiol.*, 1951, **14**: 1–21.

7. Clare, M. H. and Bishop, G. H. Properties of dendrites; apical dendrites of the cat cortex. *EEG clin. Neurophysiol.*, 1955, **7**: 85–98.
8. Fatt, P. Electric potentials occurring around a neurone during its antidromic activation. *J. Neurophysiol.*, 1957, **20**: 27–60.
9. Freeman, W. J. Oscillating corticonuclear dipole in the basal forebrain of the cat. *Science*, 1957, **126**: 1343–1344.
10. Fox, C. A., McKinley W. A., and Magoun, H. W. An oscillographic study of olfactory system in cats. *J. Neurophysiol.*, 1944, **7**: 1–16.
11. Gesell R. The driving forces of increased breathing. Pp. 221–230 in: Moulton, F. R., ed. *Blood, heart, and circulation*. Washington, D. C., Science Press, 1940.
12. Hasana, B. Ueber die elektrischen Begleiterscheinungen an der Riechsphare bei der Geruchsempfindung. *Pflug. Arch. ges. Physiol.*, 1934, 234: 748–755.
13. Kaada, B. R. Somato-motor, autonomic and electrocorticographic responses to electrical stimulation of "rhinencephalic" and other structures in primates, cat, and dog. *Acta physiol. scand.*, 1951, **24** (Suppl. 83). See Chap. III: Cortical distribution of potentials evoked by olfactory bulb stimulation in cat and monkey.
14. Landau, W. M. An analysis of the cortical response to antidromic pyramidal tract stimulation in the cat. *EEG clin. Neurophysiol.*, 1956, **8**: 445–456.
15. Liberson, W. T. Recent advances in Russian neurophysiology. *Ann. Rev. Physiol.*, 1957, **19**: 557–588. (See p. 562).
16. Lorente de Nó, R. A study of nerve physiology. *Stud. Rockefeller Inst. med. Res.*, 1947, **131, 132.**
17. Lorente de Nó, R. Action potential of the motoneurons of the hypoglossus nucleus. *J. cell comp. Physiol.*, 1947, **29**: 207–287.
18. Luse, S. A. Electron microscopic observations of the central nervous system. *J. Biophys. Biochem. Cytol.*, 1956, **2**: 531–542.
19. Marshall, W. H., Talbot, S. A., and Ades, H. W. Cortical response of the anesthetized cat to gross photic and electrical afferent stimulation. *J. Neurophysiol.*, 1943, **6**: 1–15.
20. MacLean, P. D., Horwitz, N. H., and Robinson, F. Olfactory-like responses in pyriform area to non-olfactory stimulation. *Yale J. Biol. Med.*, 1952, **25**: 159–172.
21. O'Leary, J. L. Structure of the primary olfactory cortex of the mouse. *J. comp. Neurol.*, 1937, **67**: 1–31.
22. O'Leary, J. L. and Bishop, G. H. Analysis of potential sources in the optic lobe of duck and goose. *J. cell. comp. Physiol.*, 1943, **22**: 73–87.
23. Papez, J. W. *Comparative neurology*. New York, Thomas Y. Crowell Co., 1929. xxv, 518 pp.
24. Purpura, D. P. and Grundfest, H. Nature of dendritic potentials and synaptic mechanisms in cerebral cortex of cat. *J. Neurophysiol.*, 1956, **19**: 573–595.
25. Ramón y Cajal, S. *Studies on the cerebral cortex (limbic structures)*. Tr. L.M. Kraft. London, Lloyd-Luke, 1955. xi, 179 pp.
26. Shanes, A. M. Frog nerve as generator of current and voltage. *J. cell. comp. Physiol.*, 1947, **29**: 207–287.
27. Tasaki, I. *Nervous transmission*. Springfield, Ill., C. C Thomas, 1953. 164 pp.
28. Ward J. W. Field spread potentials of the olfactory mechanism. *Amer. J. Physiol.*, 1953, **172**: 462–470.
29. Wilson, F. N., Macleod, A. G., and Barker, P. S. *The distribution of currents of action and injury displayed by heart muscle and other excitable tissue.* Ann Arbor, Univ. Of Michigan Press, 1933. vii, 59 pp.

2. *Linear Models of Impulse Inputs and Linear Basis Functions for Measuring Impulse Responses*

Next I undertook systematic exploration of the behavioral correlates of this well defined generator of background and odor-induced EEG and electrically evoked potentials in waking cats by using arrays of permanently placed bipolar electrodes (Freeman 1960a,b), first by presenting different odorants to search for changes in amplitude and frequency distributions of the EEG, then by using different unconditioned stimuli to elicit reactions relating to hunger, thirst, rage and attack, fear and flight, sexual arousal, and stages of drowsiness and sleep. The only significant variations were in the amplitude of the EEG, which increased in the low-frequency range with sleep, and in the higher frequency **gamma** range (Bressler and Freeman 1980) with the degree of arousal and motivation. There were no patterns that were specific to either the odorant stimuli, or the responses to them, or the type of motivation. An exception was the suppression of bursts in the EEG with sneezing, yawning, sniffing, or other changes peculiar to respiratory patterns. Even in sleep there was only a modest increase in slow wave delta activity. These negative results were not very exciting.

The questions emerged, how were the EEG gamma oscillations generated, how were they enhanced by arousal, and how were they suppressed by sleep and anesthesia? The single shock evoked potential also oscillated, but with a strong tendency to decay rapidly in amplitude compared with the long bursts of oscillation that accompanied inhalation. Early on I conceived the notion that when an animal learned to identify a stimulus, the mechanism of increased sensitivity to that stimulus might be expressed as a kind of "resonance" that enhanced the tendency for oscillation (Freeman 1961b). Specifically, if a waking subject was trained to respond to an electrical stimulus given to the lateral olfactory tract, the oscillatory evoked potential might last longer, or its amplitude might increase.

In order to explore this notion I needed a means for precise and reliable measurement of the evoked potential. I explored several ways to measure the degree of oscillation and of the "resonance" manifested by the evoked activity (Freeman 1961b). As mainframe and desktop computers became widely available for storing, averaging and analyzing neural potentials, I settled on the use of nonlinear regression to fit curves to evoked potentials after they had been digitized and averaged. By application of the principle of superposition to the classical physiological procedure of paired-shock testing (a "conditioning" stimulus is closely followed by a "test" stimulus), we showed that a linear domain of function existed for cortical evoked potentials that did not exceed the amplitude of the ongoing EEG (Freeman 1962; Biedenbach and Freeman 1965). By use of the Fourier transform and polar plots of the real and imaginary components (Figure 1 in Chapter 2) I showed that the sum of two damped sine waves sufficed to incorporate better than 95% of the variance in the great majority of evoked potentials from all of the experimental subjects (Figure 2 in Chapter 2). This was a major breakthrough in the acquisition of experimental data for dynamical modeling, because the same equation held for all animals and all behaviors. Only the coefficients changed. Because the basis functions (elementary curves) were linear, they conformed to the solutions to linear differential equations, opening the way to modeling the dynamics by piecewise linear approximation.

This method of analysis is widely used for developing models of neural dynamics. For example, three decades of linear analysis of the properties of axons provided the basis for the 1952 Hodgkin–Huxley nonlinear model. Linearization of neocortical models is implicit when data are analyzed by Fourier decomposition in tonotopic studies in the auditory system, and by fitting sine waves to time ensemble averages of EEG and peristimulus histograms showing oscillations in the firing probabilities of selected neurons in the visual cortex (Eckhorn *et al.* 1988; Gray *et al.* 1989). The measurements are used to estimate phase values at best frequencies, which support interesting speculations about "phase locking" and "binding" by synchronization in cortical information processing. No one seriously thinks that cortical dynamics is linear, but a linearized approximation is the best place to start (Freeman 1962, 1964b, 1972e, 1975, 1987a, 1992a).

Use of Digital Adaptive Filters for Measuring Prepyriform Evoked Potentials from Cats

A set of techniques is described for making digital measurements on averaged cortical evoked potentials by means of the IBM 7090. The basis for measurement is the selection of a set of elementary waveforms called basis functions, which are scaled by coefficients and added to replicate observed waveforms. Optimization of the coefficients is by nonlinear regression followed by cross-correlation between observed and calculated waveforms. By this means the information contained in the shape of each averaged evoked potential is incorporated into a set of numerical coefficients. Unresolved problems include the need for improved basis functions and for methods for determining confidence intervals for the coefficients.

Introduction

Cortical evoked potentials are of interest primarily as tests of changing neuronal excitabilities accompanying normal brain function. The first three steps in the analysis of these complex waveforms are proper placement of electrodes for recording, the proper choice of electrical or sensory stimulus parameters, and the establishment of behavioral control. The fourth is development of techniques for reliable measurement.

Measurement consists of comparison of an unknown entity with a set of standard scales or dimensions having numerical attributes in preassigned degree. A physical object can be described by the dimensions of size, mass, density, etc. In addition there are dimensions such as location, velocity, weight, hardness, etc. Some of these dimensions can be complex (e.g. size depends on three or more subsidiary coordinates), and some can be interdependent or nonorthogonal (e.g. specification of size and mass may determine density). In each dimension the unit is defined with reference to a standard physical entity, e.g. a unit of mass or length, and the result of measurement is expressed as an equivalence between the unknown and the sum of a specified number of units of that entity.

The dimensions of a complex waveform are elementary waveforms from which that waveform can be built by simple addition. Any finite single-valued function of time is admissible. They are called basis functions (10, 15), and they can be expressed in numeric as well as geometric form. The selection of each basis function is equivalent to the choice of a physical entity for the unit of measurement and along a dimension specified by that entity. Measurements along these dimensions are scale factors for the basic functions equivalent to the amplitudes and latency coefficients of the basis functions, and therefore to the number and time of onset of unit basis functions equal to the observed waveform.

One of the most commonly used basis functions is the unit square wave. By choosing an appropriate unit height and width and adding the units as in a histogram, the amplitude of any waveform can be specified to any desired degree of precision. By repeating the procedure any reasonable number of times and specifying the latency at those times, enough data accumulates to permit reassembling the basis functions to reconstruct the observed waveform whenever it is needed. The existence of this basis function is implicit in analog- to-digital conversion.

Another common basis function noted for its easy use is the empirical "hump" or peak. The expression of a complex waveform as a succession of peaks having specified latencies, polarities, amplitudes, and durations can permit reconstruction if the shape is given also, as it usually is pictorially. The example introduces an important modification in the form of a flexible or adaptive basis function. The shape of a "hump" can be kept essentially the same while its width is varied. This allows considerable economy of expression, for only one generic shape need be stored as the source for a whole class of basis functions, which can be stretched to fit the curve.

The usually large number of coefficients needed to match the observed waveform by the above systems of measurement implies that there is redundancy or overdetermination in the numerical results of measurement. There exists also a number of basis functions for which there is no redundancy, i.e. the dimensions are independent, and the set of coefficients required for a stated level of precision is the minimum number required to achieve that level. However, neither parsimony nor orthogonality is a necessary condition for measurement, and redundancy carries the hope of subsequent cross-check between a set of measurements.

If precision, simplicity, and economy are insufficient, what shall be the criteria for selection of basis functions? The requirements posed here are conformity of shape to the suspected time courses of real events producing the waveform, and adaptiveness along dimensions parallel to proposed changes in the state of the structure in which the events take place. More generally, the development of a system of measurement requires the choice of those dimensions which are thought to reflect degrees of essential (or desired) qualities, while rejecting those attributes of the unknown tentatively regarded as trivial or accidental, i.e. as noise. This rejection of available information is inevitable no matter what set of dimensions is chosen. Therefore, in dealing with an unknown for which uncertainty exists as to which attributes are essential, the validation of a proposed system of measurement must be based on criteria beyond mere faithfulness of reproduction. The nature and structure of the over-all system on which the observations are made must be considered in choosing useful parameters.

This report is concerned with problems of measurement in cats of the prepyriform electrical response evoked by stimulation electrically of the lateral olfactory tract, for which criteria of electrode placement and choice of stimulus parameters have been standardized (7). A set of linear basis functions is used to measure averaged responses. Preliminary validation of the choice of these dimensions and of the methods of application is based on comparison of the original data with curves calculated using the results of measurement and on comparison of the Fourier transforms of the calculated and observed waveforms. Further assay of the system as a whole is based on a derivation of these basis functions from a mathematical model for the physiology of this cortex, and on behavioral studies. The latter include testing of the ability of the system to distinguish between cortical evoked potentials recorded in differing behavioral states and estimation of covariance between the results of measurement and a concomitant measure of behavior. These more broadly based evaluations of the system will be discussed in later reports.

Methods

The techniques used for permanent electrode orientation in cats, for choice of stimulus parameters, and for behavioral control have been described (7). Signals reported in this study were recorded bipolarly across the prepyriform dipole using a Grass 5P5 difference preamplifier and a Mnemotron 400-A CAT. The averaged evoked potentials (AEP) were transmitted on-line in analog form by telephone wire to a nearby digital voltmeter and tape recorder. Each AEP was sampled 100 times at intervals of 1.25 msec over a sample duration one-quarter (0.31 msec) of the digitizing interval. The values were stored on magnetic tape in binary form and subsequently were scaled, plotted, and punched on cards by an IBM 7090. Further processing was entirely by use of FORTRAN programs written for the 7090.

The three adverse conditions encountered in recording from this cortex were time-variance, nonlinearity, and noise. The experimental conditions used for sampling the cortical response to single electrical shocks were adjusted toward a balanced reduction of these effects. The "noise" consisted primarily of the spontaneous EEG signal generated by this cortex during normal respiration, which required the use of averaging to improve the signal-to-noise ratio. The stimulus rate was set as high as possible without allowing any stimulus to fall during the detectable prepyriform response to a preceding stimulus. The time base for the averaging of the response was fixed at 125 msec, beginning simultaneously with the delivery of the stimulus to the tract.

The difficulty then arose that the frequency of spontaneous activity was usually about 40 per sec giving a wavelength of 25 msec. Five such waves precisely equaled the total duration of the time interval for averaging, and the long bursts of spontaneous activity lasting I to 2 sec were then found to give oscillatory averaged records even in the absence of electrical stimulation coherent with the time base. Therefore, the stimulus repetition rate was placed at 7.3 per sec for a duration of 137 msec (halfway between 125 msec for 5 cycles and 150 msec for 6 cycles at 40 per sec). In all cases reported here the response had terminated within that duration.

The cortical evoked potential was found to vary continuously with fluctuations in behavioral state, but because of the "noise" it was not possible to observe these changes over short periods of time. The decision was made to place each cat in a stereotyped behavioral sequence (5) consisting of anticipation (waiting in the starting box of an ergometer for the door to open), working (pulling on the rope in the ergometer), and lapping (consuming milk at the end of the runway). The AEP was taken during each of these three states over identical periods of 12.5 sec, this duration being the usual limit over which consistent work could be obtained. This interval limited the number of single-shock stimuli during any one period of observation to ninety-four.

As described previously (6) the evoked potential changes in form with increasing stimulus intensity from a low-amplitude, undamped, high-frequency, relatively sinusoidal oscillation to a high-amplitude, heavily damped oscillation at low frequency. The optimal intensity for stimulation was that which yielded oscillation at or near the frequency of spontaneous activity for each cat.

At this optimal intensity paired-shock stimulation showed that the superposition property held for the response to a good approximation, that it could be treated as a linear function. At higher stimulus intensities the amplitude of the second of two overlapping responses was generally lower, and the frequency was higher, implying the presence of stimulus-dependent amplitude and frequency nonlinearities. Moreover, the covariance of the AEP with behavior diminished with development of its functional change (7). Ease and effectiveness of analysis both required that stimulus intensity be kept as low as the other conditions would allow. In some cats the achievement of an acceptable signal-to-noise ratio required prolonged averaging at such an optimal stimulus intensity. In this study it was necessary to accept a limit on the period of averaging. Therefore in each cat the optimal intensity was determined, and, if necessary, the stimulus intensity was further increased until the initial maximum peak of the AEP was at least 10× greater in amplitude than the mean peak amplitude of ninety-four averaged traces of spontaneous activity. Stimulus current was then measured (the biphasic pulse duration firing fixed at 0.1 msec), and this same current was used for all subsequent recordings.

From 25 to 40 AEP were taken from each cat on any one day, after several weeks of daily practice in the ergometer. In association with each mean a measurement was made of the mean rate of work done by the cat during the 12.5-sec period of averaging. This set of behavioral measurements was subsequently used as the dependent variable in a multiple regression on the accompanying measurements of the AEP.

Results

The Choice of Basis Functions

The principal dimension or basis function for the evoked potential was the waveform of a damped sinusoid (complex exponential), having the four coefficients of initial amplitude (V), frequency ($\omega = 2\pi f$), decay rate (α), and phase of onset (ϕ). The sum of two of these basis functions was required for adequate representation, so that specification of each AEP required at least eight coefficients, i.e. four for each of the two dimensions (Fig. 1).

The equations for these basis functions were:

$$V_1(t) = V_1 \sin(\omega_1 t + \phi_1) e^{\alpha_1 t},$$
$$V_2(t) = V_2 \sin(\omega_2 t + \phi_2) e^{\alpha_2 t}.$$

There were several constraints on these two basis functions. Their amplitudes were zero for all times prior to and at time zero (taken as 1.25 msec after delivery of the evoking stimulus, which was approximately equal to total conduction and synaptic delay in this cortex). They could not have the same frequency. The amplitudes of each had to decay to values less than the noise level (approximately 10% of initial peak values) within the 125-msec averaging interval which restricted α_1, and α_2 to values less than zero.

Matching the initial rising negative portion of the AEP required the use of two additional basis functions, each in the form of a real exponential having the two coefficients of amplitude and decay rate.

$$V_1'(t) = V_1' e^{\beta_1 t},$$
$$V_2'(t) = V_2' e^{\beta_2 t}.$$

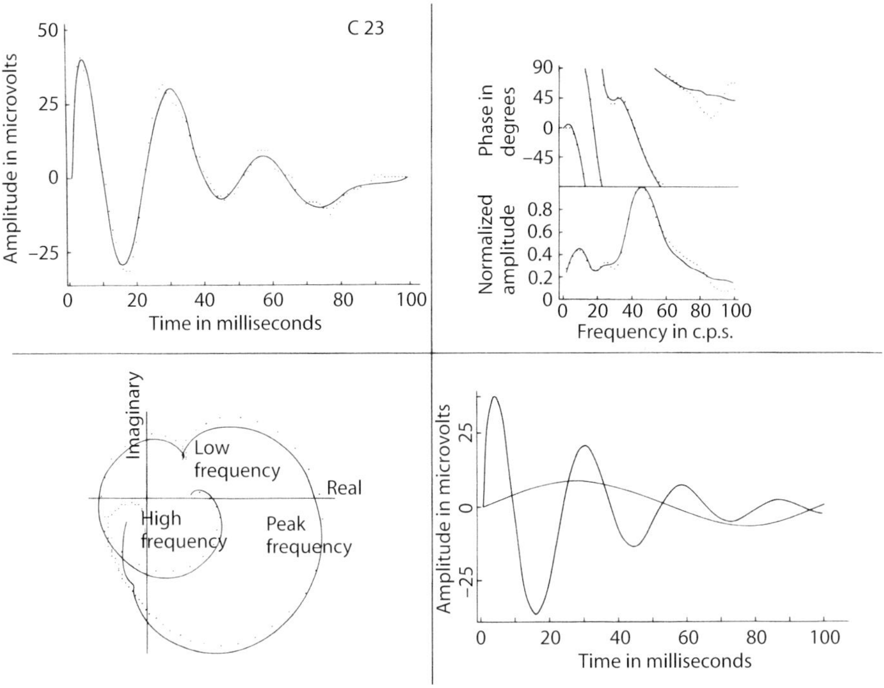

Fig. 1. Upper left: a set of digital values for an AAEP (averaged average evoked potential) as dots, with a curve representing the sum of four optimized basis functions. Lower left: polar plot of the Fourier transform of observed (dots) and calculated (curve) values. Lower right: amplitude and phase derived from the polar plot. Upper right: the two sinusoidal basis functions, each with an added real exponential function, for which the sum is shown at upper left.

The initial amplitudes, V_1' and V_2', for each of the two additional basis functions was set equal to the negative of the amplitude of one complex basis function at time zero in order to give zero amplitude.

$$\begin{aligned} V_1' &= -V_1 \sin\phi_1, \\ V_2' &= -V_2 \sin\phi_2. \end{aligned}$$

Furthermore, because the rise of the AEP to peak value took place within two to four digitizing intervals, the information available was not adequate to define effectively both the coefficients β_1 and β_2. For this reason they were set equal to a single constant, β. The sum of these four basis functions then became:

$$\begin{aligned} V(t) = V_1 &[\sin(\omega_1 t + \phi_1) e^{\alpha_1 t} - \sin\phi_1 e^{\beta t}] \\ &+ V_2 [\sin(\omega_2 t + \phi_2) e^{\alpha_2 t} - \sin\phi_2 e^{\beta t}]. \end{aligned}$$

This was the equation used to generate waveforms for comparison with prepyriform AEP. In the next two sections methods are described for evaluating these coefficients.

The Construction of a Matched Filter

Evaluation of the nine coefficients was undertaken by the method of least squares (14). That is, values for the nine coefficients were sought by means of regression, which gave the best fit between the observed and generated waveforms in a least squares sense. The two main problems encountered (once the type and number of basis functions had been determined) were uncertainty about the true waveform for each cat, and the presence of transcendental terms in the basis functions that rendered the "normal" equations nonlinear.

The uncertainty about waveform was owing to the restrictions imposed on the recording conditions, which caused the introduction of "noise" into the AEP. The remedy used was to average a set of ten to forty AEP from any one cat on any one day and take the average AEP or AAEP as the best attainable observation of the true waveform under existing experimental conditions. The AAEP was then fitted with the basis functions using nonlinear regression in the general form proposed by Gauss in 1855 (8). The first step for this was measurement of AAEP by graphic means to make a set of initial "guesses" as to what the values for the parameters might be.

Experience provided several rules of thumb. The initial deflection was always negative in voltage, but because it was displayed as upwards it was expressed as positive in amplitude. By convention both initial amplitudes were taken as positive. The larger of the two amplitudes was taken as that of the dominant basis function, V_1, and the smaller as that of the subsidiary basis function, V_2. One of the two frequencies almost always lay within a few cycle per sec of the commonest frequency for spontaneous activity for that cat (38–45 cycle per sec); the other (either dominant or subsidiary) was higher or lower. The phase of onset of the dominant basis function was always near 90° ($\pi/2$ radians), whereas that of the subsidiary was usually negative.

The dominant frequency, ω_1, was estimated by measuring the average interval between zero crossings, doubling this value to obtain wavelength, and dividing that into 2π. An initial value for α_1 was found by sketching curves through the succession of upward and downward maxima of the AEP to form a mirror-image pair of exponentials. The negative reciprocal for the time (in seconds) from zero to the point at which these fell to l/e (37%) of their initial values sufficed for α_1. Extrapolation of these curves to the ordinate at time zero gave the initial guess for V_1. The phase ϕ_1 was set to $\pi/2$. A free-hand sketch was made of the dominant basis function, which then was subtracted from the AEP. The above procedure was then repeated on the residual waveform to derive initial guesses for the four subsidiary coefficients. The coefficient β was routinely set to –400.

A valuable aid in this guesswork was provided by the Fourier transform of the AEP. The Fourier integral

$$\mathrm{AEP}(\omega) = \int_0^\infty \mathrm{AEP}(t)\mathrm{e}^{-\mathrm{i}\omega t}\,\mathrm{d}t$$

was converted by use of Euler's theorem to an expression in terms of sines and cosines. It was further modified to a summation rather than an integral to permit operation on the 100 digital values in which the AEP was coded:

$$\text{AEP}(\omega) = \sum_{n=1}^{100} \text{AEP}(t_n)\cos(\omega t_n)\Delta t - \text{i}\sum_{n=1}^{100} \text{AEP}(t_n)\sin(\omega t_n)\Delta t,$$

where Δt was the digitizing interval (0.00125 sec). This was expressed as the sum of real and imaginary parts for each frequency:

$$AEP(\omega) = \text{Re}(\omega) + \text{Im}(\omega).$$

It was convenient to calculate these sums for values of frequency from 0 to 100 cycle per sec at intervals of 2 cycle per sec. The results of this numerical integration (Fig. 1) were plotted both in polar coordinates as the locus of a frequency vector and as the amplitude and phase of this vector.

$$\text{Amp}(\omega) = [\text{Re}(\omega)^2 + \text{Im}(\omega)^2]^{1/2}$$

$$Phs(\omega) = \tan^{-1}[\text{Im}(\omega)/\text{Re}(\omega)].$$

The polar plot was useful for determining the type and number of basis functions. The amplitude curve provided the optimal basis for estimating ω_1 and ω_2.

After inserting these values into the basic equation for $V(t)$ a curve was generated and compared with the AEP. The difference between these two curves was expressed by subtracting the calculated from the observed amplitudes at each of the 100 values for time, squaring the differences, and adding to form a sum of squares of the differences:

$$\text{SSD} = \sum_{n=1}^{100} [\text{AEP}(t_n) - V(t_n)]^2$$

Optimization of the nine coefficients was undertaken in order to minimize SSD.

The curve, $V(t)$, based on initial guesses for the nine coefficients, $c_1, c_2, ..., c_9$), was used to search for a better curve, $F(t)$, and a new set of "improved" coefficients, c_1', c_2', ..., c_9'). The proposed changes in coefficients were $\Delta c_1 = c_1' - c_1$, $\Delta c_2 = c_2' - c_2$, etc. The assumption was made that the initial guesses were sufficiently close to the desired values that the improved curve could be written as a first-order Taylor series expansion of the initial curve in the altered parameters:

$$V(t) = f(t, c_1, c_2, \ldots, c_9),$$
$$F(t) = f(t, c_1', c_2', \ldots, c_9'),$$
$$F(t) = V(t) + \Delta c_1\, \partial V(t)/\partial c_1 + \ldots + \Delta c_9\, \partial V(t)/\partial c_9.$$

The values, k_i, for the set of partial derivatives of the function, $V(t)$, were found numerically, so that the differences between the initial and desired curves could be expressed as a linear equation in a set of nine unknowns, i.e. the increments in the coefficients:

$$F(t) - V(t) = k_1\Delta c_1 + k_2\Delta c_2 + \ldots + k_9\Delta c_9$$

This linear function was then fitted to the numerical values of $[\text{AEP}(t) - V(t)]$ using the usual method of least squares, i.e. the nine "normal" equations were formed by partial differentiation with respect to each of the nine coefficients, these were set equal to zero to minimize SSD, and the desired increments in parameters were found by solving the nine simultaneous equations by successive substitution.

The increments were added to the initial guesses to provide a new set of nine parameters, from which a new curve was generated and also a new value for SSD. If the new SSD was less than the previous one, the procedure was repeated; if not, the differences between the new and old coefficients were reduced until the new SSD was less than the old. The procedure was repeated as often as necessary until successive trials failed to give significant decreases in SSD or until a preset number of iterations was exceeded (in practice usually 4–10).[1]

The generated curve not uncommonly failed to converge onto the observed waveform, in which case a new set of initial guesses was supplied. The criterion for an acceptable level for SSD was somewhat arbitrary. The sum of squares of the amplitudes of the generated curve divided by 100 gave an estimate of "signal" power, and SSD per 100 estimated "noise" power. The signal-to-noise ratio in practice was required to exceed 10:1 for suitable AAEP. Values at times exceeded 150:1, although values between 20:1 to 30:1 were more common.

A useful test for comparing the generated curve with the AAEP was the application of the Fourier transform to the generated curve. Not infrequently the observed and generated curves in the time domain showed a modest but seemingly acceptable signal-to-noise ratio, but their transforms displayed prominent divergences, particularly in the low frequency cusps and loops. The differences instigated renewed search for optimal values of coefficients and often provided essential clues as to what those values should be.

The set of optimal coefficients defined a "matched filter," which was (for exponential basis functions) the best linear equivalent to the system that generated the AAEP. That is, the optimized basis functions could be used (among other things) as a set of directions for building an electrical filter network, which on impulse driving would produce as output a waveform identical to the sum of the basis functions. Apart from theoretical justification the term "filter" was appropriate, because any or all of the four basis functions could be subtracted (i.e. filtered) from the AEP to leave residues of one or more basis functions or merely the noise (unidentified residue). This became important for the next step.

The Use of Adaptive Filters

It was desired next to measure successive AEP in order to evaluate sequential changes in cortical responsiveness to electrical stimulation accompanying

1 Further technical considerations are not warranted here. The iterative computations are so laborious as to require the use of a high-speed computer. Several FORTRAN programs for nonlinear regression have been written in recent years, which are available in computer center libraries, and which differ in their detailed operations particularly with regard to recent modifications in technique (e.g. 3, 9). The program used here (written by R. M. Baer of the Berkeley Computer Center) is available from the Center under the code G2-BC-NONL.

behavior. There were two reasons for not using regression for this. First, it was believed that simultaneous optimization of all nine coefficients to fit the AEP would result in the best least squares fit of the basis functions to signal-plus-noise and not to the signal embedded in the noise. Second, nonlinear regression was found to converge fairly rapidly during the first three or four iterations but thereafter to approach a limiting value of SSD very slowly. Successive values for coefficients, on the other hand, fluctuated seemingly erratically and to a much larger degree than was reflected in changes in SSD. There were no assignable confidence limits on the optimized values, so that a prolonged series of iterations was required to make fine adjustments in the basis functions and to impart intuitive confidence in the end result. Each iteration required 20 to 30 sec of computer time, which made routine use of this procedure on AEP prohibitively costly.

The matched filter was used as the basis for identification of the "signal" in the "noise" of the successive AEP by cross-correlating the digitized AEP with the digitized curve representing the sum of optimized basis functions. However, simply identifying the AAEP in each AEP by cross-correlation was not sufficient. The additional assumption had to be made that the difference between each AEP and the AAEP was in part due to noise but in part also to variation in waveform. To seek out this variation and express it as changes in the coefficients the basis functions were changed systematically in search of a maximal cross correlation between observed and generated waveforms. That is, the matched filter from the AAEP was adapted to each AEP by changing the coefficients one at a time. Each AAEP coefficient was regarded as the mean for a set of values in the AEP, and the set of AAEP coefficients as the centroid of a hyperellipsoid distribution of values for the AEP.

The matched filter was adapted to single AEP as follows. From each AEP were subtracted ("filtered") the two real exponential basis functions and the subsidiary complex basis function generated from the optimized coefficients. This in essence constituted a pre-whitening transformation using the subsidiary basis functions as a noise filter. The mean amplitude of the filtered AEP was subtracted from all values to correct the filtered data to zero mean. Then three basis functions, $B(t)$, were generated from the coefficients of the dominant basis function, one having the frequency specified by the matched filter, the second being 1.0 cycle per sec lower, and the third being 1.0 cycle per sec higher. Each of the three was normalized to zero mean and cross-correlated in turn with $F(t)$, the filtered AEP. The cross-products between each pair of values in the two sets of data occurring at the same time, t_n, were added together:

$$\sum_{n=1}^{100} B(t_n) \cdot F(t_n).$$

This sum was divided by the square root of the sums of squares of both $B(t_n)$ and $F(t_n)$ to give the correlation coefficient, r, between the two sets of data:

$$r = \sum_{n=1}^{100} B(t_n) \cdot F(t_n) \Bigg/ \left[\sum_{n=1}^{100} B(t_n)^2 \sum_{n=1}^{100} F(t_n)^2 \right]^{1/2}.$$

If the higher (or lower) frequency curve showed the greater correlation an additional curve was generated and cross-correlated at a frequency increased (or decreased) by 1.0 cycle per sec. This was repeated until a maximal value for r was obtained. The same stepwise search was carried out for the optimal value of the decay constant α_1 in steps of 7.0. If α_1 was changed from the initial value, frequency, ω_1, was again tested and optimized.

The lag (phase expressed in units of time) of the generated curve shift was now shifted in steps of 0.125 msec (equivalent to 2° of phase for a frequency of 40 cycle per sec), and r was computed for each of 30 lags in both directions from that lag specified by ϕ_1. A new value for ϕ_1 was calculated from the lag giving the highest r. The search was returned to renewed testing of ω_1 and α_1 this time in steps of 0.5 cycle per sec and 3.5, respectively.

Values for r for the dominant component usually ranged from 0.90 to 0.99 with a median value of about 0.95. The value for β was left unchanged. The computer time required was from 3.6 to 12 sec with a median of about 6 sec.

The optimized dominant basis function and the real exponentials were then subtracted from the AEP, and the same search procedure was applied to the smaller component. These values for r usually ranged from 0.55 to 0.95 with a median value between 0.75 and 0.80. The amplitudes of the basis functions were then optimized by regression, requiring an additional 4 sec of computer time. After this stage the values for ϕ_1 and ϕ_2 were corrected for a systematic error introduced during initial averaging. Four digitizing channels were used with sequential sampling but with simultaneous display, so that the waveform taken from channel 2 had to be lagged 0.31 msec after that on channel 1, etc. The three-stage process of adapting the matched filter required usually from 12 to 20 sec of computer time for each AEP, which was 1/20 the time required using nonlinear regression.

The trial-and-error selective search used at the cross-correlation stages was a cost-cutting maneuver carrying the risk that local maxima for r might be encountered, thus terminating the search prior to its completion. Likewise, the use of discrete-interval approximations to cut costs constituted restriction of the range of variation of the coefficients, impairing to some extent the statistical properties of the results. Experience (to be discussed in a later report) showed that these methodological defects were detectable, and that their effects could be mitigated by running a set of data through the entire three-stage adapting procedure for a second time, using the "individualized" coefficients as the starting point rather than those of the AAEP. These considerations also prompted the use of fairly wide intervals for the search for values of ω and α, with a return to search using smaller intervals after specification of lag. Experience was too limited to specify precisely what the best intervals might be.

Normal Values of the Coefficients

These filter techniques were applied to the AEP of twelve cats to find mean values. The resulting mean optimized coefficients and their standard deviations are shown

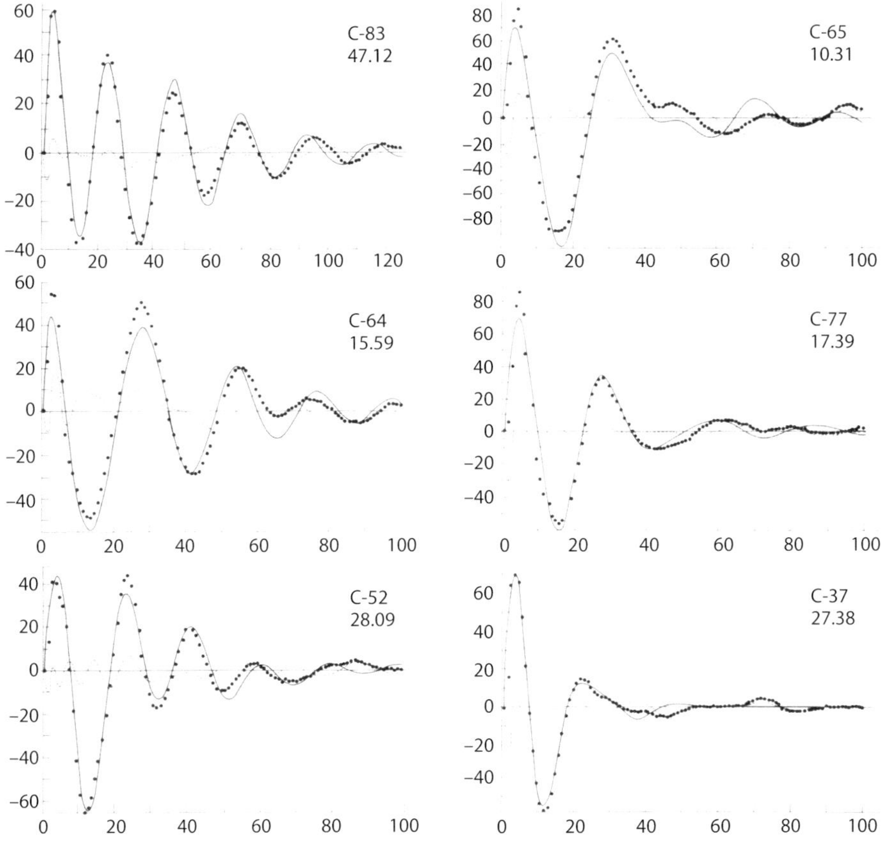

Fig. 2. The dots represent values for AAEP. The solid curves are generated from the mean coefficients (Table 1) after fitting the AEP. The dashed curves are the mean of residuals after subtracting each set of optimized basis functions from the corresponding AEP. Under the label for each cat is shown the signal-to-noise ratio for the solid curve with respect to the mean residuals.

in Table 1. Corresponding AAEP are shown for six cats in Fig. 2 as dots representing the 100 digital values of amplitude, and the sum of optimal basis functions as curves. The mean residuals for sets of AEP (Fig. 2) showed that periodicities often existed other than those two incorporated into the dominant and subsidiary basis functions. These were not explored in detail, but inspection showed that all cats had activity in the evoked potential in the range of 60 to 120 cycle per sec. In addition many cats showed evoked activity at a frequency nearly but never exactly half that of the spontaneous activity, which corresponded to an additional peak observed in the EEG power spectra of these cats.

Repeated attempts were made in several cases by means of nonlinear regression to fit generated curves to AAEP using three basis functions and thirteen coefficients. At each trial the result was a singular matrix, in the process of solving the "normal"

Table 1 Mean coefficients and SD for twelve cats

Cat	V_1 μv	ω_1 radians/sec	α_1 radians/sec	ϕ_1 radians	V_2 μv	ω_2 radians/sec	α_2 radians sec	ϕ_2 radians	S:N AEP	$\overline{S:N}$ AAEP
C37	109.4	210.49	−58.34	1.62	75.76	348.4	−81.97	−0.59	15.03	27.38
	±4.38	±19.11	±3.33	±0.15	±5.59	±7.75	±11.72	±0.06		
C51	88.75	174.18	−30.1	1.71	47.98	231.78	−21.16	1.54	10.06	11.30
	±4.06	±6.95	±2.20	±0.15	±5.64	±6.75	±4.42	±0.17		
C52	74.19	253.68	−33.71	0.27	23.82	106.34	−24.52	−2.53	21.42	28.09
	±5.82	±9.83	±5.42	0.2	±3.26	±17.33	±4.21	±0.54		
C58	64.41	178.44	−33.48	0.6	46.41	271.75	−55.95	−1.1	14.95	15.94
	±4.42	±7.41	±4.21	±0.11	±3.87	±11.69	±12.76	±0.14		
C64	113.6	183.64	−36.3	1.82	16.86	233.05	−11.46	−1.14	10.03	15.59
	±6.57	±5.42	±3.90	±0.08	±2.04	±10.53	±8.87	±0.25		
C65	114.1	150.8	−41.66	1.83	38.13	296.88	−48.52	−0.79	8.82	10.31
	±7.44	±5.62	±2.93	±0.06	±4.98	±13.27	±13.95	±0.19		
C70	142.2	240.06	−62.16	1.08	140.1	32.28	−48.99	−0.69	26.42	38.94
	±10.08	±17.40	±6.71	±0.14	±9.42	±3.11	±9.31	±0.09		
C77	69.06	177.07	−32.51	1.46	36.93	255.61	−44.16	−0.27	11.71	17.39
	±5.04	±16.80	±6.16	±0.23	±7.34	±24.67	±17.13	±0.26		
C83	55.17	280.86	−26.47	0.98	42.63	371.65	−82.29	−0.79	24.86	47.12
	±8.76	±6.62	±4.90	±0.11	±9.38	±20.90	±30.03	±0.17		
C85	113.8	202.32	−48.16	1.29	37.9	360.97	−68.54	−1.03	16.10	25.87
	±4.76	±9.75	±0.55	±0.10	±7.82	±24.99	±10.72	±0.17		
C86	90.8	183.56	−54.35	1.68	15.9	250.88	21.71	0.74	22.02	31.05
	±8.70	±13.83	±4.39	±0.09	±7.52	±7.77	±10.39	±0.18		
C89	93.79	208.57	−43.09	1.44	37.84	465.83	−156.5	−2.13	8.73	32.64
	±10.94	±22.27	±9.66	±0.16	±13.82	±108.78	±101.71	±0.81		

[a] Symbols: w = frequency values; a = decay rate; ø = phase; V = amplitudes; S:N = average signal-to-noise ratio for the set AEP; S:N = ratio for the AAEP; subscripts 1 and 2 refer, respectively, to the dominant and subsidiary components.

equations, implying that the additional dimension was not sufficiently independent of the other two to allow this, i.e. the information in the waveform of the evoked potential sufficed to fix only nine coefficients. In brief, two sinusoidal basis functions could always be fitted to an AAEP but never three.

Discussion

The mathematical techniques used in this study were selected from standard engineering procedures described in various textbooks (e.g. 4, 10, 12), so no attempt is made here to derive, justify, or explain them in detail. Emphasis has been placed on a simplified version of the theory of waveform measurement (2, 11, 13) and on the sequential steps taken to make these measurements. These same steps (but not necessarily the same basis functions) are applicable to any of the wide variety of waveform encountered in physiological studies.

There are several difficulties and sources of error in the applications reported here, which center on the unidentified residuals shown graphically in Fig. 2. These residuals were termed "noise" largely for the purpose of indicating the probable error of the results of measurement in the form of a signal-to-noise ratio. A more precise statement of the uncertainty of measurement would take the form of confidence intervals for each of the coefficients. However, nonlinear regression does not provide confidence intervals, and the multiple determinations from AEP include variance from changes in cortical state as well as variance from errors of measurement. For neither of these kinds of variance is the distribution known. Although their sums approach the Gaussian distribution in form for the separate coefficients, they are not precisely normally distributed, and a large normal distribution for one source of variance might obscure a skew or bounded deviation for another. Independent estimates for these coefficients will be required ultimately to resolve this uncertainty.

The unidentified residuals include several components. Truly random "white" noise having a flat spectrum is undoubtedly present but to a small degree. In addition there is "unwanted" signal, i.e. the sum of superimposed traces of spontaneous EEG activity, which often have dominant frequencies in the range of frequencies in the AEP, but which may have been excluded from incorporation into the signal by virtue of a deviant phase of onset or rate of decay. There are very low or very high frequency subsidiary quasi-sinusoidal components that are too small to be included in the original matched filter. Most importantly, the results of systematic errors of measurement accrue in the residuals. Partly these are from use of discrete interval approximations, but the greater proportion appears to stem from the use of linear basis functions to describe inherently nonlinear, time-variant processes.

The reason for selection of real and complex exponentials as the basis functions in this study was the finding that superposition holds to a good approximation over the linear dynamic range of cortical function being examined (6). Roughly speaking, this is a property of so-called dendritic potentials (1), which are graded and do not display refractory periods, differing in this from the highly nonlinear characteristics of muscle and nerve action potentials.

Yet there is clear evidence for a proportionality between amplitude and frequency in the AEP of many cats, such that the frequency increases as the AEP amplitude decreases over the 125-msec duration. Most cats have multimodal spectra with peaks rather close together, so that low frequency "beat" patterns may appear (e.g. C65). Distributed peaks in spectra occur, because of either multiple concomitant frequencies in the population of cortical neurons, or time-variance in frequency during averaging, or both causing too-rapid decay of AEP (e.g. C52). Much of the unidentified residue can be attributed to these phenomena, which point to the need for appropriate nonlinear basis functions to supplant those used here.

Wiener's theory for design of optimal filters (13) from which the present approach was derived (2, 12) assumes that the "noise" in which the signal is being sought is truly random. The extent to which the above types of coherence in the "noise" distort the estimates for the means and variances of the coefficients most certainly needs to be assessed as an integral aspect of a theory for measurement of cortical evoked potentials, but that is primarily a physiological rather than a statistical problem.

Finally, no attempt has been made to use other methods of curve-fitting (e.g. moments, chi-square) in order to evaluate the method of least squares in terms of cost, reliability, and precision. If digital measurements of evoked potentials are to be made routinely, as they can and must be, efforts will have to be directed also toward efficient utilization of the computer resources now available to make them.

Acknowledgements

Reprinted with permission by Academic Press Inc. from *Experimental Neurology*, Volume 10, Number 6, 475–492, December 1964.

Supported by grants (MH-06686 and NB-050350) from the National Institutes of Health, United States Public Health Service, and by the Berkeley Computer Center.

References

1. Bishop, G. H. 1958. The dendrite: receptive pole of the neuron. *Electroencephalog. Clin. Neurophysiol.* **Suppl. 10**: 12–21.
2. Bode, H. W., and C. E. Shannon. 1950. A simplified derivation of linear least square smoothing and prediction theory. *Proc. IRE* **38**: 417–425.
3. Box, G. E. P., and K. B. Wilson. 1951. On the experimental attainment of optimum conditions. *J. Roy. Statist. Soc.* **B13**: 1–45
4. Brown, R. G., and J. W. Nilsson. 1962. *Introduction to Linear Analysis.* Wiley, New York.

5. Freeman, W. J. 1959. An ergometer for measuring work from cats as an index for drive. *J. Appl. Physiol.* **14**: 1071–1072.
6. Freeman, W. J. 1962. Alterations in prepyriform evoked potential in relation to stimulus intensity. *Exptl. Neurol.* **6**: 70–84.
7. Freeman, W. J. 1963. The electrical activity of a primary sensory cortex: the analysis of EEG waves. *Intern. Rev. Neurobiol.* **5**: 53–119.
8. Gauss, C. F. 1855. *Méthode des moindres carrés; mémoires sur la combinaison des observations.* Bertrand, Paris.
9. Hartley, H. O. 1961. The modified Gauss–Newton method for fitting of nonlinear regression functions by least squares. *Technometrics* **3**: 269–280.
10. Huggins, W. H. 1960. Representation and analysis of signals. Part VII. Signal detection in a noisy world. *Johns Hopkins University Report No. AFCRC-TN-60–360.*
11. Lerner, R. M. 1959. The representation of signals. *IRE, Trans. Circuit Theory* **6**: 197–216.
12. Smith, 0. J. M. 1958. *Feedback Control Systems.* McGraw-Hill, New York.
13. Wiener, N. 1949. *Extrapolation, Interpolation, and Smoothing of Stationary Time Series.* Wiley, New York.
14. Wolfenden, H. H. 1942. *The Fundamental Principles of Mathematical Statistics.* Macmillan, Toronto.
15. Young, T. Y., and W. H. Huggins. 1961. Representation and analysis of signals. Part VIII. Representation of electrocardiogram by orthogonalized exponentials. *Johns Hopkins Univ. Report No. AFCRL-187.*

3. *Rational Approximations in the Complex Plane for Laplace Transforms of Transcendental Linear Operators*

Oscillations in a complex system such as a neuron or an area of cortex imply that recursive exchanges of ions and energy are taking place. Dynamic modeling starts with writing two coupled differential equations having at least two state variables, the output of each equation providing input to the other. These relations describe a feedback loop. One of the two equations models the forward limb, and the other models the feedback limb of the loop. In order to evaluate the coefficients in each of the two equations, it is desirable to open the loop and to record and measure the input and output of each limb. I was able to open the negative feedback loop in the prepyriform cortex by inducing deep surgical anesthesia, and to measure the resulting open loop impulse response of the forward limb by fitting to it the curve from a sum of real-valued exponential terms. My task was to restructure the equation and interpret it in terms of postsynaptic potentials from dendrites of pyramidal cells. The first question was, were the gamma oscillations the product of feedback within the individual neurons, which then synchronized their discharges by their synaptic interactions, or were they the product of interactions between excitatory and inhibitory neurons? The next question was, how could the zeros be identified, counted, and evaluated? The poles were easy to find and count, but they could not be correctly evaluated without the zeros. Finding the zeros is the hardest problem in linear analysis.

I already knew from surgical experiments isolating the olfactory system (Chapter 5) that the gamma oscillations were not driven by other parts of the brain. I had tested the competing hypothesis that the oscillations were intraneuronal in accordance with the Hodgkin–Huxley model. My two predictions from that model failed. First, the distributions of the interpulse intervals of single units did not show a peak at the wave length of the dominant frequency of the EEG in the gamma range (40–80 Hz) or its harmonics. Instead, those distributions reflected a Poisson process. Second, application to the cortex of solutions rich or deficient in calcium or containing a chelating agent did not modulate the EEG frequency in the predicted directions. In contrast, the oscillations were profoundly affected by chemicals that enhanced or blocked synaptic transmission (Biedenbach 1966). These findings gave a strong indication that the oscillations were a mesoscopic property of feedback among neural populations (Figure 1 in Chapter 3) and not the microscopic property of synchronized neuronal oscillators. Proof, however, had to come from a model of the synaptic interactions.

Landmark work had already been done mainly by Wilfrid Rall (1962) on modeling the cable properties of the dendritic arbors of single neurons. In order to model populations of neurons I sought to develop ways to surmount the incredible complexity of the neuropil that contained indefinitely large numbers of neurons and synapses (Koch and Laurent 1999) in great variety. My approach was to suppress the gamma oscillations and all unit activity by deep anesthesia, induce the open loop dendritic potential by electrical impulse stimulation of an afferent axonal path, fit the sum of exponential curves to the averaged evoked potential (Figure 3 in Chapter 3), and interpret the rate coefficients in terms of successive delays from my electrical stimulus to the extracellular dendritic response. This forward chain began with the terminal axon propagation and dispersion, followed by the synaptic delay and dendritic cable delay, and concluded with a passive membrane decay and a dendritic after-potential (Figure 4 in Chapter 3). In order to model these delays I used a rational approximation for the transcendental operators that were required to describe serial dispersive, synaptic, and cable delays.

The more difficult problem was to find a compact way to model the feedback limb of the cortex in the closed loop state of animals that were waking or under light anesthesia. By serendipity I found that a rational approximation of the cable model for a one-dimensional diffusion process was well suited to model the properties of lumped negative feedback with one delay term T_2 and one gain term K_2. This formulation did not by itself resolve the question of whether mechanisms of EEG oscillations were intracellular or intercellular, but it optimized the statistical properties of measured values of frequency and decay rate from large numbers of averaged evoked potentials. Specifically, when the values for the frequency and decay rate of damped cosines fitted to averaged evoked potentials were transformed to values for K_2 and T_2, the distributions of the coefficients were normalized, and the covariance between K_2 and T_2 was linearized. These properties of the K_2–T_2 transform enabled conventional linear multivariate statistical analyses of the parameter values (Freeman 1964c).

One of the uncertainties in using the open loop transfer function to model the closed loop responses was to determine the point of entry of the feedback path. This is essential, because the open loop poles of the feedback limb become zeros in the closed loop transfer function (Freeman 1975). The forward model of delays had to be broken into two delays: that which represented afferent axonal propagation, and that which represented intracortical synaptic and cable delays (Figure 1 in Chapter 3). In order to evaluate my digital adaptive filters I collected evoked potentials two or three times a week from animals with implanted electrodes in sets of 30 trials. During each trial I measured the rate of work done by the animal to get food. Using multiple regression I then calculated the multiple correlation coefficient between the rate of work as the dependent variable and the set of measured coefficients from the evoked potentials. I repeatedly tested and discarded sets of basis functions for measurement, using as my criterion the maximization of the multiple correlation coefficient predicting the rate of work, on the premise that the best technique of measurement would yield the highest correlation of evoked potential parameters with the behavioral assay (Freeman 1964c). In this way the lumped negative feedback model displayed in terms of root locus plots (Figures 7 and 8 in Chapter 3) served to validate the system for measuring the cortical impulse responses, so that the way was opened to reliable testing of further predictions of the model.

Induction and evaluation of open loop impulse responses and closed loop zeros are required for modeling the dynamics of neocortical feedback. Sadly, another example had not yet been published for any area of neocortex at the time of this writing.

A Linear Distributed Feedback Model for Prepyriform Cortex

A mathematical description is developed for the oscillatory prepyriform evoked potential, which is based on the concept of recurrent inhibition in a large pool of cortical neurons. A basic single-loop transfer function is presented in serial fashion, with stepwise exclusion of nonlinearities so that the over-all function can be represented in terms of Laplace transforms. Transcendental terms are eliminated where possible, and those remaining are treated by rational approximations. Estimates of parameters in the forward channel of the loop are based on measurements of a putative open-loop response obtained from cats under heavy pentobarbital anesthesia. Parameters in the feedback channel are then estimated from measurements of the phase, frequency, and decay rate of a typical prepyriform averaged evoked potential from a cat in the waking state. The model is presented in its simplest form as a linear time-invariant system based on essential characteristics of single neurons and their dominant interconnections. It is proposed as a basic schematic that can be further developed to make its predicted output compatible with the patterns of gross electrical activity recorded from this cortex.

Introduction

In recent studies of the superposition characteristics of induced activity in the spinal motoneuron pool, Granit and his co-workers (21, 22) have shown that certain excitatory and inhibitory events in the spinal cord add linearly over a wide dynamic range. Granit (21) has further called attention to the fundamental identity of his formal model with the system of linear equations proposed by Hartline and Ratliff (26) to describe interaction patterns for neurons in the crustacean eye. It is difficult to overstate the importance of these findings, for they indicate that whereas description of the properties of single cells requires the use of nonlinear differential equations such as those devised by Hodgkin and Huxley (27), the properties of neuron populations might be adequately described by linear models. Owing to the high development of linear analysis, physiologists would be free to adapt and use existing mathematical techniques for describing the dynamics of complex neuronal systems, instead of being forced either to devise and test their own nonlinear methods or to persevere in the use of intuitive logic. Nowhere might this be more advantageous than in the study of cortical potentials (6).

Since the first systematic analyses of dorsal root potentials 30 years ago (4), repeated suggestions have been made that cortical potentials might be homologous with the relatively long-lasting synaptic potentials and afterpotentials generated by differing types of spinal neurons. In recent years interest has centered on representation of evoked potentials in terms of the excitatory (EPSP) and inhibitory (IPSP) postsynaptic potentials as seen in the spinal motoneuron (12) because of the occurrence of these same patterns of electrical activity in single neurons of motor and sensory neocortex (30, 31, 35, 36), archicortex (29, 42), and cerebellum (23). The postulation that sequences of alternating peaks of cortical potential might reflect successive peaks of multineuronal IPSPs and EPSPs, while still tentative, seems now to have more experimental evidence in its support than does any alternative view (2, 12, 15, 24, 31, 35, 36, 42, 46, 47).

The prepyriform evoked potential sometimes has a particularly simple waveform composed of alternating negative and positive peaks (15). Suppose that the basis for this response were a sequence of EPSPs and IPSPs in the population of neurons generating the over-all response. Suppose in addition that the mechanism of alternation were an inhibitory reaction to initial excitation followed by rebound excitation and reinhibition, this being the gross counterpart of the lateral, recurrent, or surround inhibition observed in the activities of single cells in numerous sensory and motor systems (23, 26, 28, 47). Can a linear model be used to describe the system quantitatively? What would be its parameters, and how might they compare with those for single-neuron models? What changes in the model would be required to reproduce the varieties of waveforms the prepyriform cortex generates under differing experimental conditions?

This report begins with the derivation of a transfer function, which is a linear dynamic formulation of the cortical mechanism generating the evoked potential. This function is reduced to minimal complexity. The means are described for evaluating its parameters from measurements of prepyriform averaged evoked potentials (AEP), and for predicting the basic waveform for this cortex. The model is neither unique nor complete, nor does it offer proof of the several assumptions made. Rather it is a quantitative formulation of those assumptions with exploration of some of the implications and difficulties to which they lead.

Methods

The techniques for electrode implantation in cats, stimulation, recording, behavioral control, and averaging of evoked potentials have been described in previous reports (15, 17). Computations for measuring AEP were performed on the IBM 7090 using matched and adaptive digital filters (17). Descriptions of the Laplace transform and of the root locus techniques used here are given in numerous engineering texts of which those by Harris (25), Brown and Nilsson (9), and Smith (41) are particularly useful. Root locus techniques are used in preference to phase space or describing functions, primarily as a convenience in handling the feedback systems with distributed parameters, and also because loci for moving poles provide simple graphic display of effects of both time-variance and nonlinearity.

Results

The Basic Cortical Transfer Function

The simplified anatomical substrate for this model (34, 38) is conceived as a sheet of pyramidal cells having their dendrites oriented predominantly toward the surface and their axons toward the white matter. The dominant input path is a tract of fibers extending over the dendritic surface, with collaterals extending into the dendritic layer to form synapses with all parts of the dendrites of each pyramidal cell. Efferent axons or their collaterals terminate on interneurons and other pyramidal cells, thus establishing numerous lateral connections. The density of interconnection of any single cell with its neighbors is assumed to diminish exponentially with horizontal distance from that cell in the manner described by Sholl (40), Braitenberg (8), and others (5, 24, 45).

The proposed linear transfer function is shown in Fig. 1, which relates input to recorded output through a series of operations, $C(s)$, expressed by means of their Laplace transforms. Additional loops representing parallel paths have been omitted for simplicity. Two types of input exist, a quasirandom multichannel pulse train, $n(t)$, that fluctuates in pulse density (as does prepyriform input with such factors as respiration), and an impulse driving, $\delta(t)$, equivalent to that provided by an electrical stimulus to the lateral olfactory tract. For the fixed level and low intensity of $\delta(t)$ assumed here $n(t)$ and $\delta(t)$ are considered noninteracting or superimposable. The tract threshold need not be considered. There is a pure dead time representing

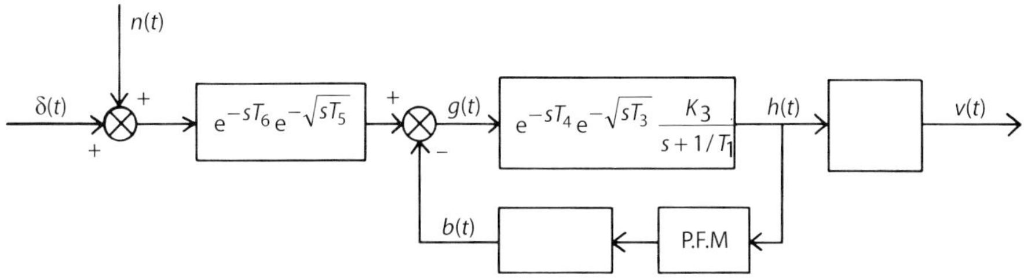

Fig. 1. The basic cortical transfer function.

conduction delay, T_6, in the tract and a distributed lag, T_5, owing presumably to a distribution of conduction velocities in the tract (7). These can be minimized by proper electrode placement and choice of the origin for time in the calculations and need not be further considered.

The signal, $g(t)$, enters a synaptic delay, T_4, followed by a distributed lag, T_3, owing to the cable properties of the pyramidal cell dendrites (37). This is cascaded into a first-order system with a time constant T_1, representing the dynamics of an average EPSP for the cells in the population. K_3 is considered equal to $1/T_1$.

Physiological evidence indicates that a pulse train should ensue at $h(t)$, which varies in frequency as a linear function of the elevation of the EPSP above a threshold value (21, 44). It is assumed formally that the noise, $n(t)$, maintains the average level of activity in the population at all times at a level such that explicit representation of a threshold element can be omitted. It is further assumed that owing to the large numbers of active cells in the pool the pulse frequency (or rather pulse density owing to the spatial dimensions also involved) can be described by a continuous function, which is directly proportional to the observed AEP analog voltage change (21). This allows omission of explicit representation for nonlinear analog-to-digital pulse-frequency modulation (PFM). Physiologically there are two outputs at this stage, one a multichannel pulse train entering the feedback loop and the other a time-dependent voltage change seen as an extracellular IR potential difference. Mathematically there is a single continuous function, $h(t)$, that is scaled by the respective gain constants, K_1 and K_2

Apart from gain, K_2, the properties of the feedback path are incorporated in a single distributed lag, T_2, which represents the sum of a set of as yet undefined synaptic or propagation and electrotonic delays, or both. It is assumed that population values for each type of sequential delay are normally distributed, so that the total set of delays is also normally distributed. For an impulse input to such a system the feedback signal, $b(t)$, therefore has the shape of a skewed Gaussian curve, the mathematical expression for which is identical with that for a cable causing the same peak delay (11, 41). It is assumed to be negative or inhibitory and therefore to produce a reduction in the pulse density [equivalent to reduced magnitude of $g(t)$] at the feedback summation point. Subsequently there must be a reduction below normal of pulse density and therefore of its analog, $h(t)$, in the forward channel.

After an additional distributed delay, T_2, there must occur a deficit of inhibitory feedback represented by $b(t)$ at the feedback summation point. Physiologically, the presence of $n(t)$ is essential to cause reexcitation in proportion to disinhibition, resulting in the start of a second and subsequent cycles. Mathematically the difficulty can be handled by equating $n(t)$ to a positive constant greater than peak negative values of $b(t)$, unless $n(t)$ must for other reasons be set near or at zero, in which case a change in configuration of the model is required (see next section). Except for some attenuation, phase lead, and slowed rise time in the initial negative (excitatory) peak, the impulse response of this system has precisely the form of an exponentially decaying sine wave (41). The decay rate and the frequency depend on the loop gain equal to the product K_2K_3 and on the several delay constants, T_1–T_4.

Two additional simplifications are made. K_1 depends in part on the magnitude of $\delta(t)$, the location of the gross extracellular recording electrodes with respect to the

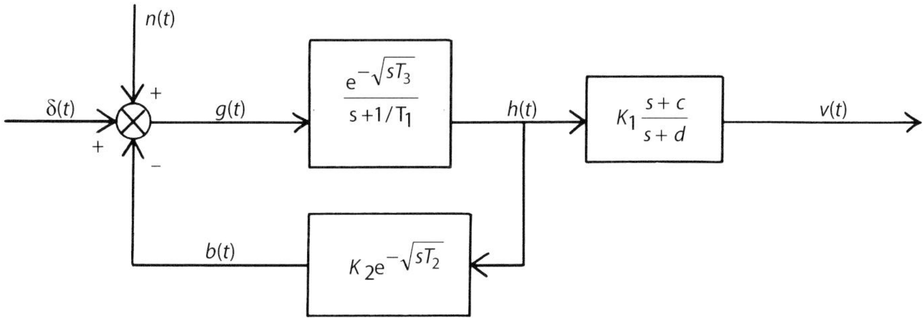

Fig. 2. The simplified cortical transfer function.

cortex, the number and position of the active cells, the gross shape of the cortex, and other factors in addition to single cellular properties. Neither K_1, K_2, nor K_3 can be estimated independently at present. K_3 can be incorporated into K_2, which redefined becomes loop gain, and into K_1 which becomes the gain constant in the opened loop. The effects of T_4 on the initial half cycle can be eliminated by adjustment of the time origin for the calculations, and its remaining effects are included in the much larger quantity, T_2, so T_4 is dropped. The simplified system configuration is shown in Fig. 2.

The numerator term, $s+c$, represents a real zero near the origin of the s-plane and the denominator term, $s+d$, represents a real pole far to the left of the origin. In a previous report (15) these were interpreted in terms of the transfer function proposed by McAlister and used by Freygang and Frank (19) to compare intracellular with extracellular microelectrode recordings across passive membrane of motoneurons. It appears likely in the present analysis that a passive membrane model is not valid here, and that the zero, c, represents mainly the fact that the controlled input to the system, $\delta(t)$, is a current pulse rather than a voltage pulse. The position of this zero on the real axis also includes the cumulative phase lag effects of those delays (T_4, T_5, and T_6) that were deleted from the model. The pole, d, represents the reciprocal of the time constant of active membrane manifested by the rise-time of an axosomatically induced postsynaptic potential, but also the cumulative rise-time effects of T_4, T_5, and T_6. The attenuation effects of T_5 (if any) are included in K_1.

The end result of model-building should be simplification. The prepyriform cortex is a part of larger systems, and its parts (the cells) are in turn composed of smaller parts. The "part" as conceived at each level of analysis may become no less complex as its size and range of function diminish. But in order to build an understanding of a neuron from knowledge of the function of cell parts and then also an understanding of an area of cortex based on the properties of its neurons it is necessary to abstract the essential features of the parts into algebraic expressions, and to combine them so that the relations between the parts are expressed by algebraic operations. The starting point of this study is the function of the pyramidal cell en masse reduced to

$$G(s) = \frac{\exp[-(sT_3)^{1/2}]}{s + 1/T_1}. \quad [1]$$

The extracellular stimulation-recording operation becomes

$$H(s) = K_1[(s+c)/(s+d)]. \quad [2]$$

The transfer function of the cortical interaction term is

$$F(s) = K_2 \exp[-(sT_2)^{1/2}]. \quad [3]$$

Each of these is given the simplest form of expression compatible with organization of the complex subsystems into a higher system, i.e. the prepyriform cortex as a whole:

$$C_s = G(s)H(s)/[1 + G(s)F(s)] \quad [4]$$

which is the closed loop transfer function between excitation and the response of the system shown in Fig. 2.

In the following two sections a means is described for evaluating the parameters in *G(s)*, *H(s)*, and then *F(s)*. A simpler form of *C(s)* is derived, which is a rational function approximation for the transcendental function given in Eq. (4). Let the input be a unit impulse ¶(s). Then the output, *V(s)*, is equal to *C(s)*, and

$$C(s) = K(s+c)/(s^2 + \alpha_1^2 \pm j\omega_1^2)(s^2 + \alpha_2^2 \pm j\omega_2^2)(s+d) \quad [5]$$

From Eq. [5] the basis functions used to measure prepyriform AEP can be derived (17). The time-variance and the inherently stabilizing nonlinearity of the model are explored with reference to prepyriform AEP in a later report (16).

The Effect of Pentobarbital

Experimentally, intraperitoneal injection of a surgically anesthetizing dose of pentobarbital (50 mg/kg) in a cat having electrodes previously implanted in the prepyriform cortex changes the normally oscillatory waveform to that form seen in Fig. 3. During the induction phase the frequency, ω, and rate constant, α, of the AEP as defined in Eq. [5] decrease, often with a decrease (13) in stability angle $[\zeta = \tan^{-1}(-\alpha/\omega)]$. As ω reaches about 40% of its initial value the second negative and all subsequent cycles disappear, and about 20 min after the initial dose the first positive peak shrinks to a very small amplitude. The first negative peak may decrease slightly in amplitude but not in latency or rise time. This form persists for several minutes and coincides with almost total suppression of spontaneous prepyriform EEG activity, and with maximal suppression of spinal and autonomic reflexes. As the peak of the anesthetic effect passes, the same sequence is seen in reverse. The positive overshoot deepens, the subsequent train of oscillations reappears along with EEG activity and in 40–60 min the AEP appears close to normal. It remains so for the 4- to 6-hour duration of surgical anesthesia.

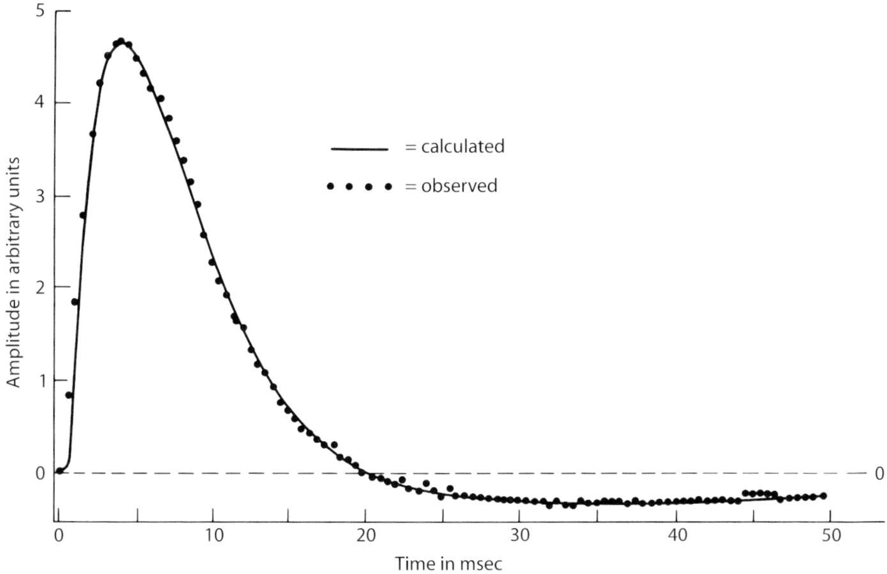

Fig. 3. The prepyriform AEP under pentobarbital.

For replication of the AEP under the condition of minimal positive overshoot the feedback gain, K_2, is set to zero. For correspondence of performance of the model to the disappearance of EEG waves, $n(t)$ is also set to zero. In the model the drop in K_2 can be attributed to a fall in $g(t)$ below an excitatory threshold in the forward limb due to the fall in $n(t)$. The transfer function then becomes

$$C(s) = K_1 \frac{\exp[-(sT_3)^{1/2}]}{s+1/T_1} \frac{(s+c)}{(s+d)}. \qquad [6]$$

The distributed lag term (11) for $\exp[-(sT_3)^{1/2}]$ can be represented in the s-plane by a line source coincident with the negative real axis. An analytic approximation is

$$e^{-\sqrt{sT}} = \frac{\left[\dfrac{(sT)^2}{(\alpha_1 T+\pi^2)^2} - \dfrac{2(\alpha_1 T-\pi^2)}{(\alpha_1 T+\pi^2)^2} sT + 1\right]}{(1+4sT/\pi^2)(1+4sT/3^2\pi^2)(1+4sT/5^2\pi^2)(1+4sT/7^2\pi^2)\ldots} \times \left[\frac{(sT)^2}{(\alpha_1 T+3^2\pi^2)^2} - \frac{2(\alpha_1 T-3^2\pi^2)}{(\alpha_1 T+3^2\pi^2)^2} + 1\right]\ldots, \qquad [7]$$

which specifies (41) a set of complex zeros in the right half of the s-plane and a set of poles on the negative real axis having values for αT of $-(\pi/2)^2, -(3\pi/2)^2, -(5\pi/2)^2$, etc. These poles represent the rate constants of a set of exponential decay terms, that can be added in the time domain to reproduce the waveform of the impulse response of a distributed lag element. In the case where the impulse response is described by a set of discrete-valued amplitudes at regular time intervals, it is possible to omit consideration of all poles having time constants much shorter than the sampling interval. The first two poles in this approximation suffice for present needs.

The locations for the complex zeros are given by the roots of the numerator terms in Eq. [7]. These zeros can be placed far to the right of the $j\omega$-axis in the s-plane by choice of the arbitrary constant, $\alpha_1 >> n^2\pi^2/T$:

$$sT = (\alpha_1 T - n^2\pi^2) \pm 2jn\pi\sqrt{\alpha_1 T}, \quad n = 1, 3, 5, \ldots, 2n+1. \qquad [8]$$

The open loop transfer amplitude effects of the zeros on the exponential terms can, therefore, be expressed by a lumped gain constant. Owing to the fact that K_1 is unspecified these effects are included in K_1.

The transfer function for the cortex under deep pentobarbital becomes

$$C(s) = \frac{K_1(s+c)}{(s+2.44/T_3)(s+22/T_3)(s+1/T_1)(s+d)}. \qquad [9]$$

The predicted impulse response for this system in the time domain is, by inverse Laplace transform:

$$v(t) = Ae^{-(1/T_1)t} + Be^{-(2.44/T_3)t} + Ce^{-(22/T_3)t} + De^{-dt}. \qquad [10]$$

The amplitudes, A, B, C, and D, of these simple exponential terms are determined by the gain (K_1) and the four rate constants ($1/T_1$, $2.44/T_3$, $22/T_3$, and d) by the method for expansion of a rational function (ratio of polynomials) into its partial fractions (9, 25).

The curve represented by the sum of four exponentials in five parameters was fitted optimally (in a least squares sense) by means of nonlinear regression (17) to the pentobarbital responses of eight cats. The AEP used were restricted to those having a detectable action potential preceding the wave response, in order to minimize the contribution of presynaptic distributed delay (T_5). Comparison between cats showed wide scatter and little conformity to physiologically meaningful values. The source of the difficulty appeared to be that the positive overshoot was deeper and more prolonged than was expected, either because K_2 had not been reduced entirely to zero, or because some low-amplitude slow positive process had been induced or uncovered by the drug in the AEP, which was not included in the model.

An alternate configuration introduces a parallel forward branch to account for the prolonged positive overshoot of the waveform as shown in Fig. 4. The transfer function is

$$C(s) = \exp[-(sT_3)^{1/2}]\left(\frac{K_1}{s+1/T_1} + \frac{K_x}{s+1/T_x}\right)\left(\frac{s+c}{s+d}\right). \qquad [11]$$

Using the same rational approximation for distributed lag and combining the fractions in the center term,

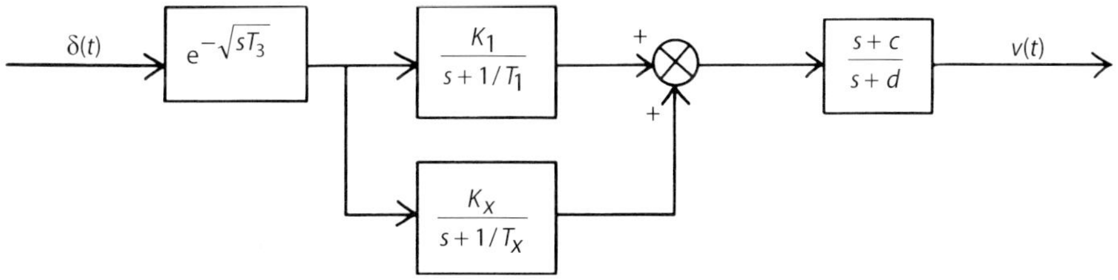

Fig. 4. The transfer function under pentobarbital.

$$C(s) = (K_1 + K_x) \frac{(s+y)(s+z)}{(s+2.44/T_3)(s+22/T_3)(s+1/T_1)(s+1/T_x)(s+d)} \quad [12]$$

where

$$y = -c, \qquad z = -\frac{K_1/T_x + K_x/T_1}{K_1 + K_x}. \quad [13]$$

Taking the inverse Laplace transform

$$v(t) = A^{-(2.44/T_3)t} + B^{-(22/T_3)t} + C^{-t/T_1} + D^{-t/T_x} + E^{-dt} \quad [14]$$

with amplitudes, A–E, determined as previously. The same pentobarbital AEP were fitted with curves generated from this equation. The mean optimal values for the parameters are shown in Table 1. The variability between cats was considerably reduced without significant change in the closeness of fit, i.e. signal-to-noise ratio, between observed and calculated curves.

The two values particularly desired from this work occur in the major forward channel; i.e. the time constants for the exponential delay, $T_1 = 4.5 \pm 0.4$ msec, and for the distributed lag, $T_3 = 3.4 \pm 0.5$ msec. The latter value represents (41) the delay equivalent to a suitably terminated RC cable of length n, resistance R per unit length, and capacitance C per unit length, $T_3 = (nR)(nC)$. The first-order exponential lag is taken to be approximated by a pole at $s = -1/T_1 = -220$ and the distributed lag is approximated by poles at $s = -2.44/T_3 = -720$ and $s = -22/T_3 = -6500$. The constants for c and d, respectively, are represented by a zero on the negative real axis at the origin and by a pole at $s = -3300$, representing a first-order exponential term with a time constant of about 0.3 msec. The exponential term for the minor loop, $s = -1/T_3 = -92$, is not considered further in the present report owing to uncertainty about its physiological significance.

Table 1 Parameters for open-loop transfer functions[a]

K_1	K_x	$1/T_1$	$1/T_x$	$2.44\,T_3$	c	d
811	−808	−223	−92	−716	−0.6	−3164
±103	±109	± 18	± 20	± 89	± 3.6	±452

[a]Means ± S.E. $N = 8$ cats. For symbols see Fig. 4.

The Prepyriform Root Loci for Feedback Gain Variation

Given the two time constants in the forward channel obtained from cats under anesthesia, it is desired to calculate K_2 (feedback gain) and T_2 (feedback delay) from AEP of normal cats using the model shown in Fig. 2. The mean values from Table 1 are used for T_1, and T_3, because it is probable that the variability from error of estimate for these parameters for each cat exceeds intrinsic differences between cats or between differing states in the same cat. Moreover, the presynaptic action potential is not seen at many recording sites, which implies that T_5 is not negligible. For development of the argument, typical values of frequency, $\omega = 250$ radians/sec, and rate constant, $\alpha =$ -30/sec, are taken from measurements of AEP previously reported (17).

The four poles representing the quantities $1/T_1$, $2.44/T_3$, and $-30 \pm j250$ are plotted in the s-plane, the complex poles being the observed closed loop poles of the distributed feedback system (Fig. 5). The total phase lag of the loop transfer function at the points in the s-plane corresponding to these complex poles must be $-180°$ $= -\pi$ radians. Part of this lag is contributed by the two poles on the real axis and is expressed as the sum of θ_1 and θ_2. The remainder $\psi = -(\pi - \theta_1 - \theta_2)$, is owing to $e^{-\sqrt{sT_2}}$ which is equivalent to a line source along the negative real axis from the

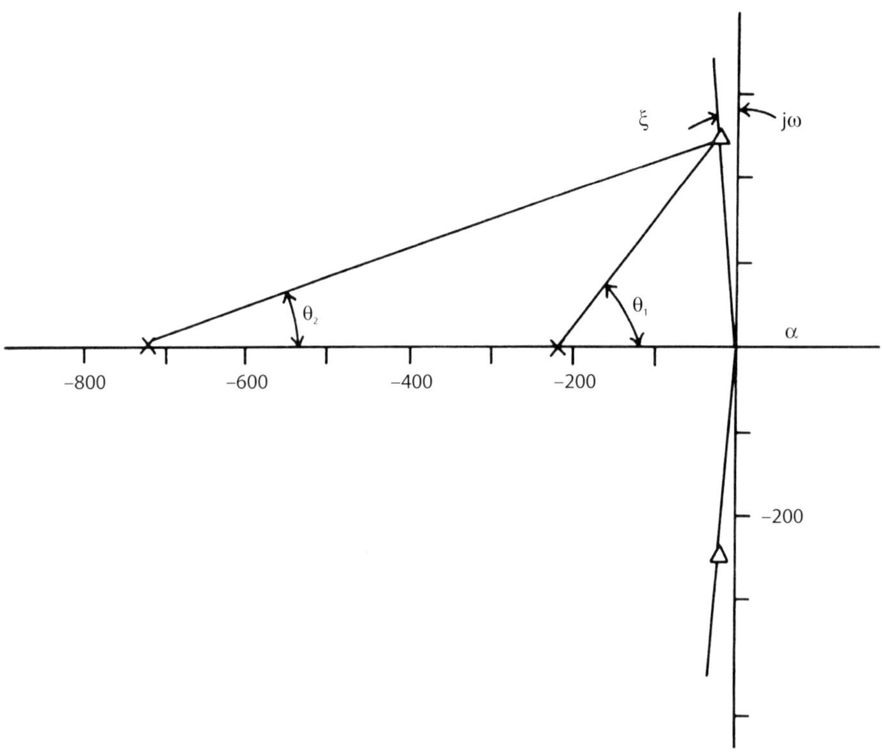

Fig. 5. The essential data required for estimation of feedback gain and delay, shown as locations of two real poles and a pair of complex poles in the s-plane.

origin to $-\infty$. The calculated value of ψ and the measured values for α and ω are used to calculate values for K_2 and T_2.

The open loop gain (db) and phase (degrees) contours in the s-plane are shown in Fig. 6 for a pure distributed lag element in a feedback loop for which the forward channel is of unity gain without delay. The transference for this element (assuming an impulse input) can be expressed:

$$F(s) = \exp[-(sT)^{1/2}]. \quad [15]$$

Taking the logarithm of both sides of the equation:

$$\ln|F(s)| + j\psi(s) = -\sqrt{sT} \quad [16]$$

Let $F = \ln|F(s)|$ and $\psi = \psi(s)$. Squaring both sides and replacing s by $\alpha + j\omega$:

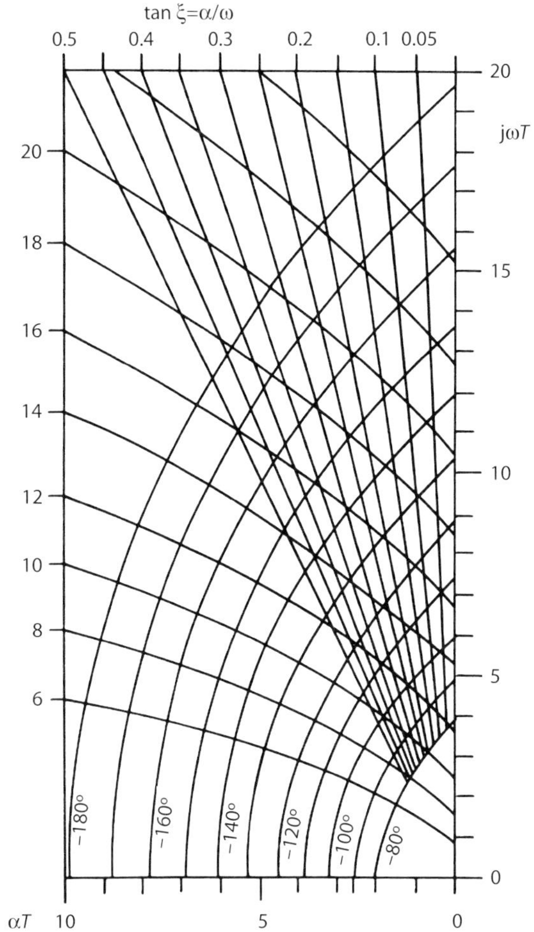

Fig. 6. Phase and gain contours in the upper left-half sT-plane for a single-loop, negative-feedback system having unity gain in the forward channel and distributed lag, $\exp[-(sT)^{1/2}]$, in the feedback channel.

$$F^2 - \psi^2 + 2j\psi F = \alpha T + j\omega T. \qquad [17]$$

Separating into the real and imaginary parts:

$$\alpha T = F^2 - \psi^2, \qquad [18]$$

$$\omega T = 2\psi F. \qquad [19]$$

These two equations are now solved for feedback gain, K_2, and feedback delay, T_2, in terms of ω and α. The variable T is first eliminated by combining Eqs. [18] and [19].

$$F_2 - \psi^2 = 2\psi F\alpha/\omega. \qquad [20]$$

The quadratic formula gives two roots for F, but only that solution is used that gives positive real-time values for T_2 (values for α are always negative)

$$F = \psi\{\alpha/\omega + [(\alpha/\omega)^2 + 1]^{1/2}\}. \qquad [21]$$

Let

$$\xi = \alpha/\omega + [(\alpha/\omega)^2 + 1]^{1/2} \qquad [22]$$

so that

$$F = \psi\xi. \qquad [23]$$

Moreover,

$$\ln(K_2) = -F. \qquad [24]$$

The feedback gain becomes

$$K_2 = e^{-\psi\xi} \qquad [25]$$

and feedback delay is derived from Eqs. [19] and [23]:

$$T_2 = 2\psi^2\xi / \omega. \qquad [26]$$

The value for representing the contribution, ψ, to phase lag of the dominant pole pair by the distributed lag element is

$$\psi = -\pi + \tan^{-1}[\omega/(220+\alpha)] + \tan^{-1}[\omega/(720+\alpha)]. \qquad [27]$$

A graphical interpretation is shown in Fig. 6, where the gain contours are calculated by solving for ωT in terms of αT and F expressed in db (decibels):

$$\omega T = 2F/8.68(F^2/8.68^2 - \alpha T)^{1/2}. \qquad [28]$$

The phase contours are

$$\omega T = 2\psi(\alpha T + \psi^2)^{1/2}. \qquad [29]$$

The normalized coordinates for ωT and αT specify that, once the stability angle, ζ, of the closed loop poles has been determined in the s-plane, changing the scale by multiplying by T will not change the angle. For a loop without forward delay the position of the closed loop pole in the sT-plane is given by the intersection of the line of angle, ζ, with the $-180°$ phase contour for the distributed lag element. In the present case, where the forward limb contributes an amount of phase lag denoted by $\theta_1 + \theta_2$, the position of the closed loop poles is given by the intersection of the stability line with the phase contour $\psi = -(180° - \theta_1 - \theta_2)$.

In the example shown in Fig. 2, the angles $\theta_1 = -50°$ and $\theta_2 = -20°$ leave a total phase lag of $\psi = -110°$ to be contributed by the feedback delay. This corresponds to an equivalent gain of −14 db, which implies that, for damped oscillation to have occurred at the specified frequency and decay rate, the open loop gain K_2 must have been 14 db, which is equal to a factor of about 5. The distributed lag time is found by dividing the value of ωT on the nearest ordinate by the observed value for ω giving in this case a value of $T_2 = 0.027$ sec. (This is converted to the peak delay T_p in milliseconds, the crest latency of an impulse response, by multiplying by 0.167×1000, to give about 4.5 msec.)

Knowing T_2 it is now possible to locate the positions of a set of poles on the negative real axis in the s-plane (Fig. 7) representing the poles of an analytic approximation of $\exp[-(sT_2)^{1/2}]$. The first six poles lie at values of $s = -95$ (from $-2.44/T_2$), $s = -850$ (from $-22/T_2$), etc. Higher order terms can be omitted. The distributed lag term $\exp[-(sT_3)^{1/2}]$ is represented as before by poles at $s = -720$ and $s = -6500$. There are three pairs of complex zeros for $\exp[-(sT_2)^{1/2}]$ and one pair for $\exp[-(sT_3)^{1/2}]$ for which locations are determined by equation [8]. The constant α_1 is set equal to 10,000 for both distributed elements. A single pole for $s = -1/T_1 = -220$ completes the pole-zero configuration (Fig. 7) approximating the transcendental expression for the feedback loop. The root loci for gain variation have been determined by conventional techniques (25).

An expanded view of the s-plane in the vicinity of the origin appears in Fig. 8. There are four poles corresponding to $-1/T_1$, $-2.44/T_2$, $-22/T_2$, and $-2.44/T_3$ for which root loci as a function of gain have been constructed in Fig. 7. The positions of the dominant closed loop poles on the right-hand locus determine the observed gain contour, and the intersection of that contour with the left-hand locus defines the position of a secondary pair of complex poles. The poles representing delay in the feedback limb become zeros in the closed loop system and contribute phase lead to the oscillatory impulse response. For that part of the response owing to the major closed loop poles the phase lead contributed by the zeros at $-2.4/T_2$ and $-22/T_2$, when added to the phase lag contributed by the remaining closed loop poles in Fig. 8, is 1.03 radians. This value lies within the normal range for phase of onset of AEP, implying that the zero, c, postulated in Fig. 1, must be far from the origin.

The transfer function for this system can now be expressed in terms of an equivalent pair of complex poles, two real zeros, and a real pole:

$$C(s) = \frac{K(s + 2.4/T_2)(s + 22/T_2)}{(s^2 + \alpha_1^2 \pm j\omega_1^2)(s^2 + \alpha_\alpha^2 \pm j\omega_\alpha^2)(s + d)} \quad [30]$$

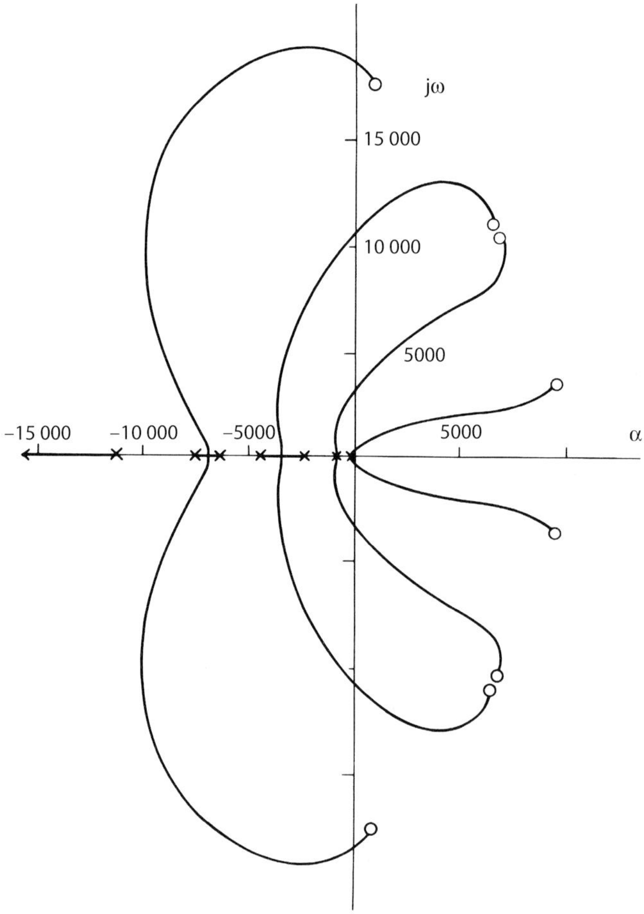

Fig. 7. A rational pole-zero approximation in the *s*-plane for the transcendental function shown in Fig. 2 and described by Eq. [4], with root loci for variation in feedback gain.

The time response to a unit impulse, $\delta(t)$, of the system with the values shown in Fig. 8 is:

$$v(t) = A[1.62 \sin(250t + 1.03)e^{-30t} + 2.01 \sin(310t - 1.62)e^{-860t} + 1.18e^{-3300t}] \quad [31]$$

where A is an arbitrary constant (Fig. 9). The transient specified by the secondary poles ($\alpha_\alpha \pm j\omega_\alpha$) has the form of a heavily overdamped second order response that approaches the time axis asymptotically without overshoot. In previous work (17) this component as a first approximation was represented by a first-order exponential, $s = -400$, while the phase was controlled by the value for the zero, c:

$$C(s) = K(s + c) / (s^2 + \alpha^2 \pm j\omega^2)(s + x) \quad [32]$$

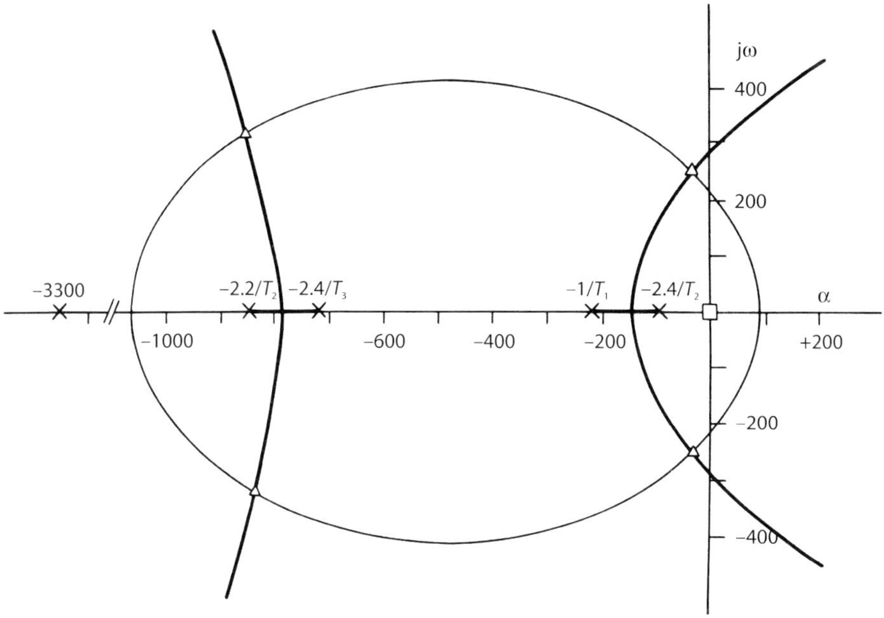

Fig. 8. A large scale representation of Fig. 7 with addition of a gain contour, two pairs of closed loop poles, and a real pole, *d*, at $s = -3300$. Closed loop zeros occur at $-2.4/T_2$ and $-22/T_2$.

and

$$v(t) = A[1.75\sin(250t + 1.12)e^{-30t} - 1.58e^{-400t}]. \qquad [33]$$

If the present formulation is the more closely correct, then an error in estimate of amplitude of the initial peak must have resulted from the previous practice, with the consequence of a systematic error of measurement as shown in Fig. 9. This pattern of error was found in the residuals from virtually all AEP (17), except those from one cat in which the error was apparently incorporated into the subsidiary basis function. That is, the rational transfer function expressed by Eq. [30] is a better approximation for cortical impulse responses than is the empirical transfer function expressed by Eq. [32], but the error is restricted to the first few milliseconds of the time-response.

Discussion

The essential features of the proposed linear model explicitly incorporate some of the outstanding properties of mammalian cerebral cortex. The model is composed of a population of relatively homogeneous elements, which are densely interconnected by transmission lines terminating randomly on one another. The density of

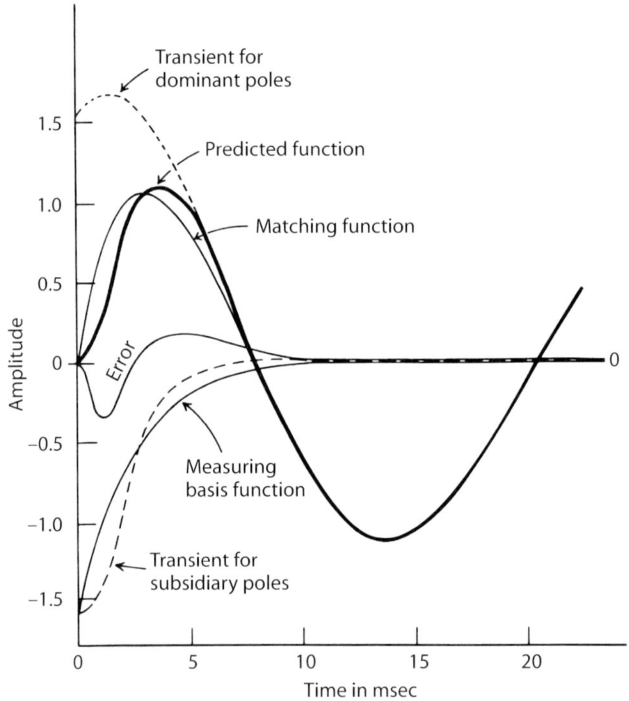

Fig. 9. Time responses predicted from Fig. 8 and calculated from Eqs. [31] and [33]. After 12 msec the predicted curve is a damped sinusoid. The "error" curve is to be compared with the "residuals" shown in Fig. 2 of Ref. (17). The matching function is the sum of the transient for the dominant poles and the first-order measuring basis function; it has been used routinely for measuring AEP (17).

interconnection of any one element with its neighbors falls exponentially with distance from that element, and the transmission delay similarly increases with distance.

Interaction of elements by virtue of these connections is assumed to be mainly inhibitory. The number of elements is taken as so great that continuous approximations can be used for both analog and digital modes of function, such as pulse trains. The characteristic standing waveform of such a randomly interconnected population is sinusoidal, with amplitude modulation of the basis carrier wave by changes in the pulse density of the input function.

Random "spontaneous" impulse activity in both input lines and intrinsic elements is a characteristic and indeed essential feature of the model, as it is of the nervous system (1, 20, 33, 39). The model predicts that EEG-type activity will vanish if afferent noise is abolished, as occurs in isolated cortex (10). Temporal (and spatial) patterns such as AEP and EEG waves can be conceived as organized deviations from the stationary, normally distributed background "noise" or "carrier signal" that is assumed to be present in the absence of patterned input, e.g. sensory or electrical stimulation. Finally, as discussed elsewhere (14) the model can be made inherently stable, particularly in the face of unusually intensive afferent stimulation, by virtue of an amplitude-dependent flattening of the curve representing the distribution of interconnection delays, which results in a nonlinear saturation of feedback gain. This is an exceedingly important property, for the problem of maintaining stability in an enormous population of interconnected and spontaneously active elements is not trivial (3, 5, 24, 32, 43, 44). This will be elaborated in a later report.

An additional feature of this model is the assumption that the population elements have impulse responses, time constants, and other properties homologous with the known characteristics of single neurons in cortex and other parts of the nervous system, which permits limitation in the choice of properties to be built into the model. That is, the seeming correspondence of values for the time constants in the forward channel of the loop with comparable known or suspected time constants of single cells elsewhere in the brain is not proof of the validity of the model. On the contrary the model was designed so that the correspondence could be attained.

The major uncertainty in the development of the model is focused at the use of a secondary parallel element in the transfer function to describe the AEP under pentobarbital. There is no direct evidence that in fact the putative cortical feedback loop was opened by this treatment, and the depth of the prolonged positive overshoot suggests that it was not, or at least not completely. The parallel-path configuration treated this overshoot as a "nuisance" parameter, which had to be dealt with in order to obtain estimates of the main forward channel properties. The necessity for making this move in order to obtain physiologically "reasonable" estimates of the other parameters served also as a reminder that no reason exists to exclude the possibility that the prepyriform AEP might consist of the concomitant superimposed responses of two or more distinct and unlike populations of neurons, although there is no reason yet to include it either. That is, the configuration shown in Fig. 4 might be real (particularly for low-frequency components in the AEP), though it was intended to represent function of the model for values of K_2 slightly greater than zero. The uncertainty introduced by this equivocal configuration into the values for the forward time constants and the real zero, c, pervaded assessment of the feedback properties, making the values presented tentative until definitive means are found for ascertaining that there is a loop as the basis for both AEP and EEG activity, and that it can be opened.

The choice of pentobarbital for this work was based on previous observation (13) of the existence of a single negative peak in the single-shock evoked potential under heavy dosage. Chloralose has also been used, particularly because it is known to suppress recurrent inhibition in the spinal cord (24) and to suppress oscillation in prepyriform AEP as well. Whatever drug is used, the requirement exists for distinguishing between the effects of feedback blockade and noise reduction. The prediction has been verified (18) that procainization or section of the lateral olfactory tract near the bulb with stimulation and recording distal to the site of blockade causes the same appearance of the AEP as does pentobarbital, which may provide the basis for experimental analysis of this nonlinear property of the cortex.

The main feature of the present model is its inherent simplicity, linearity, and time-invariance, which renders it inadequate for detailed correlation with the electrophysiological, pharmacological, anatomical, and behavioral properties of this cortex. The model can readily be modified to introduce additional complexities, but by itself it constitutes a simple key. By combining specifications for properties of single-neuron with rational approximations for the distributed properties of the population expressed by means of Laplace transforms, cortical function is stated in terms of a linear algebra, which in turn provides a basis for precise prediction, testing, and controlled measurement of both unit and multicellular potentials.

Acknowledgements

Reprinted with permission from Academic Press Inc., from: *Experimental Neurology*, Volume **10**, Number 6, 525–547, December 1964

Supported by grants (MH-06686 and NB-05035) from the National Institutes of Health, United States Public Health Service, and by the Berkeley Computer Center.

References

1. Adrian, E. D. 1950. The electrical activity of the mammalian olfactory bulb. *Electroencephalog. Clin. Neurophysiol.* **2**: 377–388.
2. Andersen, P., and J. C. Eccles. 1962. Inhibitory phasing of neuronal discharge. *Nature* **196**: 645–647.
3. Ashby, W. R. 1950. The stability of a randomly assembled nerve-network. *Electroencephalog. Clin. Neurophysiol.* **2**: 471–482.
4. Barron, D. H., and B. H. C. Matthews. 1938. The interpretation of potential changes in the spinal cord. *J. Physiol. London* **92**: 276–321.
5. Beurle, R. L. 1955. Properties of a mass of cells capable of regenerating pulses. *Phil. Trans. Roy. Soc. London, Ser. B* **240**: 699–778.
6. Bishop, G. H. 1958. The dendrite: receptive pole of the neuron. *Electroencephalog. Clin. Neurophysiol.* **Suppl. 10**: 12–21.
7. Boudreau, J. C., and W. J. Freeman, 1963. Spectral analysis of electrical activity in the prepyriform cortex of the cat. *Exptl. Neurol.* **8**: 423–439.
8. Braitenberg, V. 1963. Histology, histonomy, histologic, pp. 160–176. In *Progress in Brain Research*, Vol. 2: Nerve, Brain and Memory Models. N. Wiener and J. P. Schade [eds.]. Elsevier, Amsterdam.
9. Brown, R. G., and J. W. Nilsson. 1962. *Introduction to Linear Systems Analysis.* Wiley, New York,
10. Burns, B. D. 1958. *The Mammalian Cerebral Cortex.* Arnold, London.
11. Chu, Yaohan. 1952. Feedback control systems with dead-time lag or distributed lag by root-locus method. *Trans. AIEE, pt. 2, Applications and Industry* **71**: 291–296.
12. Eccles, J. C. 1964. *The Physiology of Synapses.* Academic Press, New York.
13. Freeman, W. J. 1960. Repetitive electrical stimulation of prepyriform cortex in cat. *J. Neurophysiol.* **23**: 383–396.
14. Freeman, W. J. 1963. A distributed delay feedback model of cortical evoked potential. *Proc. 16th Ann. Conf. Med. Biol.* **5**: 24.
15. Freeman, W. J. 1963. The electrical activity of a primary sensory cortex: the analysis of EEG waves. *Intern. Rev. Neurobiol.* **5**: 53–119.
16. Freeman, W. J. 1970. Spectral analysis of prepyriform averaged evoked potentials in cats. *J. Biomed. Systems* **1**: 3–22.
17. Freeman, W. J. 1964. Use of digital adaptive filters for measuring prepyriform evoked potential in cats. *Exptl. Neurol.* **10**: 475–492.

18. Biedenbach, M. A. 1966. Effects of anesthetics and cholinergic drugs on prepyriform electrical activity in cats. *Experimental Neurology* **16**: 464–479.
19. Freygang, W. H., Jr., and K. Frank. 1959. Extracellular potentials from single spinal motoneurons. *J. Gen. Physiol.* **42**: 749–760.
20. Granit, R. 1955. *Receptors and Sensory Perception.* Yale Univ. Press, New Haven.
21. Granit, R. 1963. Recurrent inhibition as a mechanism of control, pp. 23–37. In *Progress in Brain Research*, Vol. 1: Brain Mechanisms. G. Morruzzi, A. Fessard, and H. H. jasper [eds.]. Elsevier, Amsterdam.
22. Granit, R., J. Haase, and L. T. Rutledge. 1960. Recurrent inhibition in relation to frequency of firing and limitation of discharge rate of extensor motoneurons. *J. Physiol. London* **154**: 308–328.
23. Granit, R., and C. G. Phillips. 1956. Excitatory and inhibitory processes acting upon individual Purkinje cells of the cerebellum in cats. *J. Physiol. London* **133**: 520–547.
24. Haase, J., and J. P. Van Der Meulen. 1961. Effects of supraspinal stimulation on Renshaw cells belonging to extensor motoneurons. *J. Neurophysiol.* **24**: 510–520.
25. Harris, L. D. 1961. *Introduction to Feedback Systems.* Wiley, New York.
26. Hartline, H. K., and F. Ratliff. 1956. Inhibitory interaction of receptor units in the eye of Limulus. *J. Gen. Physiol.* **40**: 357–376.
27. Hodgkin, A. C., and A. F. Huxley. 1952. A quantitative description of membrane current and its application to conduction and excitation in nerve. *J. Physiol. London* **117**: 500–544,
28. Hubel, D. H., and T. N. Wielsel. 1963. Shape and arrangement of columns in cat's striate cortex. *J. Physiol. London* **165**: 559–568.
29. Kandel, E. R., and F. J. Brinley. 1961. Electrophysiology of hippocampal neurons. I. Sequential invasion and synaptic organization. *J. Neurophysiol.* **24**: 225–242.
30. Li, Choh-Luh 1963. Cortical intracellular synaptic potentials in response to thalamic stimulation. *J. Cell. Comp. Physiol.* **61**: 165–179.
31. Li, Choh-Luh, and S. N. Chou. 1962. Cortical intracellular synaptic potentials and direct cortical stimulation. *J. Cell. Comp. Physiol.* **60**: 1–16.
32. Mackay, D. M. 1960. Modelling of large-scale nervous activity. In "Models and Analogues in Biology." *Symp. Soc. Exptl. Biol.* **14**: 192–198.
33. Monnier, A. M., and G. Coppée. 1939. Nouvelles recherches sur la resonance des tissues excitables. I. Relation entre la. rythmicité de la résponse nerveuse et la resonance. *Arch. Intern. Physiol.* **48**: 129–144.
34. O'Leary, J. 1937. Structure of the primary olfactory cortex of the mouse. *J. Comp. Neurol.* **67**: 1–31.
35. Phillips, C. G. 1961. Some properties of pyramidal neurones of the motor cortex. *Ciba Found. Symp., Nature Sleep* **1961**: 4–24.
36. Purpura, D. P., R. J. Shafer, and F. S. Musgrave. 1964. Cortical intracellular potentials during augmenting and recruiting responses. II. Patterns of synaptic activities in pyramidal and nonpyramidal neurons. *J. Neurophysiol.* **27**: 133–151.
37. Rall, W. 1960. Membrane potential transients and membrane time constant of motoneurons. *Exptl. Neurol.* **2**: 503–532.
38. Ramón y Cajal, S. 1955. *Studies on the cerebral cortex (limbic structures).* L. M. Kraft [trans.]. Lloyd-Luke, London.
39. Rosenblueth, A., N. Wiener, W. Pitts, and J. Garcia-Ramos. 1949. A statistical analysis of synaptic excitation. *J. Cell. Comp. Physiol.* **34**: 173–205.
40. Sholl, D. A. 1956. *The Organization of the Cerebral Cortex.* Wiley, New York.
41. Smith, E. J. M. 1958. *Feedback Control Systems.* McGraw-Hill, New York.
42. Spencer, W. A., and E. R. Kandel. 1962. Hippocampal neuron responses in relation to normal and abnormal function. *Colloq. Intern. Centre Natl. Rech. Sci. Paris* **107**: 71–103.
43. Ten Hoopen, M., and A. A. Verveen. 1963. Nerve-model experiments on fluctuation in excitability, pp. 8–21. In *Progress in Brain Research*, Vol. 2: Nerve, Brain and Memory Models. N. Wiener and J. P. Schade [eds.]. Elsevier, Amsterdam.
44. Terzuolo, C. A., and Y. Washizu. 1962. Relation between stimulus strength, generator potential and impulse frequency in stretch receptor of Crustacea. *J. Neurophysiol.* **25**: 56–66.
45. Uttley, A. M. 1955. The probability of neural connexions. *Proc. Roy. Soc., Ser. B* **144**: 229–240.
46. Vastola, E. F. 1958. After-positivity in the lateral geniculate body. *J. Neurophysiol.* **22**: 258–272.
47. Von Baumgarten, R., J. D. Green, and M. Mancia. 1962. Slow waves in the olfactory bulb and their relation to unitary discharges. *Electroencephalog. Clin. Neurophysiol.* **14**: 621–634.

4. *Root Locus Analysis of Piecewise Linearized Models with Multiple Feedback Loops and Unilateral or Bilateral Saturation*

The lumped negative feedback model of cortical dynamics provided three main predictions about the relationship between the occurrence of action potentials ("units") of single neurons and the amplitude of EEG waves and averaged evoked potentials (mainly due to dendritic current). First, the probability of impulse firing of the inhibitory neurons would oscillate at the same frequency and decay rate as the oscillations in the firing probability of the excitatory neurons in the same neighborhood, whether their oscillations were measured from the EEG, the background unit activity, the averaged evoked potentials, or the post-stimulus time histograms. This prediction was verified by simultaneously constructing and measuring averaged evoked potentials and post-stimulus time histograms from both excitatory and inhibitory neurons (Freeman 1968b, 1972c, 1974b), and by calculating over long stretches of spontaneous activity the probability of unit firing conditional on the amplitude of the EEG (Freeman 1967b, 1975; Eeckman and Freeman 1990).

Second, the oscillation in unit firing probability would fluctuate in phase with the EEG for the neuron population, if the EEG was also generated by the same neurons, and a quarter cycle phase difference on the average would be found between the outputs of the inhibitory neurons and the excitatory neurons. Three contrasting examples were found. In the olfactory bulb the EEG was generated by the inhibitory interneurons, whereas the units were generated by the mitral cells, and the pulse probability waves of the mitral cells showed a quarter cycle phase lead over the oscillations of the granule cells. In the prepyriform and entorhinal cortices, the EEG was generated by the superficial pyramidal cells, and their pulse probability waves oscillated in phase with the EEG, while the pulse probability waves of the inhibitory neurons lagged the EEG oscillations at the same frequency by a quarter cycle. In the hippocampus the pulse probability wave of the pyramidal cells led the wave of dendritic EEG by a quarter cycle (Horowitz *et al.* 1973), which suggests that the hippocampal interneurons are responsible for the EEG, as they are in the olfactory bulb.

Third, with increasing stimulus intensity the unit firing probability during the first inhibitory half cycle of the excitatory neurons in the forward limb should drop to zero at the action potential threshold, so that they could not transmit the full extent of the inhibition to which they were subject. This nonlinear failure to transmit output in proportion to input would cause a reduction in loop gain, and this block would account for the well known decrease in frequency of oscillation of evoked potentials with increase in stimulus intensity (Freeman 1962).

Verification of the first and second predictions was uncomplicated in support of the hypothesis that EEG oscillations were due to interneuronal feedback, based on the use of digital adaptive filters at fixed input conditions in the small-signal, near-linear range. The results constituted strong evidence that the oscillations in the gamma range were due to interactions between neuron populations and were not the result of driving of olfactory neurons by "pacemaker" neurons in the neocortex, thalamus, or elsewhere. However, the application of the digital filters to sets of evoked potentials derived over a range of stimulus intensity yielded high-quality data showing that the lumped negative feedback model was inadequate. In order to make sense of the data I had to introduce and evaluate positive feedback loops for both mutual excitation and mutual inhibition into the model (Figure 1 in Chapter 4). In cellular terms, in a mixed population of excitatory and inhibitory neurons the excitatory cells excite each other as well as the inhibitory cells, and the inhibitory cells inhibit each other as well as the excitatory cells. This configuration of excitatory and inhibitory populations encapsulates this topology of synaptic interactions in mature neuropil, so it is the centerpiece of mesoscopic brain dynamics. It is both the conclusion of microscopic cellular studies and the starting point for modeling mesoscopic interactions.

I had to show that the reduction in nonlinear gain was due to the static nonlinearity of the threshold (Figures 2 and 5 in Chapter 4), owing to the abnormal conditions of electrical stimulation, in which the firing probabilities of the neurons in the population are seriously distorted from their normal spatial and temporal homogeneity (Freeman 1975). I had to determine whether the changes with intensity in the frequencies and decay rates (the closed loop rate constants) were due to changes in loop gain or to changes in the open loop rate constants (Figure 7 in Chapter 4). And I had to show that in the normal operating domain the nonlinear gain was bounded by an upper limit as well as a lower limit, causing "bilateral saturation" (Figure 6 in Chapter 4). Thus a more general model emerged which included what became known as the static "sigmoid" nonlinear function.

The failure of the lumped negative feedback model was shown by plots of the frequencies and decay rates of sets of averaged evoked potentials in graphs of the complex plane. These experimental values were modeled by the roots of linear differential equations, which were complex numbers. The roots were *poles* that specified the characteristic frequencies and decay rates of a linearized feedback system for particular values of the feedback gain, and *zeros* that specified forbidden frequencies. The poles gave the frequencies which appeared in the impulse responses, and the zeros appeared as notches in the spectra of the impulse responses, showing the frequencies at which the system could not oscillate. A complex conjugate pair of zeros was most easily seen in a polar plot of the Fourier transform of an evoked potential, such as that in Figure 1 in Chapter 2, as a cusp in the trajectory, most dramatically when the trajectory approached or actually enclosed the origin of the complex plane (Freeman 1970). In a model in which a parameter such as the strength of negative feedback was varied across a range, the values of the closed loop roots changed, and as they did so each root traced a *root locus* in the complex plane with frequency on the imaginary axis and decay rate on the negative real axis (Figure 7 in Chapter 3; Figure IV, the curve rising to the right in

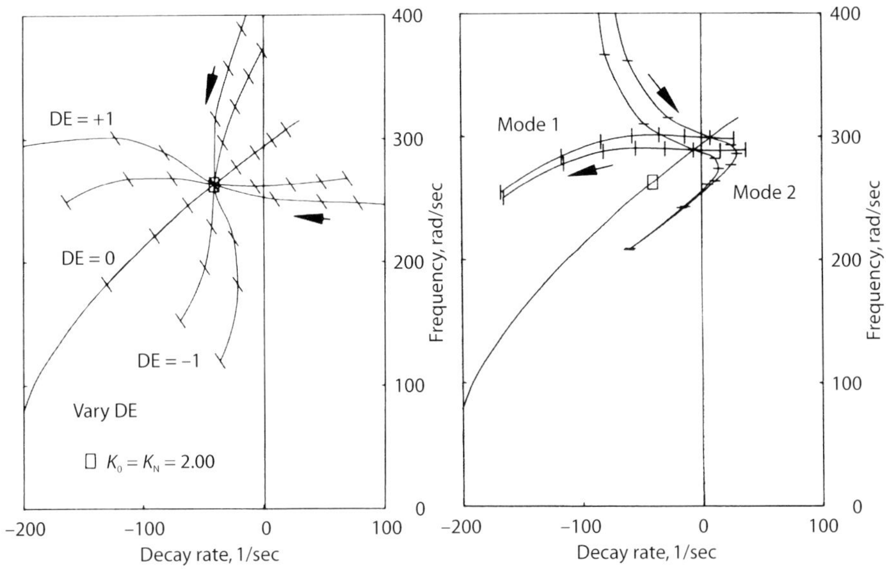

Figure IV. Each frame displays the upper quadrant of the complex plane $s = a \pm j\omega$ near the origin, $a = 0$ and $j\omega = 0$. Each point evaluates a cosine function fitted to an averaged evoked potential, where a is the decay rate in 1/sec and ω is the frequency in radians/sec of the dominant component. A succession of points designates a root locus for the positive value of a complex pole pair, as an experimental condition or parameter is changed, reflecting the dependence of frequency and decay rate on the manipulation.

Left frame: The curve sloping upwardly to the right shows the **Mode 1** root locus of the KII set for increasing gain in a single negative feedback loop. The entire root locus plot of the dependence of frequency on gain is shown in Figure 7 in Chapter 3. Blow-ups of the loci close to the origin are shown in Figure 8 in Chapter 3 and in Figure 5 in Chapter 4, where the crosses designate the values of the open loop poles. The reference feedback gain $K_0 = K_n = 2.0$ is shown by the small rectangle. The tick marks show steps of change in negative feedback gain K_n of 0.5. Decreasing gain with increasing amplitude in Mode 1 is shown by relocation of the pole downward to the left. The root locus is unchanged when positive feedback is added (mutual excitation, K_e, and mutual inhibition, K_i, in symmetry with equal gains). When the KII set receives excitatory bias, DE > 0, the root locus is rotated to a horizontal direction. The frequency is independent of amplitude, and the decay rate is accelerated with increasing input intensity, which occurs on orthodromic stimulation of the bulb through the primary olfactory nerve. When the KII set receives an inhibitory bias, DE < 0, the root locus rotates to a vertical direction, showing frequency dispersion with a fixed decay rate, which is characteristic of antidromic activation through the lateral olfactory tract. In Mode 1, the KII system is hyperstabilized by strong stimulation, up to the point of onset of seizure (Freeman 1986). From Figure 6.7, p. 360 in Freeman 1975.

Right frame: Root loci in Mode 1 at three values of Kn and DE = 0 show increased stability with increased stimulus intensity (arrow to the left). Examples of two root loci in *Mode 2* predict the destabilization of the KII set with increasing amplitude induced by input, provided that the input-induced amplitude does not exceed the background EEG or that it induces an increase in the background, as when inhalation occurs. The root loci as they are evaluated for the cat predict convergence to 250 radians/sec (40 Hz). From Figure 7.17, p. 442 in Freeman 1975.

both frames). As shown in Chapter 3 this is a theoretical root locus, in comparison to the set of points comprising an experimental root locus. The task for modeling then became not to fit a curve to a single averaged evoked potential, but to fit a root locus to the values for the frequency and decay rate from a set of averaged evoked potentials that were taken during the systematic variation of a physiological or experimental parameter, such as inducing arousal, giving a drug, or changing stimulus intensity (Chapter 5).

The most important result from this approach was the clear distinction between the two modes of cortical function shown in Figure 6 in Chapter 4 and here in Figure IV, which is a later result from Freeman (1975). Mode 1 was governed by unilateral saturation when the stimulus intensity or the level of anesthesia was varied, whereas Mode 2 was governed by bilateral saturation, embodied in the sigmoid curve, which determined the stability properties of the EEG. This article gave the main features of what I have called the KII model ("K" after Katchalsky, Freeman 1983), the KO set being a noninteractive collection of neurons, and the KI set being an interactive collection all with the same sign (+ for excitation, – for inhibition). The KII set provided the basis for understanding the population dynamics by which neuropil is destabilized by input. The root locus technique was fully developed in Chapters 6 and 7 of *Mass Action in the Nervous System* (Freeman 1975), in which I showed how an inhibitory bias from dominance of mutual inhibition served to hyperstabilize a population of coupled oscillators by dispersing their characteristic frequencies, reflecting stabilization in Mode 1 (Figure IV, left frame, vertical), whereas an excitatory bias from dominance of mutual excitation colligated the frequencies (Figure IV, right frame, horizontal). Excitatory bias also promoted synchronization at a globally shared frequency in Mode 2 (Figure IV, right frame) on input-dependent destabilization (Freeman 1992a), when the complex poles were driven to the right across the $j\omega$-axis and then to rest on the $j\omega$-axis at the real part $\alpha = 0$ and $j\omega = 250$ radians/sec (an undamped cosine at 40 Hz).

The root locus technique, when it is applied to models with multiple feedback loops, variable gains, and external bias controls, is indispensable for understanding how mixed populations of excitatory and inhibitory neurons can achieve temporal coherence over large spatial domains. These aspects of population dynamics may be necessary for solving in a novel, mesoscopic way what has become known in microscopic terms as the "binding" hypothesis (Von der Malsburg 1983; Singer and Gray 1995), in order to overcome the numerous challenges to the hypothesis that have been posed by critics (Tovée and Rolls 1992; Ghose and Freeman 1992; Rolls and Tovée 1995).

Analysis of Function of Cerebral Cortex by Use of Control Systems Theory

Abstract

The prepyriform (primary olfactory) cortex generates an oscillatory electrical potential in response to single-shock electrical stimulation of the olfactory tract. The averaged evoked potential (AEP) can be matched by the sum of 2 damped sinusoids, each specified by 4 coefficients: amplitude, V, phase, ϕ, frequency, ω, and decay rate, α. Measurements of the 8 coefficients are made by nonlinear regression and by use of digital adaptive filters. Validation of the measurements is by multiple correlation of the 8 coefficients with rate of work, W, done by cats to get food during the averaging process. Factor analysis of the data reveals certain fixed patterns of interparameter covariation. These patterns are interpreted as constraints on the neuronal mechanism generating the AEP, which are associated with specific behavioral patterns.

Analysis of the constraints is the main purpose of this work. The cortex is described as a mixed population of excitatory and inhibitory neurons. The functional characteristics of the pool of each neuron type are described with a linear differential equation based on the waveforms of AEP and on the physiological properties of single neurons. The topology and numerical constants of the model are evaluated by means of recordings of evoked potentials, EEG, and unit activity from this cortex. Three types of neuronal interaction are postulated: positive excitatory feedback, positive inhibitory feedback, and negative feedback. By appropriate definitions the strengths of interaction are reduced to functions of a single variable: feedback gain. A linear feedback system is developed to represent the neuron pools appropriately interconnected, and the output of the system in response to impulse driving is determined.

Gain is treated as amplitude-dependent, i.e. feedback gain is made to decrease with increased amplitude of input. The changes in waveform of the impulse response of the system with successive decrease in feedback gain are shown to replicate the changes in waveform of averaged evoked potentials of the cortex with successive increase in amplitude of cortical input (stimulus intensity). Additionally the cortex displays three modes of unstable operation which are found in the model as well. It is proposed that linear differential equations with variable coefficients provide a suitable basis for combining observations of unit activity with measurements of evoked potentials, EEG, and the behavioral correlates of cortical activity.

Introduction

It is generally agreed that brain function is based on organization of the activity of large numbers of neurons into coherent patterns. Much is known about the properties of single neurons and about over-all brain function in terms of animal behavior. What is lacking is understanding of the ways in which populations of neurons are organized, so that some of the main characteristics of behavior can be understood in terms of the characteristics of single neurons and their interactions with other neurons. To bridge this gap some observations are needed on the coherent activity of homogeneous populations of neurons [13]. Spontaneous EEG and evoked potentials recorded from the mammalian brain are generated by large numbers of neurons, so that measurement of these potentials would be useful toward a synthesis, provided they can be logically connected with other measurements at both cellular and behavioral levels.

This study of cortical function in the cat is therefore based on three types of data. Measurements have been given of evoked potentials generated by cell masses as recorded with macroelectrodes [14]. Patterns of change in the evoked potentials have been related to changes in the behavior of the cat [17]. And the form of the evoked potential has been related to spike potentials generated by single cells in the cortex as recorded with extracellular microelectrodes [19]. The observations on units, populations, and the whole animal are interrelated in a mathematical framework, so that these and other data can be directed with maximal effectiveness toward the development of a useful hypothesis of cortical function. The contribution in this report toward that hypothesis is the specification of a basic neuronal topology and some elementary population dynamics for a paleocortical mechanism.

Specifically, the prepyriform cortex generates an oscillatory electrical potential in response to single-shock stimulation of the lateral olfactory tract [13]. The waveform of the averaged evoked potential (AEP) can be matched [14] by the sum of two damped sinusoids, each of which is completely described by four parameters: amplitude, V, phase, ϕ, frequency, ω and decay rate, α. The dominant (initially negative) of these two basis functions (elementary waveforms used as the basis for measurement), which represents the main characteristics of an AEP, can be treated [15] as the output in response to impulse driving of a system described by a linear differential equation. Changes in AEP accompanying animal behavior can be described succinctly [16, 17] as patterns of change in the four parameters of the dominant basis function. The patterns can be replicated by changes in one of the coefficients of the differential equation.

The topology and other properties from which this equation is derived must, of course, conform to what is known or inferred about the cellular mechanism of the prepyriform cortex. A preceding analysis of the prepyriform AEP [15] is based on the postulate that a negative feedback loop exists within the neuron populations comprising the cortex.

Analogy with other neural mechanisms [4, 22, 24, 25] and the histological appearance of the cortex [32] suggests the occurrence of the following sequence of events upon electrical stimulation of the tract. The volley is thought to cause excitation of superficial pyramidal cells in the cortex, manifested by dendritic depolarization and pyramidal axonal firing, both in proportion to the degree of induced surface negativity. The pyramidal cell discharge is presumed to cause activation in the deeper layers of the cortex of interneurons, which send activity back to the superficial pyramidal cells and there cause inhibition manifested by surface positivity. Thus the AEP is thought to consist of a sequence of alternating excitatory and inhibitory postsynaptic potentials having the over-all shape of a damped sinusoid.

A formal description of this loop is based [15] on Sholl's [40] analysis of interneuronal connection densities within cortex, which suggests that a mathematical representation of interaction within a large pool of neurons can be reduced to a simple algebraic expression, which contains two constants. One constant represents the delay in transmission through the feedback limb of the loop and the other represents the interaction intensity. The constants are defined respectively as feedback delay, T_2, and feedback gain, K_2. The frequency, ω and decay rate, α, of the AEP suffice together to specify the values for K_2 and T_2 in the feedback equation.

This general formulation does not require detailed specification of the nature of the feedback loop or of the interneurons, other than that they be inhibitory. However, additional studies have shown that excitatory and inhibitory neurons interact with their own kind as well as with each other and form both positive and negative feedback loops in the cortex. Other evidence shows that the interactions are amplitude-dependent and therefore nonlinear [18, 19]. Most of this report is devoted to the elaboration of the general model to incorporate these additional properties.

But the general model has been retained as the basis for comparing the output of more complex models with the output of the cortex, because it simplifies the comparison remarkably. Calculation of K_2 and T_2 from ω and α is a conformal transformation, so that changes in ω and α can be represented as sets of points or as curves in $T_2.K_2$ coordinates. The effects of changing a coefficient in an explanatory model can be expressed as a root locus in the complex plane, which likewise transforms to a curve in $T_2.K_2$ coordinates. If a change in either the model or the cortex is equivalent to a change only in K_2, the locus is a vertical line. If the change is equivalent to a change only in T_2, the locus is a horizontal line. For a change equivalent to covariation of K_2 and T_2 the locus is an oblique line passing through the origin. Empirically [16, 17] the loci representing physiological changes in the cortex as revealed by AEP approximate horizontal and oblique lines in $T_2.K_2$ coordinates. Experimental conditions have also been found, in which these loci can be reproduced without reliance on behavioral manipulations. Therefore patterns of variation in AEP and in the output of proposed models both are expressed in terms of T_2 and K_2, so that comparison can be made in terms of the slope and intercept of two line segments. Models are considered acceptable if such segments are congruent.

The results show that properties of neuron populations can be logically derived from properties of single neurons, using well-established statistical and mathematical techniques, which include linear systems analysis, probability theory, and multivariate statistics. The main inference to be drawn from this report is that these techniques are broadly and directly applicable to the mammalian nervous system and will be required for further advances in this field.

The steps in the development are as follows: (a) A topology of interconnection is proposed, based on the histology of the cortex and on observations of its unit activity. (b) The nature of actions at the junction points is specified in terms of known principles of synaptic function [6], and the network is transformed to a more usable equivalent. (c) The function of each of two homogeneous neuron populations as revealed by responses to impulse activation (postsynaptic potentials) is expressed by a differential equation in operational form. (d) The topology and equations for the neuron populations are combined by algebraic methods. (e) The coefficients in the equations are evaluated with previously obtained physiological data [21, 22, Section II] (f) Solutions are obtained for the output of the system by graphic techniques and the inverse Laplace transform. (g) The feedback gain in the system is varied in order to simulate the changing degree of block or saturation with varying response amplitude observed during unit studies, and a family of solutions constituting a root locus is obtained. (h) The changes in the output of the model as a function of varying feedback gain are compared with the patterns of variation already derived from AEP. The comparison is made following transformation of the output of the model to $T_2.K_2$ coordinates.

Methods

Stainless steel electrodes were permanently implanted in the olfactory tract for stimulation and across the prepyriform dipole [7] for recording in each of 12 adult cats. Each AEP was the sum of 88 single-shock responses delivered at 7.3/second over a 12-second period. Stimulus intensity was raised to that level at which the peak-to-peak amplitude of the AEP was at least 10 times the mean peak-to-peak amplitude of background activity averaged in the same way. The amplitude of the AEP was measured at 100 points at intervals of 1.25 msec after the stimulus and punched onto cards.

Each AEP was regarded as the sum of noise plus two damped sinusoids as basis functions [21], each having the equation

$$V(t) = V[\sin(\omega t + \phi)e^{-\alpha t} - \sin\phi e^{-\zeta t}], \tag{0.0}$$

where amplitude V was in microvolts, frequency ω was in radians $\sec^{-1}$, the rate constants α and ζ were in $\sec^{-1}$ and the phase ϕ was in radians. The constant ζ was fixed for each cat and corresponded to the rise time of the initial peak of the AEP. Techniques for measuring the parameters have already been described [14, 16].

Values for α and ω were transformed to K_2 and T_2 as follows:

$$\begin{aligned} \delta &= (1 + \alpha / \omega)^{1/2} - \alpha / \omega \\ \theta &= \pi - \tan^{-1}[\omega / (\beta - \alpha)] - \tan^{-1}[\omega / (\gamma - \alpha)] \\ T_2 &= \delta\theta^2 / 3\omega \text{ seconds} \\ K_2 &= \exp(\delta\theta) \end{aligned} \tag{0.1}$$

where $\beta = 220 \sec^{-1}$ and $\gamma = 720 \sec^{-1}$ were previously defined [15] to represent the delay characteristics of the forward limb of the loop.

Results

1. Topology of the Cortical System.

The cortex is assumed to consist of a mixed population of excitatory (A) and inhibitory (B) neurons interacting in the form of a negative feedback loop. Studies of unit activity [19] have shown that the main events following stimulation of the olfactory tract are (a) excitation of type A cells, (b) excitation of type B cells, and (c) inhibition of type A cells. With bipolar recording of the AEP the output is that of the A cells. Diagramatically these facts are shown as:

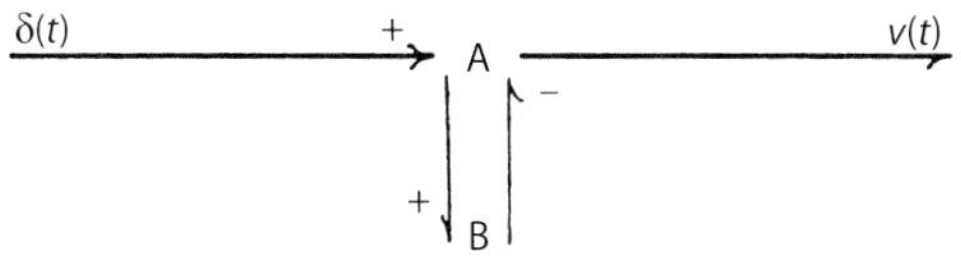

(1.0)

where the Dirac delta function, $\delta(t)$, is the input impulse equivalent to the single shock stimulus and $v(t)$ is the output equivalent to the AEP. Particularly at low stimulus intensities many type A units are not initially excited but are inhibited. Thus, the diagram becomes:

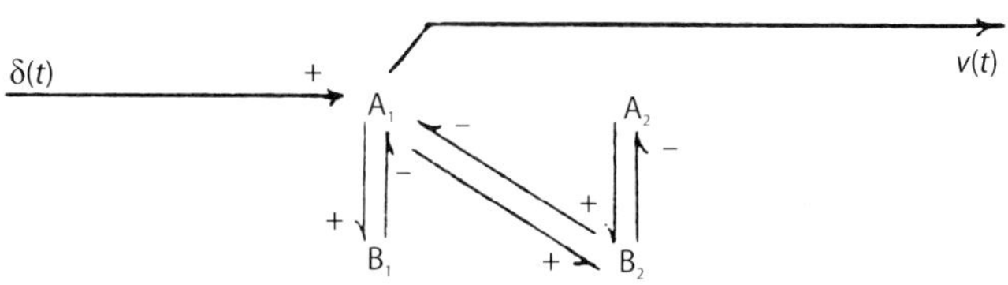

(1.1)

The diagonal connections are placed from A, to B_2 to represent the fact that long delayed excitation is very seldom the first response of a cell to electrical stimulation at any intensity. This also means that higher order interaction terms, A_3, B_3, etc., can be omitted when the input is an electrical stimulus.

Evidence has been found for believing that types A and B interact with their own kind. The introduction of this property results in the diagram shown in Fig. 1, I. Further variations in the scheme might include the addition of scaled output to $v(t)$ from A_2 and from B_1 and B_2. These would cause the initial amplitude and phase of the output in the time domain to vary in proportion to the relative amounts. These aspects are not in question here.

Equation (9.0) expresses the conclusion that a simple negative feedback loop such as that depicted in flow diagram (1.0) should have the same impulse response as the multi-loop system depicted in Fig. 1, I (provided $A = B$ and $f = g = 1$) for any value of K. This has been confirmed by analog computation, which validates the algebraic operations summarized in Fig. 1.

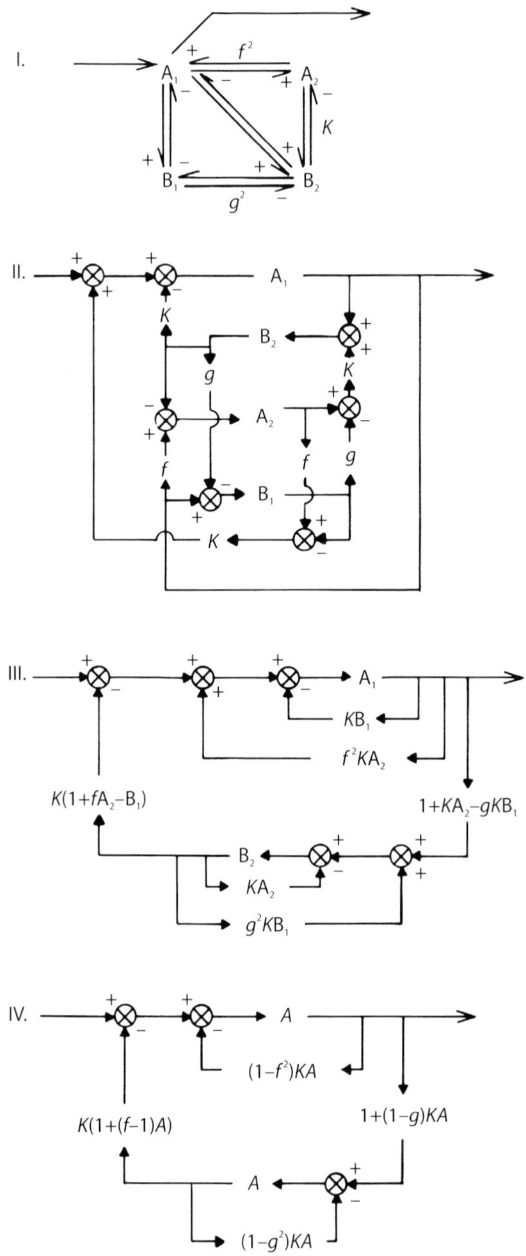

Fig. 1. I. Flow diagram for an interconnected set of excitatory (A) and inhibitory (B) neurons. II. Topological equivalent, provided the interactions are additive. III. Transformation by block substitution. IV. Simplification by setting $A_1=A_2=B_1=B_2$

2. Summation at Junction Points

The next step is specification of the nature of the interaction. It is taken to be purely additive, for the following reason. Superposition fails mainly 5–15 msec after a conditioning volley at low intensities and for progressively longer intervals at higher intensities. This is the time range in which unit studies have shown maximal saturation to take place, so that the nonlinearity is that to be introduced at a later stage in the analysis. Additionally there is failure of superposition for very short intervals (e.g. 2 msec) between test and conditioning volleys. This is attributable to the properties of the afferent tract and is not at issue here. Therefore by a topological transformation I goes into II in Fig. 1, and by block substitution can be transformed to III.

It is necessary for these transformations to specify the magnitudes of the proposed interactions. Neglecting for the present the effects of saturation it is clear that the basic determinant of these magnitudes is the anatomical interconnection density of these cells. Were the numbers of types A and B cells and also the average number of end terminals for each cell equal, then the connection density would be uniform, and the connection term for each pair of elements could be represented by a numerical constant. For unequal numbers of A and B cells the homotypic connection densities (A to A and B to B) would differ from the heterotypic connection density (A to B and B to A). Here the ratio of homotypic excitatory connection density to heterotypic connection density is denoted by f, i.e. the ratio of the number of fibers going from A cells to other A cells to the number going from A cells to B cells is f. The ratio of the number of fibers going from B cells to other B cells (homotypic inhibitory connection density) to the number going from B cells to A cells is g. If the relative numbers of connections are dependent only on the numbers of cells, then $fg = 1$. Were there differences in the branching characteristics of types A and B cells, then the product, fg, would equal a constant other than 1. It will be shown that this constant may be set equal to one by proper selection of a reference value for K (Section 5).

The feedback gain in each heterotypic (negative feedback) loop is denoted as K. Positive excitatory feedback gain is denoted as f^2K and positive inhibitory feedback gain is denoted as g^2K.

The next step is to make the assumption that type A and type B cells have sufficiently similar properties in the time domain that they can be represented by a single class of neurons, i.e. $A = B$. Apart from computational simplification the main justification for this move is that, because the evaluation of the model to be considered here concerns only the output of A_1 and not B_1 or B_2, there would be no advantage in specifying properties distinguishing A cells from B cells.

It is also assumed that the ratio of A_1 to A_2 remains constant with increasing stimulus intensity. Although for higher input the proportion of activated neurons out of the whole cortex greatly increases, as shown by unit studies, this does not mean that A_1/A_2 must change for most A_1 neurons are reactivated within 3–5 msec and are therefore A_2 neurons for some other A_1 neurons. The question is more appropriately phrased by asking what are the relative amounts of input to A_1 and A_2 and what are their relative outputs to $v(t)$. Because as before stated this relates only to the phase and amplitude of the time response, not the frequency or decay rate, it is permissible to set $A_1 = A_2$ and also $B_1 = B_2$. In Fig. 1, diagram III goes into IV.

3. The Transfer Functions of *A*.

The transference for $A = A(s)$ is derived from the "open loop" population response of type A neurons to single-shock stimulation (Fig. 3 in ref. [15]). This matter was previously considered in more detail [15]. Essentially the transient consists of a single surface-negative peak with a dead time, T, between the stimulus and the start of the response, a rise time having the rate constant $\beta - \gamma$, decay time having the rate constant, γ, and a shallow surface-positive overshoot of long duration. The overshoot [represented by the time constant, T_X, in equation (11) of ref. 15] is considered to be unimportant for present purposes. The negative peak alone can be matched by a curve having the equation.

$$\begin{aligned} v(t) &= V[e^{-\gamma(t-T)} - e^{-\beta(t-T)}], \quad t \geq T \\ v(t) &= 0 \qquad\qquad\qquad\qquad\qquad t < T \end{aligned} \tag{3.0}$$

The Laplace transform of this equation corresponds to the transform of the proposed open loop response of the population of type A neurons to single-shock stimulation and therefore to the transference for the population,

$$A = \frac{V\beta\gamma e^{-sT}}{(s+\beta)(s+\gamma)} \tag{3.1}$$

Values for β and γ have previously been obtained by fitting an equation of the form [3.0] to the open loop response (Table I in ref [15]) and are $\beta = 720$ and $\gamma = 220$. The gain is set equal to $\beta\gamma$, so that $V = 1$ and any deviation in d.c. gain from unity can be included in f, g, and K.

An estimate for the dead time, T, is taken from unit studies, which is the mean interval between the negative crest of the action potential in the tract and the start of the response of type A neurons. It is assumed here that this delay (0.87 msec) for axonal conduction through the molecular layer and the first synapse was equal to the delay for conduction to the deeper layers of the cortex plus the second synapse. The delay from the depth back to the surface is taken to be the same, i.e. $T = 0.87$ msec. Owing to the fact that this delay is much shorter than the other time constants, the expression $\exp(-sT)$ is replaced by the rational approximation given by Smith [41] as $-(s-2/T)/(s+2/T)$, which conveniently provides phase lag with negligible attenuation. Equation (3.1) becomes

$$A = \frac{-1.58 \times 10^5 (s - 2300)}{(s+220)(s+720)(s+2300)} \tag{3.2}$$

Use of equation (3.2) to represent function in the closed loop is based on three assumptions in addition to those discussed already. (a) The constants do not change with closure. This is based on the findings that incremental doses of pentobarbital can open the loop without changing the latency, rise time and initial amplitude of the surface-negative peak, and that during "spontaneous" variation of AEP or upon changes in stimulus intensity the duration of the initial surface-negative peak is constant. (b) The number of neurons involved in generating evoked potentials is so large and their individual firing patterns are so uncorrelated that the impulse output of the neurons to other neurons can be treated as a continuous function. (c) The magnitude of the continuous function is linearly proportional to the magnitude of the AEP.

The basis for assuming linearity is that the mean firing rate for many neurons is linearly related to membrane current [1, 6, 20, 23, 26, 28, 29, 39, 43], except for the limitations here expressed as saturation. Because these cells are continually firing at random and are therefore in the steady state, it is presumed that accommodation [1, 6, 20] and adaptation [1, 26] are not important. The linear relationship between membrane current and unit firing rate in single neurons has been established for steady currents. For alternating currents as in the present case it is likely that phase shift would occur between membrane current (which is the main determinant of the AEP) and membrane voltage or rate of change in voltage (which is the main determinant of firing rate). The membrane current–voltage transfer function [proposed as $(s+c)/(s+d)$ in equation (11) of ref. [15]) is not known for this population of neurons, but its significance is only for the phase and amplitude of the AEP, which are not of primary concern at present.

There is an obvious parallel between the population response and postsynaptic potentials generated by single neurons in response to impulses, because the AEP is thought to manifest the sum of extraneuronal synaptic currents of many cells. The population time constants must be consistent with the cumulative effects of synaptic delay, dispersion, passive dendritic conduction, passive decay of membrane transients, dispersal of transmitter substances, etc., but they cannot be identified with unique processes at the cellular level.

4. The Relation Between Saturation and the Gain Constants

The main nonlinear characteristic of the cortex is clearly amplitude-dependent. At low stimulus intensities the AEP evoked by the second of two volleys, which arrives during the initial surface-positivity evoked by the first, has a lower frequency than does the superimposed conditioning AEP. At higher intensities the second AEP has a higher frequency than does the first AEP, if it begins during and is superimposed on the second surface-negativity of the first [8] The frequency of the AEP is dependent on the level of background activity [17]. This is taken to mean that excessive amplitude affects the strength of interactions between cells without necessarily altering the cell time constants. The gain constants can be said to vary with the amplitude of output, and the problem is to find appropriate expressions for the dependency.

The cortical system is conceived as consisting of an interconnected population of cells, each having a linear transference between the levels of threshold and upper firing rate limit. At any moment these levels are distributed through the population, probably normally, as proposed by Rall [33] for a motoneuron pool and by others for nerve [44] and sensory systems [21], and are best described in terms of the mean and variance for each. The mean (d.c. level) and variance (r.m.s. level) for the background activity, the amplitude probability distribution for which is approximately normal [19], are the third pair of variables for the system. The test signal (AEP) has a maximum negative (upper) and positive (lower) value measured from the mean for background activity (here taken as zero, because all measurements were made using RC-coupling in the amplifiers). If the variances are Gaussian, then those for background activity may be convolved with those respectively for upper and lower limits [33] to give Gaussian distributions. Analysis then requires exploration of only two relationships: one between maximum signal negativity and the width of the upper range, and the other between maximum signal positivity and the lower range, all expressed in millivolts.

The procedure followed here is introduction of the lower limit alone (treated as a fixed clipping level in analog computation), and then of both upper and lower limits. The effects of distribution are not considered explicitly.

Clipping in a single negative feedback loop with a single unilateral clipping stage, for which the output is taken at the input to the nonlinearity, is a nondynamic nonlinearity, and the transference can be broken into two parts [41]: a linear frequency-dependent part, $F(j\omega)$, and a nonlinear amplitude dependent part, $G(a)$. In a flow diagram this is:

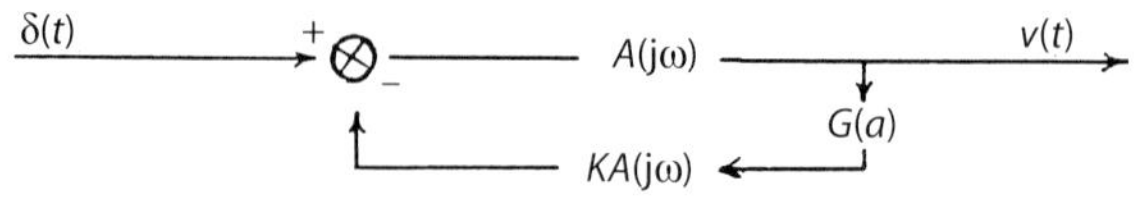

(4.0)

The effect of $G(a)$ is to reduce feedback gain as amplitude increases.

For a sinusoidal input the output for such a system consists of the fundamental frequency plus a series of higher-order harmonics [29, 31]. Harmonics are not considered here, because the methods used for measuring AEP do not give adequate information about them. The phase and amplitude of the fundamental depend on the input amplitude, so that a set of root loci are required in the s-plane to describe the response characteristics for each input amplitude. Alternatively it is possible to display the frequency-dependency in a Nichols chart showing phase, $j\phi$, and the logarithm of closed loop gain, ln G, as functions of frequency. The effect of saturation is to translate the linear locus, $F(j\omega)$, vertically downward. This change may also be represented as a vertically upward movement of the pole point [41]. This shift in pole point describes the nonlinear gain characteristic, i.e. it is the locus for $G(a)$. As shown in Fig. 2 the maximum in gain reduction owing to one-stage unilateral clipping is to 1/2 (0.69 nepers) for the fundamental frequency.

In order to measure $G(a)$ for impulse rather than sinusoidal driving, the system in flow diagram (4.0) was simulated on an analog computer, using two integrators to solve a second order differential equation representing each cell type:

$$A = \frac{0.158}{(s+0.22)(s+0.72)} \quad (4.1)$$

Dead times were omitted, and unclipped gain ($K = 4$) was adjusted to give an optimal value for Q commonly observed in the AEP. The time was scaled by about 1000. A diode was used to clip the output of the forward unit into the feedback limb. The direction of clipping was set to limit the overshoot after the initial response peak on impulse driving, and the amount of clipping was expressed by the ratio of initial amplitude to the negative of the clipping level in volts.

As for AEP [9, 11] the frequency of the output, ω, was taken from the interval between the first and second upward peaks and value for Q from the ratio of their amplitudes. The decay rate was $\alpha = \omega / 2Q$. The linear transference, $F(j\omega)$, was calculated for the system and plotted as ln G *versus* $j\phi$ (Fig. 2). For each value of ω measured from the impulse responses a conversion factor was taken from the curve, $F(j\omega)$, on this graph in millimeters/radians/sec. The corresponding value for α was multiplied by this factor, and the distance in millimeters was plotted along the normal to the curve, $F(j\omega)$, at the point α. The set of points so obtained from impulse responses for different levels of clipping (here the independent variable) described the locus for $G(a)$.

The results in Fig. 2 show that the apparent gain reduction is greater for impulse than for sinusoidal driving, and that phase lead is introduced. The latter is a reflection of the fact that the duration of the first upward peak is not affected by the

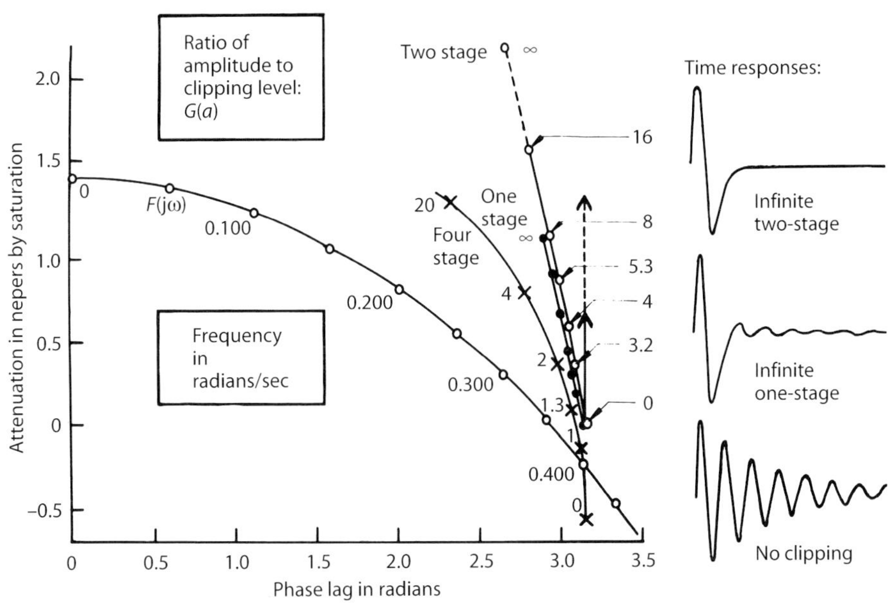

Fig. 2. Logarithm in nepers of closed loop gain versus phase lag in radians for analog solutions of flow diagrams (4.0) and (4.2) and Fig. 1, I. The effect of clipping by use of diodes (a fixed-level approximation for unilateral saturation) on the feedback system is expressed by separating the transference into a linear, frequency-dependent part, $F(j\omega)$, and a nonlinear amplitude-dependent part, $G(a)$.

clipping, so that the increasing duration of the positive overshoot gives the appearance of diminishing frequency and increasing phase lead (cf. Fig. 3 in reference [11]).

The analytic approach cannot be extended to two or more nonlinearities [41]. Yet unit recording clearly shows that saturation occurs in both the forward and feedback limbs. This was represented in the analog computation by introduction of a second diode following the feedback element:

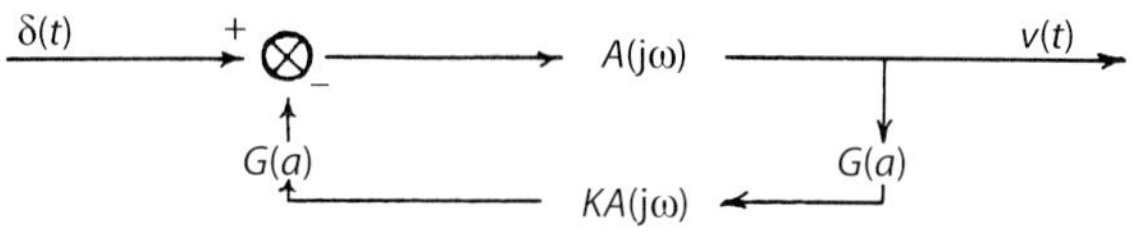

(4.2)

The locus for $G(a)$ was calculated as before (Fig. 2). The phase shift is of the same type, and the attenuation in nepers is on the average doubled. (The last point for infinite two-stage clipping could not be measured because of the lack of a second upward peak and was fixed by extrapolation). This is taken to mean that two-stage clipping results in the square of gain reduction by one-stage clipping.

With reference to Fig. 1, the results in Fig. 2 show that increasing saturation can be expressed as a reduction in K. The nature of the relation between K and amplitude of response need not be specified, save that the maximum reduction in K by clipping in a single loop is by a factor of 1/10 (2.30 nepers).

Because the limitation in firing rate over the range of response amplitudes now being considered applies only to the amount by which one cell can inhibit another, the positive excitatory loop gain, f^2K, is not altered by changes in amplitude. For any given reference value, K_0, this relation is expressed by

$$f = f_0 (K_0 \,/\, K)^{1/2} \tag{4.3}$$

where f is a constant equal to the value for f at $K = K_0$. On the other hand the positive inhibitory loop is most strongly affected. From the ratio of gain reduction for two-stage to one-stage clipping shown in Fig. 2 it is inferred that f^2K changes in proportion to the square of K.

Therefore

$$g = g_0 (K \,/\, K_0)^{1/2} \tag{4.4}$$

where g_0 is a constant equal to the fixed value for g at $K = K$. From equations (4.3) and (4.4) it follows that for all values of K.

$$fg = f_0 g_0 = \text{a constant} \tag{4.5}$$

There are only two types of cells postulated in the population so that if connection densities depend only on the relative numbers of the two types then

$$fg = 1 \tag{4.6}$$

Ideally the reference value for K_0 should be determined by that value for negative feedback gain at which there is no saturation, i.e. two-way anatomical connection density. Because this is indeterminate, K_0 is taken as that value at which f=g=1. Therefore,

$$f_0 = g_0 = 1 \tag{4.7}$$

5. The Root Locus for Unilateral Saturation

This was first done by digital computation. For each value of K, representing a level of saturation at a given response amplitude, corresponding values for f and g were obtained using equations (4.3) and (4.4). The constant K_0 representing reference negative feedback gain was set equal to one of three values (1.26, 1.78, 2.50) as shown in Fig. 5. K was varied in steps of 2 db ref. K_0 (decibels with respect to K_0 as reference level). Solutions were found by graphic techniques for the system of equations shown by the flow diagram IV in Fig. 1.

Using equation (3.2) a root locus plot was generated for the minor feedback loop in Fig. 1, IV:

$$D = \frac{A}{1+(1-f^2)KA^2} \tag{5.0}$$

Substitution of equations (3.2) and (4.3) into (5.0) gives

$$D = \frac{-1.58\times10^5(s-2300)(s+220)(s+720)(s+2300)}{[(s+220)(s+720)(s+2300)]^2 + (K-K_0)[1.58\times10^5(s-2300)]^2} \tag{5.1}$$

A root locus plot for this is shown in Fig. 3 (above) for the upper half of the s-plane. The crosses at s =–220 and s = –720 represent open loop double poles; not shown are a double pole at s = –2300 and a double zero at s =+2300. The gain contours show the logarithm to the base 10 of $(1–f^2)\ K$. For $f < 1$ the closed loop poles lie on the –180° phase contours; for $f > 1$ they lie on the 0° phase contours (including the real axis). There are 3 pairs of closed loop poles and 4 closed loop zeros, specified by equation (5.1), one each at the site of an open loop pole for each value of K, $f \neq 1$, and one at s = + 2300.

The same plot served to provide the solution for the second minor feedback loop:

$$E\frac{A}{1+(1-g^2)KA^2} \tag{5.2}$$

$$E = \frac{-1.58\times10^5(s-2300)(s+220)(s+720)(s+2300)}{[(s+220)(s+720)(s+2300)]^2 + (K-K^2/K_0)[1.58\times10^5(s-2300)]^2} \tag{5.3}$$

having the same number of poles and zeros as for D, but in differing locations in the s-plane.

The minor forward loops were

$$F = 1+(f-1)A \tag{5.4}$$

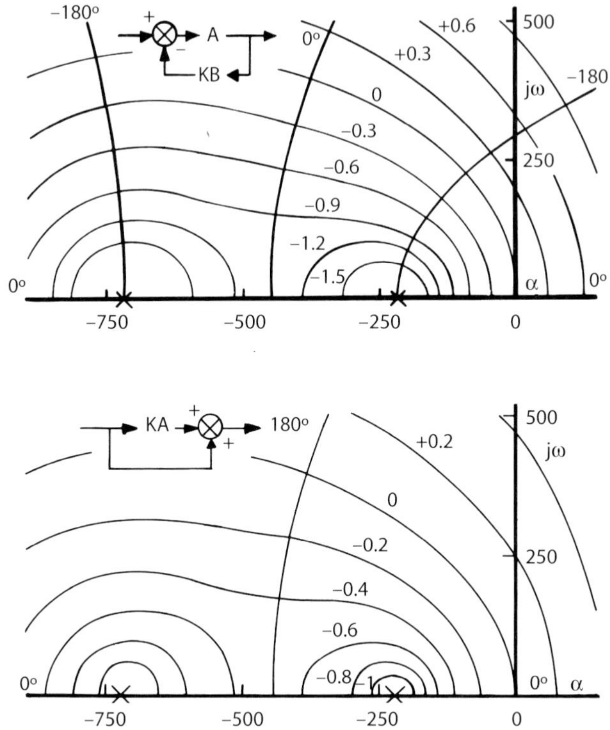

Fig. 3. Phase and gain contours in the upper half of the s-plane near the origin for the two minor feedback loops in Fig. 1, IV (above, wherein the two crosses represent each a double pole), and for the two minor forward loops in Fig. 1, IV (below, wherein the two crosses represent each a single pole). Not shown are poles at $s = -2300$, and zeros at $s = 2300$. In practice the lower diagram is used to find solutions to $1/F$ and $1/G$; the closed-loop poles located by means of the diagram are the closed loop zeros for equations (5.6) and (5.7).

$$G = 1 + (1 - g)KA \tag{5.5}$$

These were evaluated in the form $1/F$ and $1/G$ by means of the phase and gain contour plot shown in the lower half of Fig. 3. These took the form:

$$F = \frac{(s+220)(s+720)(s+2300) - 1.58\times10^5\,[(K_0 / K)^{1/2} - 1](s-2300)}{(s+220)(s+720)(s+2300)} \tag{5.6}$$

$$G = \frac{(s+220)(s+720)(s+2300) - 1.58\times10^5\,[1-(K / K_0)^{1/2}]K(s-2300)}{(s+220)(s+720)(s+2300)} \tag{5.7}$$

The solution for each minor loop had 3 zeros and 3 poles, the latter being always in the same position.

The solutions for the minor loops were then combined to find the solution for the major loop:

$$C = \frac{D}{1 + KDEFG} \tag{5.8}$$

The poles of the two minor forward loops canceled three of the zeros of each of the minor feedback loops. As indicated for two levels of K in Fig. 4 there were then 12 poles (crosses) of which 4 are shown and 8 zeros (circles) of which 3 are shown for the open major loop. For each specified value of K there was a closed-loop solution having 12 poles (triangles) of which 3 are shown and 10 zeros (squares) of which 4 are shown. The others are far to the right or left of the $j\omega$-axis.

This procedure resulted in a set of 12 root loci as a function of K. Only those in the vicinity of the $j\omega$-axis are important. These are shown in Fig. 5 for the real axis and the upper half of the s-plane near the origin. Three sets of curves are shown, one for each value of K_0. The degree of change in gain by saturation is expressed in db ref. K_0. The same family of curves results for choice of the constant $fg = 1$, showing that this constant may be eliminated by proper choice of K_0 in this model.

The computations were checked with the help of an analog computer. Two units were constructed as described by equation (4.1) and flow diagram (4.2) and were interconnected as shown in Fig. 1, I. The unilateral clipping level was varied at all 4 sites together, and the input as before was an impulse of constant height and duration. The unclipped gain (not equivalent to K_0) was set at different levels between 2 and 4 until a root locus with clipping was found, which resembled that for the digital model. Such a locus for unclipped gain of 4 is shown in Fig. 2 for in G *versus* $j\phi$. It is

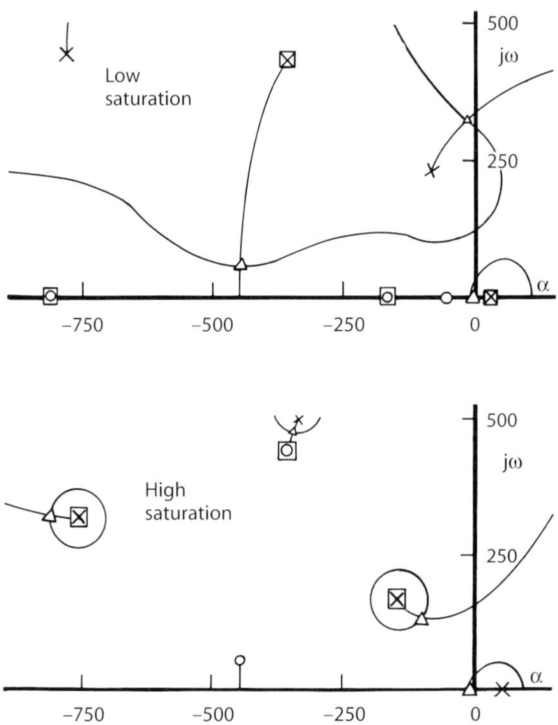

Fig. 4. Intermediate solutions are shown for K_0=2.5, and g=1.12 (above) or g=0.50 (below). Note at low saturation the closed-loop pole at the origin and the closed-loop zero to its right.

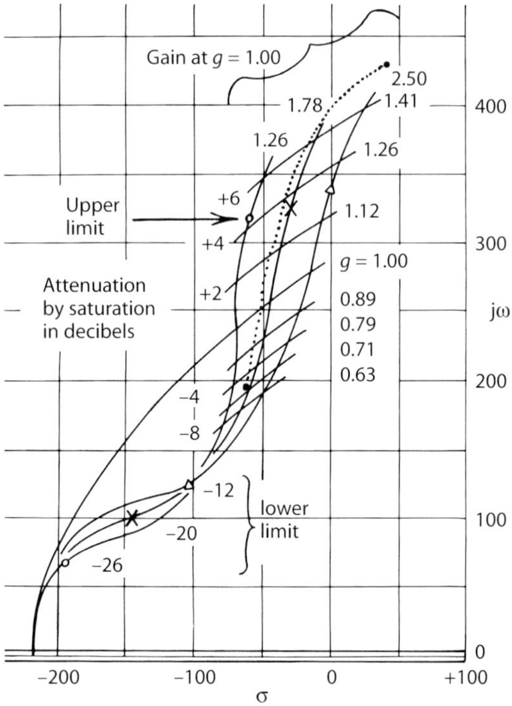

Fig. 5. Loci are shown for the poles nearest $j\omega$-axis in the condition of unilateral saturation. The dotted line represents an analog solution. The two lines parallel to the real axis represent loci for a pole that moves toward the point $s = -220$ with increasing amplitude, and then back to the right half of the s-plane The values of gain attenuation for which the pole on the real axis crosses the origin are shown for the loci of the complex pole pair by the "upper" and "lower" limits.

shown in Fig. 5 in the s-plane, where the frequency has been corrected for the differing timescales of the two systems.

The similarity of the two results validated the digital approximation. The analog device provided the additional information that the maximal gain reduction from unilateral saturation was about 1.8 nepers (16 db ref. unclipped gain or by a factor of 1/6).

6. Comparison of Model Having Unilateral Clipping with Cortex

For each value of K_0 there are three loci crossing $j\omega$-axis. Two of these are loci for a complex pole pair. With increasing amplitude these complex poles go conjugately from the right to the left in the direction of increasing stability and decreasing frequency. The third locus lies along the real axis. The closed loop pole moves from right to left along the real axis across the origin for values of $g > 1$ (low amplitude). For $g = 1$ this pole lies at $s = -220$. For values of $g < 1$ (high amplitude) the pole moves back towards and across the origin. The values for attenuation at which these two crossings occur are indicated on the other loci as "upper" and "lower" stability limits.

Comparison of these loci with those from AEP can be made in terms of the range of stability and spectrum. It is immediately clear that stability depends not merely on connection density but on the relative numbers of the two cell types. For equal numbers of A and B, $g = 1$, and without saturation the system is unstable for values of $K > 3$.

When values for g rise above about 1.32 (for appropriate values of K_0), a pole on the real axis moves across the origin into the right half of the s-plane. Such values of g are associated with an excess of inhibitory elements over the number of excitatory elements in the population, and a positive value of the pole implies a regenerative increase in the level of activity in the positive inhibitory loop for any input. Without saturation the system is unstable, as confirmed by study of the analog device. Unilateral clipping makes it stable by reducing feedback gain until the pole moves back to the origin. The presence of this pole at $s = 0 + j0$ is confirmed in the analog device by the presence of a step function in the oscillatory output of the system in response to impulse driving. The direction of the step is opposite that of the initial quarter cycle of oscillation and its amplitude is small, owing to the presence nearby on the positive real axis to the right of the $j\omega$-axis of a closed loop zero.

Evidence for the existence of this pole and zero near the origin in cortical transference is to be found in the low-frequency, initially surface-positive respiratory wave in the EEG for this cortex, and in the presence of a low-frequency initially positive transient as the subsidiary component of many AEP. The model also helps to explain the frequency modulation seen in sinusoidal bursts in the EEG, which characteristically appears as a shift in frequency from higher to lower during each burst, e.g. from 350 to 240 rad/sec [7].

The model implies that, given the lower limit firing rate of all neurons, a random network of excitatory and inhibitory neurons will be stable merely if (for non-zero background input) the latter outnumber the former by a factor appropriate for the prevailing collection density. It implies that some degree of saturation is present in normal cortex at all times as the necessary condition for stability. This has been verified by unit recording [19].

As the input intensity and response amplitude are increased to high values, a single pole on the real axis moves back again through the origin into the right half of the s-plane. There is no adjacent closed loop zero. This pole implies regenerative increase in the activity in the positive excitatory feedback loop. With increasing amplitude the pole moves still further to the right. The impulse response takes the form of a diverging exponential with the same initial polarity of amplitude as the start of the normal impulse response. (By convention in physiology, extracellular negativity of potential is displayed as "upwards" and positivity as "downwards". In the present context negativity also implies excitation as positivity implies inhibition of type A neurons and elements. Therefore both the input impulse and the start of the AEP are negative, excitatory, and upwards.)

The analog model did not become unstable with full, four-stage, unilateral clipping, unless negative feedback gain, K, was additionally reduced to half its unclipped value. This means that the amount of gain reduction possible by saturation alone is not sufficient to bring the pole on the real axis across the origin into the right half of the s-plane.

Cortex, however, is unstable for high intensities of input. In response to very intense electrical stimulation of the olfactory tract, the cortex generates a train of high-amplitude, diphasic spikes at regular intervals of 200 to 400 milliseconds for 30 to 60 seconds. The seizure closely resembles psychomotor epilepsy in humans both its EEG and its behavioral manifestations [12]. But the spikes are initially surface-positive, which is the reverse of the initial polarity of the AEP. They manifest inhibition of the type A neurons. Clearly the cortex and the model are similar, in that neither can go unstable by a regenerative increase of the activity in the positive excitatory feedback loop. Rather (in cortex) there must be run-away activity of the positive inhibitory feedback loop, during which many of the type B neurons undergo progressively strong inhibition, and many others *pari passu* are released from tonic inhibition. It is the activity of the latter bombarding the type A neurons that gives rise to the surface-positive spike.

This phenomenon will be discussed further in Section (8) in connection with the model having bilateral saturation.

Finally, both the analog and digital models generate at full saturation an impulse response that is highly damped. This is equally true of cortex in response to very strong single-shock stimulation. Both the cortex and the models become unstable, if at all, in the same way, i.e. not by ringing but with a shift in baseline.

7. Bilateral Saturation

Comparison of the loci for unilateral saturation with that described for AEP further shows that minimum stability in the oscillatory mode for the model is associated with maximal frequency, whereas for cortical function the minimum for stability lies at an intermediate frequency close to that for the EEG (cf. Fig. 6). This means that unilateral saturation cannot be used alone to account for the nonlinear characteristics of the cortex in the physiological range.

The model is modified by introducing the upper limit in firing rate. This is done by setting f = constant, so that gain in the excitatory positive feedback loop changes in proportion to K, i.e. it decreases with increasing amplitude. Owing to the flexibility in choice of $K_0, f=1$. The gain in network I in Fig. 1 is thereby made uniform and equal to K, except for the positive inhibitory feedback loop where it is K_2/K_0. Two of the minor loops in network IV of Fig. 1 disappear, and the cortical transfer function becomes

$$C = \frac{A}{1 + KAEG} \tag{7.0}$$

Upon substitution of equations (5.4) and (5.6) into (7.0),

$$C = \frac{A + (1-g^2)KA^3}{1 + (2-g)KA^2 + (1-g^2)K^2A^3} \tag{7.1}$$

Solutions each having 9 poles and 7 zeros are found as before using Fig. 3 and equations (3.2) and (4.4).

8. Comparison of Model Having Bilateral Saturation with Cortex

This system yields a zero to the right of a pole, both in the right half of the s-plane at low degrees of saturation, as in the system with unilateral saturation. Both the pole and the zero move toward the origin with increasing saturation. But the loci for the dominant complex pole pair curve away to the left to the $j\omega$-axis with decreasing saturation. The system is stable in the oscillatory mode for all levels of saturation (provided K is not too large). It is absolutely stable for high levels of input.

For the condition of zero input, as following section of the tract or administration of a large dose of pentobarbital, normal cortical EEG activity is suppressed, but there occur high amplitude, diphasic, initially surface-positive and deep-negative spikes randomly at intervals of less than one to more than sixty seconds. In the model the dominant pair of complex poles lies far to the left of the $j\omega$-axis. A single pole and zero lie on the real axis in the right half of the s-plane, so the model predicts that with no sustained input or internal activity (as observed with microelectrode recording) the output, $v(t)$, of the system in response to any random event would consist of a rapid shift in the baseline to a new level in the direction opposite to the normal impulse response. This behavior is observed during analog simulation. However, in this unstable state the output of the cortex quickly returns to its initial level after an overshoot, presumably because a regenerative burst of activity in inhibitory or excitatory neurons is self-limiting. Therefore the present model suffices to predict the existence, necessary conditions, and approximate rise time for the transients manifesting one type of observed cortical instability, but not the time course.

The random spikes in de-afferented or anesthetized cortex have waveforms indistinguishable from those occurring during seizures induced by electrical stimulation. Unit studies have shown that following the brief phase of firing induced in almost all cortical neurons by a strong afferent electrical stimulus, there is a period of suppression of 'spontaneous' or background discharge of almost all the cortical neurons, which lasts 10 to 30 milliseconds. The background input on the olfactory tract is also suppressed. During the early recovery period of background cortical unit firing many of the inhibitory neurons apparently 'escape' into a sudden burst of activity, which is necessarily self-limiting and responds to the occurrence of a pole and a zero in the right half of the s-plane, and therefore to the model having a level of saturation less than that required for stability. This means that in a qualitative way the model having bilateral saturation can account for the occurrence of both seizures and of barbiturate spikes. All that is required is the postulate of a relative refractory period for each component neuron, which is implicit in the notion of an upper limit on the firing rate of each neuron.

A rare second form of prepyriform cortical instability in the oscillatory mode has been observed in cats accidentally subjected to hyperthermia (unpublished observations), in which the output consisted of sustained high-amplitude sinusoidal oscillation expressing a limit cycle at 28/sec, which was reversibly suppressed by asphyxia, and which lasted one to two hours. This pattern of output can readily be obtained from the model by appropriate choice of K_0, f and g, but the physiological significance is unclear.

The loci for the dominant complex pair of poles curve toward the $j\omega$-axis with increasing amplitude and decreasing frequency and then away, giving rise to a minimum for stability in the oscillatory mode in the mid-frequency range. When transformed into T_2 K_2 coordinates the main locus for bilateral saturation (high 6) appears as a smooth curve rising to a sharp inflection with decreasing K, at which minimum stability for the oscillatory mode occurs. For values of K_0 between 1.78 and 2.24 the low-amplitude stability level for the monotonic mode (a single pole lies at s=(0 + j0) occurs for gain levels of +2 to +3 db ref. K_0, and the inflection at –1 to –3 db ref. K_0 Between the levels of +3 to –4 db ref. K_0 the curve can be approximated by a straight line having the equation (for three values of K_0):

$$\begin{aligned} K_2 &= 1.05 + 0.96T_2, \quad K_0 = 1.78 \\ K_2 &= 1.17 + 1.03T_2, \quad K_0 = 2.00 \\ K_2 &= 1.30 + 1.09T_2, \quad K_0 = 2.24 \end{aligned} \tag{8.0}$$

The curve shown in Fig. 6 is for $K_0 = 2.00$. (Virtually identical curves were obtained by analog computation, using K as the independent variable and setting $f^2 = 1$ and $g^2 = K/K_0$.) The result shows that with increasing saturation (diminishing K) the values for K_2 and T_2 both increase. (The anomaly that K_2 increases as K decreases results from the definition of K_2 as a normalized gain and is discussed further in Section 9).

The locus for "spontaneous" variation of AEP from cats having frequencies within the range of the EEG (Table IV in reference 2) is

$$K_2 = 1.03[\pm 0.36] + 1.06[\pm 0.11]T_2 \tag{8.1}$$

High values for K_2 and T_2 are associated with high values for the root mean square (r.m.s.) amplitude of EEG activity. Because the single-shock evoked potential is superimposed on the EEG activity, it is proposed that bilateral saturation is more likely to occur during periods of high amplitude r.m.s. EEG activity. The r.m.s. amplitude averaged over 10 to 12 second epochs is a random variable [10]; so that a sample of estimates of K_2 and T_2 derived from AEP can be expected to be distributed along the locus for bilateral saturation.

The output of the model for varying degrees of bilateral saturation consists essentially of a damped sinusoid having the four coefficients of amplitude (V), frequency (ω, transformed to T_2), decay rate (α, transformed to K_2), and phase (ϕ). Estimates for these were made at 14 levels of saturation between +3 and –4 db ref: K_0=2.00. Ordinary and partial correlation coefficients were calculated between each pair of coefficients. These are given in Table I and are compared with mean estimates derived from Table I AEP.

The dominant factor in "spontaneous", variation of AEP is described as covariance of T_2, K_2, ϕ and $-V$. The ordinary correlation coefficients derived from the model correspond to this pattern. However, two discrepancies between the

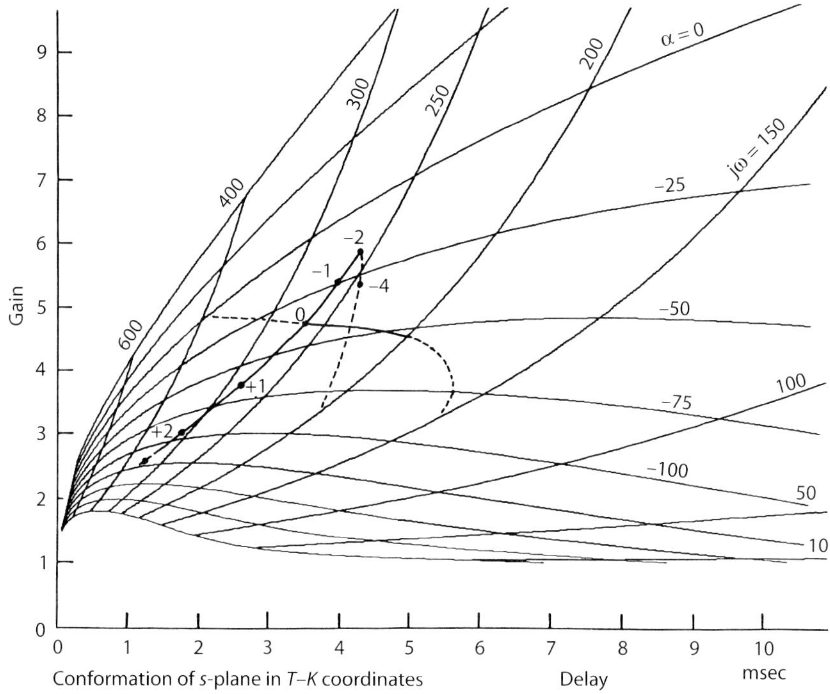

Fig. 6. The root loci for unilateral saturation (curving downward and to the right) and for bilateral saturation (moving upward to the right and then downward) are compared in T_2–K_2 coordinates. The solid segments indicate the probable limit of attenuation by saturation alone.

Table I. Interparameter ordinary and partial correlation coefficients derived from cortex, the model with bilateral saturation (vary K), and the model with a variable feedback rate constant (vary γ_B).

	VT_2	VK_2	$V\phi$	T_2K_2	$T_2\phi$	K_2f
			Ordinary			
Cortex	−.09	−.27	.01	.83	.59	.39
Vary K	−.97	−.81	−.98	.84	.99	.87
Vary γ_B	−.51	−.49	−.99	.99	.55	.52
			Partial			
Cortex	.25	−.45	−.05	.84	.59	−.42
Vary K	.15	−.51	.37	.92	−.60	.75
Vary γ_B	−.37	.37	−.99	.99	−.39	.40

model and the cortex have emerged. The partial correlation coefficients involving ϕ appear reversed in sign. Additionally the range of variation in for the output of the model is 5 to 40, whereas typically for components of the AEP having frequencies from 220 to 280 radians/second the range for ϕ is 30° to 75° [14]. It is likely that these discrepancies resulted from one or more of the simplifications introduced in Sections 2 and 3.

These results indicate that the model having bilateral saturation suffices to represent cortical function at low input amplitudes, whereas the model having unilateral saturation is more appropriate for higher input amplitudes. As previously discussed the upper limit in saturation is conceived as an average frequency of firing, and this limit no longer remains constant during an AEP having an amplitude greater than the amplitude of the background activity. For a high-amplitude AEP the firing probability may be very high during surface negativity and very low during subsequent surface positivity, when most neurons are not challenged to fire, so that the upper limit ceases to play a role. (The poststimulus time histogram for this reason is not useful for exploring the upper limit.) therefore it is desirable to represent the locus for changes in AEP with increasing stimulus intensity with two curves (Fig. 6). The low range corresponds to the model having bilateral saturation, and the high range to the model having unilateral saturation. The operating point in which the transition from one operating mode to the other takes place in cortex occurs at about the level where the amplitudes of evoked and background activity are equal. It is tentatively suggested that the transition point for the models can be set at the level where $K = K_0$. The two curves in Fig. 6 are for $K_0 = 2.00$, which cross at $K = 0$ db ref. K_0.

In normal cortex the surface-positive overshoot of the AEP in response to high intensity stimulation is very long in duration, and the apparent frequency and decay rate of the diphasic waveform are lower than values compatible with either model. A possible explanation for this is that powerful excitation of the cortex leads to secondary changes elsewhere in the brain, which react upon the cortex and delay recovery during the surface-positive phase. In support of this explanation it has been found in both undercut and fully isolated cortex that recovery time is quicker, and that values for T_2 and K_2 derived from AEP recorded from such preparations (provided recovery from anesthesia is complete) conform more closely to the predicted loci.

9. The Effect of Varying a Time Constant

In order to explore the effects of changes in gain for this system it was necessary to fix the time constants at some usable values. It is necessary now to determine the sensitivity of the system to changes in a major time constant, in order to judge how critical the choice may be. This requires that the gain, K, be fixed at some reasonable value.

From the preceding sections it is suggested that if $K_0 = 2.0$ then K should be set at 0 db ref. K_0 (2.0) in mid-range, with an upper limit of +3 db ref. K_0 ($K = 2.8$) for low amplitudes and a lower limit of –4 db ref. K_0 ($K = 1.0$) for high amplitudes. The mid-frequency is then 255 with a range from saturation alone of 315 to 200 radians/sec.

The assumption has been made that $A = B$. In particular it has been assumed that the value for the lowest rate constant, γ, in type A cells equals that in type B cells, i.e $\gamma_A = \gamma_B$. The additional assumption is now made that $K = K_0$, so that $f = g$ for AEP components having frequencies within ± 10% of the mid-frequency. In this case the flow diagram IV in Fig. 1 reduces to a single negative feedback loop having the transfer function:

$$C = \frac{A}{1 + KA^2} \tag{9.0}$$

in which $K = 2.00$ and $\gamma_B = 220$.

The validity of equation (9.0) was verified by analog computation (Fig. 1). The validity of these two assumptions was tested as follows. Eight sets of mean values for ω and α previously reported [14] were found to have the required frequency component. The values for frequency, ω, and decay rate, α, for each AEP in each set were assumed to represent a solution to the equation:

$$1 + KA^2 = 0 \tag{9.1}$$

The operational expression for A is defined by equation (3.1). Here one of the two major time constants, γ_A, and the gain, K, were treated as unknowns, the values for γ_B, β, and T being as before. There were 2 zeros and 6 poles on the real axis of the s-plane, 5 of the latter being known. The phase angles of the 7 known roots to the point $s = (\alpha + j\omega)$ were added and the total was subtracted from 180°. The remainder was the phase lag contributed by the unknown pole, which specified the position of the pole on the negative real axis. The value for K was taken from the ratio of a.c. to d.c. gain, corrected for the change in the normalized gain constant in equation (3.1),

$$K_{\text{a.c.}} = \frac{2.5 \times 10^{10} (s - 2300)^2}{(s + \gamma_B)(s + 220)(s + 720)^2 (s + 2300)^2} \tag{9.2}$$

where $s = (\alpha + j\omega)$. For $K_{\text{d.c.}}$, $s = (0 + j0)$, and

$$K = \frac{K_{\text{a.c.}}\ \gamma_A}{K_{\text{d.c.}}\ \gamma_B} \tag{9.3}$$

Mean values for K and γ_B are shown in Table II for each cat. The scatter (roughly ± 40% of the means for K and γ_B) indicates the extent to which either or both assumptions fail, and shows the range over which γ_B should be varied.

The effects on the output of the model of changing γ_B were determined. Using equation (9.0) and the types of root locus plot shown in Fig. 3 a set of solutions was found for the location of the main complex pole in the upper half of the s-plane for values of γ_B ranging from –400 to –100, $K_0 = 1.78$. On transformation to T_2–K_2 coordinates (Fig. 7) the locus appears as a straight line not quite passing through the origin (if dead time were omitted it did so). The increase in K_2 accompanying the increase in T_2 results from the way in which K is defined as a normalized gain. Alternatively if the gain ratio in equation (9.2) is not multiplied by γ_A / γ_B during computation in the s-plane, this implies a compensatory reduction in gain with reduction in γ_B, and the root locus conforms almost to a horizontal line segment in T_2–K_2 coordinates (Fig. 7).

This property accounts for the anomalous behavior of K_2 in relation to K described in the preceding section. K_2 rises as K falls (Fig. 7). If K_2 is multiplied by T_0/T_2, where T_0 is the value of T_2 for $K \neq K_0$, then owing to the intercept on the positive gain axis the curve for gain verses delay is hyperbolic, with decreasing gain for increasing delay. The linear form is clearly the more desirable as the basis for comparison of the model with the cortex.

Table II. Estimates for feedback gain, K, and lowest rate constant $\gamma_B = 1/T$, in a simple negative feedback loop, for which the impulse response is a damped sinusoid having the frequency, ω and the decay rate, α.

Cat	N	α sec^{-1}	ω rad sec^{-1}	γ_B sec^{-1}	K	T_B msec
C51	25	21	231	134	2.56	7.47
C52	38	34	253	248	1.93	4.04
C58	10	56	272	325	1.74	3.08
C64	32	11	233	121	2.73	8.30
C70	38	62	240	205	1.51	4.87
C77	10	44	256	200	2.14	5.00
C83	30	26	280	227	2.21	4.38
C86	10	22	251	161	2.62	6.21
Average	8	35	252	203	2.18	5.42
S.E.		± 6	± 6	± 23	± .17	± .63

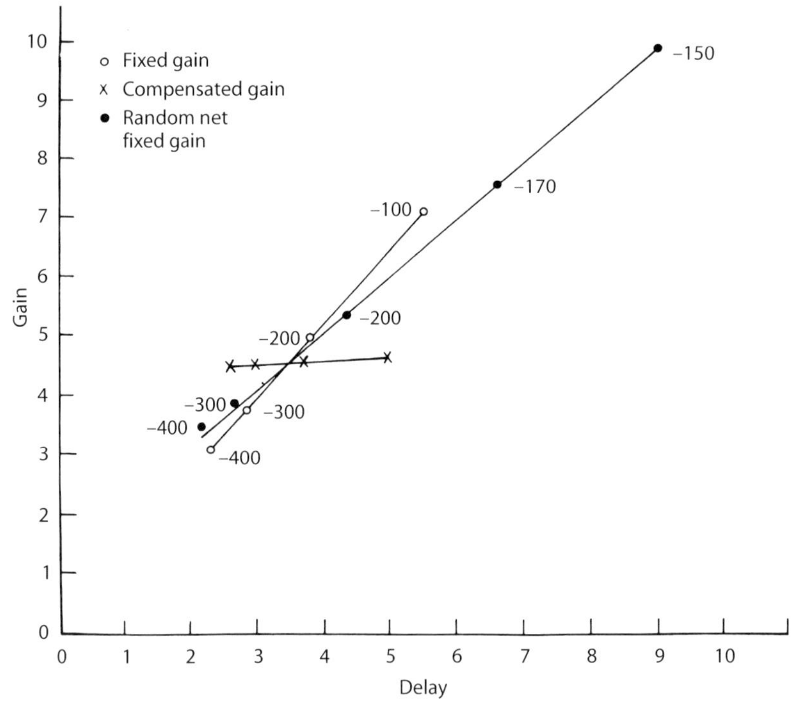

Fig. 7. Root loci are shown for a simple negative feedback loop having a variable time constant, γ_B, in the feedback limb, (o); for the same loop having both variable time constant and proportionately varying gain, (×); and for the network in Fig. 1, III having a varying time constant γ_B, (•). Conformity to the locus marked (o) implies that the impulse response of the element in the feedback limb has constant initial amplitude with varying decay rate. Conformity to that marked (×) implies that the impulse response has constant area.

It was next assumed that $f = g = 1$ and $K_0 = 1.78$, but that γ_B was variable for the system shown in Fig. 1, i.e. $A \neq B$.

The flow diagram in Fig. 1, III reduced to:

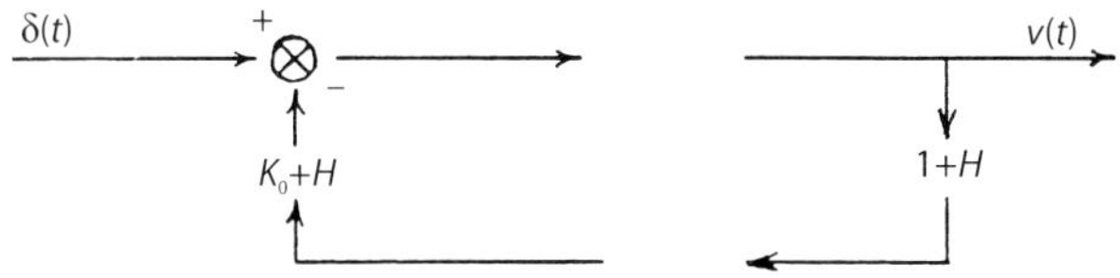

(9.4)

where:

$$H = K_0(A - B) \tag{9.5}$$

Using techniques described in Section 5 the poles and zeros were evaluated for the expression (A–B) for different values of γ_B. The four minor loops were evaluated and lastly the major loop. The locus for the complex pole nearest the $j\omega$-axis, transformed to T_2–K_2 coordinates, is shown also in Fig. 7.

A linear approximation for this curve is

$$K_2 = 1.22 + 0.96T_2 \tag{9.6}$$

Covariance of the parameters is shown in Table I (Vary γ_B). The pattern is not significantly different from that for the model having bilateral saturation with fixed time constants (Vary K).

These results show that a change in either the main gain constant or in the main feedback time constant of the model will suffice to replicate the low amplitude physiological root locus. This brings a major problem to the fore. The feedback gain (interaction strength) is a property of the population of cortical neurons and can only be estimated by measurements of population responses. But those measurements cannot be interpreted unequivocally unless the mean time constant for both forward and feedback neurons are known. Intracellular recordings might be suggested as a means for measuring the time constants, but there are numerous difficulties owing to the small size of the cells, e.g. simple bias toward the largest cells, distortions resulting from injury, etc. If the loop is opened the feedback neurons cannot be activated, but if it is closed, the input to the feedback neurons is not known. Responses to current steps given to the soma are distorted by the dendritic conductance [34], and in any case they do not provide a basis for estimating dendritic delay [5]. It seems likely that new techniques will be required to evaluate the major time constants for the system, so that the AEP can be used solely for estimation of the reference level for feedback gain, K. Until this is done K must be treated as a ratio variable.

This ambiguity raises for consideration yet another aspect of the stability of the cortical mechanism. The normal input to the cortex can be described as a sustained burst of impulses in the olfactory tract during inspiration, which causes increased synaptic activity in cortical neurons. The activity is manifested by an increase in the r.m.s. amplitude of EEG activity, which according to the present hypothesis results in an increase in saturation and thereby an increase in K_2 and T_2. According to the ionic hypothesis there should also take place an increase in conductance associated with EPSP, IPSP and after-potentials. All together should reduce membrane resistance and shorten the time constants, with the effect of reducing K_2 and T_2. The degree of effectiveness of this implied compensatory mechanism is not known. However, since *K* is presented as a ratio variable, attempts need not be made at present to distinguish the relative contributions to changes in AEP resulting from saturation or from membrane conductance changes. Both accompany changes in amplitude.

Discussion

The computational model is based on a simple physiological model. The cortex is conceived to consist of a mixed pool of excitatory and inhibitory neurons, which differ (for present purposes) only in number, geometry, and sign of output. Each neuron is capable of generating excitatory and inhibitory postsynaptic potentials governed by fixed time constants for synaptic and cable delays and passive decay of membrane transients. Firing rate is deemed proportional to membrane depolarization between a lower limit dependent at least partly on threshold (zero pulse frequency) and an upper limit probably dependent in part on afterhyperpolarization [3, 23] of single neurons following each discharge.

The large number of randomly firing neurons involved in the population response allows some simplification. Unique patterns of connection for each neuron are submerged, and uniform non-specific connections are assumed. The time constants of the system are interpreted as representing mean values for all the cells. Unit activity levels in the two parts of the pool are treated as continuous functions. The upper and lower limits on firing rate are thought to be distributed through the population, although for convenience in analog representation the lower limit has been treated as a fixed clipping level. The effect of these limits on firing rate is to reduce the strength of interaction among cortical neurons at excessive positive and negative amplitudes. This effect is described in terms of amplitude-dependent feedback gain. Certain dynamic nonlinearities (e.g. accommodation, adaptation, and variable membrane conductance) need not be introduced.

The main problem considered in this report is the translation of the physiological description into a mathematical form. The key point is the flow diagram in Fig. 1, I, which expresses the physiological hypothesis in graphic form, and which is the starting point for both analog and digital computation. The validity of the computations, including the requisite simplifications, is established by comparison of the results of analog and digital procedures applied to the system having unilateral saturation. The validity of using Fig. 1, I to represent the cortex is established by comparison of the effects of reducing feedback gain, *K*, on the system having bilateral saturation with the changes in AEP thought to accompany decreases in the strength of interaction of cortical neurons.

The computational and physiological models together suffice to describe the main features of the AEP in relation to its "spontaneous" variation [17] changes with food consumption [10], prolonged habituation [9,13], stimulus intensity [8, 11], pentobarbital [15], tetanization of the tract, and d.c. polarization of the cortex.

A basis is established for further analysis of nonlinear phenomena such as the respiratory wave [7] and psychomotor-like seizures [12]. The inherent stability of the cortex in most operating conditions is accounted for.

The main deficiency of the computational model is the lack of formal specification for the dependency of the three feedback gain parameters on response amplitude. This in turn reflects uncertainty regarding the nature of the distributions of upper and lower limits and the analytic functions that might best be used to describe them. Although for simplicity these limits have been interpreted in terms of threshold and maximum mean repetition rate for unit firing, in actuality multiple factors can contribute to amplitude-dependent nonlinearity of the cortex. Prominent among these is the reduction in amplitude of IPSP as membrane potential approaches the equilibrium potential for the IPSP, which may account for the finding in the model representing bilateral saturation that the gain in all loops (Fig. 1, I) is equal to K, except for the positive inhibitory loop, where it is K_2/K_0.

Additional deficits stem from the lack of reliable information about mean rate constants of forward and feedback neurons, the membrane current-voltage transfer function, and the statistical properties of spontaneous EEG and unit potentials, all in the intact cortex of the waking animal. These are best remedied by further observation within the existing framework of the hypothesis.

The main deficiency of the physiological model lies in its failure to account for multiple simultaneous frequency components in both AEP and the EEG, usually in the range from 20–60 cycles/second but occasionally as low as 10 cycles/second and as high as 120 cycles/second. Spectral shifts involving multiple frequencies are particularly apparent in relation to the learning process, and would appear to involve coupling of the cortex with other brain structures. The model therefore says little about what must be the main function of this cortex, i.e. the transformation of olfactory signals related to conditional responses, as the signals pass from the bulb through the cortex to tertiary olfactory structures. Analysis of this problem will require simultaneous study of multiple populations of neurons, of which the prepyriform is only one.

However, the models show that a formal relation can be established between some simple physicochemical properties of single neurons and the behaviorally correlated properties of a single large mixed population by means of linear differential equations with variable coefficients. This general approach based on the study of electrical properties of the population, appears to be suitable for use in the search for solutions to problems of greater complexity.

Acknowledgments

Reprinted with permission from: *The Logistics Review* 1967, Vol. 3, No. 12.

This work was supported by grants MH - 06686 and NB - 05035 from the National Institutes of Health, U.S.P.H.S., by the Berkeley Computer Center, and by a

Guggenheim Fellowship 1965–66. The technical assistance of G. Sugimura, J. Willey, J. Horowitz, Ben Barron and Alan Alberts is gratefully acknowledged. Analog computer facilities were provided by Professor C. Atkinson, Dept. Mech. Eng., Berkeley. I am especially indebted for valuable discussion to Professor O. J. M. Smith, Dept. Elec. Eng., Berkeley.

References

1. Araki, T., Ito, M. and T. Oshima. "Potential changes produced by application of current steps to motoneurones," *Nature (Lond.)* **191** (1961), 1104–1105.
2. Boudreau, J. C. and W. J. Freeman. "Spectral analysis of electrical activity in the prepyriform cortex of the cat," *Exptl. Neurol.* **8** (1963), 423–439.
3. Eccles, J. C., *The Physiology of Nerve Cells* (Johns Hopkins Press, Baltimore, 1957), 270 pp.
4. Eccles, J. C., *The Physiology of Synapses* (Academic Press, New York, 1964).
5. Fadiga, E. and J. M. Brookhart, "Interaction of excitatory postsynaptic potentials generated at different sites on the frog motoneuron," *J. Neurophysiol.*, **25** (1962), 790–804.
6. Frank, K. A. and M. G. F. Fuortes, "Accommodation of spinal motoneurones of cats," *Arch. Ital. Biol.*, **98** (1960), 165–170.
7. Freeman, W. J., "Distribution in time and space of prepyriform electrical activity," *Neurophysiol.*, **22** (1959), 644–665.
8. Freeman, W. J., "Repetitive electrical stimulation of prepyriform cortex in cats," *J. Neurophysiol.*, **23** (1960), 283–396.
9. Freeman, W. J., "Phasic and long-term excitability changes of prepyriform cortex in cats," *Exptl. Neurol.*, **5** (1962), 500–518.
10. Freeman, W. J., "Changes in prepyriform evoked potential with food deprivation and consumption," *Exper. Neurol.*, **6** (1962), 12–29.
11. Freeman, W. J., "Alterations in prepyriform evoked potential in relation to stimulus intensity," *Exper. Neurol.*, **6** (1962), 70–84.
12. Freeman, W. J., "Comparison of thresholds for behavioral and electrical responses to cortical electrical stimulation in cats," *Exper. Neurol.*, **6** (1962), 315–331.
13. Freeman, W. J., "The electrical activity of a primary sensory cortex: the analysis of EEG waves," *Intern. Review J. Neurobiol.*, **5** (1963), 53–119.
14. Freeman, W. J., "Use of digital adaptive filters for measuring prepyriform evoked potentials from cats," *Exper. Neurol.*, **10** (1964), 475–492.
15. Freeman, W. J., "A linear distributed feedback model for prepyriform cortex," *Exper. Neurol.*, **10** (1964), 525–547.
16. Freeman, W. J., "Problems in the measurement of cortical evoked potentials," in J. Wortis, *Recent Advances in Biol. Psychiatr.* (Plenum Press, New York, 1965) 7, Ch 24.
17. Freeman, W. J., "Correlation of goal-directed work with sensory cortical excitability," in J. Wortis, *Recent Advances in Biol. Psychiatr.* (Plenum Press, New York, 1965) 7, Ch.25.
18. Freeman, W. J., "Nonlinear gain stabilization of prepyriform cortex," (Abstr.), 1965, *Fed. Proc.*
19. Freeman, W. J., "Statistical input-output relations of single cortical neurons," *The Physiologist* **10** (1967), 172.
20. Fuortes, M. G. F. and F. Mantegazzini, "Interpretation of the repetitive firing of nerve cells," *J. Cen. Physiol.*, **45** (1962), 1163–1179.
21. Goldberg, J. M., H. O. Adrian, and F. D. Smith, "Responses of neurons in the superior olivary complex of the cat to acoustic stimulation of long duration," *J. Neurophysiol.*, **27** (1964), 706–749.
22. Granit, R., "Recurrent inhibition as a mechanism of control," 23–37. In G. Morruzzi, A. Fessard, and H. H. Jasper, *Progress in Brain Research*, Vol. 1. "Brain Mechanisms." (Elsevier, Amsterdam, 1963).
23. Granit, R., D. Kernell, and G. K. Shortness, "Quantitative aspects of repetitive firing of mammalian motoneurones, caused by injected currents," *J. Physiol. (Lond.)*, **168** (1965), 911–931.
24. Green, J. D., M. Mancia, and R. von Baumgarten, "Recurrent inhibition in the olfactory bulb. I. Effects of antidromic stimulation of the lateral olfactory tract," *J. Neurophysiol.* **25** (1962), 467–488.

25. Hagiwara, S., "Analysis of interval fluctuations of the sensory nerve impulse," *Jap. Journ. Physiol.*, **4** (1954), 234–240.
26. Hartline, H. K., and F. Ratliff, "Inhibitory interaction of receptor units in the eye of Limulus," *J. Gen. Physiol.*, **40** (1956), 357–376.
27. Hunt, C. C. and M. Kuno, "Background discharge and evoked responses of spinal interneurones," *J. Physiol.*, **147** (1959), 364–384.
28. Jenik, F., "Electronic neuron models as an aid to neurophysiological research," *Ergeb. Biologie*, **25** (1962), 206–245.
29. Katz, B., "Depolarization of sensory terminal and the initiation of impulses in the muscle spindles," *J. Physiol. (Lond.)*, **111** (1950), 261–282.
30. Kuffler, S. W., and C. Eyzaguirre, "Synaptic inhibition in an isolated nerve cell," *J. Gen. Physiol.*, **39** (1955), 155–184.
31. Levinson, E., "Some saturation phenomena in servomechanisms with emphasis on the tachometer stabilized system," *A.I.E.F. Transactions, part II*, **72** (1953), 1–9.
32. Lorente de No, R., "Action potential of the motoneurons of the hypoglossus nucleus," *J. Cell. Comp. Physiol.*, **29** (1947), 207–287.
33. O'Leary, J., "Structure of the primary olfactory cortex of the mouse," *J. Comp. Neurol.*, **67** (1937), 1–31.
34. Rall, W., "A statistical theory of monosynaptic input-output relations," *J. Cell. Comp. Physiol.*, **46** (1955), 373–411.
35. Rall, W., "Membrane potential transients and membrane time constant of motoneurons," *Exper. Neurol.*, **2** (1960), 503–532.
36. Rall, W., "Electrophysiology of a dendritic neuron model.," *Biophys. Jour.*, **2** (1962), 145–168.
37. Ramon y Cajal, S., *Studies on the cerebral cortex (limbic structures)* (L. M. Kraft Lloyd-Luke, London, 1955).
38. Rodieck, R. W., N. Y.-S. Kiang, and G. L. Gerstein, "Some quantitative methods for the study of spontaneous activity of single neurons," *Biophys. J.*, **2** (1962), 351–368.
39. Savant, C. J., *Control System Design* (McGraw-Hill, New York, 2nd Ed., 1964), 358–360.
40. Sears, T. A., " The slow potentials of thoracic respiratory motoneurons and their relation to breathing," *J. Physiol. (London)*, **175** (1964), 404–424.
41. Sholl, D. A., *The Organization of the Cerebral Cortex* (Wiley, New York, 1956).
42. Smith, O. J. M., *Feedback Control System* (McGraw-Hill,New York, 1958), 13.
43. Terzuolo, C. A., and T. Araki, "An analysis of intra-versus extracellular potential changes associated with activity of single spinal motoneurons," *Ann. N. Y. Acad. Sci.* **94** (1961), 547–558.
44. Terzuolo, C. A. and Y. Washizu, "Relation between stimulus strength, generator potential and impulse frequency in stretch receptor of Crustacea," *J. Neurophysiol.*, **25** (1962), 56–66.
45. Verveen, A. A., "On the fluctuation of the threshold of the nerve fiber," in D. P. Tower and J. P. Schade, *Structure and Function of the Cerebral Cortex* (Elsevier, Amsterdam, 1960), 282–288.
46. Wilson, V. J., and P. R. Burgess, "Disinhibition in the cat spinal cord," *J. Neurophysiol.*, **25** (1962), 392–404.

5. *Opening Feedback Loops with Surgery and Anesthesia; Closing the Loops with Noise*

One of the basic heuristic concepts of classical neurology is the idea of the tonic influence, either excitatory or inhibitory, that each part of the nervous system exerts on others. From his clinical studies of the evolution and devolution of the central nervous system, Hughlings Jackson (1884) inferred that each level from cortex to spinal cord exerted constant inhibition on the levels beneath it, and he explained many of the symptoms that he observed and charted in terms of "release" of lower centers from higher domination. Sherrington concluded, near the close of his illustrious career, that the roles of neuron pools could best be characterized in terms of "central excitatory states", which he conceived in terms of chemical activities that regulated the firings of neurons in a continuously graded manner, unlike the "all-or-none" action potential. Synaptic potentials were still unknown at the time (Sherrington 1929). In contrast, the entire system of neurophysiology in Russia was based on the 19th century work of Sechenov, who demonstrated the regulation of reflex activity in the spinal cord of the frog by steady state excitation from higher brain centers, which, if it were too strong, became paradoxical inhibition.

The concept of steady state control is implicit in the physiological concept of homeostasis, in which the body is seen to maintain its parameters at desired set points through sensory receptors and negative feedback loops. The brain somehow determines the set points. For example, according to this view, body temperature is normally kept near 37 °C by the hypothalamic "thermostat", but in disease the setting is elevated to a new set point that produces fever. Blood pressure is set by neurons in the brain stem, and the brain lowers the set point in rest and increases it in anxiety.

Most importantly, the forebrain displays a high degree of autonomy in selecting the set points of the homeostatic systems, and expressing them in its internally maintained steady states. The olfactory system presents a unique opportunity for introducing these fundamental ideas about autonomy and self-regulation in brain components into a dynamic model, in which the "tonic influence" is represented by a steady excitatory or inhibitory bias. When the olfactory system is surgically isolated from the rest of the brain, it reveals its autonomy by continuing to generate its background EEG oscillations, although they are no longer integrated into the goal-directed behavior of the remaining brain and body systems. Rather than going permanently to sleep, the olfactory system stays continually awake (Becker and Freeman 1968). Even after the receptor layer has been removed, and the olfactory bulb and prepyriform cortex are reduced to an isolated island retaining only its blood supply, the background unit activity and fluctuating dendritic potentials persist, showing that the source of the activity is internal and self-regulated. In contrast, when the prepyriform cortex is surgically separated from the bulb, bulbar activity persists, but the prepyriform goes silent – the EEG is flat, and there is no unit activity. This phenomenon has also been demonstrated by reversible inactivation of the lateral olfactory tract by local application of procaine (Freeman 1975), and by cooling with a cryogenic probe (Gray and Skinner 1988). The results showed that sub-assemblies in the brain on the order of KII sets have a high degree of autonomy, which may be suppressed, if they are deprived of an exogenous depolarizing bias, or released if the bias is hyperpolarizing. Most areas of cortex appear to require an excitatory bias, because they fall silent when isolated in slices and slabs (Burns 1958) and remain so unless artificially stimulated by any of a variety of intoxicants or by sustained electrical stimulation.

A striking outcome of this preparation was the finding that intense electrical stimulation of the stump of the severed input tract to the cortex restored the oscillations of the prepyriform evoked potential (Figure 1 in Chapter 5). The distal segments of the severed axons continue to conduct pulses, if electrically stimulated, for 10 to 14 days after sectioning, but thereafter die owing to loss of their connection with their cell bodies. These experiments replicated root loci in Mode 1 (Figure 7 in Chapter 5, upper frame), which had already been shown to hold for changes in stimulus intensity, and for changes in the levels of local and general anesthesia, except for very high tetanization with weak pulse input, when Mode 2 root loci appeared (Figure 7 in Chapter 5, lower frame). In brief, the characteristic frequency of the evoked potential is determined by the ratio of the amplitude of the input impulse to the amplitude of the pre-stimulus background activity (Figure 6 in Chapter 5). Unilateral saturation results from either increasing the pulse input or decreasing the background activity by anesthetics (Figure 5 in Chapter 4), which decreases the feedback gain and the frequency. The generality of Mode 1 root loci (Figure IV, right frame) under such diverse conditions is the strongest evidence available that the gamma oscillations arise from interactions, not from reactions to pacemaker neurons. The results showed also that the nonlinearity at the trigger zones, where dendritic current density is converted to pulse density, is far more important in determining rapid changes in loop gains than is the nonlinearity at the synapses, where pulse density is converted to wave density.

Effects of Surgical Isolation and Tetanization on Prepyriform Cortex in Cats

Single-shock stimulation of the lateral olfactory tract between the prepyriform cortex and the olfactory bulb causes both structures to generate oscillatory evoked potentials having approximately the same frequency. If the tract is cut, the potential evoked in the cortex by stimulation of the tract central to the cut is reduced to a single negative peak. The inference might be drawn that the successive peaks of the oscillatory single-shock evoked potentials are due to reverberating excitatory and inhibitory events between the bulb and the cortex.

The experiments described here are designed to show that the oscillatory mechanism of the cortex is contained entirely within the cortex. The reason that the oscillations disappear following tract section is shown to be simply that the mitral cells in the bulb, which are continually active in the normal state, deliver a depolarizing input to the neurons of the cortex and maintain them in a state of activity. When the tract is cut, the depolarizing bias is removed, and the cortical neurons become silent and can be fired only by maximal single-shock stimulation of the tract – or by sustained tetanization. In the presence of this artificial background input, the responses to superimposed single-shock stimulation of the cortex are oscillatory, just as in the intact cortex.

Methods

Three types of surgical intervention were used. Deafferentation refers to section of the olfactory tract or to removal of the bulb. Undercutting refers to severance of all connections of the cortex with the central nervous system, but with some tissue adjacent to the cortex included in the island. Isolation refers to combined undercutting and deafferentation, so that the only connection of the cortex to the rest of the animal is through the vascular system.

Completeness of isolation was checked by gross dissection and histological examination of the brains fixed in situ and carefully removed to preserve intact any remaining bulb tissue. Results from incomplete procedures are not included here. It was, in fact, astonishing how small a fragment of bulb and tract was required to sustain normal EEG and evoked potentials in the cortex.

Comparison of the responses of surgically treated cortex with those of the intact cortex in the same animal on the opposite side required precise measurement. For most of these results the techniques used (5) consisted of measuring the amplitudes, v_1 and v_2, and latencies, t_1 and t_2 in seconds, of the first two surface-negative peaks of the averaged evoked potential (AEP). The frequency, ω, was taken as $\omega = 2\pi / (t_2 - t_1)$ radians sec^{-2}. The decay rate, α, in sec^{-1} was given as $\alpha = \omega / 2Q$, where $Q = \pi / \ln(v_1 / v_2)$ (see also ref. 6).

Previous behavioral studies of the prepyriform AEP (10) have shown that its waveform changes in most physiological states in a characteristic way. The pattern is best described by postulating the existence in cortex of a negative feedback loop with a variable gain, K_2, and a variable delay, T_2. By means of the following equations, the frequency, ω, and decay rate, α, of the oscillatory AEP serve to evaluate the gain, K_2, and delay, T_2.

$$\xi = (1 + \alpha / \omega)^{1/2} - \alpha / \omega$$
$$\theta = \pi - \tan^{-1}[\omega / (B_1 - \alpha)] - \tan^{-1}[\omega / (B_2 - \alpha)]$$
$$T_2 = \xi\theta^2 / 3\omega \text{ in sec}$$
$$K_2 = \exp(\xi\theta)$$

where the constants, B_1 and B_2, are defined (7) to represent the delay characteristics of the forward limb of the loop. The pattern then consists of linear covariation of K_2 and T_2 in accordance with the equation, $K_2 = a + bT_2$, where a and b are empirically evaluated constants (10).

A feedback model has been formulated (8, 9) to represent the dynamics of the cortex, based on linear differential equations with a single variable feedback gain coefficient. The solutions to the equations for appropriate initial conditions yield damped sinusoids. Transformation of the frequency and decay rate of the sinusoid to values for gain, K_2, and delay, T_2, for differing values of the feedback gain coefficient produces a curve or locus in K_2–T_2 coordinates, that replicates the experimentally derived curve (8, 9). The feedback gain coefficient is interpreted physiologically to represent the effect of a nonlinear property of cortical neurons termed "saturation." This is the failure of mean firing rate of single neurons to follow extreme fluctuations in dendritic currents, for which experimental evidence has already been reported (11). The degree of saturation for the cortical response to any test pulse input depends on the ratio of "spontaneous" background input to test input for the cortex.

The requirement is posed in the present study, where an "artificial" background input is introduced and systematically varied, that the cortical AEP change in the same way as when the normal background input changes. Therefore the frequency and decay rate of AEP recorded in the differing conditions described in this report are transformed to estimates of gain, K_2, and delay, T_2. Conformity of the resulting curves or loci to the loci derived from normal waking cats (10) is taken to mean two things: first, the basic mechanism generating the AEP has been preserved intact in the designated experimental conditions, and second, the conclusion previously reached, i.e. this pattern of change in AEP is a manifestation of a nonlinear negative feedback loop in the cortex, is substantially correct.

Results

During the induction of barbiturate narcosis (with 35–45 mg/kg pentobarbital given intravenously in a cat having implanted electrodes), and for 1–3 hr thereafter, the prepyriform evoked potential is nonoscillatory (4). It consists of a surface-negative peak lasting 5–7 msec followed by a shallow positive overshoot lasting 30–100 msec. This has been referred to as the "open loop" response of the cortex (7, 11). A mirror-image response is observed below the molecular layer. The first stage of recovery consists of the prolongation of the negative peak by from 2 to 4 msec and the appearance of a large positive overshoot. Somewhat later, a broad second negative peak appears following the positive overshoot, so that the response may be described as oscillatory. Thereafter the frequency and duration of the oscillation progressively increase, until the response assumes its preanesthetic form (a damped sinusoid at 250 radians/sec), even though the animal remains anesthetized and responds to nociceptive stimulation of the limbs merely with flexor reflexes and without arousal. Additional doses of 20–30 mg/kg given intravenously quickly cause reversion to the nonoscillatory form. The initial amplitude is unaffected, provided respiratory embarrassment is avoided.

The occurrence of a nonoscillatory, open loop AEP coincides with the disappearance of oscillatory EEG activity and of background or spontaneous unit activity. Large initially surface-positive, deep-negative diphasic spikes occur irregularly, often in short trains ("barbiturate spikes" (1)).

The hypothesis was proposed that the alteration in cortical excitability produced by pentobarbital might be due to the diminution in background input from the olfactory bulb. Therefore a pair of electrodes placed in the bulb was used for tetanizing the tract, another pair in the olfactory tract in the temporal lobe was used for test stimulation, and a third pair in the intervening cortex across the cortical dipole was used for recording. Early in the recovery from deep narcosis, the averaged evoked potential consisted of the surface negative peak and the large surface-positive overshoot (Fig. 1). Stimulation was then applied to the bulb at a rate of 250/

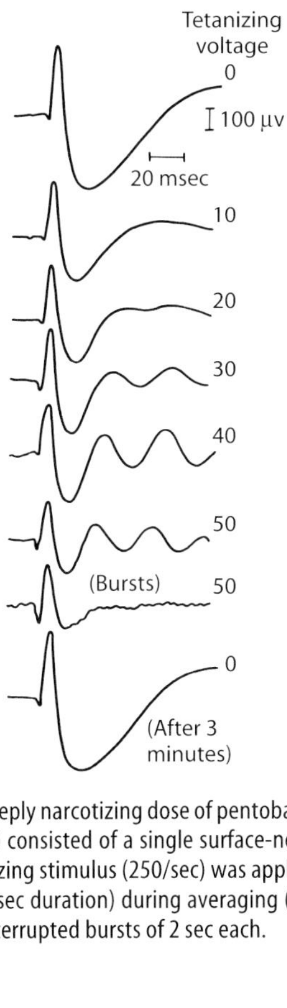

Fig. 1. During early recovery from a deeply narcotizing dose of pentobarbital, the AEP in response to low-level test stimulus (2.2 v, .01 msec, 7.3/sec) consisted of a single surface-negative peak, and a positive overshoot without a late negative wave. A tetanizing stimulus (250/sec) was applied to the olfactory tract by means of a third electrode pair (10–50 v at 0.1 msec duration) during averaging (12 sec, N=92). During the next-to-last trace the averaging was done in six interrupted bursts of 2 sec each.

sec, randomly with respect to the averaging process, and at stepwise increments in voltage. As shown in Fig. 1, the AEP became fully oscillatory, and the frequency was roughly proportional to the tetanizing voltage up to and including the normal frequency of the AEP. On cessation of stimulation the AEP reverted to its nonoscillatory form. This showed that the cortex could be returned to its normal state (insofar as the AEP is concerned) under barbiturate narcosis by supplying an artificial source of background activity. The voltages required were far in excess of threshold (e.g. 10x) for single-shock responses, and the placement of the electrodes had to be as far apart as possible, to minimize the preemption of test axons by the tetanizing input.

The lateroventral surface of the bulb, tract, and cortex of the anesthetized cat can be exposed by orbital exenteration and removal of the frontal process of the sphenoid bone. If the dura is incised carefully to avoid damage to the cortex, the AEP is normally oscillatory. Application to the olfactory tract of a 2% solution of procaine, or incision of the tract to a depth of about 0.5 mm where it emerges from the bulb, or total removal of the bulb cause the AEP to become nonoscillatory. Additional intravenous doses of barbiturate sufficient to do this on the opposite side fail to modify further the AEP on the deafferented side.

The effect of procaine is quickly reversible, but not the effect of tract section or bulb removal. In fact, recurrence of spontaneous oscillation (EEG) or of oscillatory AEP is reliable evidence for failure of complete section or complete removal of the bulb, as repeatedly confirmed on postmortem anatomical examination. The bulb has also been removed by suction through a burr hole in the floor of the frontal sinus, and electrodes have been implanted in the remaining tract and cortex. The AEP gradually disappears over a period of 5–14 days, as the olfactory tract axons degenerate, but it never becomes oscillatory. The chronically deafferented cortex does not respond to electrical stimulation.

However, if prior to tract degeneration two pairs of electrodes are placed on the olfactory tract central to the section, one for tetanizing and the other for testing, together with a third pair across the cortical dipole for recording, the effect of tetanizing is found to be similar to that in the anesthetized but surgically intact cat. As shown in Fig. 2, right, there is the difference that the maximum frequency attainable by tetanizing is about 80% of the normal level, and the responses are much less oscillatory. This is reflected in the high values for α, the decay rate.

The frequency of the AEP again depends on the tetanizing voltage (the smooth curves in Fig. 2, left, are cumulative normal distribution curves). Consistently it has been found that the maximum or asymptotic frequency of the AEP is higher for smaller test stimuli.

Tetanizing the tract running to intact cortex having a normal oscillatory AEP in response to a fixed level of test stimulation causes increased decay rate and frequency with fewer oscillations. This phenomenon is closely related if not identical to results of decreasing the test stimulus intensity in the presence of a stable level of background spontaneous activity (8, 11).

The background unit activity, which is suppressed by deep anesthesia and deafferentation (11), can be reestablished for the duration of tetanizing, but it disappears thereafter in several seconds. Figure 3 shows poststimulus time

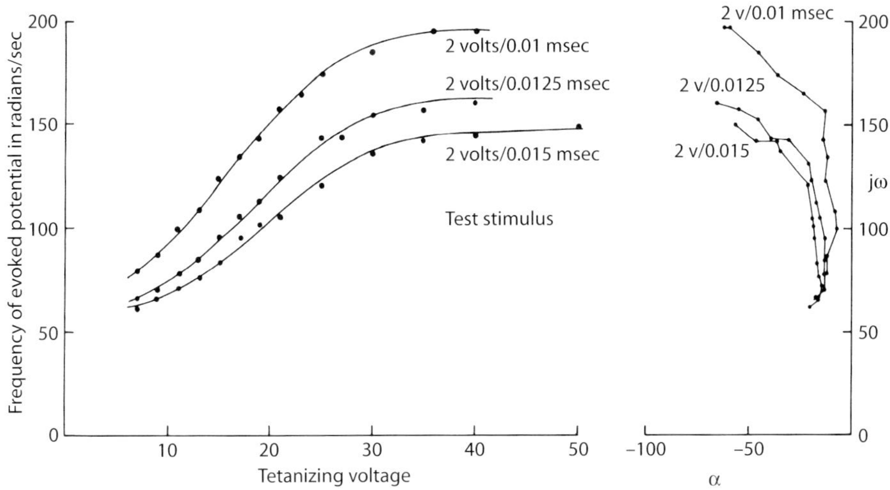

Fig. 2. About 24 hr after removal of the olfactory bulb the procedure described in Fig. 1 was repeated in a different cat for three test stimulus intensities. (At these short durations an increase in pulse amplitude and in duration are equivalent.) The frequency, ω, in radians/sec of the AEP is shown as a function of tetanizing stimulus intensity on the left. Curves are cumulative normal-distribution curves. On the right ω is plotted against α, the decay rate in 1/sec (upper left quadrant of the S plane).

histograms from a unit recorded with a microelectrode from a cortex with the bulb removed 8 hr before. The AEP above each poststimulus histogram was recorded simultaneously from a gross electrode nearby. Without tetanizing (OV) the AEP was nonoscillatory, and the unit was silent. With tetanizing at low levels (6 v) the negative peak widened at the same time as the positive overshoot enlarged, and induced unit activity appeared before background activity. (The irregularities in the traces are attributable to the tetanizing shock artifact. As previously described (9), the secondary event in the first negative peak of the AEP represents reexcitation of some neurons already excited at the same time as the inhibitory neurons are excited, before the inhibitory action can take effect.) With increasing tetanizing voltage the secondary excitation became larger than the first, sustained background unit activity reappeared, and the AEP became oscillatory. On cessation of tetanizing, the AEP and histogram reverted within 30 sec to the control patterns.

These results imply that the disappearance of the oscillations following tract section should not be ascribed to interruption of a loop between the bulb and the cortex, but to a loss of background input from the bulb to the neurons in the cortex, which normally maintains their transmembrane potential near threshold. The decrease in maximum or asymptotic frequency of AEP with increase in test stimulus magnitude or duration (Fig. 2) implies the existence of an upper limit on the effectiveness of the depolarizing input. Otherwise, the curves in Fig. 2, left, would be asymptotic to the same maximum frequency but at differing levels of tetanizing voltage.

However, it is possible that section of the tract interrupts a major feedback loop between the bulb and the cortex, which is responsible for the oscillation at 250 radians/sec, and leaves intact another loop within the cortex or between the cortex

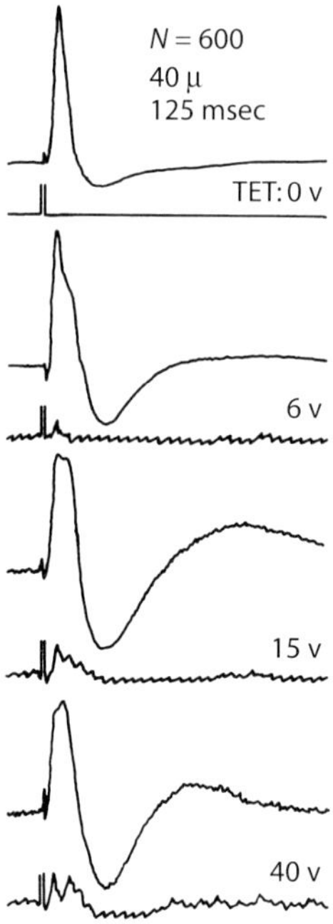

Fig. 3. Each pair of traces shows a poststimulus time histogram of the firing of a single unit in the cortex and an AEP recorded from the surface nearby. Tetanization of the stump of the severed olfactory tract caused reappearance of an oscillatory AEP with a dicrotic initial surface-negative peak and the return of unit activity. Both faded rapidly after cessation of tetanization.

and some other brain structure, which is capable of generating oscillatory potentials at some lower frequency, and then only in the presence of a depolarizing bias.

To test this possibility surgical undercutting of the cortex was undertaken. The inaccessibility and irregular curvature of the cortex made a conventional transpial approach unfeasible, particularly when the aim was to preserve the superficially located afferent pathway intact and to place multiple gross electrodes in it. An effective technique was to remove by suction all of the anterior third of one hemisphere down to the floor of the skull, but to leave intact the cortex and bulb together with some adjacent tissue. There is no question with this technique that all of the central connections were severed, and in at least 1 of 13 preparations studied for 1 or more weeks, all of the olfactory tubercle, amygdaloid nucleus, or orbital neocortex was removed. The thickness of the islands ranged from the smallest, having a maximum of 3 mm and a mean of less than 2 mm, to the thickest, with an average of 11 mm. The surface areas of the islands are shown in summary schematic form in Fig. 4.

It was expected that such islands might show patterns of spontaneous EEG activity characteristic of those normally seen here in sleep (3), because of the removal of brainstem input. On the contrary, the activity appeared in bursts at 250 radians/sec, very similar to that recorded with electrodes implanted in the contralateral intact cortex. The AEP was oscillatory at its normal frequency. It has often been noted that the AEP from the two prepyriform cortices are as similar to one another in comparison to the prepyriform AEP of other cats as are identical

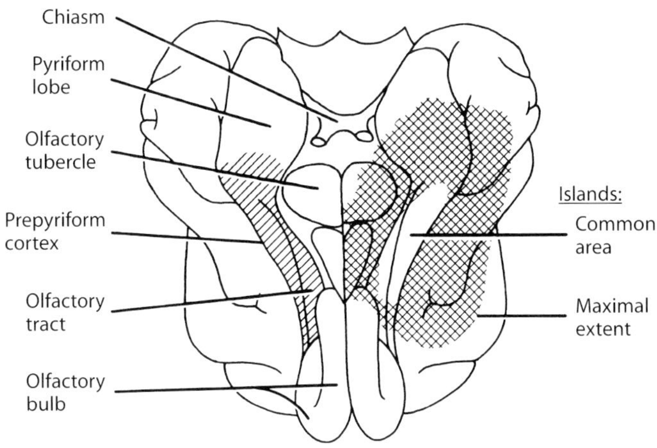

Fig. 4. Small shaded area on the left of this ventral view of the cat brain represents the surface of the prepyriform cortex. Large shaded area on the right represents the maximal surface extent of the outer borders of all of the "islands" taken together, though none was so large. The central lacuna is the area held in common by all of the islands.

twins among siblings, and this similarity was equally apparent in cats having one cortex undercut. Only in its patterns of spontaneous variation was the AEP of the undercut side uniquely different from normal (10).

Over a range of stimulus intensity from 1× to 3× threshold the AEP of the undercut cortex shows the decrease in frequency, ω, the decrease in α to a minimum, at threshold for single-shock responses, and further decrease in ω and increase in α as the single-shock response increases in amplitude above the background EEG amplitude. Following the transformation of ω and α to gain, K_2, and delay, T_2 (7), the locus for the change in values with increasing stimulus intensity is well within the normal limits for variation between cats (10). An example is shown in Fig. 5. With increasing stimulus intensity both K_2 and T_2 increase together to a minimum value for a near ω =250 radians/sec. With further increase in stimulus intensity T_2 continues to increase and K_2 diminishes.

The appearance of normal frequencies in the AEP of the undercut cortex rules out the possibility that the oscillatory AEP depends uniquely on a long loop through the thalamus or rhinencephalon, and makes it unlikely that neighboring structures are involved, unless all of them are equally competent to interact with the cortex to generate these frequencies. But it raises again the question whether the bulb and cortex might so interact, especially since the bulb generates waves at nearly the same frequency.

Therefore, in 13 cats the ipsilateral bulb was removed from 1 to 5 weeks after recovery from undercutting. The surgery was performed under ether, so that recovery from anesthesia was complete within 2 hr. As expected, an open loop response was obtained (Fig. 6, $T = 0$ v). Upon tetanizing the tract central to the cut, within the first 24 hr after bulb removal, the oscillatory AEP reappeared. The attainable frequencies were well within the range of normal values (Fig. 6, $T = 45$ v). Thus it is clear that the cortex contains within itself a mechanism for generating the AEP at

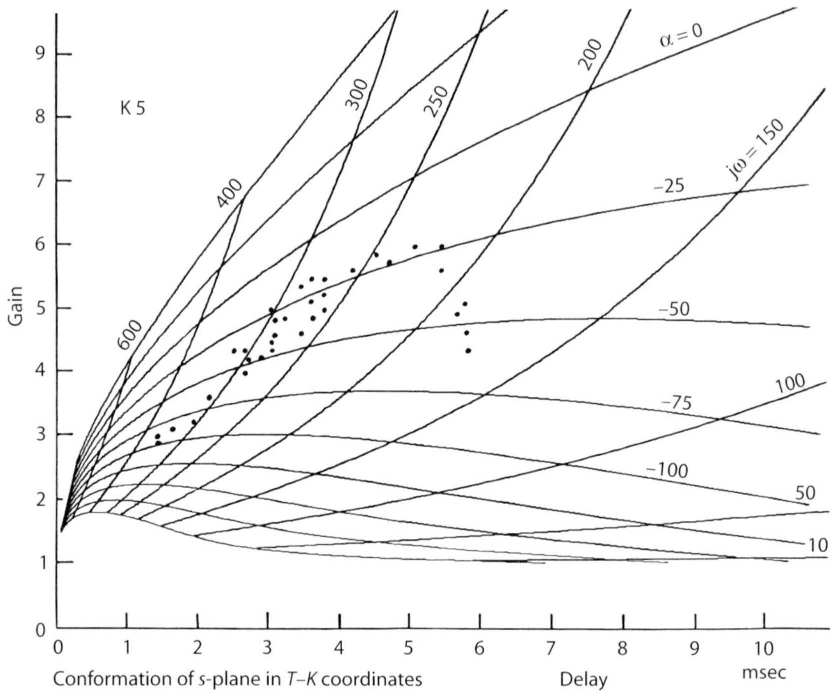

Fig. 5. An example in T_2–K_2 coordinates is shown of the frequency (ω) and decay rate (α) of AEP recorded from the undercut cortex (isolated from the rest of the brain) with olfactory bulb and tract undamaged. With increasing stimulus intensity T_2 rises monotonically, whereas K_2 increases to a maximum, at which the peak amplitude of the single-shock evoked potential reaches the peak amplitude of the EEG (a-c) activity. With still further increase in stimulus intensity, T_2 continues to increase, but K_2 either remains constant or decreases.

250 radians/sec, and the bulb contributes a background input rather than a feedback loop to sustain the oscillation.

An additional point of interest, demonstrated in Fig. 7, is that, with increasing stimulus intensity at a fixed level of tetanizing input amplitude, there is decreasing frequency, ω, in the AEP as in normal cortex. Further, the rate at which ω decreases with increasing intensity is less for high tetanizing levels than for low levels. Thus a low frequency attainable between 8 and 10-v test stimulus at $T = 25$ v is attainable with 20-v test stimulus at $T = 35$ v and is unattainable at $T = 45$ v. This confirms the proposal that the waveform of the AEP in the intact cortex must depend on the ratio of test stimulus to background activity and not on the test stimulus alone, as previously deduced (10).

Transformation of values for ω and α into K_2–T_2 coordinates reveals (Fig. 7, above) that the locus for change with increasing stimulus intensity is horizontal, i.e. that the delay, T_2, increases without change in the gain, K_2. This is characteristic of many intact animals (10) for AEP having amplitudes above the peak EEG amplitudes. On the other hand, if the test stimulus intensity is fixed and the background tetanizing level is varied, the locus is oblique (Fig. 7, below). This is characteristic of the AEP from intact cats, when the stimulus intensity is maintained constant and

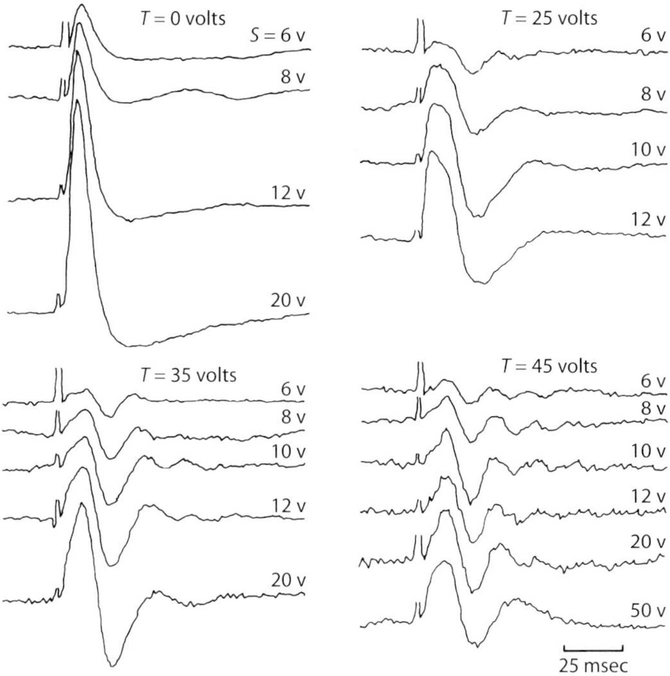

Fig. 6. Four sets of AEP are shown recorded from isolated prepyriform cortex (undercut 4 days before and bulb removed 8 hr before recording). Stimulus intensity was raised in steps, first without tetanization and then with tetanization at each of three levels.

the background activity varies spontaneously or is modified by behavioral manipulation (10). The slopes of these experimental curves are within the range of variation between normal cats. Figure 7, below, also shows replication of another previous finding: the slope of the curve is lower for higher mean values of delay, T_2.

These several common features show that the oscillatory mechanism in the isolated cortex has many properties characteristic of the mechanism of the intact cortex. It is reasonable to conclude that the mechanism is entirely contained within the cortex, though it is of course subject to extracortical modulation and control.

Discussion

The disappearance of oscillatory AEP and EEG after removal of the bulb can be attributed to loss of a steady background barrage of depolarizing impulses from the bulb to the tract, rather than to interruption of a corticobulbar loop. The loss causes the neurons of the cortex to fall below threshold for firing except for all but the most powerful stimuli. The loss can be induced by both general and local anesthesia, which bring about the same change in AEP as removal of the bulb. The deficit of

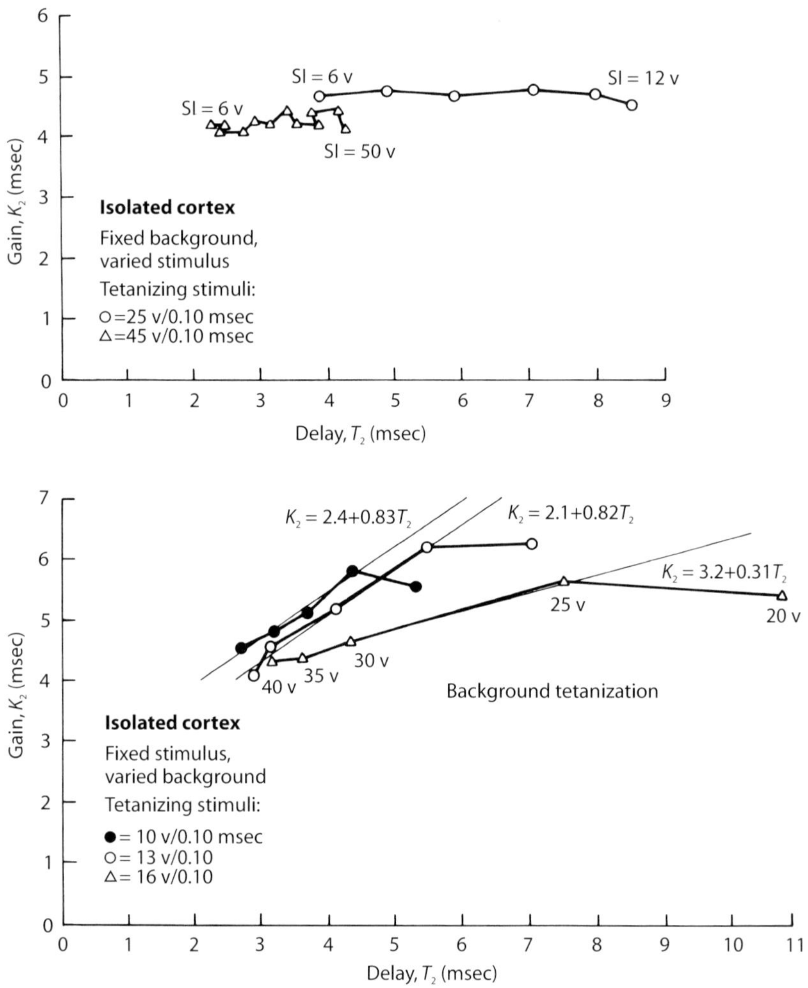

Fig. 7. Varied background tetanization with fixed test stimulus intensity yielded a horizontal locus. Varied test stimulus intensity in the presence of fixed background tetanization gave an oblique locus.

input and the distortion in AEP can be compensated by tetanization of the olfactory tract, which must (on anatomical grounds) establish an outward, depolarizing current across the axonal membranes of the superficial pyramidal cells of the cortex.

Questions arise regarding the role played by structures contiguous to the cortex in the normal function of the cortex. Tissue belonging to these structures (olfactory tubercle, orbital neocortex, and the anterior amygdaloid complex) was deliberately left intact along the margins of the prepyriform island in order to minimize direct damage and mechanical displacement of the cortex, although each of these adjacent structures was entirely removed in at least one of the 13 preparations studied. It is unlikely that the orbital neocortex plays any direct role in generating the AEP, because, unlike stimulation of the thalamus, claustrum, anterior commissure, periamygdaloid cortex, and bulb, direct electrical stimulation of it does not evoke prepyriform potentials.

The vertical thickness of the island varied from 2 to 11 mm or more. Spontaneously active units were often sampled at distances up to 2 mm above the zero isopotential surface, but activity coherent with the AEP was never found at sites more than 1,200μ from the zero (11). It is therefore unlikely that neurons lying deep to the cortex are directly and essentially involved in generating the AEP. However, the possible role of contiguous structures in modulation and control of the waveform of the AEP or spectrum of the EEG in the intact animal has not been adequately assessed.

Another problem is raised by the undercut cortex with its primary afferent tract intact. The preparation differs in this respect from the classical cortical island of Burns (2), from which all afferent connections are cut. In the absence of neural connections with the brainstem reticular formation the island shows both evoked and spontaneous activity characteristic of the waking state. Observations on the activity of undercut cortex during sleep are now being made (Becker and Freeman 1968).

The main advantage offered by the isolated cortex is the opportunity to control the level of background input, in order to test previous inferences that the form of the AEP is determined not by the magnitude of test input but by the ratio of test to background input. Indeed it is now clear that the patterns of change in the AEP previously observed by increasing the stimulus intensity or decreasing background activity (using pentobarbital) can be shown in reverse by raising the background activity, using tetanizing pulse trains to the blocked or severed olfactory tract.

The correspondence with the normal state is clearer for isolated than for deafferented cortex, in which the frequency of the AEP is consistently low. Some input of central origin to the cortex may be responsible for this, but the source and the mechanism are unknown.

The background activity in the cortex normally has both a d-c component and an a-c component, reflected respectively in measurements of transcortical d-c and EEG potentials. The tetanizing input described in this report contributed solely to the d-c bias, because of the high frequency used (250/sec). An a-c component could readily be introduced by tetanizing at lower rates (e.g. 50–80/sec). This contributed relatively little to the AEP, which tended to be somewhat more oscillatory. The difference was too small and the difficulties of establishing an equivalence between tetanizing rate and intensity too great to make this worth detailed evaluation.

Finally, during tetanization of isolated cortex the duration of the first surface-negative peak of the single-shock AEP, as well as the prominence of the second part of that peak, are greater than in the AEP of intact cortex (Fig. 6). This is partly because the tetanizing input acts directly on excitatory neurons but only secondarily on inhibitory neurons in the cortex, and mutual excitatory interaction is selectively enhanced. (As will be shown in a later report, the same phenomenon occurs during anodal polarization of both the intact and deafferented cortex.) It is, additionally, due in part to absence of the bulb. Tract stimulation normally results in antidromic invasion of the bulb by a volley on the mitral cell axons. This results in prompt and sustained inhibition of the mitral cells by bulb interneurons (12–14). Unlike the artificial tetanizing input, the spontaneous background input is transiently reduced. The reduction has the effect of superimposing a secondary response of the cortex on the primary orthodromically induced response. The addition of the secondary, initially surface-positive component having a latency of 5–8 msec shortens the duration and reduces the amplitude of the initial surface-negative peak of the AEP. Hence the bulb contributes to the form of the AEP, but it is not an intrinsic part of the oscillatory mechanism (15).

Summary

The averaged evoked potential of the normal prepyriform cortex closely resembles a damped sine wave. Local or general anesthesia, severance of the olfactory tract, or removal of the bulb will each reduce the AEP to a single negative peak, denoted the "open loop" response. The background unit and EEG activity are concomitantly suppressed, and induced unit activity occurs in response to single afferent volleys only near maximal stimulus intensity. Undercutting the cortex to sever central connections without damaging the olfactory tract does not alter the waveform, although it does alter the variability of the waveform. In each of the open loop conditions, tetanizing the remaining olfactory tract restores the AEP to its oscillatory form. The character of the oscillations closely resembles the normal AEP in relation to changing levels of test stimulus intensity and background input. It is concluded that the mechanism of the oscillatory AEP is entirely contained within the cortex and that it is subject to extracortical modulation and control. The disappearance of the oscillations following tract section or anesthesia is due to the loss of normal background input, which can be replaced by tetanizing the remaining tract. It is not the result of opening a corticobulbar loop.

Acknowledgments

The technical assistance of C. T. Mckennee and C. Becker is gratefully acknowledged.

Reprinted with permission from the *Journal of Neurophysiology*, 1968, Vol. **XXXI**, No. 3, pages 349–357

References

1. Brooks, D. C. Rhinencephalic spikes and beta waves in the cat under pentobarbital. *Am. J. Physiol.* **202**: 1221–1229, 1962.
2. Burns, B. D. *The Mammalian Cerebral Cortex*. London: Arnold, 1958.

3. Freeman, W. J. Correlation of electrical activity of prepyriform cortex and behavior in cat. *J. Neurophysiol.* **23**: 111–131 1960.
4. Freeman, W. J. Repetitive electrical stimulation of prepyriform cortex in cats. *J. Neurophysiol.* **23**: 383–396, 1960.
5. Freeman, W. J. Linear approximation of prepyriform evoked potentials in cats. *Exptl. Neurol.* **5**: 477–499, 1962.
6. Freeman, W. J. Use of digital adaptive filters for measuring prepyriform evoked potentials from cats. *Exptl. Neurol.* **10**: 475–492, 1964.
7. Freeman, W. J. A linear distributed feedback model for prepyriform cortex. *Exptl. Neurol.* **10**: 525–547, 1964.
8. Freeman, W. J. Analysis of function of cerebral cortex by use of control systems theory. *Logistics Rev.* **3**: 5–40, 1967.
9. Freeman, W. J. Analog simulation of prepyriform cortex in the cat. *Math. Biosciences* **2**: 181–190, 1968.
10. Freeman, W. J. Patterns of variation in waveform of averaged evoked potentials from prepyriform cortex in cats. *J. Neurophysiol.* **31**: 1–13, 1968.
11. Freeman, W. J. Relations between unit activity and evoked potentials in prepyriform cortex of cats. *J. Neurophysiol.* **31**: 337–348, 1968.
12. Phillips, C. G., Powell, T. P. S., and Shepard, G. M. Responses of mitral cells to stimulation of the lateral olfactory tract in the rabbit. *J. Physiol., London* **168**: 65–88, 1963.
13. Rall, W., Shepherd, G. M., Reese, T. S., and Brightman, M. W. Dendrodendritic synaptic pathway for inhibition in the olfactory bulb. *Exptl. Neurol.* **14**: 44–56,1966.
14. Von Baumgarten, R., Green, J. D., and Mancia, M. Slow waves in the olfactory bulb and their relation to unitary discharges. *Electroencephalog. Clin. Neurophysiol.* **14**: 621–634, 1962.
15. Willey, T. J. and Freeman, W. J. Alteration of prepyriform evoked response following prolonged electrical stimulation. *Am. J. Physiol.* **215**: 1435–1441.

6. *Three Degrees of Freedom in Neural Populations: Arousal, Learning, and Bistability*

Brain dynamists aim to understand the mechanisms by which neurons in brains organize and coordinate their activity in the genesis and guidance of behavior. To that end we measure the activity of parts of brains that are accessible to us, and simultaneously the behaviors of our subjects. Then we calculate statistical relationships between the two sets of data. My problem was that the prepyriform generator seemed to be involved in all forms of goal-directed behavior, but in ways that seemed non-specific. What did it contribute? Since I had learned that it could operate as a nearly autonomous subsystem in brains of small mammals, I decided to turn it loose, so to speak, and determine what it would do. I did this by using the evoked potential technique in animals trained to perform goal-directed tasks that relied on the olfactory system in preference to other senses. I used the electrical pulse to perturb the cortex, so that I could measure the path that it followed in relaxing to its pre-stimulus condition, and I inferred the degrees of freedom in respect to behavior by noting the differences in the return path that accompanied modifications in behavior.

As with unknown machines, we require a well-defined input and a good measure of output by which to define an operation that transforms the input to give the output. We know that humans and other animals select their own inputs by mechanisms of attention and expectation, and it is impossible for us to know or simulate those inputs in sufficient detail, so instead, we give pulses as sharply delineated inputs to sensory receptors and axonal tracts in order to produce cortical impulse responses, which in the most compact form reveal the characteristic frequencies that govern the neural populations comprising areas of cortex as they relax to their basal states. Those frequencies are most informative if the impulse response is well within the near-linear, small-signal range with minimal saturation, which requires that the response amplitude not exceed the peak amplitude of the on-going EEG, keeping the system in Mode 2. The signal-to-noise ratio is then less than unity, so that repeated stimulation and time ensemble averaging are required. Minimally at least 100 trials are required, which reduces the noise by a factor of 10. At least 140 msec must be allowed for the impulse response to end, limiting the repetition rate to 7/sec, so that the minimal duration of a behavioral state in which to make an observation is about 15 seconds. Thus evoked potential studies are severely constrained within limits imposed by nonlinearity, nonstationarity, and the noise contributed by the ongoing background activity.

The aim of using digital adaptive filters is to evaluate as many parameters of the basis functions and extract as much information as possible from this highly concentrated window of observation. There are two criteria of success. One is maximizing the signal-to-noise ratio by incorporating as much as possible of the variance of the data into the fitted waveform, leaving only white noise in the residuals. The other is maximizing the multiple correlation of the optimized parameters with a dependent measure of the behavior. By this behavioral measure the selection of the basis functions and the design of the filter can be optimized (Freeman 1964c). For many years I used the average power, measured in watts over the window of observation, done by cats working in an ergometer for food, or I used the frequency of pressing a bar to get water or milk. The sensitivity of the evoked potential is enhanced by training the cats to respond to the electrical stimulus as a conditioned signal. The correct response was to wait in the starting box after the gate opened for the stimulus signaling to begin work. Owing to the requirement for low intensity, the stimulus and the impulse response were very faint, like the sound of a single shot on a noisy battlefield, so that a strenuous regimen was required to train the cats to discriminate the stimulus, which otherwise never surfaced to affect behavior. During the development of a conditioned response, the variance of the train of evoked potentials was greatly enhanced (Figure 6 in Chapter 6), but after the response was formed, the evoked potential stabilized in a new form, until the response was extinguished and the cats were habituated to the electrical stimulus (Figure 1 in Chapter 6).

The outcome of each 15-second trial was a set of 8 coefficients from the digital filter: the amplitude, frequency, decay rate, and phase of each of 2 damped cosines (Table 1 in Chapter 6). The aim of this stage of modeling was to interpret the coefficients in terms of the dynamics of the neurons that generated the signals. With linear basis functions the coefficients were determined by the poles and zeros of an 8th order linear differential equation. The purpose of modeling was to establish the deterministic relation between the properties of the neurons in the populations and the characteristics of their measured behavior. In a multiple loop feedback system a change in any one parameter of the model was likely to alter the entire set of coefficients in the basis functions. Conversely, a pattern of change in the coefficients pointed to a focal alteration in the model and an identifiable neural modification, if the model was designed properly to represent the neural system. Here entered the value of the multiloop negative feedback model and its elaboration into the KII model. I based it on the premises that cortex consisted of excitatory and inhibitory neurons, each connected to others of the same and the other kind by differing synaptic gains, and that the open loop rate constants were fixed, whereas the gains at synapses and trigger zones were variable. Patterns of change in amplitudes, phases, decay rates, and closed loop frequencies were thereby interpreted as patterns of change in the strengths by which neurons acted upon each other to excite or inhibit them.

The method of choice for extracting patterns was principal components analysis of covariance matrices (or factor analysis of correlation matrices). However, preprocessing is required to normalize the distributions and linearize the covariance. These operations were done by transforming complex roots $\alpha + j\omega$ to K_2 and T_2 (Table 6.2), which linearized the correlation structure of the data (Table 6.3). Three main factors emerged that accounted for >70% of the variance (Table 6.6). The search for the neural bases and behavioral significance of these three factors guided my experimental designs for the next 20 years. The smallest factor provided the clue that proved to be essential for localizing synaptic changes with learning (Chapter 7). The next factor led to understanding of the role of excitatory or inhibitory bias given to the olfactory bulb by neurons in the forebrain to implement states of arousal (Table 6.6) or sleep (Chapter 5), the mechanism by which excitatory bias is maintained (Chapter 8), and the dependence on arousal of the steepness of the sigmoid curve and the magnitude of the input-dependent nonlinear gain (Chapter 9).

The largest factor was the most difficult to interpret, and the most important for understanding cortical dynamics. Its configuration had to be labeled "spontaneous variation", because no behavioral correlate was identified. It persisted even when all accessible aspects of behavior in waking animals were held constant, and when the olfactory system was surgically isolated. The experimentally determined complex poles from multiple sequential trials defined a root locus in Mode 2 (Figure IV in Chapter 4) that was independent of and nearly orthogonal to the root locus with changes in stimulus intensity or anesthesia in Mode 1 (Figure 6 in Chapter 6). Derivation of theoretical root loci to fit the experimental root loci in Mode 1 and Mode 2 took several years. When the requisite describing functions had been derived (Freeman 1975), the root loci in Mode 1 were found to be due to changes in the ratio of the amplitude of the evoked activity to the amplitude of the background activity (Chapter 8). The root loci in Mode 2 provided the crucial insight needed to explain the formation of gamma bursts in the olfactory EEGs in respiration during goal-directed behavior, which manifested a nonlinear input-dependent gain in the cortical mechanism (Chapter 9). The expression of this instability in the piecewise linear Mode 2 was a root locus in which the complex roots moved toward and across the imaginary axis into the right half of the complex plane with increasing response amplitude, instead of away from it as was the case for the roots upon increases in stimulus amplitude.

The form of the root locus in Mode 2 (Figure IV, right frame) and its describing function show that when the real parts of the complex roots become positive with increasing amplitude, the amplitude grows explosively, and the cortex transits to an oscillatory state, causing the gamma burst, and, further, that the frequencies are colligated to a value near 40 Hz (250 radians/sec, Freeman 1991b).

The finding of Modes 1 and 2 throws into bold relief two differences between the evoked potential and the so-called "spontaneous" EEG activity. First, the averaged evoked potential always begins at initial conditions arbitrarily set to zero, and it returns asymptotically to that level, although the zero level is obscured by residual "noise". This is because the observer does not (or should not) accept an average for which the pre-stimulus baseline segment does not average to near zero, and does not (or should not) give stimuli at such short time intervals that the next arrives before the response to the previous one has ended. By definition these conditions of observation preclude studies of instability. The interest lies in the way a perturbed system relaxes back to an assumed state of equilibrium. Second, the evoked unit activity can directly reflect only unilateral thresholding, not bilateral saturation, because the afferent electrical stimulus can augment the evoked unit activity up to levels far exceeding those achieved during EEG activity. Such strong driving can occur because, after a brief volley of firing at abnormally high rates, the synchronously activated neurons undergo recurrent inhibition, and it is during this prolonged period of suppression that their recovery processes take place. This means that the temporal distributions of firing intervals are not randomized across the neighborhood, as they are during the "spontaneous" activity.

In contrast, in the autonomous, self-regulated state the neurons individually fire at times uncorrelated with each other, their autocorrelations are noisy but on average flat, and their interval histograms conform to the Poisson distribution except for the refractory periods, which determine the upper limit in bilateral saturation. That limit is not accessible to measurement with post-stimulus time histograms, but it becomes obvious with calculation of the pulse probability conditional on EEG amplitude (Freeman 1975). Moreover, afferent electrical pulses at the low rates and intensities that are used to collect cortical impulse responses are too weak to induce instabilities, state transitions or bifurcations in cortical dynamics. They offer value as tools for the definitions of the state variables, circuitry, time constants, space constants, feedback delays and gains in cortical populations, equivalent to the "ping" of tapping an object with a hammer, but they are severely limited by the requirements for their use, which include time ensemble averaging, and the restriction to modeling with point attractors.

Patterns of Variation in Waveform of Averaged Evoked Potentials From Prepyriform Cortex of Cats

Averaged evoked potentials (AEP) obtained from different parts of the brain contain information about the state of the neuronal system generating them. Some of this information can be extracted by measuring amplitudes and latencies of successive peaks in a complex waveform. More can be retained by generating a theoretical curve and fitting it to the AEP. As the AEP changes with alterations in the state of the brain, the coefficients must be changed to preserve an optimal level of fit. Thus changes in the "form" of a neural population response can be expressed as changes in sets of numbers, which are subject to a variety of statistical and other tests.

The problems of measurement of AEP have been discussed elsewhere (14). Two features are of interest here. First, within reason the more coefficients introduced into the equation generating the theoretical curve, the better will be the fit to the AEP, and the greater will be the informational content of the set of coefficients. But the larger the number of coefficients, the greater will be the likelihood that some of the coefficients will supply information already contained in other coefficients.

This is true whether or not the system of measurement relies on orthogonal basis functions, because a change in the form of an AEP owing perhaps to variation in a single neural variable can be expected to change many or all of the coefficients used to describe the form. The result will be covariation of two or more coefficients in groups of values derived from large numbers of AEP. Patterns of covariance between coefficients can be described by tables of ordinary and partial correlation coefficients or by factor analysis. The use of these techniques for describing patterns of change in AEP is described in the present report.

Second, the existence of a unique pattern of covariance among coefficients derived from a group of AEP implies the operation of a single definable process in the neuron pool generating the AEP. Such a process can be described as a neural constraint placed on the pool by the nature of the neurons comprising it, such that the AEP can only change in a particular way and not randomly. When, as is usually the case, several factors appear to be operating simultaneously in a neuron pool, and there are also errors of measurement owing to background noise and other causes, the factors are accessible only by statistical analysis. The statistically defined patterns provide some necessary specifications for the operating characteristics of a neuron pool. With these in hand it is possible to narrow considerably the area of search for working models of the function of a neuron pool. This, in essence, is what should be expected from the study of AEP, and the present report describes some initial guidelines for this course of action.

The form of prepyriform AEP has been found to change in several behavioral circumstances, including food deprivation and food consumption (8), changes in general alertness or in attentiveness specifically to the evoking stimulus (7), and in response to changes in evoking stimulus intensity (9). Additionally there is unpredictable "spontaneous" variation in waveform from trial to trial, even when all of the above conditions are maintained constant (14, 15).

The bulk of this report is devoted to precise description of the changes in waveform constituting spontaneous variation, and to determination of the conditions in which cortical excitability (as revealed by the evoked potential) changes in just this way or in some other way. The basic premise is that this cortex as a whole operates with certain neuronal constraints or, equivalently, with a small number of degrees of freedom. It is likely that each kind of possible change will appear in more than one set of behavioral circumstances but in differing amount or direction, so that many forms of AEP may be observed despite the constraints. The results given here identify three patterns of change, two of which can be described with some clarity. The third, which is associated with long-term changes following learning, is not clearly defined.

Methods

Bipolar stainless steel electrodes were permanently implanted in the olfactory tract for stimulation and across the prepyriform dipole (11) for recording in each of 12 adult cats. Each AEP was the sum of 88 single-shock responses recorded at 7.3/sec over a 12-sec period. Stimulus intensity was raised to that level at which the peak-to-peak amplitude of the AEP was at least 10 times the mean peak-to-peak amplitude of background activity averaged in the same way. The amplitude of the AEP was measured at 100 points at intervals of 1.23 msec after the stimulus and punched onto cards. Blocks of 20–50 AEP were averaged (AAEP) for each cat and each behavioral state preliminary to measurement by digital filters.

Each AEP was regarded as the sum of noise plus two damped sinusoids used as the basis functions for measurement (9), each having the equation

$$V(t) = V[\sin(\omega t + \phi)e^{\alpha t} - \sin\phi e^{\beta t}]$$

where amplitude, V, was in microvolts, frequency, ω, was in radians $\sec^{-1}$, the rate constants, α and β, were in $\sec^{-1}$, and phase, ϕ, was in radians. The constant, β, was fixed for each cat and corresponded to the rise time of the initial peak of the AEP. Techniques for measuring the parameters have already been described (6, 12, 14).

The neurons generating the evoked potentials were conceived to consist of a layer of pyramidal cells. Synaptic excitation of pyramidal cells is manifested by a surface-negative extracellular dipole field of potential (11). These neurons receive the afferent volley from the olfactory tract fibers and transmit activity to interneurons in the cortex. The interneurons were conceived in turn to inhibit the pyramidal cells and thereby to establish a surface-positive dipole field of potential (11). The AEP was treated as a sequence of alternating EPSPs and IPSPs manifesting activity in this negative feedback loop. The lumped feedback delay, T_2, and gain, K_2, in this loop could be estimated from the frequency, ω, and decay rate, α, of the AEP by equations previously derived (13):

$$k = (1 + \alpha / \omega)^{1/2}$$

$$\theta = \pi - \tan^{-1}[\omega / (B_1 - \alpha)] - \tan^{-1}[\omega / (B_2 - \alpha)]$$

$$T_2 = k\theta^2 / 3\omega \quad \text{in} \quad \sec$$

$$K_2 = \exp(k\theta)$$

where the constants, B_1 and B_2, were previously defined (13) to represent the delay characteristics of the forward limb of the loop.

These transformations served to normalize the distributions of measurements of the variables and to simplify the patterns of covariation between parameters (15). Simplification resulted mainly from the emergence of linear covariation between K_2 and T_2. The graphic displays of covariation between most other pairs of coefficients had too much scatter to suggest any alternative to linear correlation. Ordinary correlation coefficients were calculated for the sets of each pair of the eight coefficients, and partial correlation coefficients (1) were calculated for the sets of each pair of coefficients for the dominant basis function. Additionally, a factor matrix (3) was calculated for the data from several cats in each of three behavioral states. For this purpose the variance-covariance matrix calculated from the transformed data for all cats was partitioned into an interset (between cats) matrix and an intraset (within cats) matrix. Analysis for the principal components of the variance was performed on the latter following conversion to a correlation matrix. All computations were done with the IBM 7090.

Each cat was trained to work in an ergometer (4) by pulling on a rope to get food. This apparatus was chosen because it provided a constant behavioral state lasting the 12 sec required for averaging, and because it gave a nonneuroelectrical measure of behavior by means of which to validate measurements of the AEP, as previously discussed (14). AEP were obtained during anticipation (waiting in the starting box for the gate to open), working, and lapping after 24 and 48 hr of food deprivation. Care was taken in the initial part of the work to use electrical stimulation well below the threshold for the orienting reflex (10) or for acquisition of a conditional response to the electrical stimulus. Hence the cats at this stage were regarded as naive to the olfactory tract stimulus. Subsequently four of the cats were trained to wait in the starting box after the opening of the gate until the electrical stimulus was begun (a go–no go paradigm). This was accomplished by first training the cats to press a bar for milk in response to a light, then transferring to high-intensity stimulation of the tract, then generalizing to low intensity stimulation, and finally transferring the go-no go behavior back to the ergometer. Extinction of the response contingency on electrical stimulation was achieved by allowing the electrical stimulus to run continuously throughout the period of ergometer work (10–40 runs/day lasting about 1 hr). Records during "habituation" following extinction were obtained after 10–14 days, and after 20 min of preliminary habituation on each day of measurement.

"Prolonged habituation" also was carried out by allowing each cat to remain in the starting box for 2 or more hours until all behavioral activity had ceased, the prepyriform EEG had fallen to a low level (7, 8, 10) but without the appearance of sleep waves, and the AEP had reached a stable form. (The term "habituation" as used here refers to the state of being subjected to prolonged, unchanging stimulation rather than to any neuronal change that might or might not have taken place.)

Changes in AEP, depending on stimulus intensity, were determined by use of constant-current pulses of differing magnitudes and constant duration (0.01 or 0.10 msec) monitored by an oscilloscope.

Results

Changes with Attentiveness

Examples of averages of blocks of AEP (dots) and the sum of matching basis functions (curves) are shown in Fig. 1 for four cats in each of three states: working in the ergometer and naive to the stimulus (N), working only after onset of the evoking stimulus, i.e. attentive (A), and working not contingent on the evoking stimulus, i.e. after extinction and habituation (H) to the electrical stimulus, permitting the cats to respond to a visual cue. Mean coefficients for the pairs of dominant (D) and

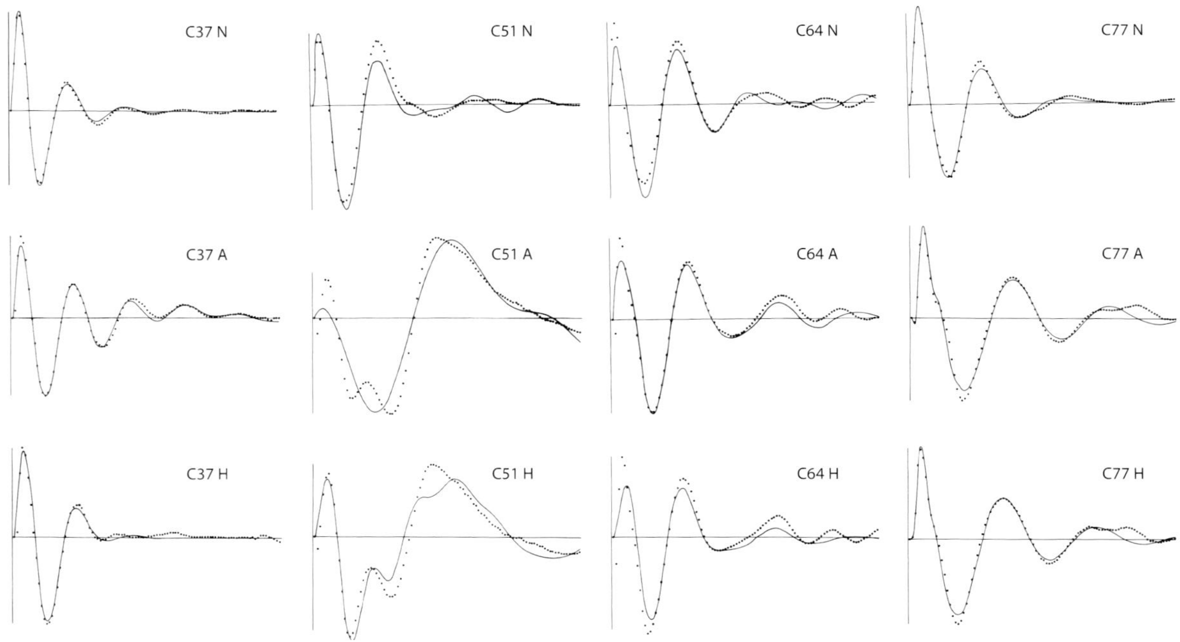

Fig. 1. Averaged evoked potentials (dots) are shown with a matching sum of basis functions (curves) for four cats in each of three behavioral states: N = naive, A = attentive, and H = habituated to the evoking stimulus. Sweep duration = 125 msec; sweep number = 88.

Table 1. AEP parameters of naive, attentive and habituated cats

	State	V_D	ω_D	α_D	ϕ_D	V_S	ω_S	α_S	ϕ_S
C37	N	121.0	240.0	71.9	1.21	42.7	472.5	180.3	−1.82
	A	128.6	255.1	40.5	0.49	29.2	57.2	6.9	−2.99
	H	137.7	257.6	49.2	0.54	43.1	84.8	30.1	−0.93
C51	N	94.3	176.6	33.2	1.62	57.8	240.0	31.1	1.28
	A	89.3	109.3	19.0	0.92	137.6	51.5	11.0	−2.29
	H	95.3	128.8	21.5	1.30	43.1	318.6	62.7	−0.80
C64	N	101.7	197.3	32.2	1.69	27.8	235.6	28.1	−1.74
	A	103.8	160.8	24.9	1.59	39.2	265.2	33.5	−0.89
	H	126.0	153.3	33.3	2.37	602.1	226.8	143.6	−1.09
C77	N	114.5	196.0	51.1	1.08	80.8	142.6	11.8	2.63
	A	102.6	144.5	26.8	1.08	103.5	715.0	187.2	3.29
	H	105.3	146.4	30.8	1.14	89.1	762.8	204.0	3.03

Amplitudes (V in microvolts), frequency (ω in radians sec^{-1}), decay rates (α in sec^{-1}), and phase (ϕ in radians) for dominant (D) and subsidiary (S) basis functions comprising prepyriform averaged evoked potentials (AEP) from four cats in three behavioral states.

subsidiary (S) basis functions are in Table 1. Mean values derived from the dominant basis function for amplitude (V), delay (T_2), gain (K_2), and phase (ϕ) with standard deviations are in Table 2.

The illustrations verify previous findings (7) that the principal change with attentiveness is an increase in the over-all duration of the AEP and the emergence or increase in amplitude of lower frequency components. Following extinction and habituation these changes are partly reversed.

For the reason discussed later (lack of a suitable physiological model to account for them) these differences between states are not at present subject to further mathematical or statistical analysis. The main points are that changes in waveform did occur between states, that these changes were reflected by changes in the coefficients (mainly decay rate, α_D, and frequency, ω_D) derived from the AEP, and that the pattern of transition summarized in Table 2 did not conform to the pattern of spontaneous variation described below, although the pattern of spontaneous variation within each state was the same in all states.

In addition, the data show what was omitted from further consideration by restricting attention to the coefficients of the dominant basis function, i.e. the residuals in Fig. 1 and the coefficients of the subsidiary basis function in Table 1. Although abstraction is mandatory in order to proceed from one level of analysis to another, it is desirable to know what is being deleted at each stage.

Spontaneous Variation

The AEP described above were measured in repeated trials on each of three widely separated days during work to obtain food. The only significant behavioral change

Table 2. Transformed parameters of dominant basis function

	N	V	T_2	K_2	ϕ
Naive					
C37	33	121.0 ± 15.4	3.23 ± 0.19	3.73 ± 0.27	1.21 ± 0.15
C51	25	94.3 ± 6.5	7.20 ± 0.39	5.90 ± 0.17	1.62 ± 0.11
C64	32	101.7 ± 5.9	6.15 ± 0.44	5.70 ± 0.32	1.69 ± 0.11
C77	30	114.5 ± 7.1	5.25 ± 0.36	4.68 ± 0.31	1.08 ± 0.20
Avg		107.9	5.46	5.00	1.40
Attentive					
C37	47	128.6 ± 20.6	3.72 ± 0.33	4.73 ± 0.24	0.49 ± 0.08
C51	21	89.3 ± 6.1	16.13 ± 1.43	8.06 ± 0.77	0.92 ± 0.20
C64	37	103.8 ± 11.9	8.77 ± 0.42	6.71 ± 0.26	1.59 ± 0.08
C77	25	102.6 ± 7.9	10.13 ± 0.54	6.72 ± 0.34	1.08 ± 0.04
Avg		106.1	9.69	6.56	1.02
Habituated					
C37	54	137.7 ± 22.7	3.44 ± 0.34	4.38 ± 0.30	0.54 ± 0.10
C51	34	95.3 ± 13.1	12.48 ± 1.58	7.53 ± 0.56	1.30 ± 0.32
C64	48	126.0 ± 12.8	8.52 ± 0.53	6.02 ± 0.26	2.37 ± 0.08
C77	24	105.3 ± 17.2	9.65 ± 1.11	6.33 ± 0.34	1.14 ± 0.09
Avg		116.1	8.52	6.07	1.33

Means and standard deviations of amplitudes (V in microvolts), peak distributed feedback delay (T_2 in milliseconds), feedback gain (K_2), and phase (ø in radians) derived from the dominant basis function fitted to the AEP (cf. Table 1).

in each day was the cumulative increase in factors related to fatigue and satiety. Yet the coefficients of successive AEP varied about their means (coefficients of variation ranging from 4% to 16%) from one trial to the next randomly with respect to the order of measurement. The coefficients of the dominant and subsidiary basis functions varied independently, except in some cats for covariation of the amplitude of the dominant basis function, V_D, and the phase of the subsidiary basis function, ϕ_S (see Table 6). The variation for each coefficient was not random with respect to other coefficients of the same basis function. In Table 3 are shown the ordinary and partial correlation coefficients for each coefficient with each of the other three. The same procedure was applied to the data from naive, attentive, and habituated cats.

No significant differences in patterns of interparameter covariation were found between states or between cats. In all sets the ordinary and partial correlations between delay and gain, $T_2 \cdot K_2$, and between delay and phase, $T_2 \cdot \phi$, were very high. The positive ordinary correlation for gain and phase, $K_2 \cdot \phi$, contrasted in all cases with the negative partial correlation for $K_2 \cdot \phi$.

The high positive correlation between delay and phase, $T_2 \cdot \phi$, was derived from a still higher negative correlation between frequency and phase, $\omega \cdot \phi$, and meant that

Table 3. Interparameter correlation coefficients

	Partial						Ordinary					
	VT_2	VK_2	$V\phi$	T_2K_2	$T_{2\phi}$	$K_{2\phi}$	VT_2	VK_2	V_ϕ	T_2K_2	$T_{2\phi}$	$K_{2\phi}$
Working												
Naive	.013	–.469	.029	.866	.721	–.463	–.213	–.341	.081	.789	.514	.245
Attentive	.520	–.676	–.213	.738	.682	–.276	.017	–.272	–.068	.828	.735	.541
Habituated	.505	–.534	–.256	.924	.720	–.548	.270	.053	.185	.937	.744	.589
Naive												
Anticipatory	.162	–.320	–.049	.831	.465	–.368	.043	–.223	–.101	.782	.268	.182
Lapping	.170	–.164	–.092	.855	.284	–.360	–.138	–.057	.037	.652	.690	.238
Attentive												
Anticipatory	.198	–.371	.340	.658	.473	–.449	–.350	–.675	.095	.725	.435	.145
Lapping	.160	–.315	–.093	.882	.680	–.425	–.250	–.268	–.185	.938	.778	.667
Average*	.254	–.450	–.047	.839	.593	–.423	–.091	–.272	.006	.832	.594	.393
Prolonged												
Habituation	.571	–.230	–.547	.789	.678	–.353	.176	.066	–.108	.776	.767	.443
Stimulated increase	.379	–.097	–.251	.905	.716	–.608	.659	.657	.019	.947	.188	.014

Interparameter ordinary and partial correlation coefficients derived from the dominant base function.
*Averages for r are by z-transform.

variations in frequency occurred without significant changes in the duration of the initially surface-negative (upward in Fig. 1) peak of the AEP.

The virtually linear relationship between gain, (K_2) and delay (T_2) was described by the line, $K_2 = a + bT_2$, where b was the slope and a was the intercept on the K_2 axis of the line representing the best fit to the data by a least-squares criterion. Mean values for the four cats given in Table 4 show that the slope, b, of the relation did not change with attentiveness, but that the intercept, a, on the K_2 axis decreased, i.e. the line was translated to the right (increased T_2) in the K_2–T_2 plane.

This relationship was examined in 12 cats (including the 4 already described) in the naive state. Mean values for gain (K_2) and delay (T_2) of dominant and subsidiary basis functions are shown in Fig. 2. The values for frequency (ω) and decay rate (α) have already been published (12). Those means enclosed in the ellipse correspond to frequencies failing within the range of spontaneous EEG activity for this cortex (2), i.e. 220–280 radians/sec. Mean values for the intercept, a, and the slope, b, were calculated for this set, and for AEP having higher and lower frequencies. Correlation coefficients were computed for the subsidiary components as well. The same pattern of interparameter covariation was found for the subsidiary as for the dominant component (this is reflected in the factor patterns in Table 6). A significant difference ($P<.01$ by t test) was found between values for b from the midfrequency and low-frequency groups as shown in Fig. 2, i.e. the slope, b, was smaller with higher values for delay, T_2. This difference held for each cat over the transition from

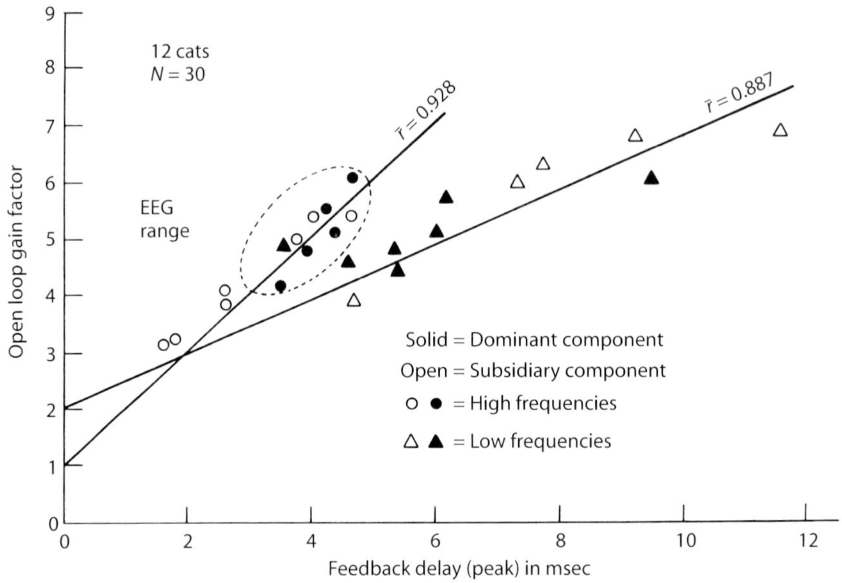

Fig. 2. Mean values are displayed in T_2–K_2 coordinates for the two components of the AEP from 12 cats. Ordinary correlation coefficients (*r*), slopes (*b*, see text), and intercepts (*a*, see text) were calculated for 10 pairs of values for each cat in 3 behavioral states: anticipation, working, and lapping. Values for *r*, *a*, and *b* were pooled for those components in the AEP having frequencies within the range of dominant EEG frequencies for this cortex; for those above the range; and for those below. Values for *r* and the lines having intercept (*a*) and slope (*b*) are from the group in the EEG range and from the group below that range.

the naive to the attentive and habituated states. For example, *C37* in all states gave relatively high values for the slope, *b*, and low values for the delay, T_2, in comparison to other cats.

Thus the main characteristics of spontaneous variation were covariation of delay and phase ($T_2 \cdot \phi$) and of delay and gain ($T_2 \cdot K_2$), and the slope, *b*, of the latter covariation was smaller for larger mean values of delay (T_2)

Food Deprivation

The effect of increasing deprivation on EEG activity and on the AEP has previously been described as an increase in amplitude in both without significant change in frequency (8). This was tested for the AEP of 7 cats naive to the stimulus by comparing means for 10 values measured on 2 successive days during anticipation, wherein the differences associated with deprivation had been found to be maximal (8). The mean values for both amplitude and subsequent rate of work increased ($P<.01$ by *t* test for each cat) respectively by 19% and 24%. Mean delay (T_2) and gain (K_2) increased by about 3% but the differences by themselves were not significant. No significant differences were found for correlation coefficients nor for values of *a* and *b*. The same results were obtained from 4 cats attentive to the stimulus, and also from comparison of the differences between the first and last halves of blocks of

Table 4. Characteristics of covariation between K_2 and T_2

Condition	No.	*a*	SE	*b*	SE
Naive, EEG frequencies	9	1.03	±0.36	1.06	±0.11
Naive, lower frequencies	11	2.03	±0.38	0.58	±0.12
Naive, higher frequencies	4	1.07	±0.52	1.14	±0.14
Working					
Naive	4	1.72		0.67	
Attentive	4	0.76		0.67	
Habituated	4	1.70		0.66	
Naive					
Anticipatory	4	1.12		0.73	
Working	4	1.46		0.66	
Lapping	4	0.91		0.75	
Average	(12)	1.16	±0.18	0.71	±0.067
Attentive					
Anticipatory	4	−0.20		0.76	
Working	4	−0.27		0.71	
Lapping	4	0.18		0.69	
Average	(12)	−0.17	±0.42	0.72	±0.084
During habitutation	4	1.24		0.66	
Stimulus increase, naive	4	1.06		0.48	
Stimulus increase, attentive	4	1.37		0.54	
C37 combined	10	0.53	±0.42	0.93	0.051
C51 combined	10	1.12	±0.36	0.59	±0.080
C64 combined	10	0.74	±0.32	0.66	±0.049
C77 combined	10	1.44	±0.31	0.57	±0.046

Slope (b) and intercept (a) on the ordinate for the relation between K_2 and T_2 in the designated conditions. $K_2 = a + bT_2$.

trials in excess of 24 on any day. By this measure the effects of satiety and fatigue were not distinguishable (except in direction) from those of deprivation.

Food Consumption

One of the more dramatic changes in prepyriform EEG activity was the abrupt decrease in amplitude at the onset of food consumption. This was clearly associated

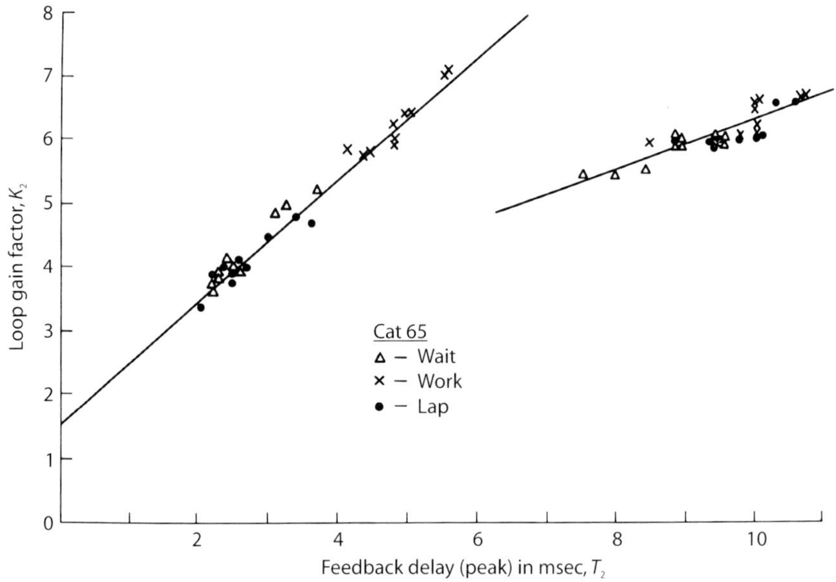

Fig. 3. Comparison of the pattern of "spontaneous" variation with the change in AEP during lapping showed no real differences.

Table 5. Parameters of AEP in relation to deprivation and food consumption

	V	T_2	K_2	φ	W
		Naive: 7 cats, deprivation			
24 hr	91.0*	6.94	4.65	1.23	369*
48 hr	108.0	7.16	4.78	1.23	457
		Naive: 12 cats			
Anticipatory	172.0	4.14	4.06	1.56	
Lapping	183.0	3.88*	3.67*	1.47	
		Attentive: 4 cats			
Anticipatory	196.0	10.52	6.88	0.91	
Lapping	198.0	10.41	6.68	0.84	

Amplitude (V in microvolts), feedback delay (T_2 in milliseconds), feedback gain (K_2), phase (φ in radians), and concomitant rate of work for food (W in milliwatts) by cats in the ergometer.

with a sharp change in respiratory patterns (8, 11), implying a reduced or desynchronized input from the olfactory bulb to the cortex. In perhaps 1 in 10 cats this decrease has not been found to occur (2, 8) and in many it has not been consistent. The AEP in the present series unquestionably changed on the average in the manner shown in Table 5 and in Fig. 3. The difficulty arose, that although the changes for 6 of the 12 cats appeared consistently in the manner shown, in 2 others they were consistently in the opposite direction. In the group of 4 attentive cats the differences shown in Table 5 by themselves were not statistically significant.

As shown in Tables 3 and 4 the correlation coefficients and values for intercept, a, and slope, b, did not differ between the states of anticipation, working and lapping for cats either naive or attentive to the evoking stimulus. As the result the values for each cat were scattered along the same regression line, as shown by example in Fig. 3. Furthermore, the means for delay (T_2), gain (K_2), and phase (ϕ) decreased together with an increase in amplitude (V) on transition from anticipation to lapping. On these several grounds the change in AEP during food consumption could not be distinguished from the pattern of spontaneous variation.

Prolonged Habituation

One of the conditions attendant on complete quiescence in a cat was the disappearance of bursts of sinusoidal activity in the EEG (7, 8). In the next set of observations each cat was stimulated continuously in a quiet environment where both EEG and comportment could be watched to avoid acquisition of AEP during sleep or grooming.

Samples of AEP were taken at intervals of 10–20 min. until the full extent of the observed change in waveform had apparently taken place. As shown in Fig. 4 there was an overall decrease in delay (T_2) and gain (K_2) along the same line as for spontaneous variation. Correlation coefficients and values for a and b are given in Tables 3 and 4. This pattern of change could be abruptly reversed by any disturbance in the environment leading to an increase in the EEG activity, almost to the full extent of its original occurrence. The only feature that distinguished this pattern of change from that seen during lapping was that here the initial amplitude (V) decreased with decrease in delay (T_2) and gain (K_2), whereas at the onset of lapping V increased.

Factor Analysis

Factor analysis is an alternative mode of description of these patterns, serving also as a convenient summary. Table 6 shows the largest three (eigenvalues > 1.00) of eight components of the pooled intraset variance among the eight variables. Data from seven cats are included under "naive," and from the same four cats under "attentive" and "habituated." The principal factor, I, in each set reflects the conjoint variation of delay (T_2), gain (K_2), and phase (ϕ) with inverse change in amplitude (V) of the dominant basis function. The same pattern is seen in the second factor but is related to the subsidiary basis function, except for phase (ϕ) in the attentive and habituated cats. The parameters of the two basis functions appear to change independently. The third factor is in each case most clearly related to the amplitude of the dominant basis function, V_D. In each set the first three factors comprise 1/3, 1/4, and 1/6 of the total variance.

This analysis was repeated for three additional naive cats, in each of which the right prepyriform cortex had been undercut, in such fashion as to sever all neural connections with the rest of the central nervous system, but to leave intact the blood

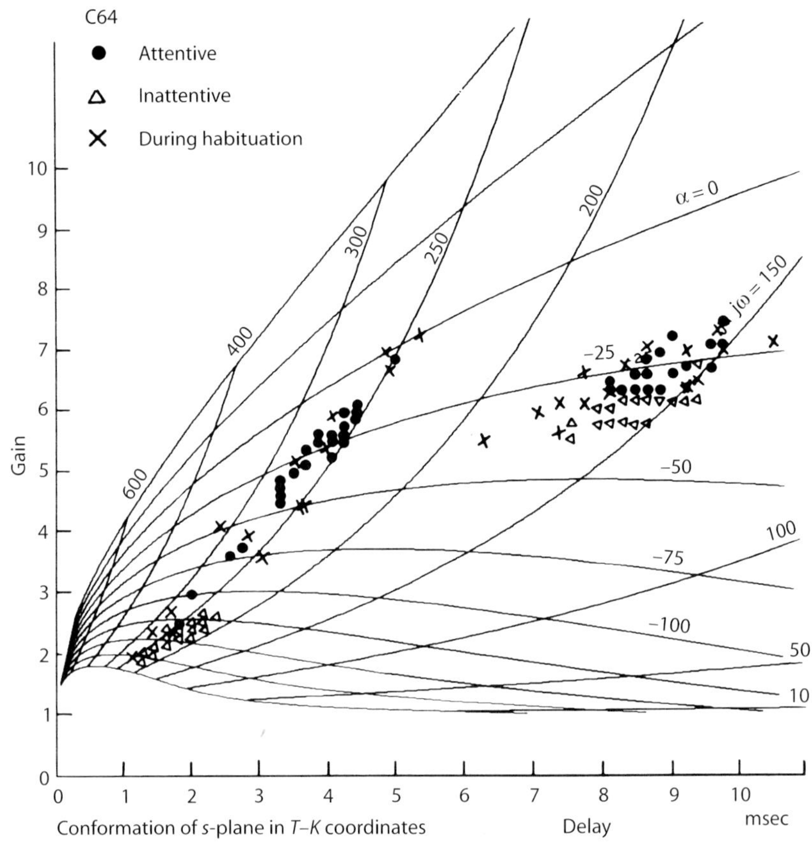

Fig. 4. Comparison of the pattern of "spontaneous" variation with that obtained by prolonged habituation showed no real differences.

supply and the olfactory bulb and tract. (Additional details including anatomical controls will be presented in a later report.) The AEP and EEG of the two sides were very similar in the waking state. Sets of 30 AEP were taken from both sides simultaneously, measured, and analyzed as described for the other data. The pattern of the principal components of the variance for measurements from the normal cortex was almost the same as that for the naive animals (except for amplitude, V_S in II). The pattern from the undercut cortex revealed a larger proportion of variance in the first two factors, and absence of a factor specifically related to the amplitude of the dominant basis function, V_D.

The linear covariation of delay (T_2) and gain (K_2) calculated from measurements of the dominant component AEP from intact and undercut cortex was not distinguishable from the patterns already described for normal cats. This shows that the pattern of change identified with factor I, denoting $-V\cdot +T_2\cdot +K_2\cdot +\phi$ depends largely (but not necessarily entirely) on peripheral events, whereas the factor identified with amplitude, V_D, is central in origin.

Table 6. Principal components for the pooled intraset variance of data from seven cats in the naive state and four cats in the attentive and habituated states*

	Naive			Attentive			Habituated		
Eigenvalues	2.185	2.058	1.153	2.718	1.848	1.325	2.614	1.790	1.314
Proportions[†]	.273	.530	.675	.340	.571	.736	.327	.551	.715
Eigenvectors	I	II	III	I	II	III	I	II	III
V_D	−.350	−.286	.728	−.211	.324	.699	−.360	.209	.728
T_{2D}	.939	−.005	.038	.900	−.206	.274	.875	.018	−.054
K_{2D}	.866	.147	−.207	.861	−.219	.181	.886	.031	−.184
ϕ_D	.611	−.216	.498	.864	−.287	−.000	.857	−.028	−.189
V_S	.183	−.626	.092	−.234	−.517	.455	.253	−.273	.803
T_{2S}	−.005	.884	.118	.130	.690	.491	−.214	.753	.178
K_{2S}	−.043	.756	.104	.405	.821	−.033	−.018	.918	.031
ϕ_S	.151	.405	.546	−.374	−.392	.529	−.298	−.510	.185

	Control Cortex			Undercut Cortex		
Eigenvalues	2.631	1.867	1.354	4.297	1.789	.877
Proportions	.329	.562	.731	.537	.761	.870
Eigenvectors	I	II	III	I	II	III
V_D	−.360	.088	.725	.879	−.288	.292
T_{2D}	.890	−.055	−.128	.735	.545	−.214
K_{2D}	.882	.096	−.158	.490	.740	−.374
ϕ_D	.638	−.158	.094	.739	.360	.157
V_S	−.429	.757	−.143	.604	.386	.630
T_{2S}	.297	.727	.317	.840	−.493	−.089
K_{2S}	.489	.240	.701	.803	−.528	−.059
ϕ_S	.108	.814	−.408	−.695	.246	.417

*Components having eigenvalues less than 1.0 have been omitted (15). Additionally, analysis is included of data from three cats, in each of which the right prepyriform cortex had been undercut but not deafferented.
[†]Cumulative proportions of total variance.

Stimulus Intensity

Values for gain, K_2, and delay, T_2, were minimal at threshold stimulation. Both K_2 and T_2 increased along loci closely similar to those for spontaneous variation, up to a region of inflexion. This region corresponded to a minimum value for decay rate, a. It also corresponded to visual threshold for single shock evoked potentials displayed on an oscilloscope, as they emerged from the background EEG activity (10). With further increase in stimulus intensity, the delay (T_2) increased with no further change or a decrease in gain (K_2). Correlation coefficients were calculated for sets of measurements ranging from the minimum at threshold to the minimum for decay rate, a, using data from four cats naive to the stimulus and the same four cats attentive to the stimulus. These showed no significant differences between cats or states and were pooled to give the values shown in Table 4. Values for the intercept, *a*, and the slope, *b*, were similarly calculated and showed that, in the range of stimulus variation between threshold for an averaged response and threshold for a single response, the mean locus for change in gain (K_2) and delay (T_2) was close to but less steep than that for the spontaneous variation.

Discussion

The outstanding feature of these results was the pattern of spontaneous variation of the AEP. The four coefficients derived from the AEP varied randomly with respect to the order of occurrence, but not with respect to each other. The pattern described by matrices of ordinary and partial correlation coefficients and by factor analysis persisted unchanged through all of the behavioral transformations known to alter the form of the AEP except sleep (5).

Insofar as the dominant component is concerned, the two main degrees of freedom for spontaneous variation are inferred to be an independent change in V and the conjoint change: $-V \cdot +T_2 \cdot +K_2 \cdot +\phi$. The same pattern of change was found upon the behavioral transition from waiting to lapping. It differed from the pattern seen after prolonged habituation only in that V changed in the same direction as $T_2 \cdot K_2 \cdot \phi$. In both of these latter two conditions there is a pronounced decrease in the EEG activity, specifically in the root-mean-square (rms) value for the alternating current component of cortical activity at 250 radians/sec. It appears that the level of the background activity may be a major factor in determining the waveform in these two conditions. Furthermore, the rms amplitude of the EEG of this cortex over 10- to 12-sec periods is known to vary randomly from trial to trial with a coefficient of variation of about 20% (8), so that the spontaneous variation in AEP can also be attributed to variation in the background rms activity.

Crucial to this interpretation is the finding that the pattern of variation in waveform for the lower range of stimulus intensity was very similar to that for spontaneous variation, although the mean amplitudes were far lower and the order of occurrence of values was not random. For this part of the locus the amplitude of the single-shock evoked potential was equal to or less than the peak amplitude of background EEG activity. If the waveform depended on the ratio of the amplitude of the impulse response to the amplitude of the background activity, it should have made little difference which was changed to give the same end result.

This parity would not hold for single shock amplitudes above the background level, where the pattern of change in the AEP differed from the pattern of spontaneous variation. Many of the data in this study were obtained from that higher range (cf. Fig. 2), mainly because the presence of a high background activity in conjunction with the requirement for a limited averaging span (to minimize time variance) required overdriving the cortex until a useable response developed. This factor appeared to account for the distribution of mean values in Fig. 2, which should be compared to the curves in Fig. 5.

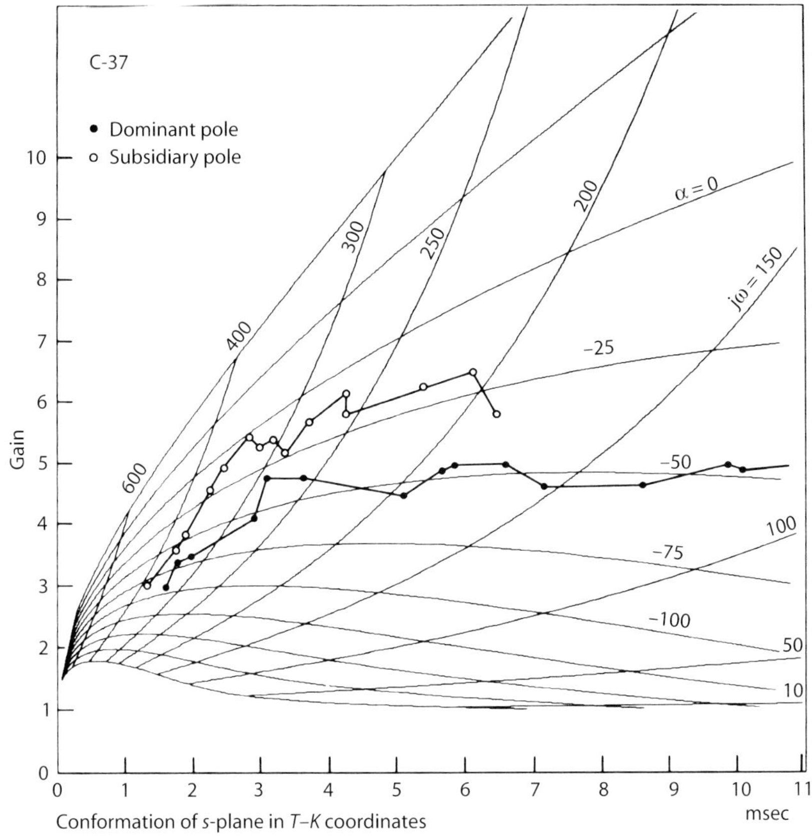

Fig. 5. Comparison of the pattern of "spontaneous" variation with the lower end of the range for response to differing stimulus intensity showed little difference.

This interpretation required that the pattern $-V \cdot +T_2 \cdot +K_2 \cdot +\phi$ persist following unilateral cortical undercutting. In this condition the amplitude of EEG activity fluctuated in company with respiration as it did on the normal side, so that the AEP were expected to change as they normally do. In fact the expected covariance of $T_2 \cdot K_2 \cdot \phi$ was found for the dominant component, but V_D changed in parallel and not inversely. The reason for this discrepancy is not clear.

The pattern of change with attentiveness specifically to the evoking stimulus followed by overtraining was not identical to the patterns associated with spontaneous variation or with changes in stimulus intensity, so that little if any of the spontaneous variation could be attributed to a fluctuating level of CC "attentiveness." This judgment was based specifically on the failure of amplitude to change in the presence of major changes in frequency, on the decrease in phase, ϕ, that accompanied an increase in gain, K_2, and delay, T_2, and the decrease in the intercept, a, rather than the slope, b, with increasing delay, T_2, in contrast to the pattern with changing stimulus intensity.

With reference to Fig. 1 and Table 1, attentiveness seemed to involve the development of activity at lower frequencies, which overshadowed but did not exclude the more normal frequencies. Such lower frequencies and prolonged actions suggest the opening or accentuation of multiple forward or feedback loops in the cortex or beyond, in addition to the simple negative feedback loop predicated in this study. This gives reason to doubt that the conversion of frequency (ω) and decay rate (α) to gain (K_2) and delay (T_2) in a simple negative feedback loop is appropriate to describe such changes.

Comparison between states involving major changes in AEP are made difficult by the requirement for two or more basis functions. As an example, a decrease in phase, ϕ, without change in frequency, ω, would imply prolongation of the initial surface-negative peak. A decrease in frequency without change in phase would imply the same. The decrease in both (Table 1) implies a rather large increase in duration of the first peak, which is apparent on direct measurement of its duration (Fig. 1). However the duration of the first peak might not result in a valid measure of a local event by the dominant basis function, if it were obscured by a parallel change in the subsidiary basis function. Perhaps the subsidiary basis function should not begin at time zero (as presently assumed) but some milliseconds later in any, or all of the behavioral conditions. Unfortunately there is as yet no theoretical basis for suggesting an appropriate lag.

Another difficulty is that the most striking changes in waveform including some dramatic high-amplitude, low-frequency fluctuations, take place during orienting, thalamic tetanization, or the middle stages of acquisition of a conditional response to electrical stimulation. The variability of the waveform is so great that averaging is often destructive. What is seen some weeks later may be a slowly developing stable residue of these major events.

An example in Fig. 6 shows the AEP of a cat before, during, and immediately after the acquisition of the bar-press response to the electrical stimulation, when the cat learned to respond reliably to high-intensity stimulation, but not yet to stimulation at the intensity used to generate the AEP. Each dotted curve shows a mean of nine AEP taken over successive trials, and each smooth curve shows the standard deviation for the nine AEP. The transient increase in variance demonstrated for the phase of response acquisition occurred in all cats and was in no case associated with an irreversible change in the mean AEP.

Studies of the performance of analogue and digital negative feedback models simulating these AEP (16) now have shown that the pattern, $-V\cdot +T_2\cdot +K_2\cdot +\phi$, can be reproduced by changing either the feedback gain or delay, but not by changing the constants in the forward limb. The pattern of change in phase, ϕ, on the other hand, occurring following overtraining in the state here called "attentiveness," required that a change in the models be made in the forward limb. Simulation of the isolated change in amplitude, V_D, required changing external forward gain without changes in loop gain or delay. These and related studies suggest that the patterns of change in AEP described here can be used in conjunction with explanatory models to localize the neuronal events in the cortex underlying the associated behavioral events.

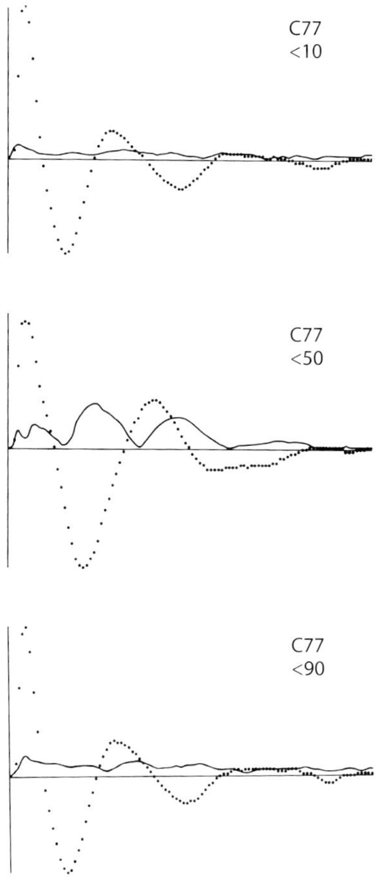

Fig. 6. Mean (dotted curves) and standard deviation (solid curves) are shown for 9 AEP in each of 3 states for 1 cat: bar-press responses to cortical stimulation less than 10% of trials: more than 50% but less than 90%; and on more than 90% of trials (criterion). The training stimulus intensity was 5.2× the stimulus used to evoke the recorded AEP. The 9 AEP for each state were recorded on every 10th trial in a sequence of 90 trials on 1 day. During the recording trials the stimulus intensity was reduced from the training level to the recording level. Bar-press responses occurred erratically during recording trials in a manner indistinguishable from intertrial bar pressing. Reliable responses were obtained at the recording level only after 2 or more weeks of additional training (10), during which the AEP changed in form (7).

Summary

A quantitative description is given of three patterns of change observed in the averaged evoked potential (AEP) in relation to behavioral changes known to influence the AEP. The most prominent pattern is that for "spontaneous" variation, which consists of well-defined concomitant changes in frequency, decay rate, phase, and initial amplitude of AEP. This is very similar if not identical to the pattern of change found during food consumption, and is similar to the patterns of change following prolonged habituation and in the lower range for stimulus intensity. It persists after cortical undercutting. It is proposed that this pattern of variation results largely from random changes in the ratio of the root-mean-square amplitude of the background EEG activity of this cortex with respect to the magnitude of the evoked response. The second pattern is an increase in amplitude of the AEP without significant other changes, which occurs with increasing food deprivation and other motivating occurrences, and which is abolished by cortical undercutting. The third pattern is an increase in the over-all duration of the AEP including the duration of the initial peak together with the emergence of low-frequency components. This takes place following development of an operant conditional on the electrical stimulation and is partly reversible following extinction and habituation. The physiological basis for this change is still obscure, but it is clear that little of the "spontaneous" variation in the waveform of AEP can be ascribed to fluctuations in cortical excitability associated with "attentiveness."

Acknowledgment

Reprinted with permission from *Journal of Neurophysiology*, 1968, **31**: 1–13.

This work was supported by Grants MH 06686 and NB 05035 from the National Institutes of Health, by the Berkeley Computer Center, and by a Guggenheim Fellowship 1965–66,

Consultation on statistical techniques with Professor Henry Scheffé and Dr. Robert Elashoff and the technical assistance of C. Tsuchitani, J. Ralph, and C. Becker are gratefully acknowledged.

References

1. Anderson, T. W. *An Introduction to Multivariate Statistical Analysis*. New York: Wiley, 1958.
2. Boudreau, J. C. and Freeman, W. J. Spectral analysis of electrical activity in the prepyriform cortex of the cat. *Exptl. Neurol.* **8**: 423–439, 1963.
3. Cooley, W. W. and Lohnes, P. R. *Multivariate Procedures for the Behavioral Sciences*. New York: Wiley, 1962.
4. Freeman, W. J. An ergometer for measuring work from cats as an index for drive. *J. Appl. Physiol.* **14**: 1071–1072, 1959.
5. Freeman, W. J. Q meter for measuring frequency specificity of cortical reactivity to electrical stimulation. *J. Appl. Physiol.* **16**: 750–751, 1961.
6. Freeman, W. J. Linear approximation of prepyriform evoked potentials in cats. *Exptl. Neurol.* **5**: 477–499, 1962.
7. Freeman, W. J. Phasic and long-term excitability changes in prepyriform cortex of cats. *Exptl. Neurol.* **5**: 500–518, 1962.
8. Freeman, W. J. Changes in prepyriform evoked potential with food deprivation and consumption. *Exptl. Neurol.* **6**: 12–29, 1962.

9. Freeman, W. J. Alterations in prepyriform evoked potential in relation to stimulus intensity. *Exptl. Neurol.* **6**: 70–84, 1962.
10. Freeman, W. J. Comparison of thresholds for behavioral and electrical responses to cortical electrical stimulation in cats. *Exptl. Neurol.* **6**: 315–331, 1962.
11. Freeman, W. J. The electrical activity of a primary sensory cortex: analysis of EEG waves. *Intern. Rev. Neurobiol.* **5**: 53–119, 1963.
12. Freeman, W. J. Use of digital adaptive filters for measuring prepyriform evoked potentials from cats. *Exptl. Neurol.* 10: 475–492, 1964.
13. Freeman, W. J. A linear distributed feedback model for prepyriform cortex. *Exptl. Neurol.* **10**: 525–547, 1964.
14. Freeman, W. J. Problems in the measurement of cortical evoked potentials. In: *Recent Advances in Biological Psychiatry*, edited by J. Wortis. New York: Plenum, 1964, vol. VII, chapt. 24, p. 235–241.
15. Freeman, W. J. Correlation of goal-directed work with sensory cortical excitability. In: *Recent Advances in Biological Psychiatry*, edited by J. Wortis. New York: Plenum, 1964, vol. VII, chapt. 25, p. 243–250.
16. Freeman, W. J. Analysis of function of cerebral cortex by use of control systems theory. *Logistic Rev.* **3**(12): 5–40, 1967.

7. *Analog Computation to Model Responses Based on Linear Integration, Modifiable Synapses, and Nonlinear Trigger Zones*

Up to this time my approach had been to identify an experimental domain of linear dynamics by means of paired-shock stimulation in order to show that proportionality and additivity held for the superimposed responses in certain ranges of the stimulus pulse interval and intensity (Biedenbach and Freeman 1965). Using measurements of cortical impulse responses by curve fitting, I constructed a series of linear dynamic models and evaluated the coefficients, extending their ranges by varying the coefficients in piecewise linear approximations. In the 1960s the use of analog hardware to model nonlinear systems had become widespread, so it seemed worthwhile to construct a network model, using operational amplifiers to simulate the linear integration performed by the dendrites of cortical populations, and diodes to simulate the static nonlinearity at the trigger zones (Figure 7 in Chapter 7), treating pulse densities as continuous variables in time and amplitude. The circuit connections were the specified by the KII model. My aim was to simulate the changes in the waveforms of evoked potentials and the root loci with increasing stimulus intensity in Mode 1, and then to simulate the patterns of change and root loci in Mode 2 (Chapter 6).

I succeeded in these aims and was pleased with the report at its publication. I now find fault with it and reproduce it here to explain why: I was confounding simulation with theory. The open loop and closed loop impulse responses (Figure 2 in Chapter 7) and the changes in the form of the impulse response with increasing stimulus intensity (Figure 3 in Chapter 7) were easy to simulate, and also to interpret by ascribing them to a reduction in the loop gains by clipping with increasing response amplitude exceeding a fixed threshold. The root loci for "spontaneous variation" (Figures 2 and 5 in Chapter 5) were also simulated by reducing the gains in all parts of the KII model (Figure 5 in Chapter 7). This, too, appeared to be consistent with the concept that with increasing dendritic current amplitude, there was increasing bilateral saturation in the trigger zones of both the excitatory and especially the inhibitory populations. The pattern of change with attentiveness, which involved a decrease in phase of onset owing to prolongation of the duration of the initial excitatory peak of the impulse response (Figure 6 in Chapter 7), was simulated by either decreasing the gain in mutual inhibition or increasing the gain in mutual excitation. This ambiguity was not resolved by simulation, as it should have been, though I came to the correct conclusion on the basis of physiological data, and confirmed it by further experiments (Emery and Freeman 1969).

Simulation of observed impulse responses, whether in analog or digital embodiments, has an appealing concreteness, and it gives us the opportunity to tweak the knobs or the values of parameters in order to change the gains and the output waveforms in unlimited varieties by trial and error, but it does not explain or prove anything. The theory ends when the device or the software has been assembled from its parts or subroutines and is put into operation. For example, students who assemble dendrites from compartments with an array of ion channels have the illusion that they are doing experiments, when in fact they are only fishing for good illustrations of known results. In the case here the analog simulation did not produce any predictions for new experiments. It was a dead end. In contrast, the piecewise linear analysis continued to yield new insights, particularly into the global stability of each model, which could be evaluated by a glance at the root locus plots, showing where and in what manner instabilities would arise by the conversion of the real parts of the solutions sets from negative to positive values. Ultimately, the extension of linear analysis led to the discovery that the analog model described here was based on the faulty assumption that the maximal gain of the system was at the resting point of the background activity, such that any increase in excitatory or inhibitory activity would reduce strengths of interactions within populations. That crucial misconception was rectified a decade later (Chapter 9).

Regarding the "long-term changes with learning" (Figure 6 in Chapter 7), evidence was already on hand that excitatory neurons excited each other in the bulb and in the prepyriform cortex (Freeman 1968b; Nicoll 1971), and that associative learning in olfaction was enabled by a small increase in the strength of synaptic connections between excitatory neurons. These findings were consistent with numerous subsequent models of associative memories based on feedback architectures (Grossberg 1975; Amari 1977; Anderson *et al.* 1977; Hopfield 1982; Kohonen 1984). The findings were in opposition to modification of the connection strength between incoming axons and the cortical excitatory cells in models with feedforward architectures as in multilayer perceptrons (Rosenblatt 1962; Rumelhart and McClelland 1990), which were the conceptual basis for physiological studies of long term potentiation and depression (LTP and LTD) in use-dependent synaptic change as a model of learning (Hebb 1949). The feedforward synapses at the input layer of the bulb are modifiable, but in normalization and logarithmic conversion of input and automatic gain control (Freeman 1975, 1992a), not in associative learning or habituation. The preponderance of excitatory interactions between excitatory neurons in cortex was not fully recognized until the 1990s, perhaps owing to a persistent reluctance to countenance "reverberatory circuits" as a basis for sustained neural activity (Chapter 8).

Analog Simulation of Prepyriform Cortex in the Cat

Abstract

The prepyriform cortex contains three types of intracortical feedback loops, negative, positive excitatory, and positive inhibitory, representing the possible modes of interaction in a mixed population of excitatory and inhibitory neurons. This system can be described by a linear differential equation with three variable coefficients corresponding to the three types of feedback gain. Solutions for the equation by analog computation for the appropriate initial conditions constitute the "single-shock evoked potential" for the model. Solutions are presented for four conditions, in each of which a single variable in the equation is modified. The changes in the output of the model are shown to replicate four patterns of variation previously described for the averaged evoked potential (AEP) of the prepyriform cortex. (1) An increase in AEP amplitude with food deprivation was replicated in the model by increasing forward gain of the system. (2) The increase in amplitude and decrease in frequency of AEP that occurs with increasing stimulus intensity was replicated by means of limiting diodes placed in the feedback limbs of the model. (3) A combined change in AEP amplitude, frequency, decay rate, and (inversely) phase, representing spontaneous" variation in AEP, was replicated by decreasing network gain. (4) A decrease in phase and frequency of AEP associated with long-term learning was replicated by a relative increase in the gain of the positive excitatory feedback loop. The model is generally stable but can be rendered unstable in each of two ways (monotonic and oscillatory) previously observed in the prepyriform cortex. The role of the positive inhibitory feedback loop is shown to be particularly important in the maintenance of cortical stability.

Introduction

The prepyriform cortex generates an oscillatory evoked potential in response to single shocks delivered to the lateral olfactory tract. The averaged evoked potential (AEP) closely resembles a damped sinusoid, which is described by the four parameters for amplitude, frequency, decay rate, and phase [1]. Four characteristic patterns of change in the AEP have been defined [2, 3]. First, in association with random fluctuations in background activity ("spontaneous" variation) and with certain behavioral states such as prolonged habituation, food consumption, mating, etc. [4], there is a conjoint change in the sinusoid, which consists of decrease in amplitude, frequency, and decay rate with increase in phase. Second, in association with long-term changes involving learning there is an increase in phase together with a large decrease in frequency [3, 5]. Third, in association with food deprivation and other motivating factors there is an increase in amplitude without other change [4]. Fourth, with increasing stimulus intensity in the range for which the amplitude of single-shock responses is higher than the peak amplitude of ongoing EEG activity, the large increase in amplitude is accompanied by a large decrease in frequency, with little change in decay rate or in the duration of the initial surface-negative peak of the AEP [6].

Studies of the spike activity recorded from single cells in the cortex [7, 8] together with histological observations on the cortex [9, 10] have led to the conclusion that the cortex consists of a mixed population of excitatory and inhibitory neurons. They are postulated to be interconnected nonspecifically, so that the neurons form negative feedback loops and both positive excitatory and positive inhibitory feedback loops. The excitatory neurons are tentatively identified as superficial pyramidal cells, which receive the impulse input from the lateral olfactory tract, and which generate the dipole field in which the AEP is recorded [11]. The inhibitory neurons are tentatively identified as interneurons lying mainly in the deeper layer of the cortex.

Both types of neurons fire randomly at mean rates (e.g. 11/sec) much lower than the characteristic EEG frequency for this cortex (40/sec). Poststimulus-time histograms show that both types tend to fire at mean rates proportional to the amplitude of the AEP (recorded from the surface with respect to the base), except that the fluctuations in firing rate for the inhibitory neurons lag the AEP by about one quarter cycle. This holds for low-intensity stimulation, where the amplitudes of single-shock responses are less than the peak amplitudes of EEG activity. For higher stimulus intensity the proportionality fails, because during surface positivity of relatively large magnitudes the firing rates of excitatory neurons go to zero and do not reflect further increases in the peak of the AEP. The same block occurs for the inhibitory neurons but is delayed by one quarter cycle [7, 8].

The diffuse cortical network can be represented by a lumped network having an equivalent topology. The time-dependent properties of the neuron types can be represented using linear differential equations, and the interaction strengths by feedback gain parameters [7]. The effect of block can be treated as a reduction in feedback gain with increased input amplitude, and can be introduced into the equations as a variable coefficient. By use of appropriate operational methods the entire system can be represented by a single linear differential equation having a single variable coefficient representing loop gain [7].

Solutions for this equation in the presence of an impulse driving function have been obtained by both digital and analog computation. The results presented here from the latter show that all four patterns of variation listed above can be replicated by appropriate changes in one or another of four gain constants, or by use of diodes for clipping to represent block of unit firing.

Method

Each neuron pool was represented by a second-order linear differential equation, which in operational form was

$$A(s) = \frac{1}{(sT_1 + 1)(sT_2 + 1)},$$

where $T_1 = 1/\beta$ and $T_2 = 1/\gamma$ were time constants representing the rise and decay times of the open-loop population response. They were given values of $T_1 = 1.4$ and $T_2 = 4.5$ sec, corresponding to values previously derived from cortex but scaled by a factor of approximately 1000.

Two neuron pools were simulated on an analog computer (Beckman EASE) and were connected into a negative feedback loop (Fig. 1). Diode clipping was introduced in the output for each simulated neuron pool, such that the overshoot after the initial peak of an oscillatory transient could be clipped. Changes were made in clipping level rather than in input magnitude to simulate the effects of increasing input. Two of the loops shown in Fig. 1 were constructed, and were interconnected

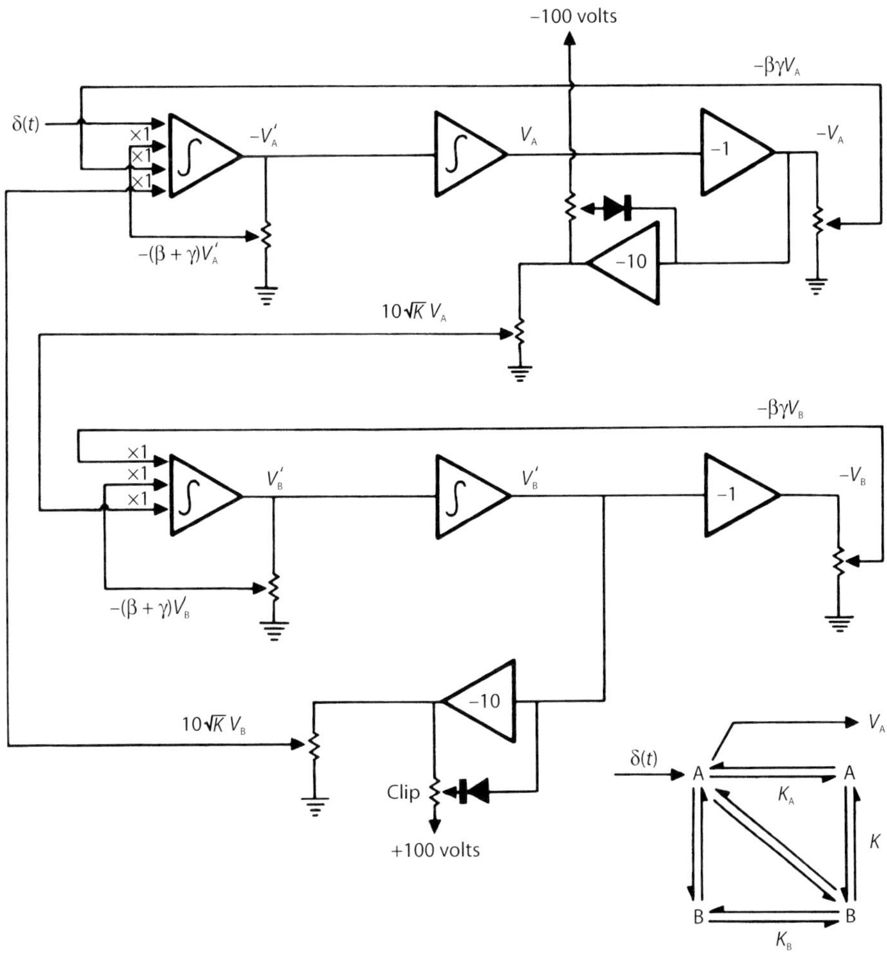

Fig. 1. Schematic diagram for simulation of a two-neuron negative feedback loop. The inset shows the connections for two such units.

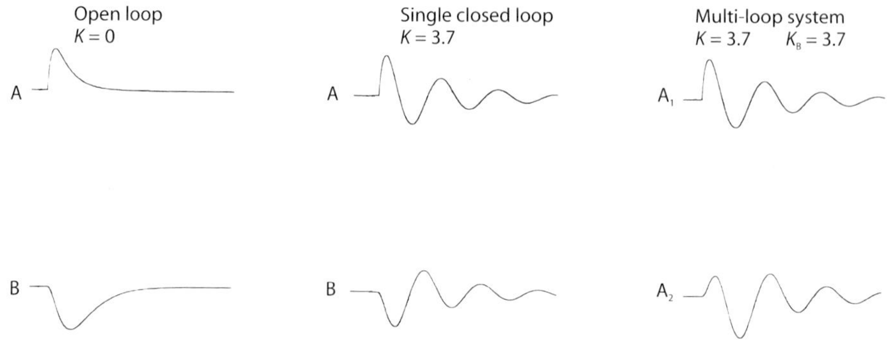

Fig. 2. Open- and closed-loop responses for impulse input. The results confirm the theory [7] that the response of A in a single negative feedback loop is identical to that for A_1 in the system with positive feedback, for all values of $K = K_A = K_B$

in the manner shown in the inset. This introduced the two types of positive feedback. Only a single diagonal pair of connections was introduced, on the basis of physiological evidence previously discussed [7]. If in this topology all the gain constants were equal, then the output of the system shown in the inset was predicted to be identical to the output of either of the simple negative feedback loops activated above in the same way. This provided a useful check on the correctness of the analog simulation.

There were four variables in this model: the negative feedback gain, K, which was identical for all three negative feedback loops in the inset of Fig. 1; the positive excitatory feedback gain, K_A; the positive inhibitory feedback gain, K_B; and the system forward gain, which was controlled by varying the magnitude of the input pulse. The symbol K_0 refers to a reference gain, which was defined [7] as that gain at which $K = K_A = K_B$.

Figure 2 shows the open-loop and closed-loop responses of the two neuron pools in a simple negative feedback loop ($K = 3.5$). Also shown are the output for unit A_1 and unit A_2 when the system was interconnected as shown in the inset of Fig. 1. A was the A unit with direct input and output, whereas A_2 was the other A unit. As predicted, the transients for A_1 and for A in either simple negative feedback loop were identical.

Results

(1) Replication of AEP Patterns

The increase in amplitude associated with food deprivation was simulated by increasing the input, either in the absence of clipping or with proportionate changes in the clipping levels.

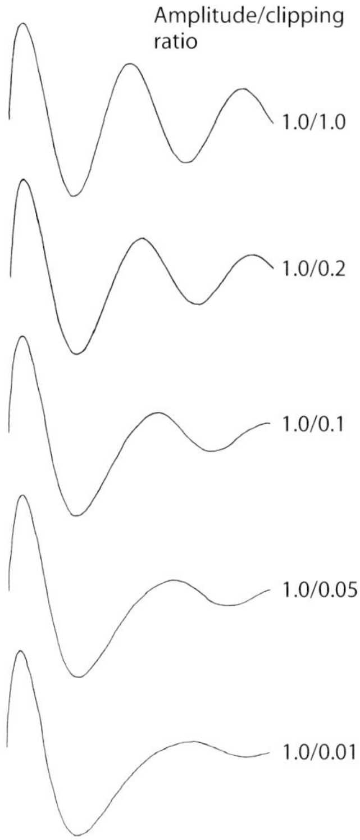

Fig. 3. The effects of clipping by diodes in the model are shown to replicate the effects of increasing stimulus intensity in cortex.

The pattern of change in AEP with increasing stimulus intensity (Fig. 3) was simulated by clipping with diodes (Fig. 1).

The pattern of "spontaneous" variation in AEP was simulated by setting $K_A = K$, $K_0 = 3.7$, and $K_B = K^2/K_0$ and varying K over the range from 2.4 to 4.8 as indicated in the right-hand column of Fig. 4. Figure 5 shows the locus. following transformation of $K_2 \cdot T_2$ coordinates [2, 7, 12].

The pattern of long-term change was simulated by setting $K_A = K_0 = 2.5$ and $K_B = K^2/K_0$ and varying K over the range indicated in Fig. 6.

(2) Network Stability

For appropriate magnitudes of gain, K, the system in Fig. 1 (inset), became unstable in either the monotonic or the oscillatory mode. Increasing $K = K_A = K_B$ all together resulted in an increase in frequency and a decrease in decay rate of the output, with onset of instability at $K = 4.1$ in the oscillatory mode. Identical results attended the same gain change in K in a single negative feedback loop (Fig. 4, $K = 4.\ 1$, left-hand column). An increase in K_B alone caused increases in both the frequency and decay rate of the oscillatory impulse response, but at $K_B = 6.1$ the system went unstable in

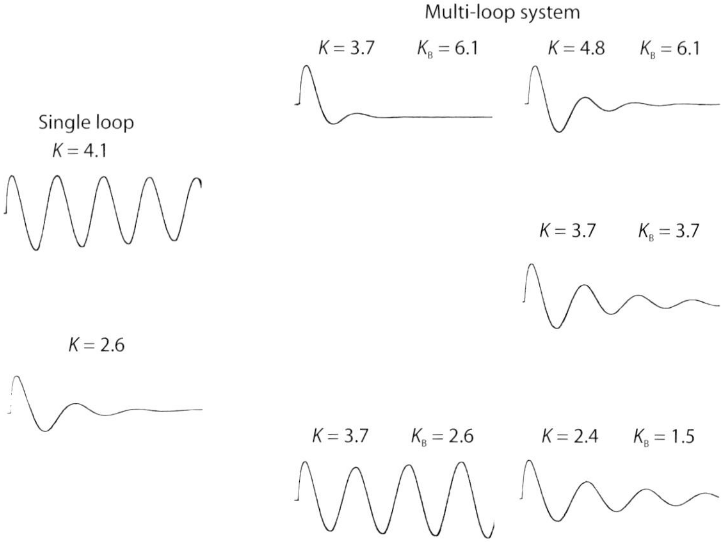

Fig. 4. Left, in a single negative feedback loop increased gain causes decreased decay rate and vice versa. Center, in the compound system increasing only the positive inhibitory feedback gain, K_B, causes increased decay rate and vice versa. Left, changing both with $K_B = K^2/K_0$ replicates cortical changes observed in several behavioral transformations, all involving changes in signal-noise ratio.

the monotonic mode (Fig. 4, middle column). If both K and K_B were increased together, the range of stability for changes in K was broadened (Fig. 4, right-hand column).

On the other hand, with decreasing $K = K_A = K_B$ the system became much less oscillatory (Fig. 4, left, $K = 2.6$). Decreasing K_B alone caused a major decrease in decay rate together with a decrease in frequency, and the system went unstable in the oscillatory mode (Fig. 4, middle, $K_B = 2.6$). Decreasing both together precluded instability (Fig. 4, right, $K = 2.4$).

The system having unilateral clipping was stable for all inputs and all levels of gain, except that for high values of K the system oscillated in a stable manner at an amplitude dependent on the clipping level.

Discussion

A derivation and detailed physiological interpretation of this model have been published with a description of the solutions by digital computation [7]. The analog model replicates the lumped cortical network generating the AEP and serves to emphasize certain important features.

Fixed-level unilateral clipping by diodes is a useful approximation for cortical function in some respects, particularly by introducing stability over wide ranges of

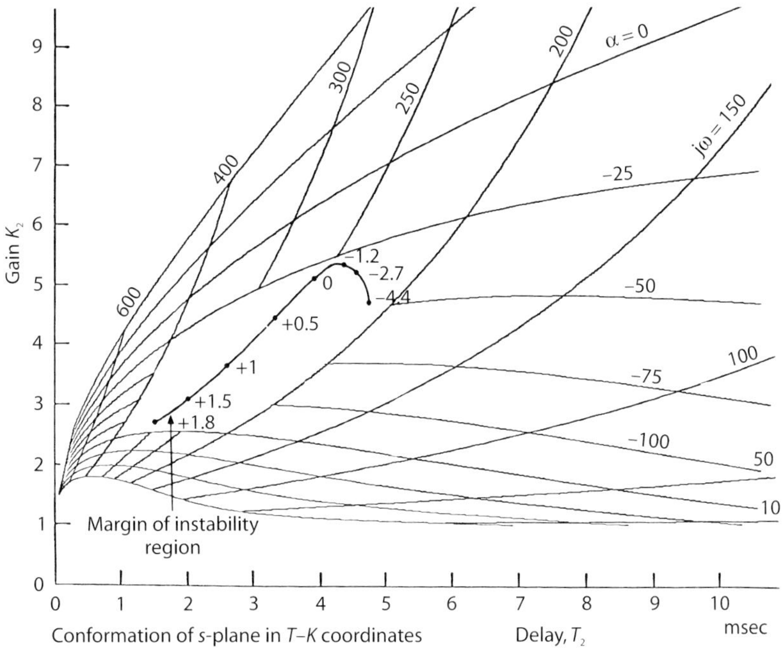

Fig. 5. The locus for changing K with $K_A = K$ and $K_B = K^2/K_0$ is shown in $K_2 \cdot T_2$ coordinates. This reproduces the locus for "spontaneous" variation of cortical AEP [3]. The values for frequency, ω, are in radians sec^{-1} and for decay rate, α, are in sec^{-1}. The numbers along the locus are values for K in decibels ref K_0. Between 1.5 db and 1.8 db the system became unstable in the monotonic mode.

input and loop gain. It is nonrepresentative in two respects, first that the blocking levels in the population of cortical neurons are dispersed in relation to the amplitude of the AEP [8], and second that clear evidence has been found for both upper and lower limits on firing in relation to surface-positive peaks of the AEP. Clipping introduces only the lower limit. However, it suffices to represent the main features of the nonlinearity manifested by cortical responses to differing intensities of input.

The level of block in unit firing in cortex is determined both by the test stimulus magnitude and the level of background activity [8]. The change in AEP amplitude with increasing food deprivation must involve an increased sensitivity of cortical pyramidal neurons not only to the test stimulus but to background input as well. It is likely that mean firing rates of most cortical neurons increase with hunger, because the rms amplitude of EEG activity increases in parallel with the AEP, both without significant change in frequency. Therefore, it seems likely that increased clipping does not take place with increasing deprivation. On the basis of this model it can be predicted that some degree of saturation must be present in all physiological states as the necessary condition for cortical stability.

An interesting feature of this model is the role played by the positive inhibitory feedback loop in the regulation of cortical stability (Fig. 4). The implication of the squared term is that the strength of interaction between interneurons is particularly sensitive to changes in the level of background input. Positive inhibitory feedback

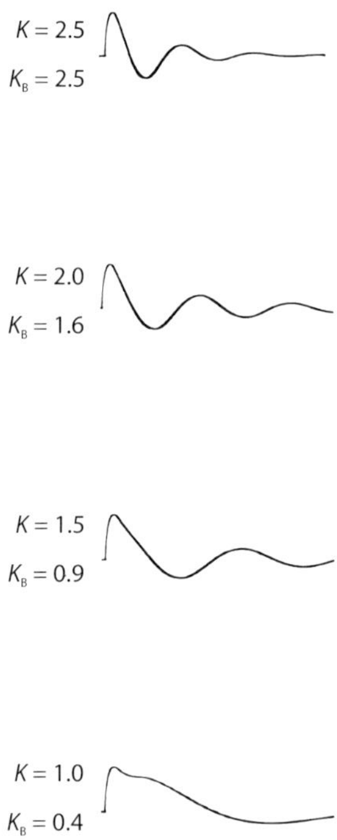

Fig. 6. Changing loop gain, K, with $K_A = K_0$ and $K_B = K^2/K_0$ replicates the pattern observed with long-term changes associated with learning [3,5].

gain is more liable to reduction with increasing test amplitude relative to background activity than is positive excitatory feedback gain representing interactions between excitatory neurons. The effect of reducing K_B is to decrease the stability of the cortex in the oscillatory mode, which is opposite in effect to reducing K. The positive inhibitory feedback loop therefore constitutes a mechanism for maximizing oscillatory interactions among cortical neurons over a wide range of interaction strengths and input levels. It normally provides for the stability of cortex, but it is probably also directly involved in the genesis of seizure activity [7].

The one difference between the model representing the pattern of "spontaneous" variation ($K_A = K$) and that for long-term variation ($K_A = K_0$) lies in the positive excitatory loop. The illustration (Fig. 6) is for a fixed value of K_A during a decrease in K_B and K, but a similar pattern can be obtained by selectively increasing K_A. This implies that the significant changes in cortical function during learning involve the interactions of superficial pyramidal neurons, which receive the primary input to the cortex.

A closely similar pattern has been observed upon delivery to the cortex of a tetanizing background input via the olfactory tract and superimposed on the test stimulus, which affects mainly the superficial pyramidal cells. The initial surface-negative peak of the AEP has two crests (is dicrotic), the first resulting from excitation by tract axons and the second from reexcitation by superficial pyramidal neurons [8]. Following tetanization of the tract the first of these peaks is potentiated, whereas following tetanization of the dorsomedial nucleus of the thalamus, it is the second (still in response to tract stimulus) that undergoes potentiation [11]. Therefore, it is in the interconnections of the superficial pyramidal neurons that changes in cortical structure and function related to learning might best be sought.

Acknowledgments

Reprinted by permission from The American Physiological Society, from *Mathematical Biosciences*, volume 2, Dr. Walter J. Freeman, "Analog Simulation of Prepyriform Cortex in the Cat", pages 181–190, Copyright 1968.

Communicated by W. Ross Adey. Supported by grants NBO5035 and MH06686 from the National Institutes of Health, U.S. Public Health Service, and by the Berkeley Computer Center. Analog computer facilities were provided by Prof. C. Atkinson, Department of Mechanical Engineering, Berkeley, and by Prof. Dr. Marko, Institut für Nachrichtentechnik der Technischen Hochschule, München, Germany. The assistance of B. Baron, B. Taskett, and H. Kling is gratefully acknowledged.

References

1 W. J. Freeman, Use of digital adaptive filters for measuring prepyriform evoked potentials from cats, *Exper. Neurol.* **10** (1964), 475–492.

2 W. J. Freeman, Correlation of goal-directed work with sensory cortical excitability. In *Recent advanc. biol. psychiat.*, Vol. 7, Chapter 25, Plenum Press, New York, 1965.

3 W. J. Freeman, Patterns of variation in waveform of averaged evoked potentials from prepyriform cortex of cats, *J. Neurophysiol.* **31** (1968), 1–13.

4 W. J. Freeman, Changes in prepyriform evoked potential with food deprivation and consumption, *Exper. Neurol.* **6** (1962), 12–29.

5 W. J. Freeman, Phasic and long-term excitability changes in prepyriform cortex of cats, *Exper. Neurol.* **5** (1962), 500–518.

6 W. J. Freeman, Alterations in prepyriform evoked potential in relation to stimulus intensity, *Exper. Neurol.* **6** (1962), 70–84.

7 W. J. Freeman, Analysis of function of cerebral cortex by use of control systems theory, *Logistics Review* **3** (12)(1967), 5–40.

8 W. J. Freeman, Relations between unit activity and evoked potentials in prepyriform cortex of cats, *J. Neurophysiol.* **31** (1968), 337–348.

9 J. O'Leary, Structure of the primary olfactory cortex of the mouse, *J. Comp. Neurol.* **67** (1937), 1–31.

10 F. Valverde, *Studies in the piriform lobe.* Harvard Univ. Press, Cambridge, Massachusetts, 1966.

11 W. J. Freeman, Distribution in time and space of prepyriform electrical activity, *J. Neurophysiol.* **22** (1959), 644–665.

12 W. J. Freeman, A linear distributed feedback model for prepyriform cortex, *Exper. Neurol.* **10** (1964), 525–547.

8. *Stability Analysis to Derive and Regulate Homeostatic Set Points for Negative Feedback Loops*

The dynamics of mutual excitation has been neglected by neurobiologists in comparison to studies of negative (homeostatic) feedback, perhaps because of a widespread misconception that negative feedback systems are intrinsically stable, whereas positive feedback systems with regenerative excitation are intrinsically unstable. Mutual excitation is widely believed to be responsible for epileptic seizures, and therefore to be avoided or suppressed in brain systems. On the contrary, each form of feedback can be stable, and with sufficiently high gain each can be destabilized to give respectively monotonic or oscillatory outputs. Feedforward excitation was proposed by Ramón y Cajal (1955) to exist as "avalanche conduction", by which the axon collaterals of a small number of mitral cells might excite other mitral cells and these yet others, in the recruitment of a large number of cells to amplify a weak olfactory stimulus (Nicoll 1971), but he did not intend by invoking this "snow-ball" effect to allow feedback, because he believed that a neuron could not function properly if it were unable to distinguish input of other neurons from its own output (Freeman 1984). The work of Cajal's student Lorente de Nó (1934) demonstrated the likelihood of feedback among cortical neurons, leading to theories of neural networks by McCulloch and Pitts (1943), and of nerve cell assemblies by Donald Hebb (1949). However, the resulting concept of "reverberatory circuits" was roundly criticized as epileptogenic, and was rejected as a basis for steady state neural activity or memory.

The existence of prolonged "after-discharge" in response to an impulse input was well known to physiologists, for example, the dorsal root potential of the spinal cord (Wall *et al.* 1962), but no one had modeled neural positive feedback to explain prolonged activity that was induced by single afferent volleys. Even more surprising was the fact that, though oscillations about set points in homeostatic negative feedback systems was well known, no one had asked how non-zero set points were established, maintained and modified by positive feedback. The olfactory bulb provided an opportunity to model this property of brain dynamic systems.

The study of the external bulbar interneurons, the periglomerular cells, by conventional cellular techniques is difficult, because they are too small for intracellular recording (the cell body diameters are about 4 microns), and their extracellular field potential of dendritic activity is a low amplitude closed field, which is very strongly masked by the powerful EEG generators in the depth of the bulb (Martinez and Freeman 1984). The cells secrete GABA on transmission, and this is widely viewed as an exclusively inhibitory transmitter in the mammalian central nervous system. Electrophysiological studies have shown a relation of periglomerular activity to attenuation (not inhibition) of the effect of bulbar input from receptors in proportion to the total amount of input. This operation is thought to provide for spatial contrast enhancement, but more importantly it performs a logarithmic transformation on bulbar input and extends the bulbar range of acceptable input intensity to four orders of magnitude, which is necessary because each glomerulus receives on the order of 20,000 axons, the input from any one or a few of which may be significant. Because GABA operates on the chloride conductance of membranes, its action can be either excitatory or inhibitory, depending on the transmembrane gradient for this ion and on the types of postsynaptic membrane receptor. Studies had not yet been done to show that chloride was concentrated in the periglomerular cells (Siklós *et al.* 1995), so that opening their chloride channels would be depolarizing.

My approach using dynamics was to measure the impulse responses of bulbar neurons to electrical stimulation of the input pathway to the bulb (orthodromic activation of the primary olfactory nerve, PON) and the output pathway (antidromic activation of the lateral olfactory tract, LOT) (Freeman 1972 a–f; 1974 a–f). My key finding was the response of a type of neuron located in the glomerular layer of the bulb, which was excited by stimuli to the PON but was neither excited nor inhibited by stimuli to the LOT. The post-stimulus time histograms of these cells showed a rapid increase in firing rate with a short start latency on PON impulse input, and then a monotonic decay to the background level without significant inhibitory overshoot. Both the amplitude and the decay rate of the response increased with increasing stimulus intensity, whereas the rate of rise decreased (Figures 1 and 2 in Chapter 8). This pattern was consistent with the properties of a positive feedback loop, in which there was amplitude-dependent saturation in the feedback path. The same effect had been already modeled as Mode 1 root loci of oscillatory impulse responses of the multiple positive and negative feedback loops in the prepyriform cortex (Chapter 4).

After testing for superposition I constructed a sum of linear basis functions and fitted these to the unit post-stimulus time histograms, yielding sets of closed loop rate coefficients (Figure 3(a) in Chapter 8.1). I took the inverse Laplace transform to get a differential equation and restructured it to conform to the topology of bulbar connections. I derived the closed loop zeros, which are essential for proper evaluation of the poles, and used the Laplace transform to generate a new set of basis functions. These gave theoretical root loci (Figure 3(b) in Chapter 8.1). I used the congruence of the experimental and theoretical root loci to estimate the dependence of loop gain on response amplitude and derived a describing function for this dependency. This function showed that, when the response amplitude was extrapolated to zero, the loop gain went to unity (Figure 3 in Chapter 8.2), giving a closed loop pole at the origin (an eigenvalue equal to zero). This led me to conclude that the periglomerular population was a self-stabilizing nonlinear feedback loop without need for internal or external inhibition to prevent runaway excitation.

The implications of these findings have been explored in several reports since that time, notably in showing that the putative transmitter of the primary nerve, the dipeptide carnosine (Margolis 1974), has a powerful excitatory action on mitral cells that can be canceled or reversed by sustained granule cell inhibition, which was modeled by pole-zero cancellation (Gonzalez-Estrada and Freeman 1980); in showing that the periglomerular cells establish an excitatory postsynaptic field of potential in the apical dendrites of mitral and tufted cells (Martinez and Freeman 1984); and in showing that a variable excitatory bias can account for the relations between unit and wave activity in the olfactory bulb (Freeman 1992a) and prepyriform cortex (Eeckman and Freeman 1991). The excitatory action of periglomerular cells on each other and on mitral and tufted cells proved to be an essential link in the model that sufficed to simulate the chaotic normal background and burst activity of the bulb (Chapter 12) and the abnormal patterns of epileptic discharge in partial complex seizures (Freeman 1986). The model supported re-examination of the actions of GABA and its mimetics and blockers on bulbar neurons, which gave electrophysiological and neurochemical evidence that the action of GABA in the outer layer of the bulb is excitatory (Rhoades 1991), and it led to confirmation of the prediction that the periglomerular neurons had a high intracellular concentration of chloride ions (Siklós *et al.* 1995). These results demonstrated the utility of dynamical modeling for prediction and explanation of the dynamics of complex neural systems.

A Model for Mutual Excitation in a Neuron Population in Olfactory Bulb

Abstract

The glomerular neurons are small cells in the olfactory bulb which form numerous synaptic connections with each other and with the output neurons of the bulb, the mitral-tufted cells. They form in the mass two anatomical feedback loops, one within the population, the other between the populations of glomerular and mitral-tufted cells. The poststimulus time (PST) histograms of single glomerular neurons on afferent electrical stimulation are monotonic, always with initial increase in induced pulse probability at varying latency of onset, implying that interactions are mutually excitatory. A differential equation is developed to represent the dynamics of afferent temporal dispersion and positive feedback in a neural population. The solutions of the equation are used to generate curves fitted to the PST histograms. The required numerical values for the coefficients are similar to those found for other bulbar neurons, which implies that the particular properties of neurons in the population arise from the topology of connections and not from the single neurons.

I. Introduction

This paper is concerned with the description of synaptic interaction within a population of neurons in the mammalian central nervous system. The anatomical circuit diagram is briefly described and the physiological responses of neurons in the population are shown in the form of poststimulus time (PST) histograms. A differential equation is developed in stages, which represents the interaction in terms of a positive feedback loop. Three solutions to the equation are shown in the form of curves fitted to the data. The outcome is evaluated and interpreted in relation to the results of others.

II. Experimental Materials

A. Anatomical Connections

Recent advances in the analysis of the neural structure of the olfactory bulb have resulted in a reasonably clear circuit diagram for its synaptic connections [1]–[9]. Although certain of their findings are still in dispute, the results of Pinching and Powell [5] and of Price and Powell [6] are adopted here as the basis for analysis of one aspect of bulbar dynamics.

The principal neurons of the bulb are projection neurons called mitral-tufted cells which send output axons to the central nervous system in the lateral olfactory tract (LOT). They receive input from incoming axons in the primary olfactory nerve (PON) which form synapses with mitral-tufted apical dendrites in a zone of specialized synaptic endings called the glomerular layer. At the simplest level of analysis, neural transmission through the bulb occurs across a single synaptic layer in a forward direction.

The mitral-tufted cells form mutual connections with two sets of interneurons. The first set, the periglomerular neurons, is located in the glomerular layer. These interneurons form receiving and transmitting dendrodendritic synapses with mitral-tufted apical dendrites. They are also interconnected with each other by axodendritic synapses. They form two feedback loops, one with each other and one with mitral-tufted cells. According to some authors [5] these interneurons receive axodendritic synapses directly from PON axons, but other observers conclude that they do not [8], [15].

The second set, the granule cells, lies in the deeper layers of the bulb. These cells have no axons. They have reciprocal dendrodendritic synapses with the basal dendrites of mitral-tufted cells, thus forming another feedback loop with mitral-tufted cells. They have no direct connections to or from the neurons in the glomerular layer.

B. Physiological Considerations

From physiological evidence [11]–[18] it is known that PON axons are excitatory to mitral-tufted cells and, directly or indirectly, to glomerular cells. Forward transmission is excitatory. The mitral-tufted cells are excitatory to the granule cells which are inhibitory to the mitral-tufted cells [12], [13], [16]. The deep loop is a negative feedback loop. On anatomical [5] and physiological [15] grounds, the actions of glomerular neurons have been inferred to be inhibitory. If so, then the glomerular-mitral-tufted connections must also form a negative feedback loop, and the mutual connections among glomerular cells must form a positive (inhibitory) feedback loop.

Mutual connection implies the possibility of interaction. The strength of interaction can be represented by a dimensionless factor, the feedback gain. For appropriate levels of gain, the expected output for the negative feedback loops in response to impulse input is oscillation of neural activity at some frequency characteristic of the bulb. The expected output for the positive feedback loop with impulse input is an abrupt increase or decrease in level of activity followed by a monotonic return to the resting level.

In physiological studies [19] single-shock electrical stimuli were delivered either to the PON or the LOT. Units were recorded from the glomerular layer and from deeper layers. All units displayed background activity consisting of pulses at irregular intervals resembling a Poisson process with a dead time. PST histograms [20] were constructed while stimulating at each site at one of several different stimulus intensities.

Units in the deeper layers responded to both inputs. The PST histograms were oscillatory at the frequency of the concomitantly averaged evoked potential (AEP). They began with either excitation or inhibition. The frequency of the oscillation decreased with increasing stimulus intensity [19]. These units were called "mitral-tufted units." (Granule cells have no axons and do not generate extracellularly detectable action potentials.)

On single-shock stimulation units in the glomerular layer tended to fire in short bursts lasting up to 50 ms at rates varying from 40/s to over 250/s. The later spikes in such trains tended to have lower amplitudes than the earlier spikes. These characteristics have already been described [12], [14], [15], [18].

Most units in the glomerular layer responded only to PON stimulation and not LOT stimulation. If they did respond to LOT stimulation, their PST histograms were oscillatory on PON stimulation. If they did not, their PST histograms on PON stimulation [19] showed a sustained increase in pulse probability with subsequent exponential decay in pulse probability to the background level. The crest amplitude (maximum induced pulse probability) and decay rate both increased with increasing stimulus intensity (Figs. 1 and 2). There was no clear evidence of induced decrease in pulse probability below resting level at any time during the PST histograms, irrespective of the locations of the units near the center or near the margins of the focal response fields in the bulb [21]. These were called "glomerular units."

C. The Approach to the Model

The oscillatory PST histograms from mitral-tufted units were fitted by means of nonlinear regression with damped sine waves. These curves were generated by the

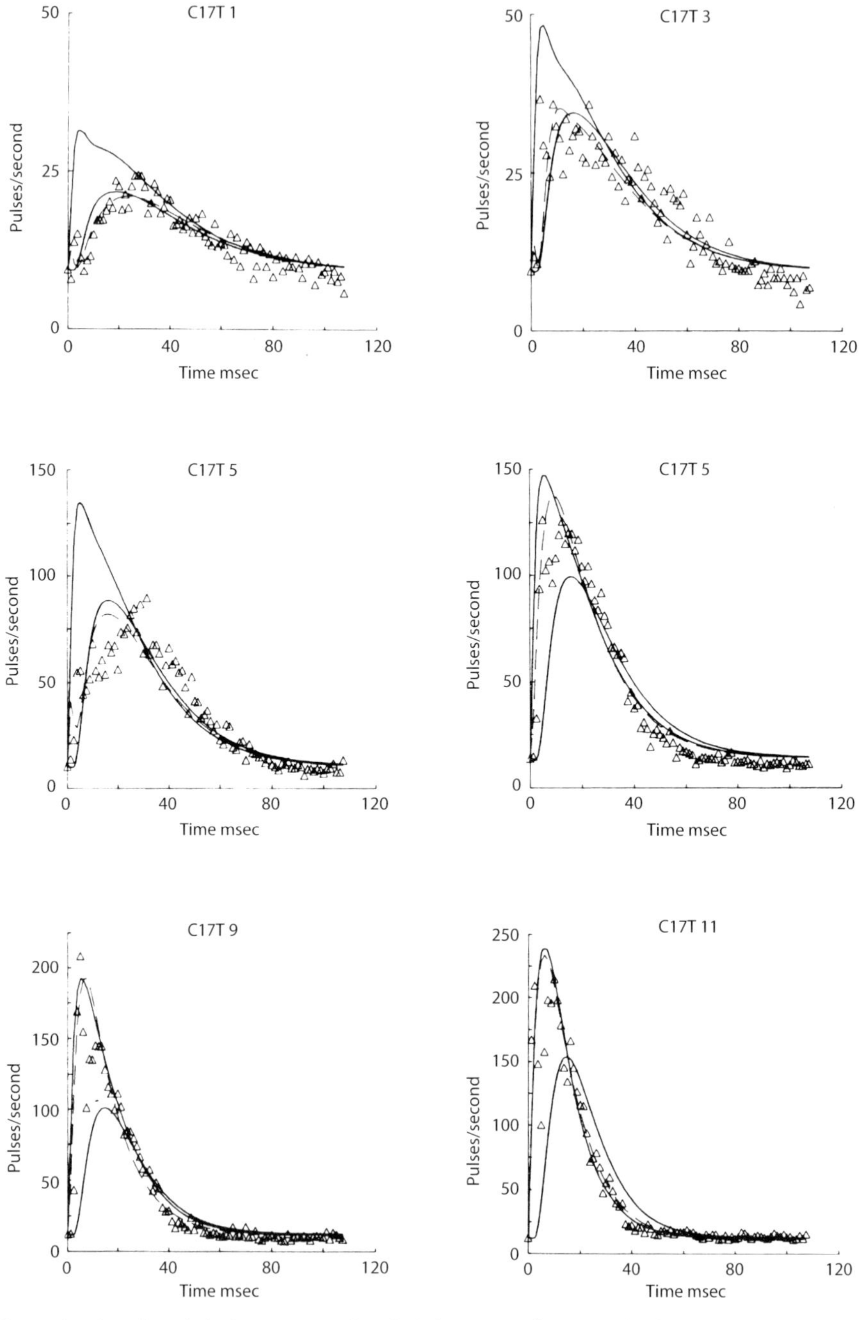

Fig. 1. The plotted symbols show six examples of PST histograms from a single glomerular unit train on repeated single-shock PON stimulation. From top to bottom the stimulus intensity was increased, and the number of stimuli was decreased (see Table I). With increasing stimulus intensity, the crest amplitude of the PST histogram increased, and so did the decay rate from the crest. The slow rise time implies that the neuron was a member of the feedback subset at low stimulus intensity. The dashed curves were generated by (1.5) and Table I. The solid curves were generated by (4.1), (4.3), and Table III.

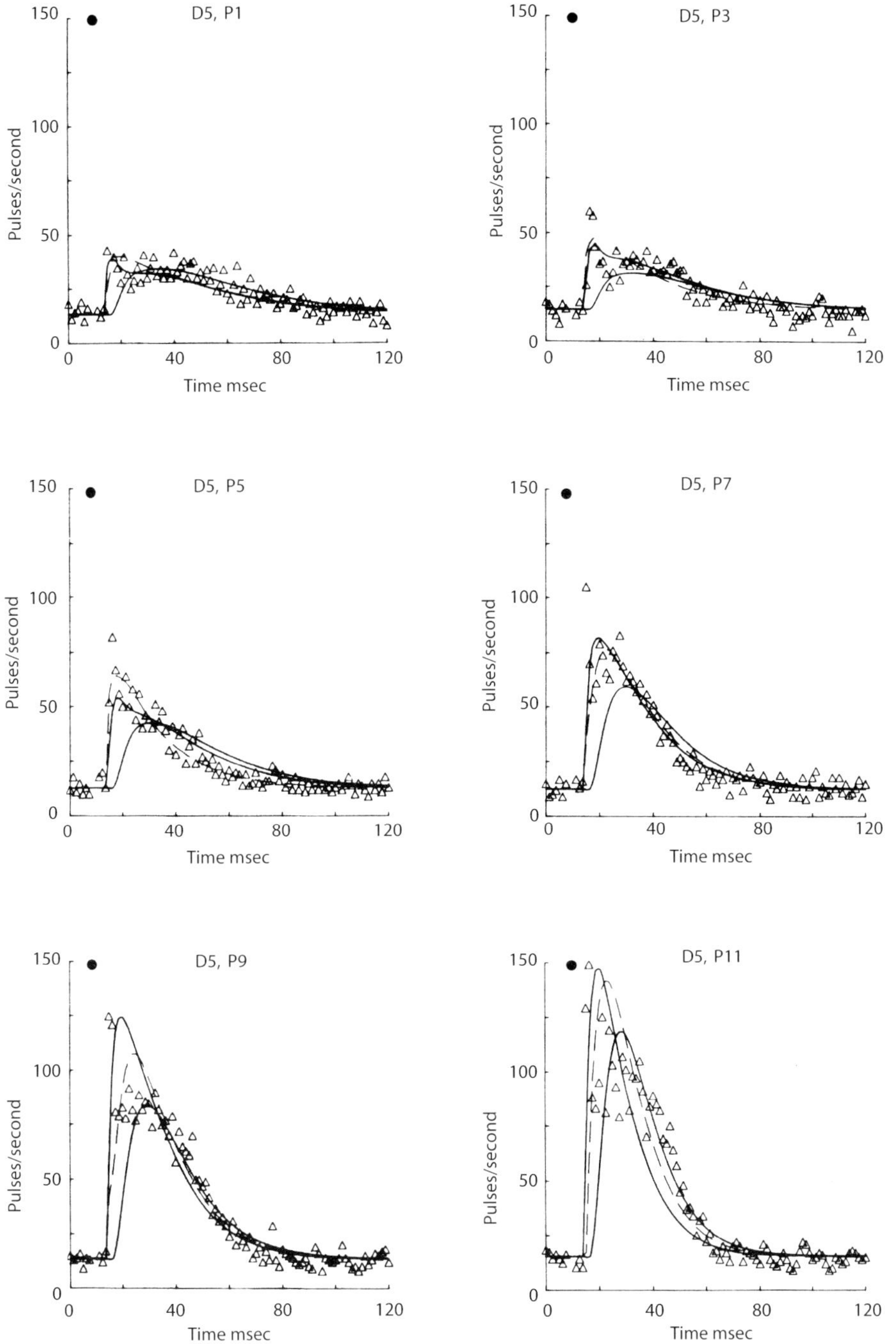

Fig. 2. See caption for Fig. 1. The fast rise time implies that the neuron was a member of the forward subset at low stimulus intensity. See Tables II and III.

solutions to a differential equation, which represented the mitral-tufted neuron population in the forward limb (excitatory) of the deep negative feedback loop, and the granule cell population (inhibitory) in the feedback limb [19].

The nonoscillatory impulse responses of PST histograms from glomerular units (Figs. 1 and 2) were not compatible with negative feedback. A new differential equation was required to represent the interaction of glomerular units in terms of a positive feedback loop.

The equation was constructed, solved, and evaluated so that curves could be generated and fitted to glomerular unit PST histograms. The main difficulty encountered was the unavailability, due to bulbar conditions already described [22], of direct measurements of the open loop rate constants of the population, such as have been made for other bulbar [19], [23] and cortical populations [24], [25]. The afferent dispersive delay in the PON [19], [21], [23] could not be separated easily from the delays within the population.

The approach used was to set up and solve an equation for a model without afferent temporal dispersion, and to find how that model was incompatible with positive feedback in the neural population. Dispersive delay was then introduced in a form already derived [19], and the PON dispersion and glomerular open loop rate constants were both evaluated. The values were checked against those found for other populations.

III. Derivation of the Model

A. Assumptions

The following assumptions were made. First, the glomerular neurons were assumed to form a homogeneous population of neurons identical in all respects save one: their relation to input. There were two subsets, both having connections to the PON. An electrical stimulus excited only a fraction of PON axons. One subset of the population called the forward subset received impulses from the PON, and the other called the feedback subset did not. Second, each subset acted on the other but not on itself (due to the refractory period of neurons in the subsets). Third, the two subsets were uniformly distributed over the surface, so that the interaction strength was constant with respect to the locations of the two subsets of the population. Fourth, the ensemble average of the performance of a single neuron over N repeated events (as shown in a PST histogram) was the same as the performance of N' neurons in its subset during any one event. Fifth, the distribution of input was uniform over the region of interest, and the dependence of conduction delays on distance was negligible. The system was treated as a lumped circuit. Sixth, the interaction strength between the subsets was expressed by a dimensionless gain coefficient, which had a fixed value during any one event, but which varied for events in different experimental conditions, such, for example, as change in stimulus intensity. The feedback gain was assumed to be equal to the square of the forward gain of each subset. Seventh, the transfer function of each subset was described by means of a linear ordinary differential equation having constant rate coefficients. While the output of any one neuron is a discrete pulse train, the output of a subset consists of a sufficiently large number of pulse trains to be represented as a continuous pulse density [26], [27].

B. The Empirical Equation for the Impulse Response

The system for analysis consisted of a delay line $A_m(s)$ in the PON, in series with a positive feedback loop $P(s)$ among periglomerular neurons. The open loop impulse response of the periglomerular neurons was assumed to have the form

$$p(t) = P_0\{\exp[-a(t-\tau)] - \exp[-b(t-\tau)]\} \quad (0.1)$$

where a, b, and P_0 were positive real numbers for amplitude and rates of change and τ was synaptic dead time. The transfer function was defined as

$$A(s) = \frac{K_e^{0.5} ab(c-s)}{(s+a)(s+b)(s+c)} \quad (0.2)$$

where $e^{-s\tau}$ was replaced by a rational approximation $(c-s)/(s+c), c = 2/\tau$, and $K_e^{0.5}$ was the square root of feedback gain. The closed loop transfer function for output from the feedback subset was

$$P_c(s) = \frac{A^2(s)}{1-A^2(s)} \quad (0.3)$$

The operation of the PON on the electrical stimulus $I\delta(t)$ to yield the dispersed bulbar input was represented, as it was previously [19], [23] by the expression

$$A_m(s) = 1 + K'\exp[-(sT)^{1/2}] \quad (1.0)$$

which specified the input as a pulse followed by a surge with a fast rise and a slow decay. The amplitude of the surge relative to the pulse was given by K'. The rational approximation for this operation [23], [28] was

$$A_m(s) = \frac{(s+z_1)(s+z_2)}{(s+p_1)(s+p_2)} \quad (1.1)$$

in which $p_1 = (\pi/2)^2/T$ and $p_2 = (3\pi/2)^2/T$, and z_1 and z_2 were determined by the quadratic equation [28, appendix I]. The combined transform for the PON and the periglomerular population was

$$P_p(s) = P_c(s)A_m(s). \quad (1.2)$$

In expanded form using (0.2), (0.3), (1.1), and (1.2)

Table I Glomerular PST histograms – values for coefficients in (1.5) and dashed curves in Fig. 1

Set	I	N	P_1	β	γ	P_2	ω	α	ϕ	P_3	T
	μamp			1/sec	1/sec		rad/sec	1/sec	rad		msec
C17T1	75	1055	8.8	44	881	1.6	326	331	−2.44	−9.9	37
C17T3	90	500	97	60	816	32	289	361	−2.51	−96	36
C17T5	105	500	98	63	837	14	255	350	−2.65	−132	32
C17T7	130	500	58	70	904	15	228	325	−2.48	−47	26
C17T9	160	250	287	91	744	112	227	375	−2.95	−189	23
C17T11	200	120	436	97	738	34	210	368	−3.13	−167	18

Table II Glomerular PST histograms – values for coefficients in (1.5) and dashed curves in Fig. 2

	I	N	V_1	β	γ	V_2	ω	α	ϕ	V_3	T
	μamp		p/sec	1/sec	1/sec	p/sec	rad/sec	1/sec	rad	p/sec	msec
D5P1	25	200	84	40	927	27	318	337	−2.47	−64	46
D5P3	27	200	159	49	807	65	328	371	−2.74	−117	45
D5P5	29	200	205	57	845	85	245	381	−2.82	−130	41
D5P7	31	200	276	67	908	146	208	401	−2.78	−209	26
D5P9	33	200	944	79	899	265	218	402	−3.11	−844	25
D5P11	35	200	919	95	1110	186	201	412	−3.17	−889	17

Note: $N = 200$.

$$P_p(s) = \frac{K_p K_e a^2 b^2 (s+z_1)(s+z_2)}{(s+\gamma)(s+\beta)(s+\alpha+j\omega)(s+\alpha-j\omega)(s+p_1)(s+p_2)} \tag{1.3}$$

where, for simplicity $\tau = 0$. K_p is forward gain and incorporates input amplitude I. The inverse transform was

$$\begin{aligned} p_p(t) &= P_a \exp(-\beta t) + P_b \exp(-\gamma t) \\ &+ P_c [\sin(\omega t + \phi)\exp(-\alpha t)] \\ &+ P_d \exp(-p_1 t) + P_e \exp(-p_2 t). \end{aligned} \tag{1.4}$$

The values of the zeros $s = -z_1$ and $s = -z_2$, were not known, so the amplitude coefficients in (1.4) had to be treated as independent variables in fitting the specified curve to PST histograms by nonlinear regression. However, the 11 degrees of freedom were unmanageable. Therefore, the equation was broken into three identifiable parts or components, each with. zero amplitude at time zero. This was done by setting $P_b = -P_a - P_c \sin\phi$, $P_1 \triangleq P_a$, $P_2 \triangleq P_c$, $P_3 \triangleq P_d = -P_e$. Component P_1 represented the main shape of the transient, especially the falling phase. Component P_2 represented the shape of the rising phase as determined by the complex pair of roots. And component P_3 represented the effect of the input dispersion on both rising and falling phases.

Equation (1.4) was thereby written in a new form with nine degrees of freedom.

$$\begin{aligned} p_p(t) &= P_1 [\exp(-\beta t) - \exp(-\gamma t)] \\ &+ P_2 [\sin(\omega t + \phi)\exp(-\alpha t) - \sin(\phi)\exp(-\gamma t)] \\ &+ P_3 [\exp(-p_1 t) - \exp(-p_2 t)]. \end{aligned} \tag{1.5}$$

The curves generated from this equation are shown as dashed curves in Figs. 1 and 2. The values for the parameters are given in Tables I and II. The positions in the s plane of $s = -\beta$, $s = -\gamma$, $s = -\alpha + j\omega$, and $s = -\alpha - j\omega$, for Fig. 1 and Table I are shown in Fig. 3(a).

C. Evaluation of the Closed Loop Transfer Function $P_c(s)$ and $A_m(s)$

The desired result at this stage was the separation of the total delay in the system into two parts. One signified by $A_m(s)$ and by component P_3 represented the dispersion in the PON and terminal synapses. The other signified by $P_c(s)$ and by components P_1 and P_2 represented the delay properties of the glomerular neurons.

The residue for $P_c(s)$ from each set of PST histograms consisted of four roots, there being two real poles at $s = -\beta$ and $s = -\gamma$ and one pair of complex poles at $s = -\alpha - j\omega$ and $s = -\alpha + j\omega$. The six pairs of complex roots over the observed range of stimulus intensity were roughly aligned with the $j\omega$ axis [Fig. 3(a)] in what are described [26] as a pair of empirical root loci. The average of the values for α, the decay rate, were roughly equal to the mean for the real roots $s = -\beta$ and $s = -\gamma$. This entire configuration was compatible with the pattern of closed loop poles for a system having open loop double poles at approximate values of s = 200/s (for a) and s = -600/s (for b).

The complex roots lay on curving segments. This implied that synaptic delay $T \neq 0$ had to be included in the open loop transform. Equation (0.2) was confirmed as the basis for the open loop transfer function, with c = 2300/s. Several pairs of values for a and b were introduced, and for each pair a root locus plot was constructed by standard techniques [28], as shown in Fig. 3 (b). The values for a and b giving root loci for the complex poles most closely approximating the empirical root loci were a = 230/s and b = 550/s. With these values for a, b, and c (0.2) became

$$A(s) = \frac{K_e^{0.5} \cdot 2.9\times10^8 (2300 - s)}{(s+230)(s+550)(s+2300)}. \qquad (2.0)$$

From the feedback equation (0.3) the closed loop transfer function was

$$P_c(s) = \frac{K_c K_e 8.5\times10^{16} (2300 - s)^2}{(s+230)^2 (s+550)^2 (s+2300)^2 - K_e \cdot 8.5\times10^{16}} \qquad (2.1)$$

For each value of K_e there was a gain contour intersecting the real axis of the s plane to the right of s = -230/s. The value for $-s$ at this intersection corresponded to the rate constant β which was the decay rate of the PST histogram from its crest.

The relation between β and the logarithm of K_e was determined from the root locus plot in Fig. 3(b). Over the observed range of variation in β indicated by the gain contours, the relationship was almost linear, so that to a good approximation

$$K_e = \exp(-0.017\beta) \qquad (2.2)$$

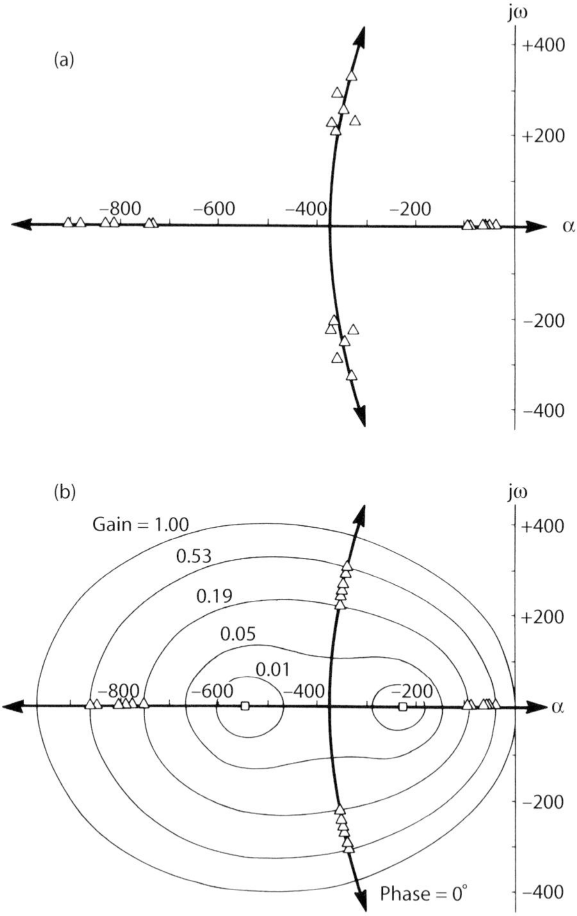

Fig. 3. Four sets of roots are plotted in the s plane (truncated at $s = 0$, $s = -1000/\text{s}$, and $s = \pm \text{j}400/\text{s}$). (a) "Empirical" root loci are indicated by the triangles representing the values obtained by fitting (1.5) to six PST histograms (Table I). The solid curves are root loci calculated from (2.3). (b) Root loci and gain contours for K_e are shown as calculated from (2.3). The triangles are the poles and the squares are the zeros introduced into (4.3). Not shown are two poles at $s = -2300/\text{s}$ and two zeros at $s = +2300/\text{s}$, which gives curvature to the complex root loci.

where 0.017 is an empirical constant in seconds. A value for K_e was determined by using β from each PST histogram (Tables I and II) in (2.2) and was introduced into (2.1). The denominator was expanded and factored into six roots as follows:

$$P_c(s) = \frac{K_c K_e 8.5 \times 10^{16} (2300 - s)^2}{(s + \alpha + \text{j}\omega)(s + \alpha - \text{j}\omega)(s + \beta)(s + \gamma)(s + \delta)(s + \varepsilon)} \quad \textbf{(2.3)}$$

The value for K' in (1.0) was estimated from the locations of the two zeros $s = -z_1$ and $s = -z_2$ in (1.1). The root locus for the zeros lay on the real axis between the two poles $s = -p_1$ and $s = -p_2$ and along a vertical line in the s plane midway between them. The negative values in Tables I and II for P_3 in comparison to the positive values for P_1 implied that the zeros did not lie between the real roots at $s = -\beta$ and $s = -p_1$ or between $s = -\gamma$ and $s = -p_2$. The relatively small values for P_2 (on the average $P_1/P_3 = 3.6$) in comparison to both P_1 and P_3 (on the average $P_1/P_3 = -1.3$) in Tables I and II implied that $s = -z_1$ and $s = -z_2$ lay closer to the complex roots at $s = -\alpha \pm j\omega$ than to the two real roots. The negative values for phase ϕ in Tables I and II implied that z_1 and z_2 were also complex roots, and that the positive imaginary part was less than ω. These conditions required that K' for glomerular cells lie between 1.5 and 1.8.

The axonal transfer function (1.0) was evaluated by setting $K' = 1.8$ and taking the value of T for each case from Tables I and II.

$$A_m(s) = 1 + 1.8\exp[-(sT)^{1/2}]. \tag{3.0}$$

In (1.1), the values of the two zeros $s = -z_1$ and $s = -z_2$ were determined by K' and T. This completed the evaluation of input dispersion.

D. Computation of the Closed Loop Solutions

The overall transfer function $P_q(s)$ was given by the product of $K_q P_c(s)$ and $A_m(s)$. From (1.1), (1.2), and (2.3)

$$P_q(s) = \frac{K_g K_e 8.5\times10^{16}(s+z_1)(s+z_2)(2300-s)^2}{(s+\alpha+j\omega)(s+\alpha-j\omega)(s+\beta)(s+\gamma)(s+\delta)(s+\varepsilon)(s+p_1)(s+p_2)} \tag{4.0}$$

where K_q is defined to include stimulus intensity I. The inverse transform of (4.0) was obtained by partial fraction expansion.

$$\begin{aligned} P_q(t) = K_q\,[&P_1\exp(-\beta t) + P_2\exp(-\gamma t) \\ &+P_3\exp(-\delta t) + P_4\exp(-\varepsilon t) \\ &+P_5\exp(-p_1 t) + P_6\exp(-p_2 t) \\ &+P_7\sin(\omega t + \phi)\exp(-\alpha t)\,. \end{aligned} \tag{4.1}$$

The coefficients P_1 through P_7 and ϕ were determined by the four fixed parameters a, b, c, and K' and by the two variables K_e and T, derived from Tables I and II. K_q was evaluated by fitting curves from (4.1) to the PST histograms. Examples are shown as the solid curves in Figs. 1 and 2 which have the slower rise time.

The output specified by (4.1) was for the feedback subset of the loop. The transfer function for the forward subset directly receiving PON input was

$$P_r(s) = K_r P_c(s) A_m(s) A^{-1}(s)\,. \tag{4.2}$$

In expanded form

$$P_r(s) = \frac{K_r K_e^{0.5}\, 2.9\times10^8 (s+z_1)(s+z_2)(s+230)(s+550)(s+2300)(2300-s)^2}{(s+\alpha+j\omega)(s+\alpha-j\omega)(s+\beta)(s+\gamma)(s+\delta)(s+\varepsilon)(s+p_1)(s+p_2)}\,. \tag{4.3}$$

Table III Closed loop coefficients for PST histograms for (4.1)[a]

Case	K_r p/sec forward limb	or	K_q p/sec feedback limb	K_e feedback gain	T msec PON delay
C17T1	.24		.23	.47	37.
C17T3	.47		.63	.35	36.
C17T5	1.49		1.97	.34	32.
C17T7	1.57		2.22	.30	26.
C17T9	2.05		3.32	.21	23.
C17T11	4.99		3.96	.19	18.
D5P1	.20		.32	.50	46.
D5P3	.35		.38	.43	45.
D5P5	.51		.75	.37	41.
D5P7	.80		1.15	.31	26.
D5P9	1.39		2.12	.26	25.
D5P11	1.67		3.72	.19	18.

[a]Fixed coefficients: $a = 230/s$; $b = 550/s$; $c = 2300/s$; $K' = 1.8$

The inverse transform was identical to that given by (4.1), but the values for P_1 through P_7 and ϕ were different, and the forward gain coefficient K_r had to be specified independently (Table III).

The waveforms for $p_r(t)$ are shown in Figs. 1 and 2 also as solid curves. They have the faster rise time and sharper crest. The conformance of the solid curves from (4.3) to the data was taken as evidence that the positions of the closed loop zeros at $s = -z_1$ and $s = -z_2$ were substantially correct. Therefore, the values for the open loop poles at $s = -p_1$, $s = -p_2$, and the value for dispersion gain K' were considered to be valid. The conformance of PST histograms to the inverse transforms of either (4.1) or (4.3) was taken as evidence that the open loop transfer function for A(s) in (0.2) was correctly evaluated. This completed the model for PON afferent dispersion cascaded in series with glomerular population excitatory interaction.

IV. Evaluation and Interpretation

A. Deviations of Predicted Curves from Observed PST Histograms

At low stimulus intensity some glomerular units have short-latency increases in pulse probability (Fig. 2) and others have longer latency (Fig. 1). Either is compatible with the model, the short-latency units being members of the feedforward subset and the longer latency units being members of the feedback subset. At higher stimulus intensities, almost all become members of the feedforward subset, owing to the increased probability that any given neuron will receive PON input at high intensity.

The crests of the PST histograms are typically serrated, due to the tendency of single neurons to fire repetitively at regular intervals. Such intervals are not equal across neurons, so the population response must be smoother. The goodness-of-fit of the curves to the data could be improved by allowing K' to take lower values at higher stimulus intensities, implying that the input then conforms more closely to an impulse. However, this elaboration does not seem worthwhile at this time.

The falling phase of the PST histograms (at about 40 ms) often shows an excess in observed over predicted pulse probability (cf. C17T5, Fig. 1). This results from the tendency of the neurons to fire in short bursts at relatively constant burst duration [14], [15]. The later spikes of such trains, especially at higher stimulus intensity, tend to become smaller in amplitude. Because of the reduced effectiveness (transmitter release/impulse) of smaller spikes arising from a depolarized baseline [29], [30] it seems likely that the excess numbers of spikes at this time in the PST histograms is not a significant deviation from the predicted effectiveness of the output of the neuron.

B. Interpretation of the Coefficients

The derived values for glomerular population open loop rate constants are similar to those obtained for other populations (Table IV). This implies that despite large differences in cytogeometry and transmitter chemistry, the passive membrane rate constants (a), cable delays (b), and synaptic delays (c) are not greatly different among these populations.

The dispersion time for PON input to mitral-tufted cells was previously found from measurements on AEP's to average $T = 32$ ms, in both open loop [23] and closed loop [19] states. Because the average dispersion time for the two glomerular units is the same $T = 31$ ms, it is reasonable to infer that both populations receive substantially the same input.

The AEP measurements were made at fixed input intensity near the middle of the range used here. The decrease in T with increasing intensity requires explanation. Voronkov and Gusel'nikova [31] have described a form of presynaptic inhibition [4] in the glomeruli of the bulb. Careful measurement of the magnitude and the time of onset of this partial block of synaptic transmission [31] have shown that it peaks between 10 and 30 ms after the initiating stimulus. The magnitude of the partial

Table IV Comparison of open loop rate constants[a]

Neural mass	a 1/sec	b 1/sec
Glomerular	230	550
Mitral-tufted	219 ± 11	612 ± 27
Granule	219 ± 11	796 ± 38
Cortical:		
superficial pyramidal granule	223 ± 18	716 ± 89

[a]Mean ± S.E.

block to a test input increases with increasing stimulus intensity to over 80 percent at high intensity. These data imply that the leading edge of a dispersed afferent volley initiates a partial block of the trailing edge of the volley, in proportion to input magnitude. The effect is to reduce the degree of effective temporal dispersion of the input. Therefore, T must decrease with increasing stimulus intensity.

The values for feedback gain K_e representing the strength of mutual action within the population, lay within the range between 0 and +1, implying stability of the population over the observed range [32]. The values for forward gain K_q and K_r increased more rapidly than in linear proportion to input intensity. This was also found to be true for the amplitude of the PON compound action potential [21]. The positive values for K_e, K_q, and K_r imply that the positive feedback is excitatory and not inhibitory. That is, glomerular neurons must be mutually excitatory in the main, and not mutually inhibitory.

C. Consideration of the Results of Others

The conclusion that glomerular neurons are predominantly mutually excitatory is at odds with previously published inferences.

Pinching and Powell [5] observed in electron micrographs that periglomerular and glomerular short-axon cells have flattened synaptic vesicles and asymmetric synaptic membrane thickening, whereas mitral-tufted cells have spherical vesicles and symmetric membrane thickenings. In reviews by Uchizono [34] and others, flattened vesicles have been identified with inhibition and spherical vesicles with excitation. However, they noted that there are many exceptions to the rule and that it "cannot be regarded as proven" [5, p. 391].

In electrophysiological studies Shepherd [15] has shown by paired shock excitation of the PON that test responses of glomerular neurons may be facilitated or defacilitated by preceding conditioning responses. The paired shocks were set at a threshold intensity, such that firing occurred in response to the conditioning (first) stimulus on 50 percent of trials. Units recorded deep in the glomerular layer where they were most common were reported to undergo prolonged inhibition to the test (second) shock, provided they had fired in response to the conditioning input, but otherwise to fire repeatedly. This apparent block was attributed to inhibition by the glomerular cell of that mitral-tufted cell, through which the glomerular cell was inferred to have received its input.[1] At high intensity of PON paired shock stimulation the test responses of glomerular units were "generally suppressed." In a recent review, Shepherd concluded that periglomerular neurons might "provide for

1 The topology of connections was described by White [8], who distinguished two types of dendrite in the glomeruli: those receiving synapses from PON axons (Stage 1) and those forming synapses with Stage I and other dendrites but not PON axons (Stage II). He proposed that Stage I were mitral-tufted dendrites and Stage II were periglomerular dendrites. He stated that "it was not possible in this study to identify the origins of dendrites. However, a tentative identification based on comparison of synaptic arrangements in the glomerulus with those of the external plexiform layer (as described by Reese and Shepherd [10]) is possible" [8, p. 78].

inhibitory feedback onto the principle relay neurons" and "for additional inhibition of neighboring glomeruli" [16, p. 881].

An alternative interpretation of his data is that, because he set his stimulus strength at the level where each observed unit fired with short latency on 50 percent of trials, then on half the trials the stimulated neuron received direct input from the PON and lay in the feedforward subset with short response latency, and on the remaining trials it lay in the feedback subset with longer latency. The failure to fire subsequently to the first spike in the former case can be attributed to augmented presynaptic partial block at the appropriate time in the local region.

This is because if the cell fired early, then the local input on that trial had to be at greater intensity, and so also did the level of partial block. Hence his data can be construed to support the concept of mutual excitation among glomerular neurons.

The present results require that if glomerular and mitral-tufted cells interact directly, as their anatomical synaptic relations suggest they do, the interaction is also mutually excitatory. An alternative is that the positive (excitatory) feedback occurred through direct axodendritic connections, whereas (inhibitory) dendrodendritic connections of glomerular to mitral-tufted cells were nonfunctional in the anesthetized animal. Another alternative is that the properties of the glomerular neurons as shown in Figs. 1 and 2 are the open loop properties, and that none of their output synaptic connections were functional. Apart from the difficulty of explaining such massive selective block, both the alternatives deny the possibility of inhibitory action in the same conditions where inhibitory action has been purported to occur [15], [16].

Acknowledgments

Reprinted with permission by the Institute of Electrical and Electronics Engineers, Inc.

First Published in: *IEEE Transactions On Biomedical Engineering*, Vol. **BME-21**, No. 5, September 1974, pp. 350–358

This work was supported by the National Institute of Mental Health Grant MH 06686.

Computations for this paper were done at the Berkeley Computer Center.

References

[1] K. H. Andres, "Der Feinbau des Bulbus Olfactorius der Ratte unter besonderer Berficksichtigung der synaptischen Verbindungen," *Z. Zellforsch. Mikrosk. Anat.*, vol. **65**, pp. 530–561, 1965.

[2] J. W. Hinds, "Reciprocal and serial dendrodendritic synapses in the glomerular layer of the rat olfactory bulb," *Brain Res.*, vol. **17**, pp. 530–534, 1970.
[3] A. J. Pinching and T. P. S. Powell, "The neuron types of the glomerular layer of the olfactory bulb," *J. Cell Sci.*, vol. **9**, pp. 305–345,1971.
[4] A. J. Pinching and T. P. S. Powell, "The neuropil of the glomeruli of the olfactory bulb," *J. Cell Sci.*, vol. **9**, pp. 347–377, 1971.
[5] A. J. Pinching and T. P. S. Powell, "The neuropil of the periglomerular region of the olfactory bulb," *J. Cell Sci.*, vol. **9**, pp. 379–409, 1971.
[6] J. L. Price and T. P. S. Powell, "The synaptology of the granule cells of the olfactory bulb," *J. Cell Sci.*, vol. **7**, pp. 125–155, 1970.
[7] W. Rall, G. M. Shepherd, T. S. Reese, and M. W. Brightman, "Dendrodendritic synaptic pathway for inhibition in the olfactory bulb," *Exp. Neurol.*, vol. **14**, pp. 44–56, 1966.
[8] E. L. White, "Synaptic organization in the olfactory glomerulus of the mouse," *Brain Res.*, vol. **37**, pp. 69–80, 1972.
[9] F. Valverde, *Studies on the Pyriform Lobe*. Cambridge, Mass.: Harvard Univ. Press, 1965.
[10] T. S. Reese and G. M. Shepherd, "Dendrodendritic synapses in the central nervous system," in *Structure and Function of Synapses*, G. D. Pappas and D. P. Purpura, Ed. New York: Raven, 1972, pp. 121–136.
[11] A. V. Minor, "Integral evoked potentials and activity of single neurons in the frog olfactory bulb" (in Russian), *Neurophysiologia*, vol. **1**, pp. 269–278, 1969.
[12] F. Orrego, "The reptilian forebrain, II–Electrical activity in olfactory bulb," *Arch. Ital. Biol.*, vol. **99**, pp. 446–465, 1961.
[13] W. Rall and G. M. Shepherd, "Theoretical reconstruction of field potentials and dendrodendritic synaptic interactions in olfactory bulb" *J. Neurophysiol.*, vol. **31**, pp. 884–915, 1968.
[14] G. M. Shepherd, "Neuronal systems controlling mitral cell excitability," *J. Physiol.*, vol. **168**, pp. 101–117, 1963.
[15] G. M. Shepherd, "Physiological evidence for dendrodendritic synaptic interactions in the rabbit's olfactory glomerulus," *Brain Res.*, vol. **32**, pp. 212–217, 1971.
[16] G. M. Shepherd, "Synaptic organization of the mammalian olfactory bulb," *Physiol. Rev.*, vol. **52**, pp. 864–917, 1972.
[17] R. von Baumgarten, J. D. Green, and M. Mancial "Slow waves in the olfactory bulb and their relation to unitary discharges," *Electroencephalogr. Clin. Neurophysiol.*, vol. **14**, pp. 621–634, 1962.
[18] C. Yamamoto, T. Yamamoto, and K. Iwama, "The inhibitory systems in the olfactory bulb studied by intracellular recording," *J. Neurophysiol.*, vol. **26**, pp. 403–415, 1963.
[19] W. J. Freeman, "Measurement of oscillatory responses to electrical stimulation in olfactory bulb of cat," *J. Neurophysiol.*, vol. **35**, pp. 762–779, 1972.
[20] G. L. Gerstein and N. Y.-S. Kiang, "An approach to the quantitative analysis of electrophysiological data from single neurons," *Biophys. J.*, vol. **1**, pp. 15–28, 1960.
[21] W. J. Freeman, "Spatial divergence and temporal dispersion in primary olfactory nerve of cat," *J. Neurophysiol.*, vol. **35**, pp. 733–744, 1972.
[22] W. J. Freeman, "Depth recording of averaged evoked potential of olfactory bulb," *J. Neurophysiol.*, vol. **35**, pp. 780–796, 1972.
[23] W. J. Freeman, "Measurement of open loop responses to electrical stimulation in olfactory bulb of cat," *J. Neurophysiol.*, vol. **35**, pp. 745–761, 1972.
[24] W. J. Freeman, "A linear distributed feedback model for prepyriform cortex," *Exp. Neurol.*, vol. **10**, pp. 525–547, 1964.
[25] W. J. Freeman, "Relations between unit activity and evoked potentials in prepyriform cortex of cats," *J. Neurophysiol.*, vol. **31**, pp. 337–348, 1968.
[26] W. J. Freeman, "Linear analysis of the dynamics of neural masses," *Annu. Rev. Biophys. Bioeng.*, vol. **1**, pp. 225–256, 1972.
[27] W. J. Freeman, "Waves, pulses and the theory of neural masses," *Progr. Theoret. Biol.*, vol. **2**, pp. 87–165, 1972.
[28] O. J. M. Smith, *Feedback Control System*. New York: McGraw-Hill, 1958.
[29] J. C. Eccles, *The Physiology of Synapses*. Berlin: Springer, 1964.
[30] B. Katz, "The transmission of impulses from nerve to muscle, and the subcellular unit of synaptic action," *Proc. Roy. Soc.*, vol. **155B**, pp. 455–479, 1962.
[31] G. S. Voronkov and K. G. Gusel'nikova, "Inhibition in glomeruli of the olfactory bulb," *Neurosci. Transl.*, vol. **4**, pp. 425–426, 1968.
[32] W. J. Freeman, "Attenuation of transmission through glomeruli of olfactory bulb on paired shock stimulation," *Brain Res.*, vol. **74**, pp. 77–90, 1974.

[33] W. J. Freeman, "Stability characteristics of positive feedback in a neural population," *IEEE Transactions On Biomedical Engineering*, Vol. **BME-21**, No. 5, September 1974, pp. 358–364.
[34] K. Uchizono, "Synaptic organization of the Purkinje cells in the cerebellum of the cat," *Exp. Brain Res.*, vol. 4, pp. 97–113, 1967.

Stability Characteristics of Positive Feedback in a Neural Population

Abstract

Glomerular neurons were driven by single shock stimulation of the primary olfactory nerve (PON). The poststimulus time (PST) histograms showed a rapid increase in pulse probability to a crest, followed by an exponential decay to background level P_0. The crest magnitude and decay rate both increased with increasing stimulus intensity. The decay rate was extrapolated to zero at threshold. This implies that loop gain is unity at threshold and below threshold.

Based in part on this finding, an equation is derived to express the dependence of loop gain K_E on the instantaneous value of amplitude, which is given by the normalized value of induced pulse rate $(P - P_0)/P_0$ in PST histograms. The mean value of loop gain $\overline{K}_E$ is dependent on the normalized value of the mean-induced pulse rate $(\overline{P} - P_0)/P_0$. The values for $\overline{K}_E$ at different levels of stimulus intensity are shown to approximate the estimates of loop gain K_e previously derived from the decay rates of the PST histograms by means of a positive feedback model.

It is concluded that the neural population with positive (excitatory) feedback is self-stabilizing without inhibitory input, and its steadystate output provides a depolarizing bias, which maintains the neural negative feedback loops in the olfactory bulb and cortex in a linear dynamic range.

I. Introduction

In [1] the performance characteristics were described for a population of neurons in the mammalian olfactory bulb in response to impulse activation. The impulse responses were nonoscillatory and initially excitatory, which indicated that the population consisted of mutually interconnected excitatory neurons.

A model was developed to describe the dynamics of this population in terms of a positive feedback loop. It was assumed that the population of neurons could be divided into two mutually exclusive subsets. The forward subset received the controlled input, the impulse, and the feedback subset did not. The neurons in the two subsets were assumed to be uniformly distributed over the population, so that interaction strength did not depend on location. The dependence of conduction delay on distance between subsets was assumed to be negligible. The impulse input was assumed to be uniform over the region of interest, so a lumped circuit approximation was used. Each subset was renewed from the other subset.

The time-dependent output for each subset was assumed to conform to a linear differential equation with fixed coefficients for the duration of each response.

The transfer function for the system under study had two parts. The first part represented the afferent dispersive delay in the input nerve carrying the impulse. The second part was the feedback loop in the population. The coefficients in the two parts were of two kinds: first-order rate coefficients and dimensionless gain coefficients representing the interaction strengths between subsets. The study in [1] was concerned mainly with the determination of the rate coefficients. This paper is concerned with further evaluation of the feedback gain coefficient K_e. The dependence of K_e on loop output is shown, and the resulting stability properties of the population are discussed.

II. Relation of Feedback Gain to Output Amplitude

A. Physiological Observations

Unit potentials identified as coming from neurons belonging to the glomerular neuron population were characteristically active in the absence of afferent electrical stimulation. They tended to occur in short bursts lasting 30–50 ms, at burst rates up to 250/s, although single spikes were more common. Mean frequencies varied between cells from less than 1/s to over 80/s. The average for 54 cells was 9.2/s. Whether or not bursts occurred, the mean firing probability was higher during each inspiration.

Interval histograms of the background activity were constructed from records of unit trains 2 to 4 min in length. An example from a glomerular unit (Fig. 1, right) shows that there was a minimum interval of 2 to 4 ms within which no pulses occurred after each pulse. The maximum likelihood of firing occurred (for this unit) at an interval of 25 ms. Thereafter the likelihood of occurrence of an interval diminished exponentially with the length of interval. The expectation density calculated for the same pulse train (Fig. 1, left) shows a refractory period followed by a relatively high likelihood of firing from 25 to 75 ms (reflecting the presence of short bursts), and thereafter decreases to a steady level in likelihood of firing. Such results showed that pulse trains from single glomerular units (and, in this respect, deeper lying units as well) superficially resembled the output of a Poisson process with a refractory period or dead time.

On electrical stimulation of the primary olfactory nerve (PON) there was an increase in poststimulus probability of firing, which began with a latency of a few

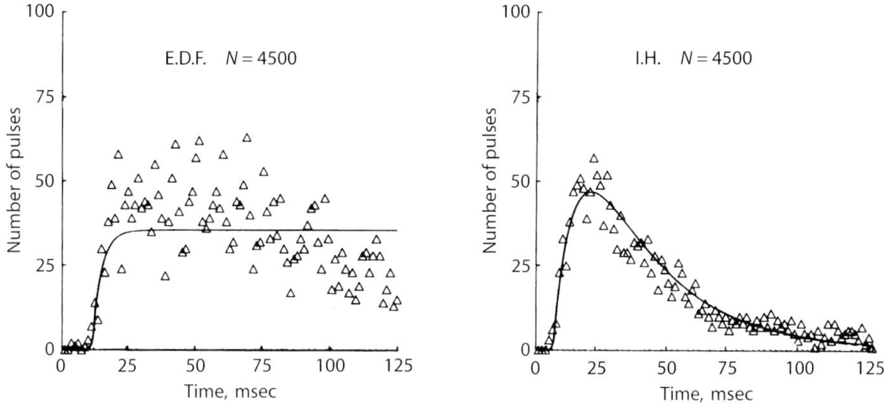

Fig. 1. The statistical properties of the "spontaneous" or background pulses of a single "glomerular" unit are shown by the expectation density (left) and the interval histogram (right). The number of pulses occurring in each time interval (milliseconds) after the occurrence of a preceding pulse are shown by the triangular symbols. The expectation density shows the absolute and relative refractory periods, and the tendency to fire in bursts. The interval histogram shows the exponential distribution of firing intervals. The curves were generated by empirical equations. For the expectation density $p(t) = 35\{1 - \exp[-(t - 8.8)/2.7]\}$, $t > 8.8$, where $p(t)$ is in pulses/unit lag time and t is in milliseconds. For the interval histogram $p(t) = 111\{\exp[-(t - 4.8)/22] - \exp[-(t - 3.8)/6.5]\}$, $t > 4.8$. The constants have no theoretical significance, but they help to evaluate empirically the length of the refractory periods.

seconds, depending on the distance between the stimulus and recording sites and on the PON conduction velocity. The pulse probability then rose rapidly to a crest and decayed monotonically to the background level of activity. With increasing stimulus intensity the rate of rise decreased and crest amplitude of pulse probability increased. The rate of return of pulse probability to baseline values also increased. These properties have already been described [1].

B. Dependence of Coefficients on Stimulus Intensity

The model consisted of two stages in series. The first part $A_m(s)$ represented temporal dispersion in the PON [1, eq. (3.0)].

$$A_m(s) = 1 + 1.8\exp[-(sT)^{1/2}]. \tag{0.1}$$

The forward gain coefficient 1.8 represented the relative importance of dispersion. The delay coefficient T in milliseconds represented the distributed delays in the PON by a lumped parameter. This decreased with increased stimulus intensity [1, table III] from 46 ms to 18 ms. A physiological explanation has been given.

The second part $P_c(s)$ represented the mutual excitation of glomerular neurons.

$$P_c(s) = \frac{A^2(s)}{1 - A^2(s)} \tag{0.2}$$

where

$$A(s) = \frac{K_e^{0.5}\, 2.9 \times 10^8 (2300 - s)}{(s+230)(s+550)(s+2300)} \,. \qquad \textbf{(0.3)}$$

The three rate constants [1, eq. (2.0)] *a*, *b*, and *c*, represented respectively, passive membrane, cable, and synaptic delays of neurons in the forward and feedback subsets of the population. The transfer function for PON electrical stimulation and for output from the feedback set was

$$P_q(s) = K_q P_c(s) A_m(s) \,, \qquad \textbf{(0.4)}$$

and for output from the forward set it was

$$P_r(s) = K_r P_c(s) A_m(s) A^{-1}(s) \,. \qquad \textbf{(0.5)}$$

The forward gain coefficients for the forward subset K_r and feedback subset K_q increased with increasing stimulus intensity [1, table III]. The feedback gain K_e decreased with increasing stimulus intensity [1, table III].

The solution to (0.4) and (0.5) for each value of K_e yielded a set of closed loop roots. The root lying on the real axis of the *s* plane closest to the origin β represented the decay rate of the PST histograms. The relation between K_e and β was approximated by

$$\log_{10}(K_e) = -0.0075\beta \qquad \textbf{(0.6)}$$

over the range of observed values for β, indicated by the arrows in Fig. 2.

The observed values for β were verified by fitting the PST histograms with curves generated by the inverse transforms of (0.4) and (0.5). β increased with increasing

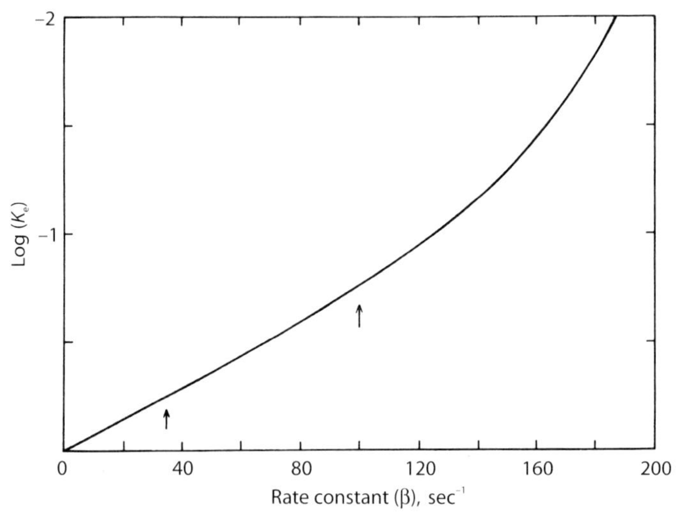

Fig. 2. The relation between the logarithm (base 10) of K_e and the rate constant β is shown as derived from [1, fig. 3(b)]. It is approximated by (0.6). At $\beta = 0$, $K_e = 1$.

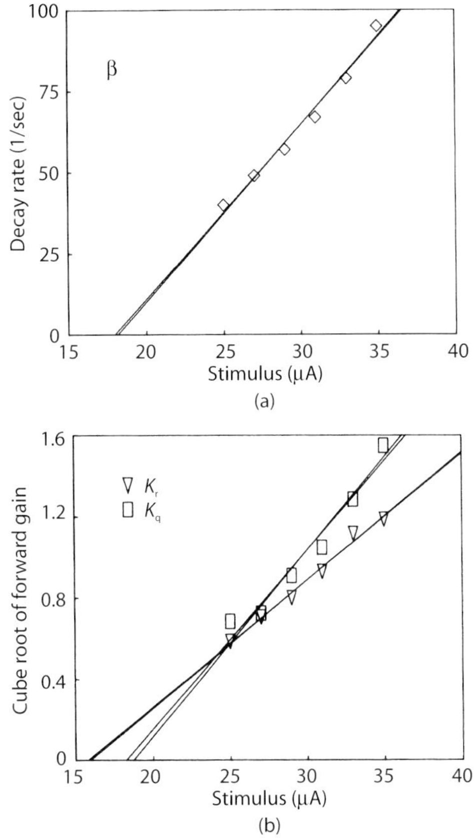

Fig. 3. Both the decay rate β (top) and the cube roots of the amplitudes K_q and K_r (bottom), of the PST histograms of set (*D5P*) decreased with decreasing stimulus intensity. The extrapolated threshold range is shown by the intersections of the regression lines with the abscissa. (For values see [1, fig. 2, tables II, III].)

stimulus intensity (Fig. 3, top) . In the other direction β was extrapolated to zero near threshold for the impulse response.

The response threshold could not be determined from the set of measurements of crest amplitude because the rate of rise to the crest and the decay rate were also dependent on stimulus intensity. An alternative approach was based on a previous empirical finding [2] that the amplitude of the PON compound action potential increased with the cube of stimulus current (Fig. 4) over the same stimulus range used for the PST histograms [1]. The cube root was taken for each value of K_r and K_q, which were defined in [1] to include a constant I representing input intensity. The results are shown in the bottom portion of Fig. 3 in which the cube roots of forward gain are plotted against stimulus intensity. Linear regression lines were plotted for both sets of gain coefficient against stimulus intensity. The intercepts of the four regression lines on the abscissa were used to define a range, in which probable value for threshold current lay.

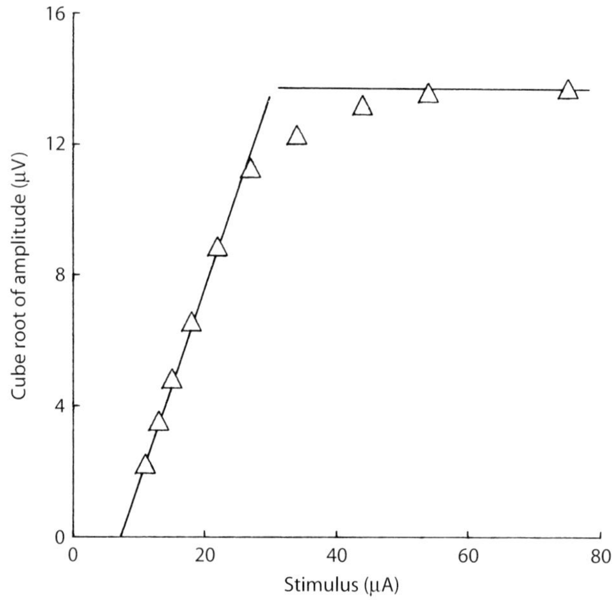

Fig. 4. The peak-to-peak amplitude of the PON compound action potential [2] increased in proportion to the cube of stimulus current, over the range of stimulus intensity from 1 to 4× threshold.

Regression lines were also computed for β against stimulus current. The segment of the abscissa between the intercepts of the two regression lines lay within the threshold range. It was concluded that within the limits of experimental error, the value for β could be extrapolated to zero at the threshold for the response.

The data in Fig. 3 implied by (0.6) that K_e decreased with decreasing input intensity to $K_e = 1$ at threshold. This implied that for zero-stimulus input on the PON, each glomerular unit transmitted as many pulses as it received on the average in time. It implied further that if a means were found to introduce an inhibitory pulse into the population, K_e would increase to some value above 1.0, and the activity in the population would return to its prestimulus level. This implication could not be tested directly, because the required input was not yet found. However, "spontaneous" activity of glomerular units did return during recovery from silence under deep anesthesia, and it was maintained after surgical isolation of the bulb. Positive (excitatory) feedback can explain the background activity in this population. A similar conclusion has been reached by Anninos *et al.* [3].

C. Characterization of Neuronal Forward Gain

The loop gain K_e has been defined in the present analysis as a fixed coefficient holding over the duration of observation of an event. The results show that K_e varies in relation to input intensity and therefore to response amplitude. Because the response amplitude varies during the course of the response as well as in relation to input intensity, one must allow the possibility that loop gain, if defined in a different way, should vary during an event, in which instantaneous amplitude varies with time. Then K_e should conform to the mean value $\overline{K}_E$ taken by the redefined loop gain K_E over the measured period of the PST histogram.

An alternative definition for K_E introduced elsewhere [4] is based on the concept that each subset of neurons is a two-stage transducer. An incoming pulse density P_d is converted to waves of dendritic current V in the dendrites, and the sum of dendritic current is converted to an outgoing pulse density P_q at trigger zones. For each stage the conversion is nonlinear but can be described by piecewise linear approximation.

Normally a subset of neurons is in some background steady state of activity, in which its mean wave activity is at some level taken arbitrarily as zero, $V = 0$, and its mean pulse density (total pulses/unit time) input is taken as $P_d = P_0$. Because by assumption the forward and feedback subsets are identical except with respect to the test impulse, the output of each subset is equal to the input to the other subset. Therefore, at rest ($V = 0$), $P_a = P_0$. The wave activity V is not the same as transmembrane potential. It is the mean level of dendritic activity, which can be measured extracellularly only in some conditions [5]. V is dependent on input pulse density P_d and it determines output pulse density P_q. The amplitude dependencies of P_d and P_q on V are now derived.

For each increment of excitatory input P_d the level of V increases by an increment in proportion to the input by a constant z_e times the difference between the prestimulus level, which is treated as zero, and the level V_e at which transmembrane potential equals the equilibrium potential of the excitatory synaptic mechanism [6]. Thus

$$\frac{dV}{dP_d} = \zeta_e (V_e - V), \quad V \geq 0. \tag{1.0}$$

If the subset is assumed to be excitatory and to receive only excitatory input, inhibitory actions need not be considered. For the range of function to be dealt with V_e is much greater than any value of V, so (1.0) is simplified as follows:

$$\frac{dV}{dP_a} = \zeta_e V_e , \quad V \geq 0. \tag{1.1}$$

For increasing dendritic excitatory current in a subset of neurons, there is increasing probability of pulse occurrence per unit time for each neuron and increasing pulse density of output for the subset.

For each increment in wave input dV there is an increment in pulse density output dP_a. which depends on rate constants $\gamma_e > 0$ for excitation and $\gamma_i > 0$ for inhibition, and the prevailing pulse density output.

$$\frac{dP_a}{dV} = \gamma_i P_a , \quad V < 0 \tag{1.2}$$

$$\frac{dP_a}{dV} = \gamma_e (P_a - P_m), \quad V > 0 \tag{1.3}$$

where P_m is a limiting value for pulse density of the subset.[2]

The solutions to the differential equations are

$$P_a = P_0 \exp(\gamma_i V), \quad V < 0 \tag{1.4}$$

$$P_a = P_m - (P_m - P_0)\exp(-\gamma_e V), \quad V > 0 \tag{1.5}$$

$$P_a = P_0, \quad V = 0 \tag{1.6}$$

The value for P_m can be determined by the condition that P has a continuous derivative with respect to V at $V = 0$

$$\frac{dP_a}{dV} = \gamma_i P_0 \exp(\gamma_i V), \quad V \le 0 \tag{1.7}$$

$$\frac{dP_a}{dV} = \gamma_e (P_m - P_0)\exp(-\gamma_e V), \quad V \ge 0 \tag{1.8}$$

At $V = 0$ (1.7) and (1.8) yield the result

$$P_m = P_0 (\gamma_i / \gamma_e + 1). \tag{1.9}$$

Measurements on other bulbar and cortical neurons have shown [4] that $\gamma_i / \gamma_e = 2.0$. Equation (1.8) is evaluated as

$$P_a = P_0 [3 - 2\exp(-\gamma_e V)], \quad V \ge 0. \tag{1.10}$$

The derivative of P_a in (1.10) is

$$\frac{dP_a}{dV} = 2\gamma_e P_0 \exp(-\gamma_e V), \quad V \ge 0. \tag{1.11}$$

Because there is no demonstrated inhibitory event under consideration, the condition in which $V < 0$ need not be considered.

---

2 The limiting value P_m is not the same as the maximum pulse rate to which any one neuron can be driven for shorter or longer periods of time. In accordance with the ergodic hypothesis, it is the average long-term maximal pulse rate, including recovery periods from high-frequency bursts, for any one neuron, or for the subset. To the extent that strongly synchronous firing develops, the ergodic hypothesis does not hold, as shown by Anninos *et al.* [3]. In the conditions holding for the observation reported here, where stimulus intensity did not exceed 3× threshold, synchrony was at a low level, except during the rising phase of the PST histograms, when, as noted earlier [1], the computed curves failed to match the observed responses.

The glomerular neurons are interconnected by axodendritic synapses, so that the postulated conversions between waves and pulses can be observed by recording pulses on the axons. The periglomerular neurons are reciprocally connected with mitral-tufted cells by intraglomerular dendrodendritic synapses [7]. It is clear that transmission across large numbers of such synapses in parallel is graded, but it is not known whether the individual synapses transmit according to an all-or-none law, which cumulatively would cause the whole set of synapses to have the dynamics specified for axodendritic transmission.

D. Characterization of Instantaneous Feedback Gain

The forward gain $K_E^{0.5}$, of such a subset can be defined as the product of these two ratios because each specifies the magnitude of output for a given input.

$$K_E^{0.5} = \frac{dV}{dP_d}\frac{dP_a}{dV}, \quad V \geq 0. \tag{2.0}$$

The loop gain between two subsets is the square of $K_e^{0.5}$. From (1.1), (1.11), and (2.0)

$$K_E = (2\gamma_e P_0 \zeta_e V_e)^2 \exp(-2\gamma_e V), \quad V \geq 0. \tag{2.1}$$

From the results summarized in Fig. 3, when $V = 0$, $K_E = 1$. Therefore,

$$(2\gamma_e P_0 \zeta_e V_e)^2 = 1 \tag{2.2}$$

and

$$K_E = \exp(-\gamma_e V), \quad V \geq= 0. \tag{2.3}$$

The value for V in (2.3) is not accessible to measurement. Upon integration (1.1) becomes

$$V = \zeta_e V_e (P_d - P_0), \quad V \geq 0. \tag{2.4}$$

It is permissible to identify P_0 in (2.4) with P_0 in (2.1), because the output for each subset in the steady state is equal to the input for the other subset. On substitution of (2.4) into (2.3), and multiplying the exponent by P_0/P_0, the result is

$$K_E = \exp[-2\gamma_e P_0 \zeta_e V_e (P_d - P_0) / P_0], \quad V \geq 0. \tag{2.5}$$

The PST histogram is measured from P_a, the output of a single cell or a small number of cells. From the uniformity assumption, it also represents the output of other cells in the same subset. Each cell in the receiving subset is assumed to receive from each other cell not refractory and therefore in the transmitting subset, but weighted according to the density of connection, which must vary from zero up to some maximum. The weighting must affect both P_a and P_0, so that normalized output $(P_a - P_0)/P_0$ can be summed over each subset, without regard to weighing. The normalized output for one subset must be equal to normalized input for the second subset $(P_d - P_0)/P_0$.

The difference between P_a and P_d becomes important when the two subsets are distinguished by which receives the input from the PON. Their outputs will differ only in latency and rise time. For simplicity the PST histograms can be treated for present purposes as corresponding to P_a, P_d, or their mean, P. The subscript is therefore dropped. From (2.2) and (2.5)

$$K_E = \exp[-(P - P_0) / P_0], \quad V \geq 0. \tag{2.6}$$

Thus loop gain can be estimated directly from the normalized pulse density as determined from the PST histograms.

Table 1 Estimation of gain K_E from pulse densities

Case	P_0 background p/sec	P_C crest p/sec	$\overline{P}$ mean p/sec	$\overline{K}_E$ average feedback gain
C17T1	9.4	11.9	14.5	.58
C17T3	9.6	26.1	18.5	.39
C17T5	11.4	42.6	31.5	.17
C17T7	13.9	85.1	41.8	.13
C17T9	12.1	119.6	42.6	.08
C17T11	12.9	173.9	45.4	.08
D5P1	13.2	35.2	22.9	.48
D5P3	14.4	39.6	23.2	.54
D5P5	12.7	59.9	25.0	.38
D5P7	12.4	71.5	29.5	.25
D5P9	13.4	92.6	37.3	.17
D5P11	15.3	108.6	41.7	.18

In order to estimate a mean value for $\overline{K}_E$ comparable to K_e the value for P in (2.6) is set equal to the mean pulse rate $\overline{P}$ across the entire duration of each PST histogram beginning from the foot of the response. The value for P_0 is taken from the prestimulus baseline segment in set ($D5P$), and from measurements on background rate between stimulus trains for constructing PST histograms in set ($C17T$). The values are shown in Table I for ($D5P$). For both sets of PST histograms the results are summarized in Fig. 5. The values for $\overline{K}_E$ are plotted on the ordinate; those for K_e are shown on the abscissa. The line at 45° represents the predicted correlation for the 12 pairs.

From the values shown in Table I for the crest amplitudes P_c, of the PST histograms, it is apparent that at the crest of the larger responses K_E approaches zero, implying that when actions reach a high level, interactions within the population reach a very low level. Presumably most cells are already strongly depolarized and are firing frequently, but their refractory periods do not permit them to respond to additional input.

The convergence of the results of these two approaches to estimation of feedback gain shows that it is permissible to define the gain as a fixed parameter over the duration of a response to impulse input. It implies further that if the glomerular neurons are mutually excitatory, they have amplitude-dependent input-output and gain characteristics, which closely resemble those of other bulbar and cortical populations [4].

III. Implications Concerning Physiological Function

A. Stability and Background Activity

If the glomerular units are generated by mutually excitatory cells with excitatory input from the PON and no input from within the bulb, must they not have among

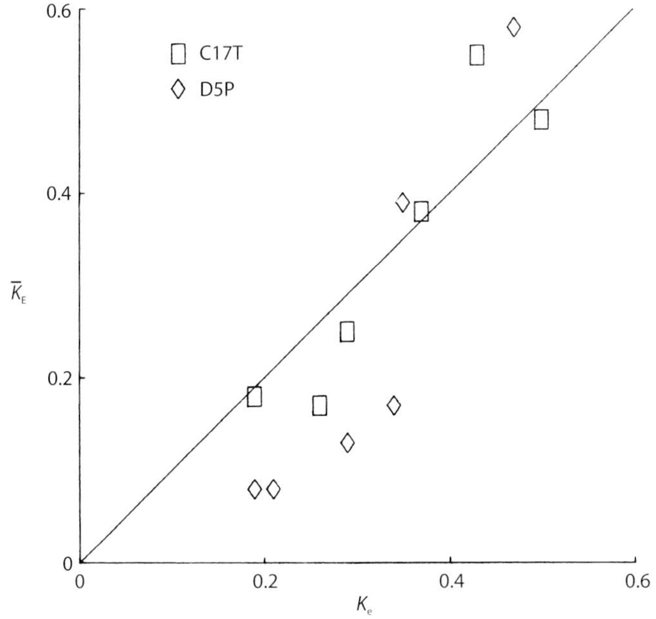

Fig. 5. Results are compared of two independent methods for estimating loop gain from PST histograms. K_e is from (0.6). K_E is from (2.6). Values are given in Table I (and in [1, table III]).

them some inhibitory cells to prevent runaway excitation? Given the amplitude-dependent nonlinearity of the forward gain characteristic of the cells, the answer is no. Suppose that a mass of such neurons is so densely interconnected that each transmits to many others. For some minimal level of activity, each pulse transmitted by one cell might so strongly affect neighboring cells that in effect more than one pulse is received back from those cells. The loop gain exceeds unity. If in any brief time period every cell continually transmits more pulses than it receives, then the activity of the mass must increase. However, as it increases, according to the model more cells are in a refractory period from a preceding pulse and cannot respond. This reduces the number of pulses transmitted by each cell, in comparison to the number received, and the gain is reduced. The activity of the mass must continue to rise until the gain is reduced to unity, at which level the activity is stabilized, each cell sending as many pulses as it receives.

If an additional pulse volley is delivered by way of an afferent path, the induced increase of activity must reduce gain below unity, because each cell is already at a steady-state limit in firing rate. Owing to the increase of input, each cell must transmit more pulses but nevertheless fewer pulses then it receives, because it is more likely to be refractory. The induced activity level must then decay slowly to the background level. This accounts for the prolonged response duration of the mass, for which the rate constant, may be much less than the rate constant of any single neuron. There is no overshoot.

If the afferent volley is increased in magnitude, the activity is driven to correspondingly higher levels, and the gain is reduced still further. The rate of return of activity down to the background level must be increasingly rapid, because at low levels of gain, the number of pulses transmitted compared to those received is low. This accounts for the dependence of the decay rate of induced responses on the input intensity.

The proof for this hypothesis is that the value for the decay rate β extrapolates to zero for threshold inputs. It is proposed that the anatomical connection density provides for a possible feedback gain K_E well in excess of unity, and that the average duration of the refractory periods of the cells determines the background level P_0 at which activity is maintained. The level may be under central control by means of the centrifugal afferents terminating on periglomerular and short axon cells [7]. The proposal accounts not only for the existence of background activity but for the Poisson-like character of the unit trains, their persistence following bulbar isolation and deafferentation (unlike cortical unit trains which are terminated by deafferentation), and the possibility for stabilization at any level over a range of amount of "spontaneous" activity, by variation in the mean time of recovery of the cells after firing.

It is concluded that inhibitory neurons are not required to stabilize an excitatory population. Were they scattered throughout such a population, as are short-axon cells among periglomerular cells, they would merely augment the required level for anatomical excitatory connection density. It is, however, possible that they are present, either to serve some local function in relation to particular cells, or to serve as sites of termination for centrifugal axons, and to amplify centrifugal input acting to regulate the background activity [7]–[9].

B. Output-Dependent Gain in Another Neural System

Amplitude-dependent gain has been described for another neural system in the retina (light adaptation) by Rushton [10] and others [11]. He denoted light input intensity as a logarithmic variable by I and retinal output by the normalized variable V and found the general relation:

$$I = V\exp(V).$$

If the loop gain K_E is defined as the ratio of output P to input I and if the output is taken as the normalized pulse density $P = (P - P_0)/P_0$ then from (2.6):

$$\hat{P} / I = \exp(-\hat{P})$$

and

$$I = \hat{P}\exp(\hat{P}).$$

In both cases the equations serve to describe the mass actions of large numbers of neurons without explicit reference to the cellular mechanisms responsible for them.

C. Functional Significance of Positive (Excitatory) Feedback

Mutual excitation among periglomerular neurons is significant to the extent that the set of neurons transmits its activity to other sets of neurons in the bulb, particularly the mitral-tufted cells, which are the principal projection neurons of the olfactory bulb. This is a complex question which cannot be considered here in full detail. However, two brief comments are in order.

Periglomerular neurons, on combined anatomical and physiological evidence, have seven distinct input channels and six identifiable output channels. Only two of each are considered here. Each periglomerular neuron receives axodendritic synapses from PON axons and dendrodendritic synapses from mitral-tufted apical dendrites in the glomeruli [7]. Each periglomerular cell transmits output to mitral-tufted cells through dendrodendritic synapses with the apical dendrites in the glomeruli, and through axodendritic synapses on the shafts of the apical dendrites just outside the glomeruli [7]. If all four of 'these connections are excitatory, two conclusions follow.

First, the steady-state background activity of periglomerular neurons must be transmitted to mitral-tufted cells. Experimental evidence that glomerular neurons excite mitral-tufted cells in this way is that background activity persists in the mitral-tufted population following deafferentation of the bulb. This is in contrast to the prepyriform cortex, to which the mitral-tufted cells project, in which the excitatory neurons are rendered silent by deafferentation [12], even though the excitatory neurons are interconnected and mutually excitatory, as are the mitral-tufted cells [13]. It is inferred that the periglomerular excitatory population provides the background depolarizing bias, which is required to maintain the negative feedback loops in the bulb and vortex in an optimal dynamic range [4], independently of the level of afferent input on the PON.

Second, in their anatomical observations, that periglomerular axons form axodendritic synapses on the apical dendritic shafts of mitral-tufted cells, Pinching and Powell [7] described the frequent occurrence of synaptic complexes in the glomeruli, in which a PON axon terminal formed synapses with two adjacent dendrites, one periglomerular and the other mitral-tufted, and in which each dendrite formed a synapse with the other. This implies that the periglomerular cell and the mitral-tufted cell are simultaneously given an excitatory impulse by the PON axon. If the time-locked activity of the pair of cells were also transmitted to the apical dendritic shaft of the mitral-tufted cell by the periglomerular axon, the likelihood of transmission of excitation to the mitral-tufted cell body must be enhanced for that external input. That is, an arrangement of parallel feedforward excitation might serve as a coincidence detector, serving to minimize the transmission of random fluctuations in glomerular activity of internal origin. This may explain the finding that the transfer function $A_m(s)$ required to describe PON input to glomerular and mitral-tufted cells [1], [14], [15] had two components in parallel [1, eq. (1.0)]. The output specified by this transfer function for impulse input (electrical stimulation of the PON) was an impulse followed by a dispersed volley. The dispersed volley can be accounted for by differences among PON axons of conduction rates and distances [2], but dispersion would exclude the possibility of singular impulse transmission by the PON. The fact that bulbar neurons responded as if they did receive an impulse must be attributed to the synaptic mechanisms of the bulb. Coincidence detection might play that role.

Acknowledgment

Reprinted by permission from *IEEE Transactions on Biomedical Engineering*, Vol. **BME-21**, No. 5, September 1974, pp. 358–364

Manuscript received April 30, 1973; revised February 1, 1974. This work was supported by the National Institute of Mental Health Grant MH 06686.

Computations for this paper were done at the Berkeley Computer Center.

References

[1] W. J. Freeman, "A model for mutual excitation in a neuron population in olfactory bulb", *IEEE Transactions On Biomedical Engineering*, Vol. **BME-21**, No. 5, September 1974, pp. 350–358.
[2] W. J. Freeman, "Spatial divergence and temporal dispersion in primary olfactory nerve of cat," *J. Neurophysiol.*, vol. **35**, pp. 733–744, 1972.
[3] P. A. Anninos, B. Beek, T. J. Csermely, E. M. Harth, and G. Pertile, "Dynamics of neural structures," *J. Theoret. Biol.*, vol. **26**, pp. 121–148, 1970.
[4] W. J. Freeman, "Linear analysis of the dynamics of neural masses," *Annu. Rev. Biophys. Bioeng.*, vol. **1**, pp. 225–256, 1972.
[5] W. J. Freeman, "Depth recording of averaged evoked tential of olfactory bulb," *J. Neurophysiol.*, vol. **35**, pp. 780–796, 1972.
[6] J. C. Eccles, *The Physiology of Synapses*. Berlin: Springer, 1964.
[7] A. J. Pinching and T. P. S. Powell, "The neuropil of the periglomerular region of the olfactory bulb," *J. Cell Sci.*, vol. 9, p. 379–409, 1971.
[8] B J. Dennis and D. I. B. Kerr, "An evoked potential study of centripetal and centrifugal connections of the olfactory bulb in the cat," *Brain Res.*, vol. **11**, pp. 373–396, 1968.
[9] G. M. Shepherd, "Synaptic organization of the mammalian olfactory bulb," *Physiol. Rev.*, vol. **52**, pp. 864–917, 1972.
[10] W. A. H. Rushton, "Visual adaptation: The Ferrier lecture, 1962," *Proc. Roy. Soc. B.*, vol. **162**, pp. 20–46, 1965.
[11] R. Shapley, C. Enroth-Cugell, A. B. Bonds, and A. Kirby, "Gain control in the retina and retinal dynamics," *Nature*, vol. **236**, pp. 252–253, 1972.

Part II

Designation of Contents as Meaning, not Information

9. Multichannel Recording to Reveal the "Code" of the Cortex: Spatial Patterns of Amplitude Modulation (AM) of Mesoscopic Carrier Waves

The changes in the state of a system that are the subject of dynamics stem from mechanisms of varying complexity. In the present context I choose to describe five levels of complexity in cortical function. At the first level I assume that a cortical system is stationary and autonomous, so that its operation can be described just by a set of state variables. The state of the system is assumed to be at rest when there is no input, and its change in state is determined by observing its output at a succession of points in time following a specific input. That state is expressed in numbers by measurement of the state variables. The operation of the system is discovered by continuing to collect input–output pairs until there are no more surprises. Then a rule is derived to specify how the system transforms the input to give the output. The rule is stated with a differential equation or its equivalent. The fixed parameters in the equation are evaluated from measurements of the input and output. This is not a causal science. The output is not "caused" by the input but is related to it in accordance with the equation viewed as a rule or law. This level of description is exemplified in Chapters 1 to 3, wherein the input is an electrical stimulus, and the cortical impulse response serves to define a filter, according to which the cortex transforms the input pulse into an oscillation. By linear extrapolation the EEG is described as the response of the filter thus defined to an input of steady state trains of action potentials from a sufficiently large number of neurons to approach white noise.

At the second level the cortical system is given the impulse input repeatedly at differing intensities. The output is observed to change with stimulus intensity, but this does not necessarily mean that the system is altered. It usually means that the cortical system is still stationary, but that changes in the input intensity alter the performance of the system, owing to the presence of amplitude-dependent nonlinearities. If the nonlinearities are static and not time-varying, curves representing the impulse response from the same equation are fitted to the collection of observations by allowing the parameters to vary. The linear model is extended beyond the small-signal domain by use of subsidiary equations describing the dependence of its parameters on the input intensity. This approach is exemplified by the root locus technique introduced in Chapters 3 and 4. Piecewise linear approximation with describing functions is used to express the nonlinearities.

At the third level the impulse input is fixed, and the cortical system is changed by behavioral manipulation, such as by food or water deprivation or exposure to conditions that induce sleep or emotional reactions of rage, fear or sexual arousal, and by surgical or pharmacological manipulations that induce anesthesia or that alter the strengths of synaptic interactions among the neurons comprising the system. These system changes are structural. Analysis at this level is exemplified in Chapters 5–7 in terms of the changes in impulse responses with arousal by unconditioned stimuli and learning under reinforcement, which are modeled by changes in gain coefficients that represent synaptic strengths.

At the fourth level the concept of the state transition is developed, according to which the cortical system displays discontinuous jumps in its performance, such as the transition between a state of zero activity under anesthesia and an awake state with background activity (Chapter 3). Physicists use the term "phase transition" to describe these jumps to differing conditions, as for example the freezing or boiling of water, but engineers prefer to use the word "phase" to describe time relations between oscillating state variables. In brief, the same neurons perform in strikingly different ways, depending on the ways in which they are connected and interact. These system changes are topological as well as structural, and they lead to identification of multiple domains of operation, each of which must be evaluated in terms of its stability, and to description of the manner in which transitions can be induced from one stable domain to another. This level is exemplified in Chapters 5–7 and in more depth in Chapter 8.

The fifth level of dynamics is directed to signal processing, where a behaviorally relevant input is defined as a signal, and the cortical response is used to determine how the cortical system can transform the input or is induced to change its state and, ultimately, enable an appropriate behavioral response. This level of analysis is crucial for neurodynamics, because the algorithms devised by input-output analysis of impulse responses have no biological meaning, unless and until the dynamics they reveal has been adapted to describe the operations that are performed on behaviorally effective stimuli. One step in this direction has been to train animals to respond to the input pulse as a conditioned stimulus (Chapters 6 and 7), but it turns out that a cortical response to an odorant stimulus differs fundamentally from its impulse response. The difference is somewhat analogous to that between the ringing of a bell evoked by striking it with a hammer with the ringing induced by exposing it to vibrations at its natural frequency that destabilize it. In the latter case the response is non-local and uncertain in time of onset.

The most widespread view among neurobiologists at this level, a view which I shared at the time the work in this chapter was done, is that stimuli are vectors of information, which is delivered to sensory receptors, transduced to action potentials, and processed by central circuits to give representations of the stimuli. In the olfactory system the information processing hypothesis is based on cellular evidence for a spatial code. Among the 10^8 olfactory receptors in the nose, genetic studies using cloning reveal at least 10^3 types of receptor (Ressler *et al.* 1994; Axel

1995; Wang *et al.* 1998). Cells of each type are distributed in the lining of the nose to facilitate capture of odorant molecules brought by sniffing, so with each inhalation an odorant stimulus is transduced to a spatial pattern of action potentials that is transmitted in parallel to the olfactory bulb. Only a small number of the receptors tuned to each odorant is activated on each sniff, perhaps 10^2 out of 10^5, so the principal task of the bulb is to generalize over each pattern in order to identify the class of the odorant represented by the sample inhaled. Classification is facilitated by convergence of axons of each chemical type to the glomeruli that attract that type during growth and development of the brain, but owing to the low specificity of receptor types for a range of odorants, multiple glomeruli as well as multiple neurons must be involved with each odorant classification in a combinatorial code (Matsunami and Buck 1997; Malnic *et al.* 1999). Coding must also be combinatorial in the olfactory cortices to which the bulb transmits, because the responses to odorants of the neurons in those areas show no greater specificity or selectivity than do the receptors or the bulbar neurons (Haberly and Bower 1989). The information processing hypothesis predicts that a spatial pattern of bulbar neural activity must exist for each distinctive odorant, whether or not there is also temporal structure in the form of precise timing relationships among the action potentials of the active neurons to "bind" the neurons into an assembly to represent the odorant to other parts of the brain. If there is temporal structure, it must have strong components in the gamma band, which contains the natural frequencies of the olfactory neural populations.

My first experiment to detect the spatial code was to record simultaneously from a set of ten microelectrodes in a prefabricated array, for which I built a miniature stereotaxic drive that could be fitted onto a cylindrical well cemented into an opening in the skull over the dorsal bulb. This allowed me to insert the array into the external plexiform layer of the bulb parallel to the mitral cell layer in a waking cat or rabbit. It was a challenging operation to position the array so that multiunit activity appeared on all ten electrodes simultaneously, monitor the activity on ten oscilloscopes, adjust and maintain the ten amplitude windows to accept a small number of units from each channel, digitize, store the incidence times in computer memory, and collect data from the 1 to 5 neurons on each channel in 30 second epochs of recording during presentation of different odorants to a thirsty and restless subject. The offline data analysis proved to be equally formidable. I demonstrated the existence of pulse probability waves at the frequency of the EEG, as I had for single cells, on all ten channels at the same frequency, but the spatial patterns of amplitude and phase of these waves were as variable on successive trials of the same odorant as between odorants (Freeman 1975). I concluded that a spatial sample of less than 50 neurons from the 10^5 neurons comprising the mitral cell population was hopelessly inadequate.

I turned to multichannel EEG recording, for which I had the option of 64 channels, each reflecting a weighted sum of the activity of at least 10^4 among the roughly 10^7 granule cell interneurons. The correlations between mitral pulses and granule waves had shown that the pulse probability wave amplitudes were strongly correlated with the EEG wave amplitudes in different times and conditions at the same location, at the same times in different locations, and for any two units separated by pulse amplitude windows on the same channel, all at the common instantaneous gamma frequency (Freeman 1975). I concluded that the local gamma amplitude was an imperfect but still adequate indirect measure of the activity of the local neighborhood of mitral cells, so that a set of 64 amplitudes would give an estimate of the spatial pattern of activity in a sample of roughly 20% of the population of mitral cells, oscillating at the same gamma frequency, but leading the granule cell activity on the average by one quarter cycle of phase at the shared frequency.

The spatial analysis of EEG activity in chronically implanted animals presented technical as well as theoretical challenges. I designed and built prefabricated arrays of 64 electrodes to place onto the brain surface in a single surgical procedure, requiring that it fit into a surgically accessible window onto the brain, with wires to a connector fixed on the dorsal skull and a cutaneous opening small enough to minimize infection. The device had to withstand autoclaving and be reusable for new subjects (Eastman 1975). The number of electrodes was limited by the available multiplexing technology that allowed me to sample at 1 msec intervals minimally on each electrode. This limit was set by the analog-to-digital conversion equipment then available and by the maximal core-to-disk data transfer rate with double buffering, which would allow me to record continuously for up to 10 sec on each trial. The spacing between electrodes was limited at the upper end by the size of the surgical window on the bulb or cortex to 4×4 mm and 4×7 mm respectively, and at the lower end by the optimal Nyquist sampling frequency that I derived from spatial spectral analysis of the EEG (Freeman 1978b). The interelectrode distance (which was equivalent to the digitizing interval in time series analysis) that I fixed upon was 0.5 to 0.8 mm.

I chose to display bursts over a fixed time interval of 100 msec as 64 traces and found that almost invariably there was a common number of peaks, troughs and zero crossings on all traces, despite variations in their amplitudes, phases and details of rise and decay (Figure 1 in Chapter 9). Therefore I adopted as my standard the display of contour maps of root mean square (rms) amplitude in microvolts and phase in radians to express the spatial pattern of each burst (Figures 2 and 3 in Chapter 9). On the basis of the information processing hypothesis I confidently expected to find differences in the spatial patterns of amplitude with different odorants. Instead I found that the gamma patterns were like handwritten signatures and facial expressions: no two bursts were identical, but each was easily recognizable as unique for each animal. From inspection of hundreds of contour plots I found also that no two patterns were identical among either the control or the odorant bursts, and that the variances among the control bursts and among the odorant bursts were about the same as the variance between the control and odorant bursts. There was no specificity of AM patterns on casual presentations of odorants. Later studies showed that the spatial patterns of amplitude modulation (AM patterns) did change consistently and significantly with odorants, but only when the animals were given an unconditioned stimulus paired with an odorant (Freeman and Schneider 1982), or when norepinephrine was injected into the bulb (Gray *et al.* 1986), but obviously, the AM patterns were endogenous. They came from within the animals, and they could not be derived from the stimuli by any form of filter or logical algorithm. Furthermore, unlike the information that could be

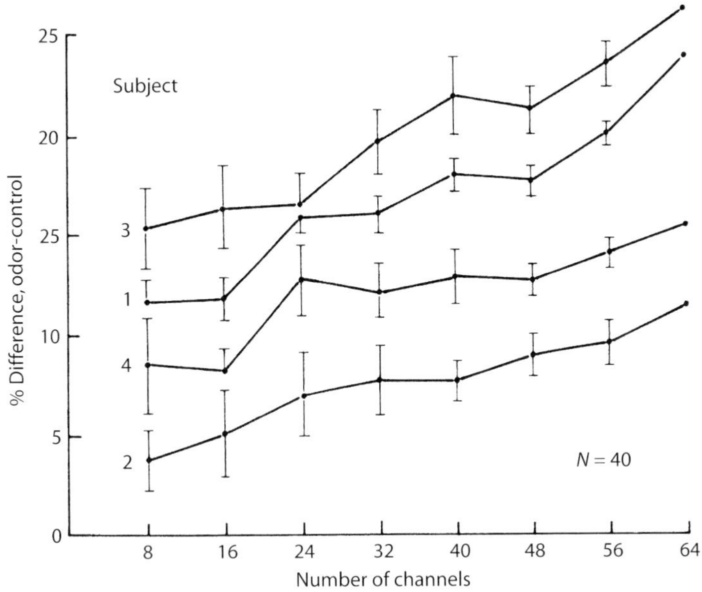

Fig. IX. A behavioral assay was used to search for localized activity that might indicate the sites of processing of odor-specific information. The assay was to determine the proportion of bursts that could be correctly classified between two odorant presentations in randomly alternated trials in a single session (Viana Di Prisco and Freeman 1965; Barrie *et al.* 1996 in Chapter 13). The null hypothesis was that there were no differences among all channels of the array, whether they had high or low amplitudes, high or low variances, and so on. Channels were selected for deletion by a random number generator on 40 trials at each number of deletions in 8 steps of 8 channels for each rabbit. The Mean and Standard Error of classification efficacy are shown for each number of channels remaining. This procedure provided the basis for the search for individual channels that consistently contributed to high or low classification values. No differences between channels were found, showing that classificatory information density was uniform over the array. The result supported the conclusion that in contrast to microscopic sensory patterns that are spatially localized, mesoscopic patterns are dispersed in space. From Freeman and Baird (1987, Figure 16, p. 405).

assigned to neural firing in sensory coding, which is spatially discrete and nonuniform, the information that could be used to characterize perceptual patterns was uniform in spatial density, though the intensity or amplitude of activity varied as a function of location (Figure IX).

These early results revealed a striking difference between bulbar and prepyriform dynamics, the significance of which was not appreciated at the time. In essence, the bulb appears to have the capacity for creating novel patterns; the prepyriform cortex does not. The difference was found by comparing the amplitude and phase patterns of the evoked potentials with those of the EEG. The phase patterns of the prepyriform EEG and the evoked potential on stimulation of the lateral olfactory tract both conformed to the direction and gradient predicted from the conduction velocities of the axons and their terminal branches in the tract (Figures 7 and Figure 8 in Chapter 9). The same conformity held for the amplitude patterns of the EEG and the evoked potential (see Figure 9 in Chapter 1). In contrast, the amplitude patterns of the initial peaks of the bulbar evoked potential to either orthodromic stimulation of the primary olfactory nerve (PON, Freeman 1974a; Figure 4.49, p. 267, in Freeman

1975) or antidromic stimulation of the lateral olfactory tract (LOT) depended on the location in the nerve or tract of the stimulating electrode (Figure 4.50, p. 268 in Freeman 1975), but the AM patterns of the gamma activity in the EEG did not conform to the pattern of the initial peak of the evoked potential. Instead they appeared to conform to the spatial patterns of secondary ringing triggered by the initial peak of the evoked potential (Figures 4 and 6 in Chapter 9). The difference in bulbar phase patterns was even more dramatic. Whereas the phase of the orthodromic evoked potential conformed to the direction and conduction velocity of the axons in the PON, the phase patterns of the bulbar gamma bursts had the form of a cone (Figure I in the Prologue), thus providing solid evidence that the AM patterns are self-organized. Evidence for phase cones was sought in the prepyriform cortex; none was found (Freeman *et al.* 1995).

This pattern difference between evoked and induced gamma activity indicates that the bulb is poised to create new AM patterns by self-organization when it is destabilized by sustained receptor input, whereas the prepyriform cortex operates on its input passively. The difference in dynamics appears also in the continuing endogenous activity of the de-afferented and isolated bulb, in contrast to the deep silence of the isolated prepyriform cortex (Chapter 5). It appears that the bulb and cortex are coupled as master and slave, using similar oscillatory dynamics to perform very different operations underlying perception (Kay and Freeman 1998). These findings from 20 years ago are pertinent to present studies of the sensory neocortices (Barrie *et al.* 1996), in which evidence is found for the self-organization of AM patterns under conditioning stimuli. If the primary receiving areas maintain a master-slave interaction with their targets of transmission, where are their equivalents to the prepyriform cortex, and what are their roles? Does the prepyriform cortex act as a controller of the chaotic dynamics in the olfactory bulb? These are some of the questions raised by spatial AM pattern analysis of gamma activity on the millimeter scale, for which the full answers have yet to be realized.

Spatial Properties of an EEG Event in the Olfactory Bulb and Cortex

Most theories of olfaction hold that some aspects of olfactory coding are spatial (Adrian 1950a; Le Gros Clark 1957; Moulton and Tucker 1964; Wenzel and Sieck 1967). For at least some discriminable odors there should exist differing spatial patterns of neural activity that represent the odors in the olfactory bulb and cortex. This hypothesis has been tested by recording the unit responses to selected odors of receptors and bulbar neurons in different parts of the mucosa and bulb. The results have been inconclusive. Each neuron responds to an almost unique array of odors, and its response profile broadly overlaps that of others. This had led to the further postulate that, if the locations of activated neurons are important for olfactory information, then the neural images of odors may be spatially distributed, either in a matrix of glomeruli (Hainer *et al.* 1954; Leveteau and Macleod 1966) or in an array of neurons, perhaps analogously to the hologram (Lettvin and Gesteland 1965.).

If this were so, then the only feasible experimental approach would require simultaneous recording at multiple sites in the bulb and cortex, because the code would have to be read on a neural surface in parallel channels at each significant moment (Hainer *et al.* 1954). The technology of unit recording is not yet adequate for this task. Approaches to multiple-channel EEG recording (Moulton 1963) have seemed promising but have not been fully developed.

This report consists of a description of the spatial properties of a prominent component of the bulbar and cortical EEGs, as a prelude to the use of the EEG for the study of olfactory information processing by the bulb and cortex.

The pioneering studies of Adrian (1950a,b) on olfactory bulbar EEG and unit activity were done in rabbits under urethane. Adrian found that stimulation with odors caused an increase in the rate of firing of some bulbar neurons and the onset of a high-amplitude sinusoidal oscillation that he called the 'induced wave.' On cessation of the stimulus the EEG returned to low-amplitude, relatively high-frequency activity that he called 'intrinsic.'

Subsequent studies of the temporal patterns of the EEGs of the bulb and cortex in animals with chronically implanted electrodes (Boudreau and Freeman 1963; Boudreau 1964; Hughes *et al.* 1969; Freeman 1975) revealed a strong correlation between the EEG and the respiratory cycle. With each inspiration there is a surface-negative slow wave followed by a surface-positive wave during expiration. Superimposed on this wave is a sinusoidal burst resembling the 'induced wave' of Adrian that begins near the crest of inspiration and extends into the expiratory period. The frequency within bursts ranges from 40 to 80/sec in the bulb and from 38 to 58/sec in the cortex. The burst duration is about half the respiratory period, ranging from 50 to 300 msec or more depending on respiratory rate. When recorded monopolarly at the surface of the bulb and cortex the peak-to-peak amplitude of the burst varies with location from a few microvolts to several hundred microvolts. In the depth of the bulb the amplitude may reach several mV.

In animals with chronically placed electrodes the bursts appear only in waking motivated states (hunger, fear, curiosity, etc.) and occur whether or not odors are deliberately administered. In anesthetized preparations under urethane or chloralose-urethane, bursts may appear repeatedly for seconds to minutes at a time in response to odors, but they are usually found to have a lower frequency than the bursts in waking animals. For this reason it is unlikely that the induced waves in waking and anesthetized animals have the same neural mechanism, so the present studies of spatial distributions of EEG were carried out in waking, minimally restrained animals.

Preliminary studies of the waking induced wave have been reported (Freeman 1978), mainly concerning the spatial frequency spectrum.[1] Briefly, measurements were made of the root mean square (rms) amplitude of single bursts of induced wave activity in waking cats and rabbits, that were recorded from the bulbar surface with linear arrays of 64 electrodes at intervals of 40 microns. The Fourier transform was taken of the data. There was found to be a sharp cut-off in the spectrum at about

1 cycle/mm. This conformed to the cut-off predicted from volume conductor analysis of the bulbar EEG generator (Freeman 1975). This result led to the adoption of 0.5 mm as the interval between electrodes for much of the work to be reported.

Methods

Multi-channel recording was from chronically implanted, prefabricated arrays of electrodes (Eastman 1975). Three types were used. The first was a rectangular array (6×10) of 60 wires (300-μm stainless steel coated with Formvar) at spacings of 0.8 mm center-to-center (4.0×7.2 mm), that was sufficiently large to cover the lateral aspect of the bulb in the cat and rabbit, or a substantial part of the anterior olfactory nucleus (AON) or prepyriform cortex, yet small enough to fit within the boundaries of these structures.

The second type was an array of 64 wires (35-μm stainless steel) in a line (1×64) at spacings of 40 μm (2.5 mm in length). This was implanted over the lateral aspect of the bulb in cats and rabbits in order to determine the spatial spectrum of the EEG.

The third type was a square array of 64 wires (300-μm stainless steel (8×8) with intervals of 0.5 mm (3.5×3.5 mm), that was designed to fit over the dorsal or lateral aspect of the bulb in rabbits, and with an optimal spacing with respect to the spatial spectrum of the bulbar EEG.

The wires were embedded in a block of dental cement shaped by filing to conform approximately to the recording surface. The proximal ends were inserted into a block connector pre-shaped to fit on the calvarium.

The arrays were placed directly on the intact dura over the dorsal bulb or onto the lateral bulb or cortex following removal of the orbital contents and resection of part of the medial wall of the orbit with a drill.

An air jet for cooling was used during drilling. The medial, anterior and ventral bulbar surfaces were not accessible. The orbit was filled with agar made with Tyrode's solution containing Zephiran (1%) and was closed with dental cement. Reference and ground leads were placed in the orbit and on the back of the skull. In some animals a bipolar stimulating electrode was placed stereotaxically from above in the lateral olfactory tract (LOT) at the level of the (AON), and another in the bulb

---

1 The spatial spectrum is precisely analogous to the temporal spectrum. Suppose, for example, that the EEG is sampled from a single channel at intervals of Δt msec for 64 samples. Or suppose that the autocorrelation of an EEG trace is calculated for 64 lags. The Fourier transform of the function gives the temporal spectrum, which shows the characteristic rates of change in potential with respect to time. Now suppose, instead, that the EEG is sampled at a single time from 64 electrodes on a line at intervals of Δx mm. The Fourier transform of this function gives the spatial spectrum, which shows the characteristic rates of change in potential with respect to distance along the array. The properties and uses of those spectra are quite similar. For example, by using a very high digitizing rate Δt or Δx one can demonstrate the absence of significant activity at temporal or spatial frequencies above a certain cut-off value f_t cycles/sec or f_x in cycles/mm. The reciprocal gives the wavelength in msec or mm and half the wavelength yields the optimal digitizing rate in time or the optimal electrode interval in space.

or cortex for EEG monitoring in that structure not covered by an array. These were omitted in most animals to minimize adventitious surgical damage.

The EEGs were amplified with 64 fixed gain (10K) pre-amplifiers (0.1 Hz to 3 kHz band-width) and monitored 4 at a time with oscilloscopes and a polygraph. They were filtered at 10 and 300 Hz (3 dB fall-off) to remove thermal noise and the respiratory slow wave, which was often confounded by the electroolfactogram, eye movements and cable noise. Recordings were monopolar with respect to one or two electrodes at the back of the head.

The EEGs were multiplexed and digitized sequentially at 10-μsec intervals across the array (6 bits/sample with a resolution of 17 μV/step). Readings were repeated at 1 msec intervals for block durations of up to 900 msec and stored in an Interdata 716 computer. At the completion of a recording block the records were inspected channel by channel for artifacts. The times of onset and termination of bursts were noted and the block (57,600 words) was written onto magnetic tape for off-line processing by a CDC 6400.

Missing points (up to 6) due to faulty or polarized electrodes were evaluated by the average of the signals from two adjacent points.

Three types of display were used. First, the time series of single bursts were drawn by the Cal-Comp plotter for the 60–64 channels. The degree of likeness was estimated by forming an ensemble average and computing the correlation coefficient for each channel with the ensemble average. Second, the root mean square (rms) amplitude was computed for each point over the duration of the burst, and the results were displayed as a contour plot with second-order extrapolation. Third, the cross correlation was repeated at each channel with time lags of ± 1 and ± 2 msec. The 5 correlation coefficients for each site were fitted with a vertical parabola. The time difference between real time and the apex of the parabola was used to estimate the time lag of each record from the ensemble average. The peak frequency for the average was found by the fast Fourier transform and the time lag was expressed as a phase at that frequency. The 60–64 values for phase were displayed in a contour plot.

The use of contour plots depended on the presence in each trace of a common waveform at whatever amplitude and phase. Occasionally at the margins of an array or due to faulty editing this was not the case. The phase then took widely deviant values and was essentially undefined. For display purposes such points were deleted. The system was designed to compute amplitude and phase distributions over two or more regions within arrays having locally coherent waveforms, but this approach was found to be needed in very few instances.

As a check on the computations an alternative method was devised. An analytic curve was fitted to the ensemble mean waveform of a burst using nonlinear regression. Then the optimized basis function was fitted to each of the 60 individual waveforms, allowing the amplitude and phase coefficients to vary. The resulting contour plots were virtually identical to the plots derived by the first method.

The system was calibrated by placing an array in a normal saline bath and applying a sinusoidal current to the bath with a pair of electrodes. A 100-msec burst of a sine wave at 40 Hz and 70 μV was used. When one of the two current generating electrodes was close to the reference electrode and the other was close to the ground electrode, and both were far from the array, the phase and amplitude values were expected to be uniform. Amplitude differences were found with a standard deviation of ±1.1% due mainly to varying channel gains. A phase gradient of 0.158 rad was found along the 1×64 array (0.063 rad/mm) due to the successive sampling at 10-μsec intervals. A correction was made for the sampling sequence, and the standard deviation of the phase measurements was reduced to ±0.012 rad.

Recordings were made beginning 3–5 days postoperatively with the animals placed in a restraining box to minimize body movements. A cone was placed over the nose with a constant flow of charcoal-filtered air, into which odors could be introduced without interruption of flow. When it was present at all, burst activity was most reliably induced by depriving each animal of food or water for 48 h before recording. When this failed the animal was removed from the restraining box and allowed to explore the recording cage. When this also failed to elicit burst activity the animal was exposed to urine, vaginal secretions, feces of the same or other species, various foodstuffs, smoke from burnt paper and tobacco, etc. If the EEGs were not isoelectric (indicating surgical or electrode problems), the animal was examined at intervals over several weeks. In no case was burst activity seen to emerge later than 4–5 days postoperatively.

At postmortem the animals were perfused with formalin, and the arrays were removed to verify by gross inspection the location and orientation of the array, and to determine whether damage had occurred due to hemorrhage, infection, or the overgrowth of bone or scar tissue. The brain was removed and sectioned by hand to determine depth electrode sites. Instances where pathology was found were tabulated as surgical failures and not as absence of a focus.

Results

(1) The Shape, Size and Location of Bulbar Induced Wave Foci

In healthy, motivated cats and rabbits with implanted bulbar surface arrays, the bursts were seen (if at all) at all recording sites (Fig. 1). The oscillation usually had the same frequency over the array, and as the frequency changed during each burst, it did so in the same way at every site. The amplitude of the induced wave varied greatly between sites and the times of peaks and zero crossing (latencies) varied slightly. The rate of rise and decay of the envelope of the induced wave varied over sites, but the monotonic increase and decrease was common to all sites. Superimposed activity at low amplitude other than that of the induced wave was seen over local collections of sites at both low frequencies (residues of the slow respiratory wave that was not completely removed), intermediate frequencies (commonly at about half that of the induced wave), and high frequencies (corresponding to the intrinsic activity of Adrian 1951).

Observations over successive bursts showed that the induced wave frequency, duration, and mean amplitude varied unpredictably, as well as the precise shape of

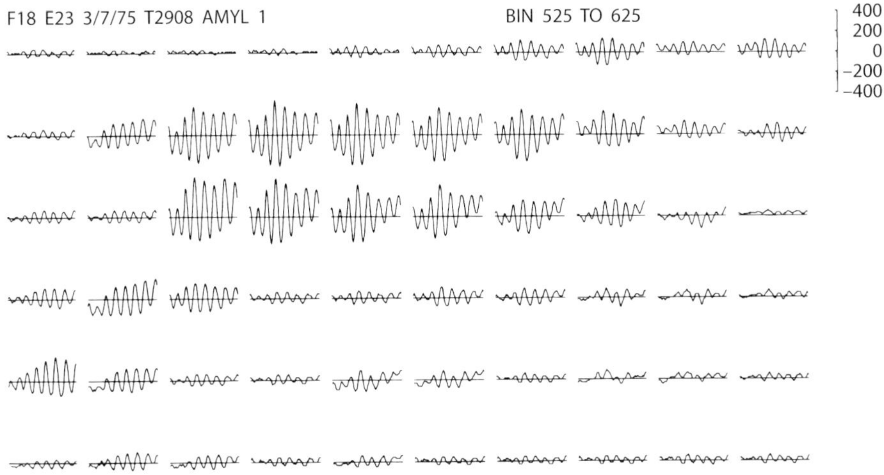

Fig. 1. A single unaveraged burst was recorded from 60 electrodes (4×7 mm) on the lateral bulb of a rabbit. Duration: 100 msec; amplitude scale in μV; digitizing interval: 1 msec; electrode interval; 0.8 mm. Negative upward. Odor: amyl acetate 1:10 in filtered air. EEG was filtered at 10 and 1000 cps. The posterior edge of the bulb lay close to the junction of the center and right thirds of the array. The lateral olfactory tract (LOT) extended posteriorly to the right across the center of the right third of the array.

the envelope. However, the relative amplitudes and latencies (with respect to the mean amplitude and latency across the array) were fairly constant for each animal. This constancy served to define an active focus in the bulb or cortex for the induced wave.

The spatial distribution of a focus was measured with the root mean square (rms) amplitude of the induced wave over the approximate duration of a burst, thus deleting information about the shape of the time envelope. Then the size, shape and location of the focus were displayed in a contour map (Fig. 2A).

An active focus was found in 4 of 6 cats and 3 of 4 rabbits with a 6×10 array over the lateral bulb, in 10 of 16 rabbits with an 8×8 array over the lateral bulb, but only 1 of 6 rabbits with an 8×8 array over the dorsal bulb. The foci varied in shape along a continuum from nearly elliptical to scattered and choppy. In 13 of the 18 positive cases the highest amplitudes occurred at contiguous sites, and the amplitude decreased monotonically with distance in all directions from the center of the focus. Five of the 13 contiguous foci were smooth in outline (Fig. 2A), and the other 8 were highly irregular with sharply serrated outlines. In the remaining 5 cases the sites with highest amplitudes were scattered over the array, suggesting that an active region could consist of two or even multiple more or less confluent small foci (Fig. 3).

The mean size of a focus was expressed as the equivalent half-amplitude diameter. This was found by measuring the area enclosed within the half-amplitude contour of each focus and computing the diameter of a circle with the same area. A fraction of a circle was used for foci in which the half-amplitude contour intersected the array border. From 11 rabbits with 8×8 arrays on the lateral bulb the diameter was 2.56 ± 0.64 mm^2. The active area enclosed within the array and the half-amplitude contour was 4.30 ± 1.67 mm^2.

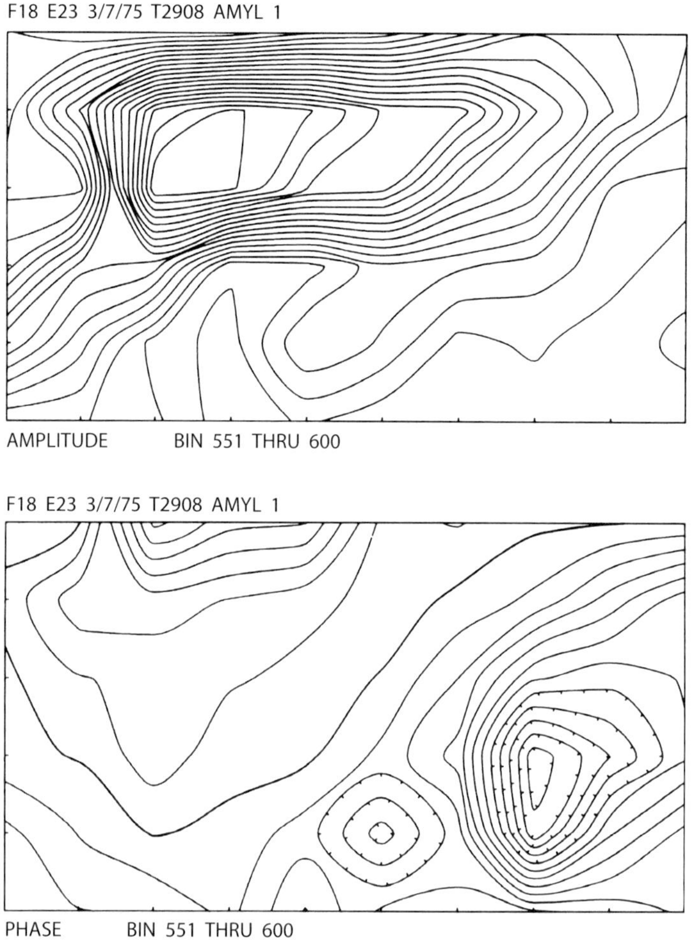

Fig. 2. A: Upper frame shows a contour plot of the rms amplitude from the center 50 msec of the burst shown in Fig. 1. B: Lower frame shows a contour plot of the phase. The amplitude maximum was 74μV; the minimum was 12μV; the contour interval was 6.2μV. The phase maximum lead was $\pm$0.40 rad; the minimum was $-$0.27, rad; the contour interval was 0.067 rad (1.0 rad $= 57°$).

The proportion of the bulbar surface, enclosed within a focus was estimated by assuming that the occurrence of a focus was equiprobable over the whole bulb, including the dorsal and lateral areas actually observed with 8×8 arrays in 22 rabbits. The area of the 8×8 electrode array was 12.25 mm^2, but the average area actually placed on the bulb was 94% of this or 11.5 mm^2. From the gross dimensions of the bulb the surface area was estimated to be 74.2 $\pm$ 7.4 mm^2, so that the electrode sampled 15.5% of the bulbar surface. Assuming that the ratio of 4.3 mm^2 active area to 11.5 mm^2 array aperture held for the whole bulb, the half-amplitude area was calculated to be 27.8 $\pm$ 7.6 mm^2. Further, 11 of the 22 rabbits showed no focus, so the estimate was reduced by half to 14 mm^2 or about 20% of the bulbar surface area.

Apart from the greater likelihood of finding a focus on the lateral rather than the dorsal bulbar surface, there was no relation between the size or shape of a focus and its position in the bulb or the location of the array.

(2) Phase Patterns Within Bulbar Induced Wave Foci

The phase distribution over the bulb was analyzed in 6 cats with 6×10 electrodes wholly on the bulb and aligned with its long axis, and in 4 rabbits with 6×10 electrodes partially over the bulb and aligned with the LOT. The average phase gradient for each animal was found to be in the anteroposterior direction roughly parallel to the trajectory of the PON (Fig. 2B). A mean value for each animal was calculated from the average phase difference in radians between the column at the anterior and posterior borders of the array (or the column in the array found at post-mortem to be closest to the posterior edge of the bulb), divided by the distance between the two columns in mm^2. The average value for the 10 animals was 0.07 ± 0.03 rad/mm, at a mean frequency of 322 ± 17 rad/sec. The expected value, if the phase gradient were dependent on PON conduction velocity for initiation of the induced wave (Fig. 4.32 on p. 230 in Freeman 1975), would be 0.4 rad/mm. Hence the observed mean phase gradient was 6 times too small to be accounted for by the known PON afferent delay.

The phase maxima and minima almost always lay within the array and were usually found on the slopes of amplitude maxima rather than at the crests (Fig. 2B). The maximal phase difference (average from 10 animals) was 0.31 ± 0.10 rad (less than 1 msec at the mean frequency of 322 rad/sec or 5% of one cycle) over an average distance of about 2 mm. The results showed that there was a high degree of synchrony within foci, and that spread did not occur radially from the center nor parallel to the PON.

The maximal phase gradient between adjacent sites in the 10 animals was seldom aligned with the PON. It averaged 0.21 ± 0.10 rad/mm. This is equivalent to 0.33 cycles/mm. For the frequency of 322 rad/sec (19.5 msec/cycle) the phase velocity was 1.5 m/sec. Since the velocity of an oscillatory event observed at the surface is 1.7 times faster than the velocity of the spread of the activity in the generator in the bulb (Freeman 1975), the implied deep velocity was 0.90 m/sec. Allowing that the measurements of phase difference at intervals of 0.8 mm spacings may have been attenuated by 50% due to volume conduction (Freeman 1977) the deep conduction velocity was closer to 0.45 m/sec. This was close to the conduction velocity of unmyelinated axons such as in the PON (0.42 m/sec), so that the maximal rate of spread was consistent with the limitations on an available mechanism of transmission of neural activity within the bulb.

(3) The Spatial Rates of Change in Amplitude

The maximal rms amplitude in the foci of 20 cats and rabbits varied between animals from 37 to 168 μV (averaged over 6 to 10 bursts each) with a mean over

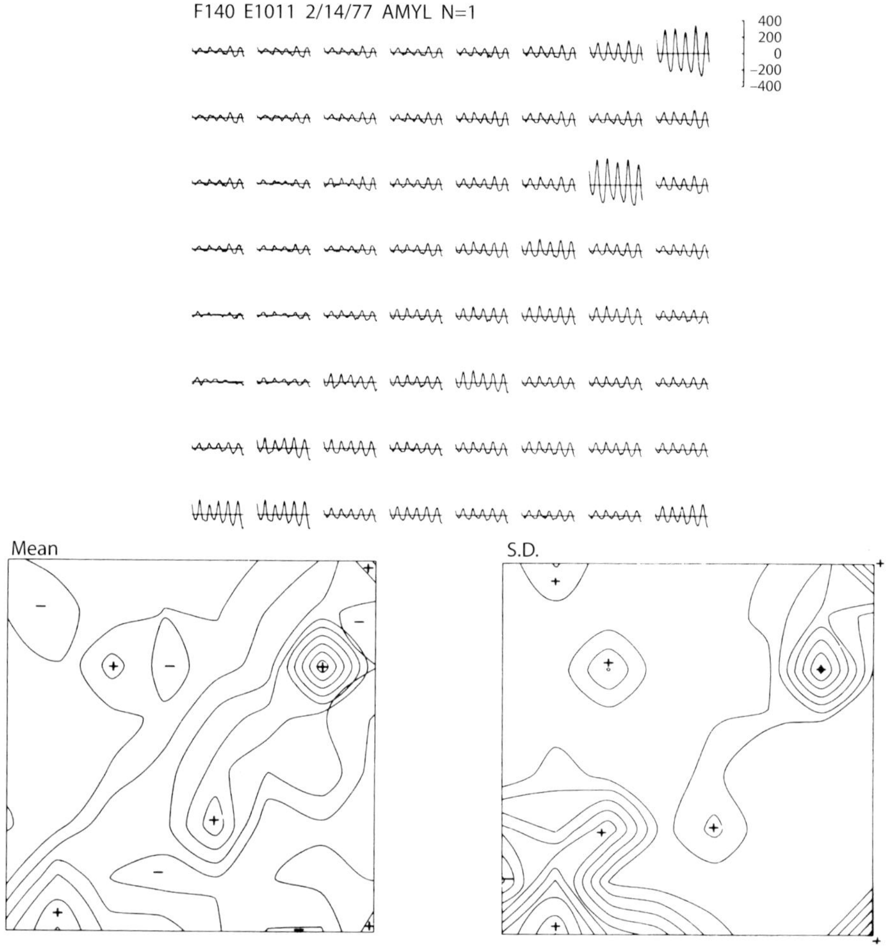

Fig. 3. Upper frame shows the 64 traces from a single burst (100 msec recording duration, scale in μV) from an array (3.5×3.5 mm) on the lateral bulb of a rabbit. Negative upward. The lower frame at left is the mean of 10 maps on the same day during brief exposures to amylacetate 1:10. Maximum: 182 μV; minimum: 26 μV; interval: 16 μV. The (–) signs indicate minima or troughs (not negativity) to distinguish them from the maxima or peaks. The lower frame at right shows the contour plot of the standard deviation of the normalized amplitudes. The mean standard deviation (SD) of the non-normalized data was 36 μV from the entire frame, with a coefficient of variation of 0.42. The grand mean and SD of the normalized data were of course 0 and 1. The range of the 64 normalized means was 3.67 SD to −1.23 SD with a contour interval of 0.49 SD. The range of the 64 normalized SD was 1.22 to 0.07 SD with a contour interval of 0.13 SD.

animals of 84 ± 32 μV. The minimal rms amplitude outside the foci ranged from 15 to 44 μV with a mean of 27 ± 7 μV. The ratio of maximum to minimum was 3.1 ± 1.2.

The rate of change in amplitude was examined in 11 rabbits with 8×8 arrays (0.5 mm spacing) and in 3 cats and 4 rabbits with 1×64 arrays (40-μm spacing). In most of the 8×8 group and all of the 1×64 group the rate of reduction in rms amplitude did not exceed 0.75 from one site to a neighboring site at 0.5 mm distance. However, in 4 of the 8×8 group there were from 1 to 3 pairs of sites separated by 0.5 mm, between which there was a steep amplitude fall-off (e.g. Fig. 3). In the mean contour plot over a set of 10 bursts in each animal the lowest amplitude at a site 0.5 mm distant from a high amplitude site was 0.51 ± 0.07 times the higher amplitude. That is, the average maximum observed decrease in rms amplitude with distance was 0.51/0.5 mm.

The maximal rms amplitude decrease between 2 adjacent electrodes was still faster for individual cases, averaging 0.41 ± 0.09. There were several cases in which a difference <0.3 was observed and one instance of <0.2.

Volume conductor analysis of the bulbar field potentials (Freeman 1975, 1977) has shown that the pattern of decrease of potential at the bulbar surface from the epicenter above a small subset of bulbar neurons (equivalent in diameter to one glomerulus) is nearly exponential for distances 0.5 mm. The rate constant depends on the depth of the zero isopotential surface of the dipole field. For a depth of 1000 μm (as in the cat) the fall-off is at 0.86/mm, so that the difference at 0.5 mm is by 0.65. For a depth of 600–800 microns (as in the rabbit) the fall-off at 0.5 mm is respectively by 0.36 (at a rate of 2.04/mm) to 0.50 (at a rate of 1.36/mm).

This result has several implications. First, the maximal observed rate of change in potential with distance was compatible with a model in which an active region in the bulb may be sharply bounded by an inactive region. One need not postulate that an induced wave focus is bordered by regions acting strongly out of phase. This is necessary for the use of the rms as a measure for the focus. The extreme instances of local differences may be explained as due to local variations in the depth of the granule cells generating the surface potentials, but this cannot be confirmed without measuring depth profiles of potential with penetrating electrodes. Second, it reveals a substantial redundancy in the signals from the 64 electrodes, because the event at each site appears at lower amplitudes on many surrounding sites, simply by volume conduction. Third, it implies that although the change in potential with distance across a focus may be gradual, the focus may consist of a collection of 'on–off' elements involved in the induced wave, at a spatial grain much less than the grain of the rectangular electrode arrays, and is perhaps at the level of the glomeruli (having a spatial frequency of about 4/mm).

(4) The Intrinsic Variance of Amplitude in Bulbar Foci

During the first few days of observation after implantation the focus in each animal underwent changes in its size and shape, generally enlarging and increasing in overall amplitude. Thereafter, if behavioral conditioning was not used, the focus stabilized into a recognizable spatial pattern that persisted for weeks. The median period of observation in 20 animals with 6×10 arrays that did not undergo training was 3 weeks, with a maximum of 24 weeks. The 18 animals with 8×8 arrays were so observed for shorter periods. However, from one burst to the next and from day to day the precise shape of each focus varied around its stabilized norm.

The daily amplitude variation was determined by constructing a matrix of 60–64 rms values for each of 6–10 induced bursts and making a contour plot of the mean and one of the standard deviation (SD) across bursts for each animal on one day. The plots for the mean and SD were similar in form, implying that the SD was generally proportional to the mean.

Then the 6–10 sets of rms values were normalized as follows. The mean for each set or map was subtracted from each of the 60–64 values in order to remove the variance across bursts that was introduced by activity at the monopolar reference electrode. The 60–64 values were divided by their own standard deviation to remove the variance in burst amplitude from one burst to the next. The mean and normalized SD (NSD) for each point was calculated as before across 6–10 bursts. The plots of the normalized means were not changed, but the plots of the NSD showed that the maximal intrinsic variation occurred on the slopes of the contiguous focus and not at the center (Fig. 4), and at the peaks of the choppy foci (Fig. 3). The maximal variance was seen at half a dozen scattered sites around each focus, usually singly but occasionally in pairs.

The distributions of the amplitude values at each of the 60–64 sites were examined for normalcy by calculating the 3rd and 4th moments. These were well within the limits for the normal distribution for both the raw and the normalized data, independently of the local values of the mean and SD. In particular, there was no evidence to indicate that the amplitude at a site with a large SD was switching between high and low values ('on' vs. 'off'), which would lead to a skewed, platykurtic or bimodal amplitude distribution at that site.

The distribution of mean amplitudes across the 64 sites was nearly normal for each of the contiguous foci. For the choppy foci the distributions were skewed with a small number of excessively large values forming a high amplitude tail.

This between-burst variance can be seen to have a high spatial frequency that exceeded the resolution of the arrays with 0.5 mm spacing, i.e. 1/mm. An insufficient number of sets was taken with the 1×64 electrode to evaluate its detailed character.

A different form of variation with a low spatial frequency was revealed by subtracting each of 10 sets from the mean for the set. This took the form of a shift in rms amplitude over a large portion of the array by an amount usually less than 0.5 SD. The location, size, and direction of the shift varied unpredictably from one sampled burst to the next. Commonly the spatial frequency was 0.25/mm (Fig. 5).

The variation from one recording session to the next (usually several days apart) was also examined. The patterns of mean amplitude and SD as seen in contour maps were easily reproducible over days and even weeks. The high (spatial) frequency variation representing fluctuations in the details of the shape of the focus between days was well within the SD on any one day. The low (spatial) frequency variation was more prominent across days and had the same range of amplitude as the variation of single maps from the mean on any one day (Fig. 5, right frame). This component of the variance was visually imperceptible in comparing contour maps of the means for two days, but it was clearly shown by maps of the differences between two sets of means (Fig. 5, left frame).

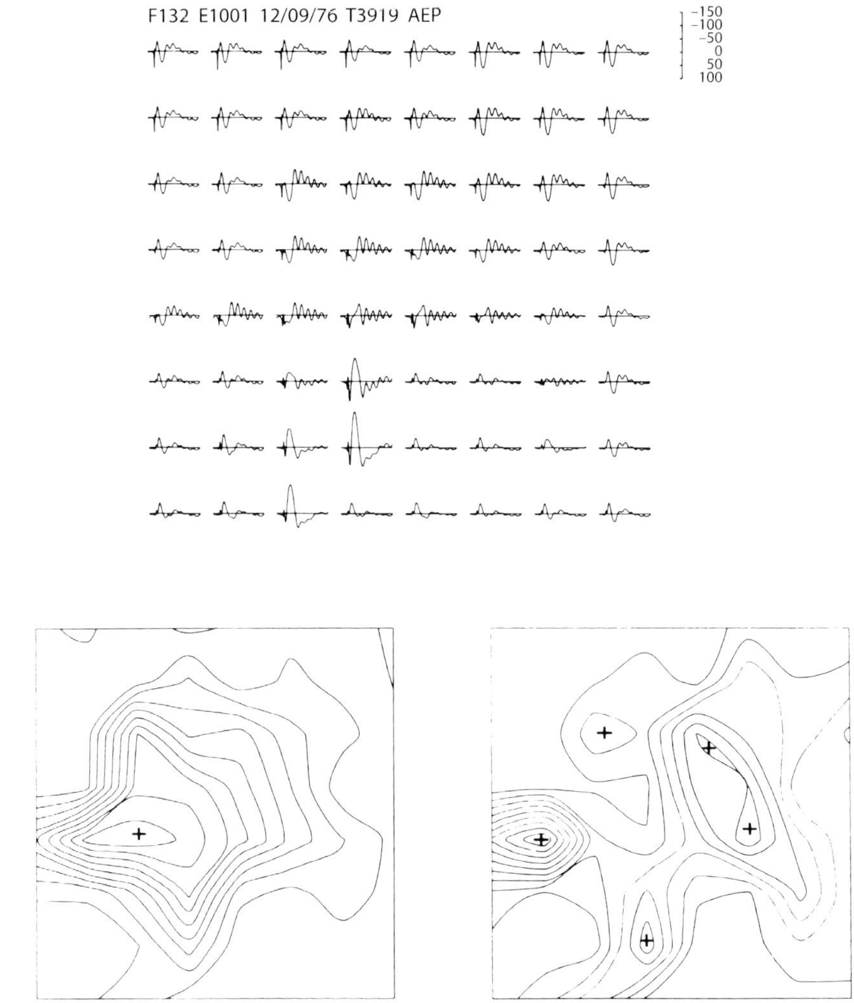

Fig. 4. Upper frame shows a set of 64 averaged evoked potentials (AEP) from an array (3.5×3.5 mm) on the lateral bulb of a rabbit, on stimulation of the LOT at 9 V/0.01 msec, 7/sec. Negative upward. Duration is 100 msec. Scale is in microvolts. $N = 130$. Left lower frame shows the mean of 10 sets of normalized rms amplitudes. The spatial distribution burst amplitude is virtually identical to that of the oscillatory after-burst in the AEP. Lower right frame shows the normalized SD. The maxima are located around the central peak of rms amplitude. The variance is focused at selected sites (see also Figs. 3 and 6) and therefore has a high spatial frequency; that is, the pattern is rough rather than smooth as in Fig. 5.

(5) Comparison of Bulbar AEPs with Induced Wave Foci

In rabbits with 8×8 arrays on the lateral bulb, bipolar stimulating electrodes were implanted stereotaxically in the ipsilateral LOT and prepyriform cortex for monitoring the cortical EEG. One or both pairs of these electrodes served to activate the bulb antidromically. In bulbs without foci the 64 averaged evoked potentials (AEPs) were uniform in appearance across the array, though varying in amplitude. The AEP

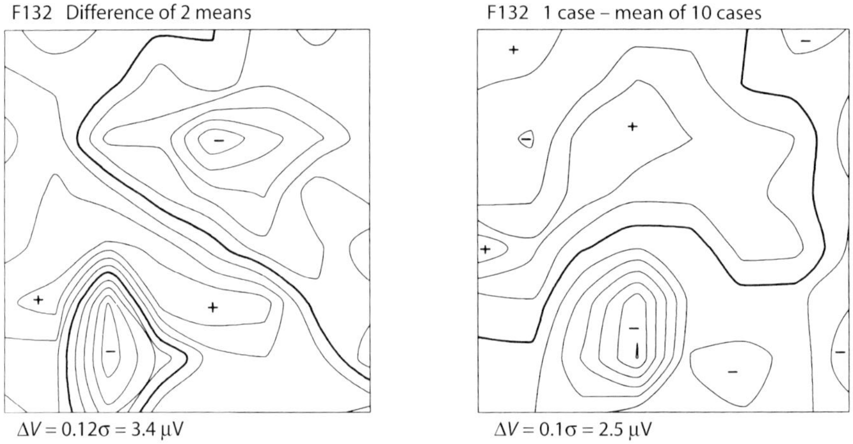

Fig. 5. A different form of variance from that shown in Figs. 3, 4 and 6 was revealed by subtracting one mean map from another (left frame) or by subtracting a single map from the mean of 10 maps (right frame). This low-amplitude component of the variance had a low spatial frequency. A map of the mean and of the normalized SD for this focus are shown in Fig. 4.

consisted of a brief mitral cell compound action potential followed by a negative peak lasting 5–7 msec, leading into a damped sine wave back to baseline.

In bulbs with foci the AEPs were more complex (Figs. 4 and 6). Typically, the initial peak of the AEP following the compound action potential was initially positive in the vicinity of the focus. The focus and the positive region broadly overlapped but were not congruent. The positive peak usually had a double, dicrotic, crest, and the regions of positivity for the two crests were overlapping but not identical. This peak was followed by a negative overshoot, on which was superimposed a sinusoidal oscillation decaying slowly to baseline. The amplitude distribution of the oscillation was identical to that of the focus. This part of the AEPs clearly resulted from entrainment or phase re-setting of induced wave activity (Scott and Aaron 1977) by the electrical stimulus.

In the bulbar region surrounding the focus the AEPs had the conventional form with initial negative peaks, and their amplitude distribution was unrelated to the focus. The spatial rate of change between the initially negative and positive regions was much too great to allow interpretation of the positive regions as 'inactive' potentials accompanying 'active' regions in the adjacent or opposite sides of the bulb (Shepherd and Haberly 1970). Moreover, the magnitude of positivity often greatly exceeded the absolute magnitude of negativity in nearby regions.

In 6 of 11 rabbits with foci and with two stimulus sites the AEPs in or near the focus were initial positive for both sides. In 3 rabbits they were positive for only one of the 2 sites and in 2 rabbits they were negative for both sites, indicating a complex dependency of the phenomenon on stimulus location.

This incidental finding is additional confirmation of the existence of the induced wave focus as a local region of specialized neural dynamics in the bulb and suggests that conventional electrophysiological analysis may be required in behaving animals to determine the nature of the specialization.

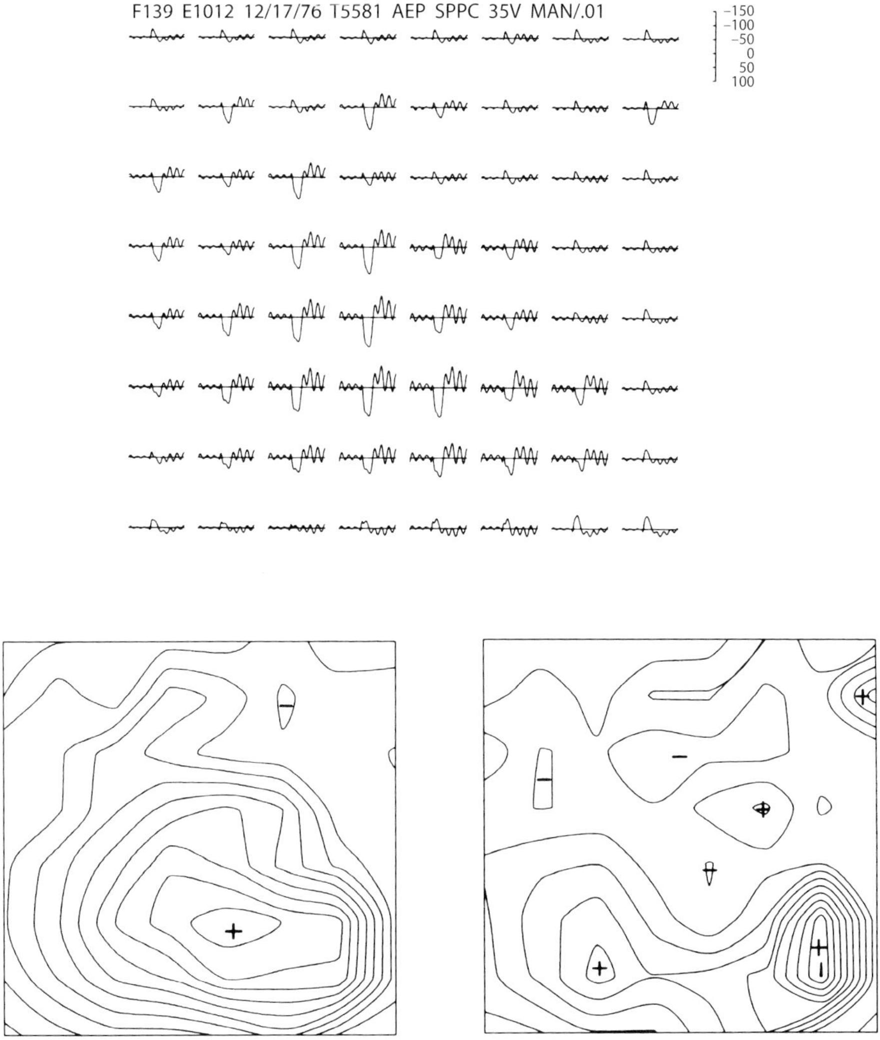

Fig. 6. The AEPs are shown from a 3.5×3.5 mm array on the lateral bulb of a rabbit, on stimulation of the LOT at 35 V 0.01 msec, 7/sec. Duration: 100 msec; scale in microvolts. Negative upward. Through and around the elliptical focus the initial peak is surface-positive and diphasic (see also Fig. 4). The surrounding AEPs are initially surface-negative as under anesthesia or in the absence of a focus. The oscillatory component of the AEP has the same spatial distribution as that of the bursts (lower left frame, mean of 10 maps). The maxima of the normalized, variance (lower right frame) are eccentric to the peak of the focus.

(6) Induced Wave Foci in the AON and Prepyriform Cortex

The 8×8 arrays were implanted over the anterior olfactory nucleus (AON) or prepyriform cortex of 3 cats and 9 rabbits. The same procedures as for the bulb were applied to derive contour maps of amplitude and phase.

The maximal rms amplitude for the AON and cortex over the 12 animals averaged $72 \pm 28\,\mu V$. The mean minimal rms amplitude was $27 \pm 9\,\mu V$ and the ratio between the extremes was 2.7 ± 0.9. The amplitude plots gave the impression of a scattered collection of foci that were randomly distributed over the AON and cortex on both sides of the LOT. In some animals the foci were sparse and small, reflected in the activity at only one site or two adjacent sites (Fig. 7, frontal prepyriform). In others the foci were confluent into a deeply serrated and irregular shape (Fig. 8, frontotemporal prepyriform). As in the case of the bulb those patterns tended to enlarge over the first 1 or 2 recording sessions and to remain stable thereafter for several weeks.

The phase plots showed phase lead (maximum) over the LOT, particularly at its anterior end (to the left in Fig. 7), with increasing lag to either side of the LOT and with a shallower gradient toward the posterior LOT. The mean phase gradient for 10 animals (excluding that in Fig. 8 where the array lay just dorsolateral to the LOT) was 0.037 ± 0.004 rad/mm. At the mean induced wave frequency this was equivalent to 7.36 m/sec phase velocity. Assuming that the same or a similar ratio (1.7 : 1) between surface and deep velocity held in the bulb and PPC, the deep velocity was 4.3 m/sec. This is comparable to the LOT conduction velocity (5 m/sec) over the relevant portion of its trajectory (Freeman 1975).

Off the LOT the phase plots were highly irregular although fairly reproducible on successive trials for each animal. Phase maxima and minima often coincided with peak amplitudes of foci. The average maximal phase difference for the 11 animals was 0.48 ± 0.22 radians or about 8% of one cycle. This small value and the lack of a clear phase gradient off the LOT showed that the cortical induced wave more closely conformed to a standing wave with some phase dispersion than to a travelling wave.

The variance patterns of the AON and cortical foci were quite similar to those already described for the bulb.

(7) Effects of Odors and of Pentobarbital

Each of the animals with implanted arrays over the bulb or cortex was presented with a number of odors either by hand or an olfactometer and nose cone. The odors included a variety of water-soluble compounds (e.g. acetone, butanol, pyridine), esters (e.g. amyl acetate, methyl salicylate), oils (e.g. clove, cedarwood) and 'natural' odors (e.g. food pellets, sawdust, urine and feces of rabbits and cats, etc.). Care was taken to record one of the earliest bursts occurring after introduction of the odor to minimize the effects of adaptation. Visual inspection of the 60 to 64 EEG traces of bursts recorded during sniffing or breathing these odors showed no significant differences in the foci from the patterns observed in room air or with filtered air, either in phase, frequency or amplitude.

For most animals a more stringent test was used. Two records for phase and amplitude plots were made in each of 3 conditions: breathing room or filtered air, or brief presentation of a chemical such as amyl acetate, or presentation of a food odor.

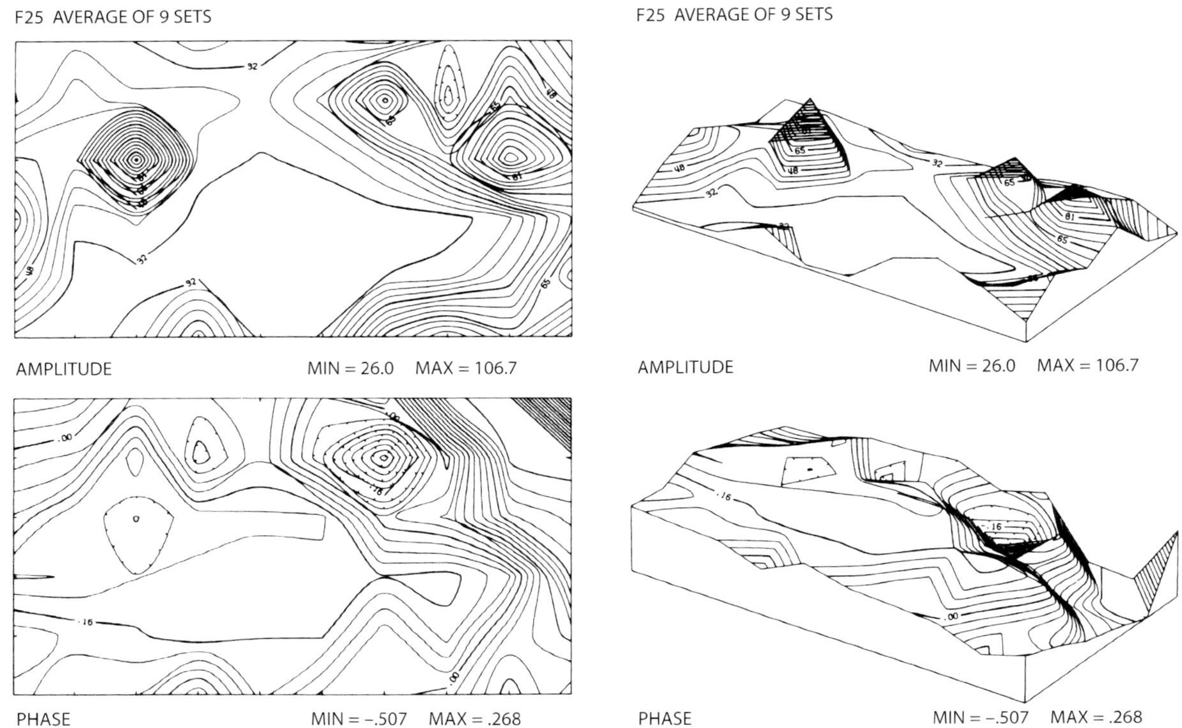

Fig. 7. Left upper frame: amplitude distribution of 9 single bursts from a 4×7 mm array placed on the frontal prepyriform cortex of a rabbit. Scale in microvolts. Left lower frame: phase distribution in radians of the same burst. The frames at right show the same data in perspective. The LOT ran from left to right parallel to the long axis of the array at the junction of the lower and middle thirds.

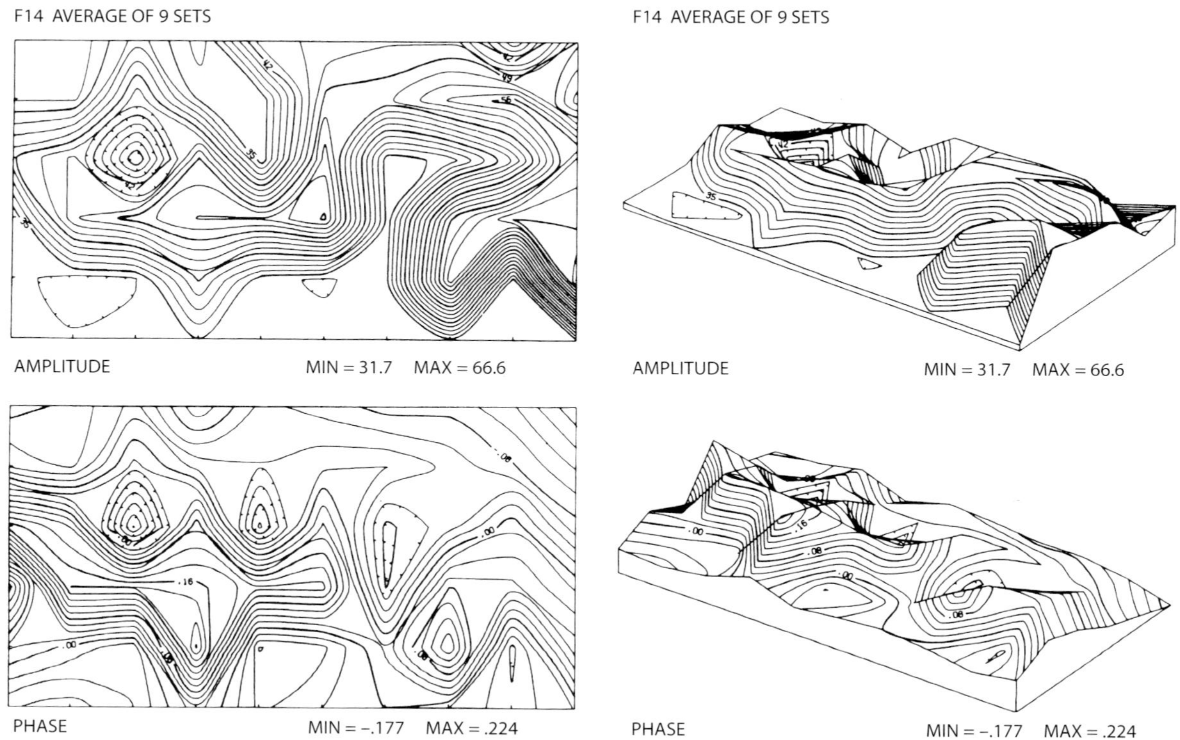

Fig. 8. Left upper frame: amplitude distribution of 9 single bursts from a 4×7 mm array placed on the temporal prepyriform cortex of a cat. Scale in microvolts. Left lower frame: phase distribution in radians. The frames at right are perspectives of the same data. The LOT ran from left to right along the lower edge of the frame to the middle of the lower edge and then turned downward.

The plots were compared by visual inspection. Differences were found between the plots in accordance with the variation already described, but they were not related to the 3 conditions. Hence it was concluded that, in the condition of presentation of a brief stimulus, whether novel or familiar, to a food-deprived animal, the spatial pattern of the induced wave bore no relation to the odor.

Records were made under barbiturate anesthesia from arrays at the time of implantation to determine whether a focus was present. Even when wave activity was present having some relation to the respiratory cycle, the records were of no predictive value in assessing whether a focus would appear on recovery. Six of the implanted animals with well characterized foci were re-anesthetized with pentobarbital in the hope that some clue could be found to guide placement in other animals at the time of surgery. In all of these animals the induced waves promptly disappeared. Several hours later there was recurrence of weak wave activity. In 2 of the 6 the amplitude and phase patterns were similar to those in the waking state, but in the other 4 the patterns were different and rather more variable.

Discussion

In the attempt to understand the coding of sensory information that is distributed over a neural ensemble the chief requirement is for a large number of observations made concomitantly on the neural activity in the ensemble, so that the spatial structure of the activity can be deduced. The most feasible approach at present is EEG recording from arrays of electrodes placed on the brain surface. The two main problems here are the management of the large numbers of data that rapidly accrue and the interpretation of the recordings in terms of what EEG events signify for neural activity. This report is mainly concerned with the first problem, the second having been dealt with elsewhere (Freeman 1975).

Four steps were used here for data reduction. First, an EEG event, the burst or induced wave, was identified by pen recordings as a manifestation of distributed neural activity and it was isolated by use of analog filters and time windows of observation. Second, it was digitized at time and space intervals determined from the spectral properties of the event. The unitary character of the event was quantitatively evaluated by cross-correlating the trace from each recording site with the ensemble mean. Third, measurements were made of its principal attributes, i.e. the distributions of amplitude, frequency and phase, and some secondary aspects such as the shape of the envelope and the modulation of frequency and phase within bursts. Fourth, the values for rms amplitude were normalized for each burst, so as to remove the overall variability of the amplitudes from one burst to the next and to highlight the spatial variability in the structure of successive bursts.

The first of two major findings in this study is that the capability for high amplitude burst activity in the waking state is restricted to a relatively small fraction of the bulb. This appears not to be an artifact of surgical damage or the placement of the arrays onto the bulbar surface, because the foci were as likely to have maximal amplitudes at the dorsal, ventral or anterior margins of the arrays as at the center. Furthermore, in unit recording from the bulbs of 12 rabbits each with a well implanted over the bulb for recording in the waking state (Freeman 1975 and unpublished data), the phasic fluctuation in firing rates with respiration and with the induced wave was found only in limited regions of the bulb not much over 2 mm in diameter. Foci were found only in the lateral bulb and ventromedial bulb, usually those two in each rabbit. (The anterior and anterodorsal parts were not accessible with the available angles of penetration.) This method minimized the risk of surgical damage to most of the bulb prior to observation, but it did not allow evaluation of the precise location, shape and stability over time of the foci. It did show that the slow wave, induced wave, and phasic variation in mean unit firing rate were all localized in the foci.

The proportion of the bulb capable of this form of burst activity is uncertain. An estimate was made that about 20% of the bulbar surface is enclosed within half-amplitude contours of foci. This was in part based on the assumption that none of the rabbits without foci had atrophic rhinitis, so that estimate may be too small by a factor up to 2. The relation between the half-amplitude width and the size of a non-normally distributed subset of active granule cells and the effect of bulbar curvature on this relation are both unclear. The results from unit recording indicate that a proportion of 1:5 is not unreasonable.

The second major finding is that in each animal the focus is stable in form and location, within certain limits of high and low spatial frequency variation. This holds over both days and weeks during breathing of room air or filtered air and on presentation of a variety of odors (but without conditioning).

Concerning olfactory coding, the results thus far clearly disallow the use of the EEG to confirm Adrian's postulate (1950a) that different odors activate neurons in grossly different parts of the bulb. Further, if odor related information is reflected in the surface EEG, it is unlikely to be found in the phase or frequency of bursts, because the ranges of variation of these variables are too small, and the amount of variation between bursts without odors is as great as that between bursts with different odors. The EEG amplitude patterns do have the range and complexity of variation that are required to reflect responses to odors or perceptual states, but they have not as yet been found to correspond to odors.

A more detailed study of the relations of burst patterns to odors may be helped by better understanding of the neural mechanisms that cause them to appear (Freeman 1975). Briefly, the bulb consists of an interactive population of excitatory and inhibitory neurons. Such a neural mass, when it maintains mutual excitation and inhibition as well as negative feedback, comprises a KII set (Freeman 1975). When it has a relatively weak level of interaction, the activity levels of the neurons tend always to fluctuate about a non-zero equilibrium. This state is manifested by the intrinsic waves of Adrian. A potential evoked by single shock stimulation decays to the equilibrium with the pattern of a damped sine wave.

When the interaction level is increased sufficiently, the response to even a small perturbation is an incremental oscillation. Owing to the saturation nonlinearities in the neurons the oscillation approaches a stable limit cycle. The interaction level depends on the amplitude of background activity in the KII set. Because the background activity is increased by receptor input, the bulbar KII set tends to switch from the stable equilibrium state to the stable limit cycle state with each inspiration and back again with each expiration. This accounts for the repetitive form, envelope, frequency modulation and duration of the bursts of induced waves, and also for the slow respiratory wave and the phasic variation in mean bulbar unit activity with respiration (Macrides and Chorover 1972).

The dependence of limit cycle activity on receptor input is shown by its absence on removal of phasic input, either temporary by closing the nostrils or permanent by surgical deafferentation of the bulb. The dependence on centrifugal input is shown by the absence of burst activity in a waking but presumably non-motivated animal. A nonolfactory stimulus such as pinching the tail of a resting cat suffices rapidly to elicit bursting. Bursts are seen to occur regularly in the undercut bulb with an intact PON (Becker and Freeman 1968), suggesting that the induction of burst activity may be due to removal by arousal of a tonic centrifugal suppression, but it is equally likely that surgical undercutting removes tonic inhibition between the bulbs by way of the anterior commissure (Callens 1967) and centrifugal input (Dennis and Kerr 1968), which may normally be blocked or overridden by an excitatory centrifugal command on arousal.

The observed spatial patterns of phase within bulbar foci are consistent with a limit cycle origin of the induced wave and not with "ringing" caused by impingement of PON input to the bulb. The local phase gradients need not reflect conduction delays within foci, although they are consistent with them. In a collection of oscillatory subjects of neurons with different frequencies, that are constrained by interaction to a common frequency, those subjects with higher natural frequencies would have relative phase lead, and vice versa. In the cortex it remains unclear whether cortical foci are resonating as tuned oscillators to bulbar input, as implied by the often high coherence between bulbar and cortical EEGs (Boudreau 1964), or have as well the capability for limit cycle activity at the same or lower frequencies.

Anesthetics with the exception of urethane or chloralose and urethane tend to suppress background activity, but especially under urethane and occasionally even under barbiturates an induced wave appears, usually in response to odor stimulation (Adrian 1950). There are multiple ways in which a KII set can go into a limit cycle state, and the fact that the burst frequency under urethane or barbiturate is usually much lower than in waking, motivated animals indicates that the mechanisms of induced waves in waking and anesthetized animals are different. For example, a high frequency limit cycle occurs with partial suppression of inhibitory neurons. Hence studies on induced waves under anesthesia may have limited applications to waking animals.

Regarding the synaptic mechanisms, the existence of limit cycle foci implies that the strength of transmission of some types of synapses must be modified in comparison to the strength of synapses in the surrounding areas, but it does not show which ones. The AEPs at sites within foci are more oscillatory directly in proportion to the limit cycle amplitudes at those points. This could mean that the mitral-tufted and granule cell negative feedback synapses are strengthened within foci, but it could also mean the PON or periglomerular synapses onto mitral-tufted apical dendrites are strengthened in foci, so that the background activity is greater, and both the EEG and AEP are more oscillatory with or without strengthened negative feedback synapses.

The tendency for AEPs in some animals at some sites to be initially positive at the surface may reflect a strengthening of the synapses of centrifugal axons in the LOT from the prepyriform or other parts of the forebrain, which establish EPSPs at the deep ends or IPSPs at the superficial ends of the granule cells. Alternatively, the initial surface positivity of the AEPs may result from activation by LOT collaterals of neurons in the anterior olfactory nucleus (AON), that in turn transmit to the granule cells and cause EPSPs at their deep ends. That is, the focus may reflect strengthening of a feedback loop between the bulb and the AON, at synapses either in the bulb or in the AON or both. The modification might instead be on centrifugal synapses in parts of the AON, that vary the level of motivation. More detailed electrophysiological analysis in waking animals and large-scale modeling will be required to sort out these and other possible mechanisms. The main obstacle at present is the difficulty in making conventional electrophysiological studies, such as depth profiles, focal PON and LOT stimulation, drug effects, and selective ablation, when the phenomenon appears robustly only in motivated, behaving animals.

The significance of the induced wave for olfactory information processing remains obscure. The limit cycle activity may have little or nothing to do directly with olfactory coding. Perhaps it is a semi-pathological but generally harmless instability that arises in a mass of interactive excitatory and inhibitory neurons, due to local aberrations in synaptic connection densities. Or perhaps it is a source of highly coherent and therefore informationally impoverished neural activity, which is used by other parts of the bulb and the cortex for purposes of timing, testing, and calibrating both within and between the bulb, nucleus and cortex. The output of a focus is apparently (on anatomical grounds) delivered uniformly to the LOT projection areas and may signal, for example, the onset, termination, and average intensity of receptor input, and include compensations for the levels of as yet unspecified centrifugal controls. In this view one would search for coding in unit and slow wave activity outside of foci and would study the foci as subsidiary elements in a complex pattern recognition device.

On the other hand, the limit cycle foci may do the bulk of odor information processing, perhaps in a holographic mode (Lettvin and Gesteland 1965; Chapter 7 in Freeman 1975), while the rest of the bulb remains unused. However, the limit cycle activity may constitute a specific form of information that does not represent the sensory input. For example, it may represent an image of the input that the animal expects to receive or is searching for. In this view one would present odors in combination with unconditioned stimuli, in order to create and modify expectations of odors in an attempt to modify the spatial patterns of foci.

These possibilities are not mutually exclusive, and there may be others, but whichever is entertained, it will be useful to be aware of the spatial nonuniformity as well as the spatial patterns of the variance of bulbar EEG activity.

Summary

Simultaneous records of the EEG were made from rectangular arrays of 60–64 electrodes successfully implanted over the surfaces of the olfactory bulb and cortex of 9 cats and 34 rabbits. Analysis was focused on bursts of sinusoidal activity at frequencies of 38–80 Hz that accompanied respiration in waking motivated animals and resemble the induced wave of Adrian. The bursts occurred in focal areas of the bulb and cortex at peak rms amplitudes averaging about 80 μV and rising by a factor of about 3 times above the relatively inactive areas. The foci sometimes appeared as randomly scattered small active zones but often were confluent into irregular shapes or occasionally a smoothly elliptical shape. From 20–40% of the bulbar surface is enclosed within the half-(peak) amplitude contours of bulbar foci.

The foci differed greatly in size, shape, location and amplitude between animals, but for each animal the shape was relatively constant over days and weeks. Maximum variation occurred at scattered sites on the slopes of foci and there was also low rms amplitude, low (spatial) frequency variation from burst to burst and from day to day. The phase differences between sites within bursts were a small fraction of a cycle (5–8%). The foci were not significantly altered in shape or location by presenting odors to the animals but were abolished or changed under barbiturate anesthesia. Preliminary results from bulbar evoked potentials suggest that foci may result from specialized conditions created by centrifugal input or from interactions between the bulb and more central structures.

The results fail to support the hypothesis that different odors are represented by activity in grossly different parts of the bulb. If the spatial distributions of the bursts reflect specific information, that information cannot depend on the odor being received but may depend on internal states including motivation and perhaps expectancy. The EEG amplitude patterns are more likely to contain this information than the frequency or phase distributions of bursts, for which the ranges of variation appear too small. The main problem for analysis is the reduction of the vast amount of data from multichannel recording. Some solutions used here include calculation of rms amplitudes and normalization of amplitude patterns for single bursts.

Résumé

Proprietes spatiales d'un événement EEG dans le bulbe olfactif et le cortex

Des enregistrements simultanés de l'EEG ont été effectués au moyen de faisceaux rectangulaires de 60–64 électrodes implantés avec succés sur les surfaces du bulbe olfactif et du cortex chez 9 chats et 34 lapins. L'analyse est centrée sur les bouffées d'activité sinuso"dale aux fréquences de 38–80 Hz, qui accompagnent la respiration chez des animaux éveillés et actifs et qui ressemblent aux ondes induites d'Adrian. Ces bouffées surviennent dans des aires focalisées du bulbe et du cortex avec des amplitudes de pics rms d'environ, en moyenne, 80 μV, et environ 3 fois plus amples que dans les aires relativement inactives. Ces foyers apparaissent parfois comme un ensemble aléatoire de petites zones actives mais souvent confluent en formes irréguliéres ou occasionnellement en une forme 1égérement elliptique. 20–40% de la surface du bulbe est contenu à l'intérieur des contours du foyer bulbaire d'amplitude correspondant à la moitié de l'amplitude de pics.

Ces foyers différent grandement en dimension, forme, localisation et amplitude d'un animal à l'autre, mais pour chaque animal la forme est relativement constante d'un jour à l'autre ou d'une semaine à l'autre. La variation maximale survient à des points répartis sur les pentes du foyer, ou on note également des amplitudes rms basses, une variation faible (spatiale) de fréquence d'une bouffée à l'autre et d'un jour à l'autre. Les différences de phase entre les localisations à l'intérieur des bouffées sont d'une petite fraction d'un cycle (5–8%). Les foyers ne sont pas significativement modifiés dans leur forme ou leur localisation lorsqu'on présente des odeurs aux animaux, mais sont abolis ou changent sous anesthésie barbiturique. Des résultats préliminaires concernant les potentiels évoqués bulbaires suggérent que ces foyers peuvent résulter de conditions spécialisées créées par les afférences centrifuges ou venant des interactions entre le bulbe et des structures plus centrales.

Ces résultats ne confirment pas l'hypothése suivant laquelle différentes odeurs sont représentées par la mise en jeu de parties trés différentes du bulbe. Si les distributions spatiales des bouffées reflétent l'information spécifique, cette information ne peut pas dépendre de l'odeur qui est reçue mais peut dépendre d'états internes comprenant la motivation et peut être l'attente. Les patterns d'amplitude EEG sont plus susceptibles de contenir cette information que la fréquence ou la distribution de phase des bouffées, pour lesquelles les marges de variations apparaissent trop faibles. Le probléme essentiel pour l'analyse est la réduction de la vaste quantité de données venant d'enregistrement à plusieurs canaux. Certaines solutions utilisées dans ce travail comportent le calcul des amplitudes rms et la normalisation des patterns d'amplitude pour des bouffées isolées.

Acknowledgements

Reprinted with permission from Spatial properties of an EEG event in the olfactory bulb and cortex; *Electroencephalography and Clinical Neurophysiology*, 1978, **44**: 586–605.

(Accepted for publication: October 6, 1977)

Supported by a grant MH06686 from the National Institute of Mental Health.

References

Adrian, E.D. The electrical activity of the mammalian olfactory bulb. *Electroenceph. clin. Neurophysiol.*, 1950, **2**: 377–388.

Adrian, E.D. Sensory discrimination, with some recent evidence from the olfactory organ, *Brit. Med. Bull.*, 1950a, **6**: 330–332.

Becker, C.J., and Freeman, W.J. Prepyriform electrical activity after loss of peripheral or central input or both. *Physiol. Behav.*, 1968, 597–599.

Boudreau, J.C. and Freeman, W.J. Spectral analysis of electrical activity in the prepyriform of the cat. *Exp.*, 1963, **8**: 423–439.

Boudreau, J.C. Computer analysis of electrical activity in the olfactory system of the cat. *Nature*, 155–158, 1964, 201.

Callens, M. *Peripheral and central regulatory mechanisms of the excitability in the olfactory system.* Brussels: Presses Academiques Europeennes, 1967.

Dennis, B.I. and Kerr, D.I.B. An evoked potential study of centripetal and centrifugal connections of the olfactory bulb in the cat. *Brain Res.*, 1968, **11**: 373–396.

Eastman, C. Construction of miniature electrode arrays for recording cortical surface potentials. *J. Electrophysiol. Tech.*, 1975, **5**: 28–30.

Freeman, W.J. *Mass Action in the Nervous System.* New York, Academic Press, 1975.

Freeman, W.J. (1978) Spatial frequency analysis of an EEG event in the olfactory bulb. *Multidisciplinary Perspectives in Event-Related Brain potential Research*: 531–546. Ed: D.A. Otto. U.S. Govt. Printing Office, EPA-600/9-77-043.

Hainer, R.M., Emslie, A.G. and Jacobson, A. An information theory of Olfaction. *Ann. N.Y. Acad. Med.*, 1954, **58**: 158–174.

Hughes, J.R., Hendrix, D.E., Wetzel, N.S., Johnston, J.W., Jr. Correlations between electrophysiological activity from the human olfactory bulb and the subjective response to odoriferous stimuli. In C. Pfaffman, (Ed.), *Olfaction and Taste III.* New York, Rockefeller Press, 1969.

Le Gros Clark, W.E. Inquiries into the anatomical basis of olfactory discrimination. *Proc. Roy. Soc.*, 1957, **146B**: 299–319.

Lettvin, J.Y. and Gesteland, R.C. Speculations on smell. *Cold Spring Harbor Symp. Quant. Biol.* **30**: 217–225, 1965.

Leveteau, J. and MacLeod P. Olfactory discrimination in the rabbit olfactory glomerulus. *Science*, 1966, **153**: 175–176.

Macrides, F. and Chorover, S.L. Olfactory bulb units:Activity correlated with inhalation cycles and odor quality, *Science*, 1972, **175**: 84–87.

Moulton, D.G. Electrical activity in the olfactory system of rabbits with indwelling electrodes. In Y. Zotterman (Ed.), *Olfaction and Taste*, Pergamon, Oxford, 1963, pp. 71–84.

Moulton, D.B., and Tucker, D. Electrophysiology of the olfactory system. *Ann. New York Acad. Sci.*, 1964, **116**: 380–428.

Scott, J.W. and Aaron, D.D. Average odor induced waves in the olfactory bulb of the rat. *Exp. Neurol.*, 1977, **55**: 654–665.

Shepherd, G.M., and Haberly, L.B. Partial activation of olfactory bulb: analysis of field potentials and topographic relation between bulb and lateral olfactory tract. *J. Neurophysiol.*, 1970, **33**: 643–653.

Wenzel, B.M. and Sieck, M.H. Olfaction, *Ann. Rev. Physiol.*, 1966, **28**: 381–434.

10. Relations Between Microscopic and Mesoscopic Levels Shown by Calculating Pulse Probability Conditional on EEG Amplitude, Giving the Asymmetric Sigmoid Function

The first significant new insight from gamma spatial analysis emerged on re-examination of the sigmoid function representing bilateral saturation. The AM patterns of gamma activity show the capacity of the olfactory system to construct novel responses that are determined by the internal dynamics, and only indirectly by the properties of the stimulus. Therefore, there must exist in the bulb a mechanism by which the input on each inhalation serves to initiate the construction of an AM pattern to be transmitted centrally from the bulb. Most dynamical systems give outputs that are determined by processing of inputs, and that serve as "representations" of the inputs (Freeman 1983; Freeman and Skarda 1985, 1989). We know that for each olfactory burst an input is required, because gamma bursts are prevented by the simple act of closing the nostrils with the fingers to force mouth breathing by awake animals. The most likely mechanism of burst formation is sensitization of the bulb by input, such that the volley of receptor input to the bulb lowers the threshold of certain bulbar neurons to interact among themselves. I concluded that each inhalation might induce an abrupt increase in the feedback gain among bulbar neurons, causing the entire bulb to transit from an equilibrium state under noise during exhalation (Mode 1) to a limit cycle state during inhalation (Mode 2) (Figure IV in Chapter 4).

The location at which I sought evidence for this input-augmented gain was the trigger zones of bulbar neurons, as measured by the sigmoid curve. A valuable clue for this search had been provided by one of my students (Sunday and Freeman 1975), a mathematician who had analyzed my data by taking numerical derivatives of the experimental pulse probability curves. He found that the maximal slope was not at the point of the curve where the input was zero, as I had previously supposed (Freeman 1975), but that the maximum was displaced to the excitatory side (Figure 1 in Chapter 10), and more prominently in awake animals (Figure 2 in Chapter 10) than in anesthetized animals. This meant that excitatory input to the mitral cells not only increased their output; it also increased their sensitivity to fresh input while they were still excited. That new input was provided predominantly by their own axon collaterals to each other as described by Ramón y Cajal (1955). The sensitivity was enhanced up to 40,000-fold among neurons in which learning had strengthened the mutually excitatory connections to form Hebbian nerve cell assemblies (Freeman 1979b). The equation that I derived used the concept of voltage-dependent sodium permeability expressed as the *m* factor in the Hodgkin–Huxley equations (Figure 3 in Chapter 10). Using this asymmetric sigmoid function (Figure 4 in Chapter 10) and taking its derivative to get the nonlinear gain at the population trigger zones (Figure 5 in Chapter 10), I replicated my collection of evoked potential data (Freeman 1979b) and showed how an array of coupled oscillators, each with the asymmetric sigmoid curve bounding its excitatory and inhibitory populations, could replicate the main features of learning appearing in the EEG data (Freeman 1979c). This model explained the neural mechanism for the bistability of the olfactory system manifested in Mode 1 and Mode 2 (Chapter 4), as well as the mechanism by which state transitions could be induced by receptor input during inhalation, but not by electrical impulse stimuli, thereby explaining the difference between the bulbar responses to inhalations versus the impulse responses (Chapter 9).

In my discussion I pointed out the utility of the technique of correlating dendritic membrane current density with axon firing frequency of single neurons as a powerful tool in cellular physiology, and its adaptation to cortex by calculating the conditional nerve impulse probability from simultaneous recordings of trains of action potentials and the local EEG amplitudes (also in Eeckman and Freeman 1990). Several unanswered questions are listed regarding unexplained details in the data. The technique is simple, powerful, and informative, yet since its inception (Freeman 1967b, 1975) it has been underutilized. It provides a powerful rationale for global and circular causality (Freeman 2000) in the relations it reveals between the microscopic and mesoscopic levels of activity in the form of pulse density waves of oscillations in the gamma band. This is because the linear causal action in the bulb is the excitation of granule cells by mitral cells, so that the input variable (mitral pulse density) should appear on the abscissa, and the output variable (EEG amplitude of granule cells) should appear on the ordinate. If plotted in this relation the curves in Figure 2 in Chapter 10 would appear as if rotated 90° counterclockwise and then flipped right to left as a mirror image. Linear causality was the basis on which I interpreted the sigmoid curve of prepyriform data, because the upper asymptote was not apparent (Figure 3.18c on page 158 in Freeman 1975). This was a serious mistake, which was rectified only after it became clear that the sigmoid curve is a population property described by circular causality, not a neural property that is regulated by direct forward synaptic transmission and described by linear causality (Freeman 1995, 1999a). The upper limit on mitral trigger zones holds irrespective of which measure is chosen as the independent variable. The high nonlinear gain in the driven prepyriform populations prevents the upper asymptote, Q_m, from being observed directly (Eeckman and Freeman 1991).

Nonlinear Gain Mediating Cortical Stimulus–Response Relations

Abstract

Single isolated neurons show a nonlinear increase in likelihood of firing in response to random input, when they are biased toward threshold by steady-state depolarization. It is postulated that this property holds for neurons in the olfactory bulb, a specialized form of cortex in which the steady state is under centrifugal control. A model for this nonlinearity is based on two first order differential equations that interrelate three state variables: activity density in the pulse mode p of a local population (subset) of neurons; activity density in the wave mode u (mean dendritic current); and an intervening variable m that may be thought to represent the average subthreshold value in the subset for the sodium activation factor as defined in the Hodgkin–Huxley equations. A stable steady state condition is posited at (p_0, u_0, m_0). It is assumed that: (a) the nonlinearity is static; (b) m increases exponentially with u; (c) p approaches a maximum p_m asymptotically as m increases; (d) $p \geq 0$; and (e) the steady state values of p_0, u_0, and m_0 are linearly proportional to each other. The model is tested and evaluated with data from unit and EEG recording in the olfactory bulb of anesthetized and waking animals. It has the following properties. (a) There is a small-signal near-linear input–output range. (b) There is bilateral saturation for large input, i.e. the gain defined as dp/du approaches zero for large $|u|$. (c) The asymptotes for large input $\pm u$ are asymmetric. (d) The range of output is variable depending on u_0. (e) Most importantly, the maximal gain occurs for $u > u_0$, so that positive (excitatory) input increases the output and also the gain in a nonlinear manner. It is concluded that large numbers of neurons in the olfactory bulb of the waking animal are maintained in a sensitive nonlinear state, that corresponds to the domain of the subthreshold local response of single axons, as it is defined by Rushton and Hodgkin.

Introduction

Sensory systems are remarkable for their combined sensitivity and stability. A stimulus that is sufficient to excite one or a few receptors in a prepared animal can give rise to widespread autonomic, postural and purposive locomotor changes within 200–300 ms, which implies that the initial activity has been rapidly spread widely through the forebrain in a controlled manner. It further suggests the existence at an early stage in each sensory system of neural mechanisms for rapid increase in the numbers of neurons responding to a stimulus from one or a few neurons to many, say 10^4 or more, as the basis for dissemination through the CNS.

One plausible mechanism requires that receptor input to a small number of 2nd order sensory neurons not only excite them but increase their transmission strength. If these neurons were excitatory to each other as well as others in their neighborhood, then regenerative neural activity would result. That is, the stimulus action would be to increase the feedback gain in a local ensemble of neurons that was prepared to receive the stimulus. But the ensemble must be stable in three respects. The response should not occur or should be distinctly different in the presence of adventitious events (noise). It should be self-limiting in its maximal amplitudes. And it should be rapidly self-terminating, so that a new stimulus sample can be taken through the same ensemble within a few hundred ms.

The key to understanding this mechanism is the nonlinear relationship between input and output for a subset of neurons in parallel within an ensemble (Freeman, 1975). Each neuron performs two time-dependent nonlinear operations. Incoming axonal impulse trains (the "pulse" mode) are converted to continuously varying dendritic current (the "wave" mode) at synapses, and the current serves to control the generation of impulses at trigger zones. The distributions of pulses for single neurons tend to be Poisson, and the wave amplitude distributions to be Gaussian. Over a subset of neurons, in which the time scale of a sensory event is one or two orders of magnitude longer than the time scale of a nerve impulse, the pulses give rise to a continuous pulse density, and the pulse-wave and wave-pulse transformations can be treated as static nonlinearities (Baker, 1978). The input-output operation for the subset can be expressed as the product of two nonlinear functions in pulse and wave amplitude. The derivative of the nonlinear function with respect to wave amplitude gives the nonlinear gain of the operation.

An expression for the nonlinear function has been derived experimentally (Freeman, 1975) from the relation of cortical neural firing probabilities to the EEG. The frequency of firing of a neuron or a cluster of neurons recorded with a microelectrode is taken as a manifestation of the pulse density of a neural subset around the electrode. In the appropriate neural geometry (Freeman, 1975, p. 233) the EEG recorded at or near the subset is a manifestation of the average membrane potential of neurons in a subset. A set of simultaneous measurements of EEG amplitude and pulse occurrence sampled at a high rate (e.g. 1 ms) over an extended time span (e.g. 100 s) yields an EEG amplitude histogram and the mean pulse rate. Then the empirical probability of cell firing is determined conditional on the EEG amplitude. This pulse probability distribution is an expression of the nonlinear operations of pulse-wave or wave-pulse conversion. Its numerically computed derivative with respect to wave amplitude is a pulse probability density[11] that is equivalent to the nonlinear gain. This gain curve is the focus of interest in this report.

1 Confusion should be avoided between pulse density p and conditional pulse probability density $dP(V)/dV$ The pulse density denotes the pulse rate/unit volume of tissue enclosing a neural subset. It can be estimated by ensemble averaging over unit trains as in constructing a post-stimulus time histogram. The pulse probability density is estimated by constructing the pulse probability distribution conditional on amplitude V and then computing the derivative with respect to V.

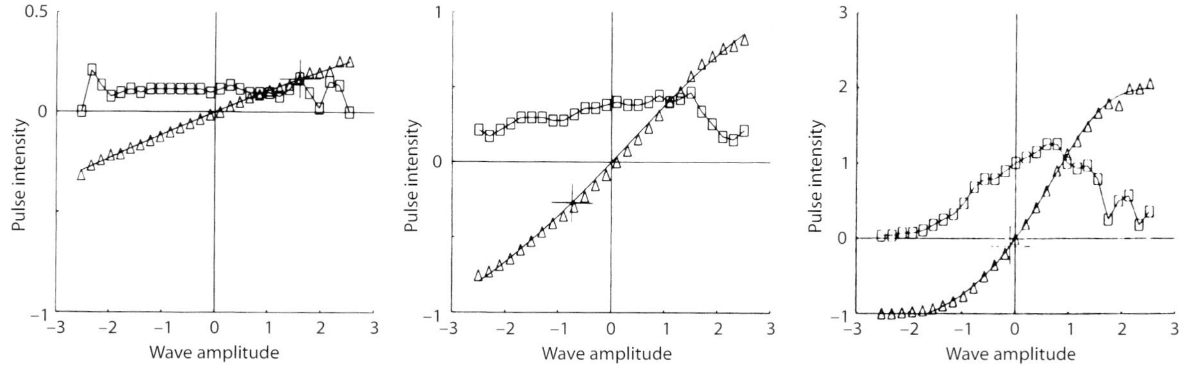

Fig. 1. Three examples are shown of the normalized conditional pulse probability distribution (Δ) and the numerically computed pulse probability density (□) from anesthetized rabbits. The curves fitted to the pulse probability distribution (Δ) are from Eqs. (5.0), (5.1), (10.0), and (10.1). The parameters are:

	Q_m	σ	Δp	Δv
left	1.00	0.10	−0.16	−0.16
middle	2.02	0.55	0.52	0.42
right	2.33	1.10	0.09	0.09

The experimental origin is at $(\hat{P}(\hat{V})-\hat{\bar{p}}, \hat{u}-\hat{\bar{u}})$. The small cross in each frame shows the estimated true origin from evaluation of Δv and Δp.

Conditional Pulse Probability Distribution and Density

The nonlinear function has been derived experimentally by analysis of the relation of nerve impulses to the EEG in the olfactory system of cats and rabbits. Details have already been published (Freeman, 1975). Units were recorded singly or in clusters from a microelectrode in cortex $p(t)$ while the EEG was recorded from a surface electrode $v(t)$. After appropriate amplification and filtering of the two signals, the experimental pulse probability conditional on EEG amplitude and time lag was calculated as follows. The EEG was band-pass filtered to remove unit activity and the low-frequency respiratory wave of the olfactory EEG and was then digitized over a time segment of 10–100s; the N values were normalized to zero mean and unit standard deviation, σ. The amplitude range was divided into 60 bins 0.1 in width from +3 to −3. For each occurrence of an amplitude $\hat{V}$ at time t the question was asked, did a pulse occur at each time lag T in $p(t + T)$, $-25 \text{ ms} \le T \le +25$ ms? If so, a count of 1 was added to the appropriate bin of a table over $\hat{V}$ and T. The resulting histogram was divided by N and divided by the amplitude probability density to give the conditional pulse probability $\hat{P}(\hat{V}, T)$.

Cross sections of the table at fixed $\hat{V} \neq 0$ gave $\hat{P}(T)$ which was oscillatory at the frequency of the EEG. Cross sections at $T = T_{max}$ where the value of $\hat{P}(T)$ was maximal gave $\hat{P}(\hat{T})$, which was normalized by dividing the function by the mean firing rate $\hat{\bar{p}}$.

The main conclusions from analysis of experimental curves from the olfactory bulb in anesthetized animals were as follows.

a) The maximally varying nonlinear functions were found at T_{max} values for which the average of T_{max} was equal to one quarter cycle of the dominant frequency of the EEG. This reflected the fact that the EEG was generated by inhibitory neurons and the pulses by excitatory neurons forming a negative feedback loop in the bulb.
b) For low amplitude EEG activity $\hat{P}(\hat{T})$ at $T = T_{max}$ was nearly linear over the interval $\pm 3\sigma$.
c) During high-amplitude EEG activity the curves became nonlinear with (on the average) monotonically increasing pulse probability P with increasing wave amplitude and horizontal asymptotes (a "sigmoid" curve as shown in Fig. 1). The sigmoid curve was interpreted to show wave-pulse conversion (wave being the independent variable).
d) The sigmoidal curves were asymmetric about the mean firing rate $\hat{\bar{p}}$ with a lower asymptote at $\hat{p} = 0$ and an upper asymptote at $3\hat{p}_0$.

Analysis of the numerically computed derivatives of the experimental curves (Sunday and Freeman, 1975) showed that

e) the dominant peak of the pulse probability density usually occurred in the interval $\hat{V} = (\sigma / 2, \sigma)$, and its magnitude increased linearly with $\hat{\bar{p}}$.
f) The experimental curves from waking rabbits, in which EEG amplitudes were much greater than in anesthetized animals, revealed with increasing EEG

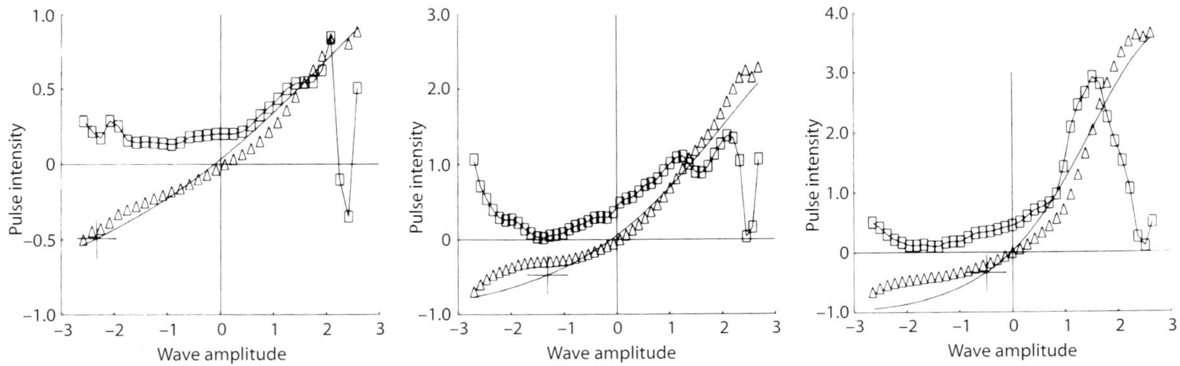

Fig. 2. Three examples are shown of the normalized conditional pulse probability distribution (Δ) and density (□) from waking rabbits; no clear upper asymptotes are seen. The fitted curves (see legend for Fig. 1) have the parameters:

	Q_m	σ	Δp	Δv
left	4.84	0.33	0.97	0.77
middle	6.83	0.56	0.91	0.74
right	14.86	0.52	1.62	1.05

amplitude a marked increase in the steepness of the nonlinear curve, the degree of asymptotic asymmetry, and the displacement of the peak of the density curve toward higher amplitude (Fig. 2). The newer findings (e–f) were incompatible with an earlier model (Freeman, 1975), in which it was assumed that the maximal nonlinear gain was at v=0, and that the degree of asymptotic asymmetry was fixed at 2, so the construction of a new model was undertaken.

Derivation of the Model

The element of cortical function is taken to be an ensemble of neurons in a local neighborhood that can be lumped without regard to spatial dimensions. The ensemble is described in terms of its topology (the network of connections) and the operations performed by the subset of neurons at each node in the network. The operations of the subset occur in parallel at the same time and are the statistical averages of the operations of the neurons comprising the subset. This report deals with the operations; the topology need not be considered here.

Each subset has two state variables: activity density in the pulse density mode $p(t)$ in axons and activity density in the wave mode $v(t)$ in dendrites. There are four operations. Dendritic activity is converted to axonal activity at trigger zones by a nonlinear operation $G_a(v)$. It is linearly delayed and dispersed in time by axonal propagation and multiplied by axonal branching $F_a(p)$. Axonal activity is re-converted to dendritic activity at synapses by a nonlinear operation $G_d(p)$. The synaptically induced activity is integrated and delivered to the trigger zones by a linear operation $F_d(v)$. In the order of operation the sequence is

$$G_a(v)F_a(p)G_d(p)F_d(v).$$

When, in the physiological range of operation, the axonal nonlinearity $G_a(v)$ dominates and thereby restricts the dendritic operation $G_d(p)$ to a linear range, $G_d(p)$ can be replaced by a constant ζ. Then the two linear operations can be combined into a single operator $F(p)$ in the sequence

$$G_a(v)\zeta F(p).$$

The linear operation $\zeta F(p)$ need not be considered further here.

The proposed model of the nonlinearity contains three state variables. The first two p and u (Fig. 3) represent activity density respectively in the pulse and wave modes. Alternatively q and v are normalized from p and u to zero in the steady state and are used for observation, measurement and computation. The third variable m is to be defined.

In the steady state wherein $\mathrm{d}p/\mathrm{d}t = 0$ and $\mathrm{d}u/\mathrm{d}t = 0$, $m = m_0$, $u = u_0$ and $p = p_0$. The variables for computation and experimental measurement are

$$v \triangleq \beta(u - u_0), \tag{1.0}$$

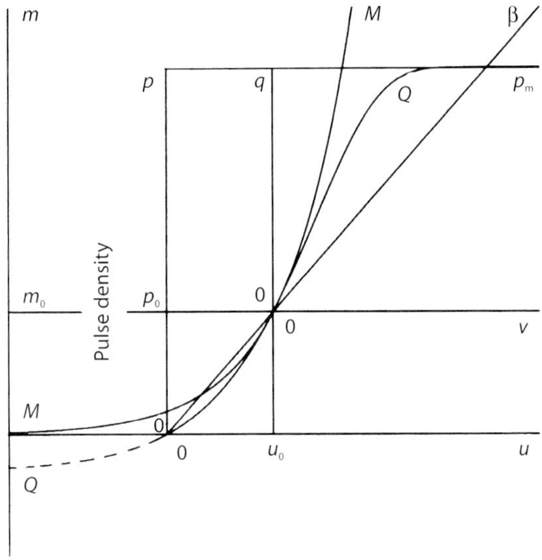

Fig. 3. Normalized pulse density Q is a function of wave amplitude u. The steady state values are $u = u_0$ and $p = p_0$. The normalized variables are $q = p - p_0$ and $v = u - u_0$. The intervening variable m is orthogonal to p and u but is shown in the p–u plane for convenience. Curve M is from Eq. (2.2). Curve Q is from Eq. (5.0)

$$q \triangleq p - p_0 , \tag{1.1}$$

where $\triangleq$ means "is defined as equal to" and β is a scaling factor. The dimensionless normalized variables are

$$Q \triangleq (p - p_0) / p_0 , \tag{1.2}$$

$$M \triangleq (m - m_0) / m_0 . \tag{1.3}$$

Three properties are assigned to the neural subset in the local ensemble. First it is assumed that the m variable is regeneratively voltage-dependent. Corresponding to this property,

$$\frac{dm}{du} = \alpha m \tag{2.0}$$

For any change in u from u_0 and m from m_0

$$\int_{m_0}^{m} \frac{dm}{m} = \alpha \int_{u_0}^{u} du . \tag{2.1}$$

From Eqs. (1.0) and (1.3)

$$M = e^{v} - 1 \tag{2.2}$$

Second, it is assumed that p increases with increasing m, but that there exists an upper limit on pulse density p for any given steady state p_0 and u_0. There are many factors in this limitation, some (such as voltage-dependent saturation of Na-activation, potassium conductance and sodium inactivation) depending on membrane potential, others (such as hyperpolarizing after–potentials) depending on previous activity, and these factors are expressions of nonlinear and time–varying relationships. Over the subset and over the longer time scale of its operation (compared to the time scale of pulses) it is assumed that the nonlinearities become static, and that the rate of increase in pulse density $\mathrm{d}p/\mathrm{d}m$ is proportional to the difference between p and the prevailing maximum p_{m} and is inversely proportional to the range p_0–p_m.

$$\frac{\mathrm{d}p}{\mathrm{d}m} = \beta' \frac{p_m - p}{p_m - p_0} \tag{3.0}$$

where β' is a scaling factor.

Integration from p_0 to p and from m_0 to m,

$$\int_{p_0}^{p} \frac{\mathrm{d}p}{p_m - p} = \frac{\beta'}{p_m - p_0} \int_{m_0}^{m} \mathrm{d}m, \tag{3.1}$$

gives

$$p = p_0 + (p_m - p_0)(1 - \mathrm{e}^{-\beta'(m - m_0)(p_m - p_0)}). \tag{3.2}$$

To simplify the notation the prevailing maximal pulse density p_{m} is expressed as a dimensionless number from Eq. (1.2).

$$Q_m \triangleq (p_m - p_0) / p_0. \tag{3.3}$$

From Eqs. (1.1), (1.3) and (3.2)

$$Q = Q_m \left[1 - \exp\left(-\beta' \frac{M}{Q_m} \frac{m_0}{p_0} \right) \right] \tag{3.4}$$

The third property of the system is taken from the empirical observation that for most neurons over a physiological range the steady state pulse rate is proportional to steady state transmembrane current or level of depolarization (Granit *et al.*, 1963; Stein, 1967; Calvin, 1975; Heyer and Llinàs, 1976).

$$p_0 = \gamma u_0, \tag{4.0}$$

where γ is a scaling factor. This linear relationship cannot be derived from the Hodgkin–Huxley system (Agin, 1964; Cooley and Dodge, 1966), although it can be extracted by appropriate modifications of the system (Dodge, 1972; Fohlmeister *et al.*, 1974). The relationships between m_0 and both p_0 and u_0 must also be specified. Intuitively it seems that m_0 should increase monotonically with u_0 over the physiological range of operation of a neural subset, and p_0 should increase monotonically with m_0. Equation (4.0) requires that each monotonic function be the inverse of the other. In the absence of experimental or theoretical evidence to the contrary, it is assumed that

$$p_0 = \beta' m_0 \tag{4.1}$$

Since m is not measured, $\beta' = 1$.

The relationship between u_0 and Q_m is based on the empirical property that $p \geq 0$ and therefore $Q \geq -1$. The value for $u = 0$ is set at $Q = -1$. From Eqs. (1.0), (2.2), and (3.4)

$$M = e^{-\beta u_0} - 1 \tag{4.2}$$

$$e^{-M/\gamma} = 1 + \frac{1}{Q_m} \tag{4.3}$$

Because u_0 is not measured, $\beta' = 1$ and

$$u_0 = -\ln\left[1 - Q_m \ln\left(1 + \frac{1}{Q_m}\right)\right]. \tag{4.4}$$

On combining Eqs. (2.2), (3.4), (4.0), and (4.4) (see Fig. 3)

$$\begin{aligned} &Q = Q_m (1 - \exp[-(e^v - 1)/Q_m]), \quad v > -u_0 \\ &Q = -1, \quad v \leq -u_0. \end{aligned} \tag{5.0}$$

From Eqs. (1.2) and (4.0) the pulse output is

$$p = \gamma u_0 (Q + 1). \tag{5.1}$$

The derivative of Eq. (5.0) is equivalent to the experimental pulse probability density curve. It is found by treating the two implicit operations [(2.0) and (3.0)] as serial

$$\frac{dQ}{dv} = \frac{dM}{dv}\frac{dQ}{dM}. \tag{6.0}$$

From Eqs. (2.2) and (3.4)

$$\frac{dQ}{dv} = \exp[v - (e^v - 1)/Q_m]. \tag{6.1}$$

At $v = v_0 = 0$,

$$Q_0 = 0 \tag{6.2}$$

$$\frac{dQ}{dv} = 1 \tag{6.3}$$

Equation (6.1) describes the forward gain in the operation $G_a(v)$. The gain approaches zero as v approaches $\pm\infty$ and goes to a maximum at $v \underline{\underline{\Delta}} v_0$. The second derivative

$$\frac{d}{dv}\left(\frac{dQ}{dv}\right) = \frac{dQ}{dv}(1 - e^v/Q_m), \tag{6.4}$$

is zero at $v = v_g$, so that

$$v_g = \ln(Q_m). \tag{6.5}$$

The peak of the gain curve shifts to the right and its height increases with increased Q_m and [by Eq. (4.4)] U_0. The third derivative is

$$\frac{d^2}{dv^2}\left(\frac{dQ}{dv}\right) = \frac{dQ}{dv}(1 - 3e^x / Q_m + e^{2v} / Q_m^2), \tag{6.6}$$

and is zero at

$$e^v = Q_m(1.5 \pm \sqrt{5}/2). \tag{6.7}$$

When $v = 0$, there is a maximum in the second derivative for

$$Q_m = 2.62. \tag{6.8}$$

This value for Q_m provides a convenient dividing point between high and low sensitivity. For $Q_m < 2.62$ the acceleration of gain for $v \geq 0$ is negative, whereas for $Q_m >$ 2.62 the acceleration is positive.

The nonlinear gain for wave-pulse conversion is

$$\frac{dp}{dv} = \gamma u_0 \exp(v - e^{-(v-1)/Q_m}), \tag{6.9}$$

and the maximal gain at $v = v_g$ is

$$\left(\frac{dp}{dv}\right)_g = \gamma u_0 Q_m e^{-1+1/Q_m} \tag{6.10}$$

For $Q_m \leq 1$, $v_g \leq v = 0$, and for $Q_m < 0.46$, $v_g < V = -u_0$ where $p = 0$, so the gain is bounded for small values of Q_m. For $Q_m >> 1$ there is no upper limit on maximal gain other than that set by Q_m and u_0, but peak displacement is limited to $v_g < V = +u_0$.

This completes the definition of the nonlinear function. The parameter γ is equivalent to the sensitivity of wave-pulse conversion and to the slope of a straight line approximating the relation between firing rate and transmembrane current (the so-called *f*/*I* curve). The steady state wave amplitude u_0 may be thought of as the controlling parameter for both p_0 and p_m (the sensitivity and range of wave-pulse conversion). It is inaccessible to measurement. However, it can be evaluated from measurement of the dimensionless parameter Q_m.

Examples of the nonlinear function are shown in Fig. 4. The main features are:

1. A near-linear range for the domain of small wave amplitudes about the origin.
2. A sigmoidal shape with large wave amplitudes that represents bilateral saturation.
3. Asymmetry of the asymptotes about the origin.

Examples of the nonlinear gain curves are shown in Fig. 5. The main features are:

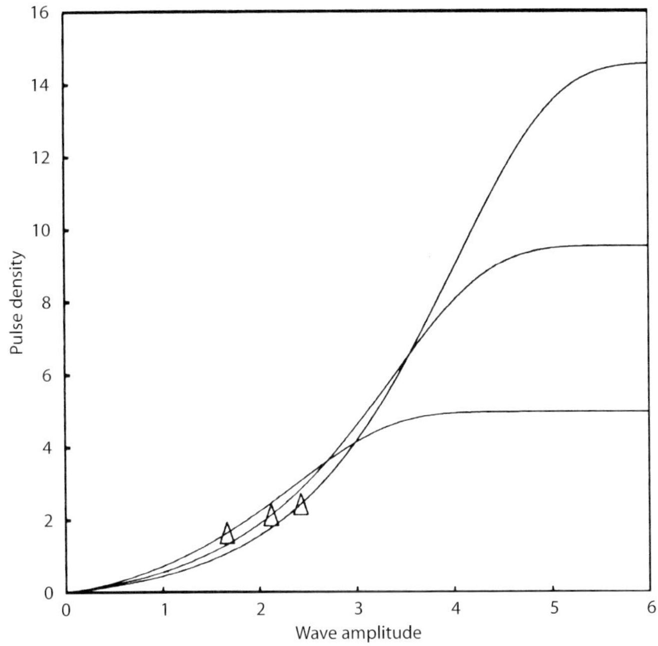

Fig. 4. Three examples of nonlinear functions $p = G_a(v)$ are shown for $Q_m = 2.0, 3.5, 5.0$. The triangles show the steady state values of (p_0, u_0) [see Eqs. (5.0) and (5.1)]. Note the sigmoid shape

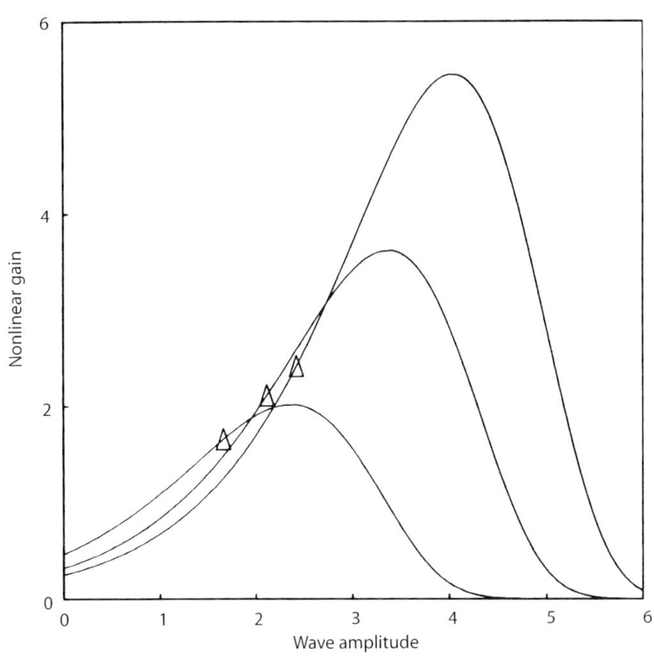

Fig. 5. Three examples of nonlinear gain curves are shown for the curves in Fig. 2 [see Eq. (6.1)]. Note that the gain increases with wave amplitude increasing from the steady state origin at (p_0, u_0)

1. Diminished gain approaching zero at both extremes of wave amplitude.
2. Increasing gain with increasing positive wave amplitude from the origin (steady state).
3. Increasing gain with increasing steady state wave amplitude u_0 and therefore with asymmetry of the asymptotes.

Evaluation of the Model

The experimental data on which this approach is based are the time series of the EEG and pulse trains recorded from cortex. The absolute measures in volts and pulses/s are arbitrary, depending on the local geometry of the recording arrangements and the clipping level in the pulse derivation system. The wave and pulse measurements must therefore be normalized.

Use of the proposed model requires that the neural activity density $\hat{u}(t)$ be identified with the state variable $v(t)$ by the scaling factor β,

$$v(t) = \beta(\hat{u}(t) - \hat{u}_0). \tag{7.0}$$

The EEG time series that manifests the neural variable $\hat{u}(t)$ is passed through RC-coupled amplifiers. Over a time epoch of sufficient length, say 10–100 s, the time average is zero and the standard deviation is $\hat{\sigma}$. The normalized EEG has had an unknown mean value $\hat{\bar{u}}$ subtracted from it. In order to maintain the equivalence specified by Eq. (7.0) it is necessary to specify the relation between the average operating value of the wave mode in the neural system u during the time epoch of observation and the steady state value of the wave mode in wave-pulse conversion $\hat{u}_0$.

$$\Delta\hat{u} \triangleq \hat{\bar{u}} - \hat{u}_0 \tag{7.1}$$

Corresponding to $\Delta\hat{u}$ there is a quantity $\Delta v = \bar{v}$ that represents a possible offset in the model representing the neural system. If the operating and steady state values correspond, both $\Delta\hat{u}$ and Δv are zero.

Otherwise, on normalizing both sides of Eq. (7.0),

$$\frac{v(t)}{\sigma} = \frac{\beta[\hat{u}(t) - \hat{\bar{u}} + \Delta\hat{u}]}{\beta\hat{\sigma}}. \tag{7.2}$$

Then the normalized EEG signal is

$$\hat{V}(t) = \frac{\hat{u}(t) - \hat{\bar{u}}}{\hat{\sigma}} \tag{7.3}$$

and the equivalent normalized state variable in the model at any time t is

$$V = \frac{v - \Delta v}{\sigma} \tag{7.4}$$

The equivalence between model and experiment in the pulse mode is by a scaling factor β''

$$p(t) = \beta'' \hat{p}(t), \tag{7.5}$$

$$\bar{p} = \beta'' \hat{\bar{p}}, \tag{7.6}$$

where p is mean firing rate of units or unit clusters. The normalized conditional pulse probability is

$$\hat{P}(\hat{V}) = \hat{p}(\hat{V}) / \hat{\bar{p}}. \tag{7.7}$$

If the mean operating value of the pulse mode p is equal to the steady state value p_0, then the difference

$$\Delta\hat{p} \triangleq (\hat{\bar{p}} - \hat{p}_0) / \hat{p}_0 \tag{7.8}$$

is zero, and

$$\hat{Q}(\hat{V}) = \hat{P}(\hat{V}) - 1. \tag{7.9}$$

If both Δp and Δu are zero, then it is predicted that

$$\hat{Q}(\hat{V}) = 0, \qquad \hat{V} = 0. \tag{7.10}$$

The converse does not hold. If $\Delta\hat{u} \neq 0$ and $\Delta\hat{p} \neq 0$,

$$\hat{Q}(\hat{V}) = \hat{P}(\hat{V})(1 + \Delta\hat{p}) - 1. \tag{7.11}$$

and $\Delta\hat{p}$ must be evaluated experimentally. In the model an equivalent normalized pulse density Q' is given by

$$Q' = (Q+1)/(\Delta q + 1) - 1 \tag{7.12}$$

where $\Delta q = \overline{Q}$ and $\overline{Q} = (\bar{p} - p_0)/p_0$ by Eq. (1.2).

The results from measurement of $\hat{P}(\hat{V})$ in data from anesthetized animals (Freeman, 1975) has shown that

$$\hat{Q}(\hat{V}) = -0.02 \pm 0.06, \qquad \hat{V} = 0,$$

so it might be assumed that in this condition $\Delta\hat{u}$ and $\Delta\hat{p}$ are neither significantly different from zero. Then the upper asymptote Q_m can be estimated by fitting the curve from Eq. (5.0) to the data, by use of nonlinear regression with minimization of the sum of squares of the differences.

Two alternative approaches are based on measurement of the normalized gain curve

$$\hat{\dot{Q}} = \mathrm{d}\hat{Q} / \mathrm{d}\hat{V}. \tag{8.0}$$

First, from Eqs. (6.1) and (7.3) at V=0,

$$\dot{Q}_0 = \sigma. \tag{8.1}$$

From Eqs. (6.3), (7.3), and (8.1),

$$Q_m = \exp(\dot{Q}_0 V_g). \tag{8.2}$$

The second approach is based on the finding that the height of the peak in the gain curve Q_g at V_g increases with Q_m. From Eq. (6.1)

$$\dot{Q}_g = \sigma e^{\sigma V_g} \exp[-(e^{\sigma V_g} - 1)/Q_m]. \tag{8.3}$$

From Eq. (5.0)

$$\frac{Q_m - Q_g}{Q_m} = \exp[-(e^{v} - 1)/Q_m]. \tag{8.4}$$

On combining Eqs. (8.2)–(8.4) and solving for Q_m,

$$Q_m = Q_g + \dot{Q}_g / \dot{Q}_0. \tag{8.5}$$

To recapitulate, under the assumption that $\Delta\hat{u} = \Delta\hat{p} = 0$ is measured by curve-fitting,

$$Q_m^a = \hat{Q}_m \tag{9.0}$$

and four additional measurements are made:

$\hat{V}_g$ = wave amplitude at peak of gain curve,

$\hat{\dot{Q}}_g$ = peak value of gain curve at $\hat{V}_g$,

$\dot{Q}_0$ = value of gain curve at $\hat{V} = 0$,

$\hat{Q}_g$ = value of nonlinear function $\hat{P}(\hat{V})$ at $\hat{V}_g$,

where

$$\hat{Q} = \hat{P}(\hat{V}) - 1 \tag{9.1}$$

$$\hat{\dot{Q}} = \mathrm{d}\hat{P}(\hat{V}) / \mathrm{d}\hat{V} \tag{9.2}$$

From these measurements

$$Q_m^b = \exp(\hat{\dot{Q}}_0 \hat{V}_g), \tag{9.3}$$

$$Q_m^c = \hat{Q}_g + \hat{\dot{Q}}_g / \hat{\dot{Q}}_0. \tag{9.4}$$

The results of fitting curves by nonlinear regression to 24 normalized conditional pulse probability curves from anesthetized animals have given (mean and standard deviation)

$$Q_m^a = 1.77 \pm 0.53,$$
$$\sigma^a = 0.63 \pm 0.27.$$

Measurements on the peak of the gain curve and its intercept at $V = 0$ have given

$$Q_m^b = 2.05 \pm 0.65,$$
$$Q_m^c = 2.14 \pm 0.55,$$
$$\sigma^{bc} = 0.62 \pm 0.28.$$

This outcome replicates previous results by a similar method (Freeman, 1975) in giving a value for Q_m near 2. However, the correlation coefficients for the 3 sets of measurements of Q_m and the t-values ($df = 22$) for the differences between means are

$$r_{a \cdot b} = 0.60,\ p < 0.01 \quad t_{a \cdot b} = 1.60,\ \text{n.s.},$$
$$r_{a \cdot c} = 0.79,\ p < 0.01 \quad t_{a \cdot c} = 2.32,\ p = 0.02,$$
$$r_{b \cdot c} = 0.72,\ p < 0.01 \quad t_{b \cdot c} = 0.51,\ \text{n.s.}$$

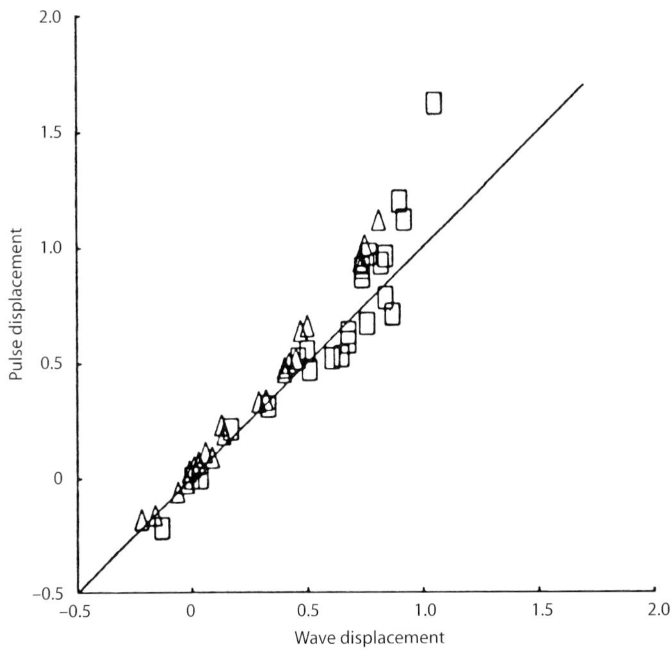

Fig. 6. The results of fitting Eq. (5.0) to conditional pulse probability distributions from anesthetized (Δ) and waking (□) animals have yielded estimates of Δp (pulse displacement) and Δv (wave displacement) that appear to be linearly related

The correlation coefficient for the 2 sets of measurements of σ is

$$r_{a \cdot bc} = 0.98, \quad p < 0.01.$$

The results show that the correlations predicted by the model between the asymptotic asymmetry of the nonlinear function and the amplitude and positive displacement of the peak of the gain curve exist in the experimental data, but there is an unresolved discrepancy between Q_m^a and Q_m^c.

Equation (5.0) could not be fitted to the experimental curves from waking animals under the assumption that $\Delta\hat{u} = 0$, although the average value for $\hat{Q}(0) = -0.02 \pm 0.05$ suggested that $\Delta\hat{p} = 0$. The alternative measures for Q_m gave widely disparate values: $Q_m^b = 1.50 \pm 0.41$ and $Q_m^c = 6.69 \pm 3.77$ with a correlation coefficient $r_{b \cdot c} = 0.78$ ($p<0.01$). The upper asymptote Q_m^a was not delineated.

The assumption that $\Delta\hat{u} = 0$ was dropped, and both Δq and Δv were introduced into Eq. (5.0) for the fitted curve. From Eq. (7.4)

$$v = \sigma V + \Delta v, \tag{10.0}$$

and Eq. (7.11) was used to fit Q' to the data. Equation (8.2) was used to derive the initial guesses for Q_m from the probability density curves for the nonlinear regression. The curve was fitted to 24 experimental conditional pulse probability distributions from anesthetized animals (see Fig. 1), giving the following means, standard deviations and correlation coefficients.

$$\begin{aligned} Q_m &= 2.64 \pm 1.22 \qquad & r(Q_m \cdot \sigma) &= 0.67, \ p < 0.01, \\ \sigma &= 0.64 \pm 0.34 & r(Q_m \cdot \Delta v) &= 0.80, \ p < 0.01, \\ \Delta v &= 0.26 \pm 0.30 & r(\sigma \cdot \Delta q) &= 0.53, \ p < 0.01, \\ \Delta q &= 0.34 \pm 0.37 & r(\Delta q \cdot \Delta v) &= 0.99, \ p < 0.01. \end{aligned}$$

The means for Δv and Δq were slightly but significantly different by matched t-test (difference = 0.08 ±0.08; t = 4.93, p<0.001). The same curve was fitted to 23 experimental conditional pulse probability distributions from waking animals (see Fig. 2), giving

$$\begin{aligned} Q_m &= 5.31 \pm 2.77 \qquad & r(Q_m \cdot \sigma) &= 0.37, \ \text{n.s.}, \\ \sigma &= 0.46 \pm 0.18 & r(Q_m \cdot \Delta v) &= 0.67, \ p < 0.01, \\ \Delta v &= 0.62 \pm 0.28 & r(\sigma \cdot \Delta q) &= 0.26, \ \text{n.s.}, \\ \Delta q &= 0.67 \pm 0.39 & r(\Delta q \cdot \Delta v) &= 0.93, \ p < 0.01. \end{aligned}$$

The means for Δv and Δq were not significantly different by matched t-test. The values for Δq and Δv for both sets of data are plotted in Fig. 6.

As shown in Fig. 2 the experimental and fitted curves differed consistently, in that the former were more sharply inflected (concavely upward) to the right of the origin, and the pulse probability at strongly negative values of v was higher than predicted by the curves. The residual discrepancies have reduced the precision of measurement of the four coefficients, so that the values in the waking state have more uncertainty than those from the animals under anesthesia. Comparison of the two groups shows that the mean values for three of the four coefficients increased in the waking state. These increases can largely account for the change in slope of the normalized conditional pulse probability distributions. Under the assumption that the deviations from the means were random, the differences between the means for Q_m (t = 4.23), Δq(t = 2.94), and Δv(t = 4.21) were significant by pooled t-test at $p<0.01$, and the difference for Δq(t = –2.24) was significant at $p<0.05$.

The next step was evaluating the coefficient γ in the two states. The approach used was based on the property of near-linearity of the model for input and output near u_0 and p_0. For $\Delta u << u_0$, and $\Delta p << p_0$ Eq. (4.0) was expanded into a linear approximation

$$p_0 + \Delta p \approx \gamma(u_0 + \Delta u). \tag{10.1}$$

By definition $\Delta q = \Delta p$. From Eq. (1.0) $\Delta v = \beta \Delta u$. Because the choice of scaling p and u after normalization was arbitrary, $\beta = 1$ and

$$\gamma \approx \Delta q \,/\, \Delta v. \tag{10.2}$$

The mean values for u_0 in the two sets were 1.82±0.37 and 2.39±0.43. The mean ratios $\Delta v \,/\, u_0$ were 0.14 and 0.25 with an overall range of –0.14 to +0.38. The ratios were not sufficiently small for the linear approximation.[2]

A correction was made for the values of Δq by use of Eq. (5.0). Wave displacement was expressed in units of u_0 by

$$v_c \triangleq \Delta v \,/\, u_0 \tag{10.3}$$

and predicted pulse displacement was calculated for each case. The corrected values Δq_c and γ_c were given by

$$\Delta q_c = \Delta q \cdot v_c \,/\, q_c\,, \tag{10.4}$$

$$\gamma_c = \Delta q_c \,/\, \Delta v. \tag{10.5}$$

The experimental values (± S.D.) for the two sets were

anesthetized	waking
$\Delta q_c = 0.32 \pm 0.33$	$\Delta q_c = 0.59 \pm 0.34$
$\hat{\gamma} = 1.14 \pm 0.95$	$\hat{\gamma} = 1.04 \pm 0.32$
$\hat{\gamma}_c = 1.08 \pm 0.94$	$\hat{\gamma}_c = 0.94 \pm 0.31$

The corrected values $\Delta \hat{q}_c$, when plotted against $\Delta \hat{v}$ as in Fig. 6, were close to the line at 45°, i.e. the upward departure from 45° seen at the upper end of the cluster of points in Fig. 6 disappeared. The product moment correlations between $\hat{\gamma}_c$ and $\hat{u}_0$, $\hat{Q}_m$, $\Delta \hat{p}$, and $\Delta \hat{v}$ were not significantly different from zero.

2 One case was deleted for which $\Delta v = 0$ ($Q_m = 1.25$)

Over 98% of the variance in the estimates of γ_c came from the ratios of 15 pairs of values nearest zero. A best estimate for γ_c was obtained by pooling the ratios from both sets (see Fig. 6)

$$\bar{\gamma}_c = 1.01 \pm 0.03 \ (\pm\text{S.E.}, N = 32, \Delta v \geq 0.20),$$
$$\bar{\gamma}_c = 1.02 \pm 0.41 \ (\pm\text{S.E.}, N = 14, \Delta v < 0.20).$$

The ranges of Q_m for these estimates were respectively 1.69–14.86 (waking) and 1.00–3.29 (anesthetized). Analysis of the subgroup of 32 cases revealed the following means and standard deviations.

	anesthetized	waking
N	12	20
$\hat{Q}_m$	3.53 ± 1.07	5.79 ± 2.65
$\bar{\gamma}_c$	1.14 ± 0.06	0.94 ± 0.17

Thus $\bar{\gamma}_c$ was slightly (14%) but significantly ($t = 8.08, p < 0.001$) greater than unity under anesthesia, but the difference was not apparent in the waking state. The coefficient of variation of $\bar{\gamma}_c$ (S.D./mean) was 1/3 that of $\hat{Q}_m$. It was concluded that the question could not be answered whether γ was an independent variable, because the errors of measurement may have exceeded the possible range of variation. For interpretation of the present data its value was fixed at unity.

It was now feasible to estimate the average maximal gain for the two sets of data. From Eq. (4.0) and the two mean values for $\hat{Q}_m$, $\hat{u}_0$ increased from 1.88 under anesthesia to 2.48 in the waking state. By Eq. (6.10) the maximal gain $(dp/dv)_g$ increased two-fold from 2.65–5.85. From the overall ranges of $\hat{Q}_m$ (1.00–14.86) and $\hat{u}_0$ (1.18–3.44) it was inferred that maximal gain varied by more than an order of magnitude (1.18–20.08) due to the relatively small variation in $\hat{u}_0$.

Discussion

The main purpose of this work has been to construct a nonlinear gain curve that replicates experimentally derived gain curves from the mammalian olfactory system. An earlier model (Freeman, 1975) incorporated three main features: a near-linear small-signal range; bilateral saturation (the gain approached zero with both extremes of wave amplitude), which provided stability for neural feedback systems having this operation, and a 2:1 asymmetry of the asymptotes, which was predicted from studies of the EEG and evoked potentials of the olfactory bulb and cortex. The present model retains these three and adds two new features: the gain increases with positive (excitatory) input, and the gain is modifiable in a pattern that depends on background or steady state activity, which in turn is presumed to be under centrifugal control (Freeman, 1975).

This model by no means exhausts the analysis of conditional pulse probability distributions. Other types of curve having *J*-, *U*-, or *V*-shapes have been found for many bulbar and cortical neurons in addition to the predominant sigmoid shapes described here. Analysis of these forms is still in progress and is beyond the scope of this report. Moreover, note should be taken of the residual discrepancies between experimental and theoretical curves as shown in Fig. 2, which suggest either that the data were not processed properly (e.g. perhaps the low frequency EEG respiratory wave was not completely filtered out) or that one or more aspects of the model is deficient (e.g. perhaps the assumption of a static nonlinearity is incorrect; or perhaps pulse-wave conversion is not kept within a linear range when EEG amplitude is very high; or perhaps multiple nonlinearities are reflected in the experimental data – to reiterate, the EEG wave amplitude and the pulse trains in the olfactory bulb stem from two different subsets of neurons in a negative feedback loop). However, these limitations appear not to vitiate the five main properties presented here.

The present model is based on a third state variable m, in addition to p and u that can be related directly to observables in pulse and wave activity of subsets. Although it is not necessary to do so, one can think of m intuitively as equivalent to an average value in a subset for the voltage-dependent sodium activation factor of nerve membrane in the Hodgkin–Huxley system (Hodgkin, 1964). At or near threshold the voltage–dependence is approximately exponential in the form of Eq. (2.2). Activation has been introduced as a dimensionless variable intervening between u and p for the reason that membrane polarization influences firing probability indirectly through the subthreshold electrochemical states of membrane. Neurons in a subset are subthreshold about 99% of the time, because pulses lasting about 1 ms occur for each neuron at mean rates of about 10/s. Pulse density p can be interpreted as sodium current in a subset, because each pulse depends on a quantal movement of sodium across membranes. Wave amplitude u is directly proportional to average transmembrane potential, when it is measured in the appropriate extracellular geometry. Then the defining relations in the model can be expressed as $\mathrm{d}m/\mathrm{d}u$ in $1/V$ and $\mathrm{d}p/\mathrm{d}m$ in amperes, γ in mhos at trigger zones and ζ in Ω at synapses.

The model was developed by invoking a non-zero steady state condition expressed in u_0, that constitutes sustained subthreshold depolarization of neurons in subsets, and that determines steady state and maximal firing rates, maximal gain, and the positive displacement of maximal gain on u with respect to u_0. The meaning of u_0 remains unclear unless the topology of the ensemble in which the neural subset is embedded is defined. For the present u_0 can be interpreted as a steady state imposed on the subset under observation by a non-zero equilibrium in the larger system that is largely under contrifugal control (Freeman, 1975).

The relation between u_0 and p_0 was assumed to be linear on the basis of empirical observations of single neurons and receptors. A linear relation between p_0 and m_0 was found to be consistent with this. No other relation was found to be consistent, but the relations were not shown to follow one from the other necessarily.

The principal advantage of the model appears to arise from relating the maximal pulse density p_m to p_0 and to the intervening variable m, instead of directly to the membrane potential or wave amplitude, as might be done in extensions of the classical "2-factor" theories. The maximum p_m and its influence are clearly not simply related to u or even dominantly so, but more so to the recent past pulse activity engendered by m and the more remote past embodied in p_0. To give detailed expression to those origins would require integration over multiple time-varying nonlinear relationships, and yet further integration over the ensemble, i.e. a statistical mechanical generalization of the Hodgkin–Huxley system in standard or modified forms (e.g. Rinzel, 1978). Considering that a pulse maximum must exist in some form in each subset, it seems preferable at this stage of exploration to make it a dimensionless parameter Q, that is dependent on u_0.

Although u_0 cannot be measured directly, it can be evaluated by measurement of the asymmetry of the nonlinear function, provided that three other coefficients are evaluated. The standard deviation σ is a scaling factor for the domain of u in a particular observation. A low value of σ with a low value of Q_m is found when $G_a(v)$ is restricted to the near-linear range. This suggests that the input to the system is then restricted to a small range. Most neurons and clusters of neurons in anesthetized animals give normalized conditional pulse probability curves in the very near-linear range, so that the nonlinearity cannot be evaluated from them. A low value of σ with a high value of $Q_{>m}$ is often found in curves from neurons in waking animals. It implies that the domain of u is bounded by one or more nonlinearities in the system outside the subset under observation.

The displacements Δp and Δv represent deviations of $\hat{p}$ and $\hat{u}$ from $\hat{p}_0$ and $\hat{u}_0$ on the average over a given period of measurement. It is obvious that in the absence of input the state variables would remain at $\hat{p}_0$ and $\hat{u}_0$ and, further, that because the input that leads to oscillatory EEG waves in the bulb is solely excitatory, one can expect the mean values of $\hat{\bar{p}}$ and $\hat{\bar{u}}$ to be greater than $\hat{p} = 0$ and $\hat{u} = 0$. However, it is not obvious a priori that $\Delta\hat{p}$ and $\Delta\hat{u}$ should exceed zero, because u_0 and p_0 are defined as values in the steady state without reference to the possible existence of steady input. Failure of the attempt to evaluate the model while fixing $\Delta v = 0$ implies that u_0 is independent of (but not orthogonal to) the mean level of time-varying input over an extended time span. This is one reason for invoking centrifugal control.

Remarkably, over a wide range of variation in Q_m (and u_0) the values for Δq and Δv were closely related. This result was obtained in two measurements on the same data by nonlinear regression with independent starting guesses.

A tentative interpretation is based on the property of near-linearity in the model for small input and output. The experimental values for the sensitivity factor $\hat{\gamma}_c$ indicate that γ is not significantly different from unity, that it is not related to Q_m or u_0, and that it does not account for the variation in shape of the conditional pulse probability distributions. This indicates that, unlike u_0, the sensitivity of neurons in a subset, as measured by their f/I curves, is not a significant variable under centrifugal control in relation to behavioral arousal. The finding that $\gamma_c = 1$ accounts for the observations that $\hat{Q}_0$ is near zero in both data sets, by virtue of the symmetry of displacement of the origin, and it also explains the finding that $\sigma^a = \sigma^{bc}$. Considering that measurements of pulse and wave activity density must be interrelatable, it shows that the procedures described here and previously in more detail (Freeman, 1975) for the normalization of EEG and unit data, and for evaluation of σ, have been optimally designed.

This interpretation must remain tentative pending further study. The wave amplitude $\hat{V}$ and pulse density $\hat{p}$ in these data came from different neural subsets in the bulb. The degree to which the pulse–wave nonlinear operation may have affected the measurements of $\bar{\gamma}_c$ and Q_m cannot at present be adequately assessed. Moreover, experiments should be done to determine whether $\bar{\gamma}_c$ and Q_m vary in relation to many types of behavioral and pharmacological manipulation, so that if γ represents a variable, the values can be driven over a range exceeding the probable error of measurement. However, its relative constancy here is consistent with the stability of threshold and sensitivity of single neurons (Katz, 1940). It is concluded that many subsets of neurons in the bulb, comprising about 20% of its surface area (Freeman, 1978), can be maintained in a nonlinear excitable state that is equivalent to the subthreshold domain of the "local response" of single neurons (Rushton, 1937; Hodgkin, 1938), with the degree of excitability being under centrifugal control. If a local ensemble containing sensitized subsets that are mutually excitatory is excited, the basis exists for a regenerative increase in activity in response to an adequate stimulus. This meets the requirement posed in the introduction for sensitivity. Owing to the near linear range of function about the origin the system can be made stable for small inputs. The bilateral saturation property bounds the system within a range determined by threshold and u_0. Thus two of the stability requirements are met. Examination of the properties of stability in the presence of high amplitude noise and the self-termination of an event induced by time-delimited input requires study of a larger system.

Acknowledgements

Reprinted with permission by Springer-Verlag, from "Nonlinear Gain Mediating Cortical – Response Relations", *Biological Cybernetics* **33**, 237–247 (1979).

Supported by a grant MH 06686 from the National Institute of Mental Health, and by a Research Professorship from the Miller Institute.

The author is grateful to Dr. Daniel M. Sunday, Dr. Soo-Myung Ahn and Brian C. Burke for assistance in various aspects of this work.

References

Agin, D.: Hodgkin–Huxley equations: logarithmic relation between membrane current and frequency of repetitive activity. *Nature* **201**, 625–626 (1964)

Baker, C.L.: Nonlinear systems analysis of computer models of repetitive firing. *Biol. Cybernetics* **29**, 115–123 (1978)

Calvin, W.H.: Generation of spike trains in CNS neurons. *Brain Res.* **84**, 1–22 (1975)

Cooley, J.W., Dodge, F.A., Jr.: Digital computer solutions for excitation and propagation of the nerve impulse. *Biophys. J.* **6**, 583–599 (1966)

Dodge, F.A., Jr.: On the transduction of visual, mechanical, and chemical stimuli. *Int. J. Neurosci.* **3**, 5–14 (1972)

Fohlmeister, J.F., Poppele, R.E., Purple, R.L.: Repetitive firing: dynamic behavior of sensory neurons reconciled with a quantitative model. *J. Neurophysiol.* **37**, 1213–1227 (1974)

Freeman, W.J.: *Mass action in the nervous system.* New York: Academic Press 1975

Freeman, W.J.: Spatial properties of an EEG event in the olfactory bulb and cortex. *Electroencephalogr. Clin. Neurophysiol.* **44**, 586–605 (1978)

Granit, R., Kernell, D., Shortess, G.K.: Quantitative aspects of repetitive firing of mammalian motoneurons caused by injected currents. *J. Physiol.* **168**, 911–931 (1963)

Heyer, C.B., Llinàs, R.: Control of rhythmic firing in normal and axotomized cat spinal motoneurons. *J. Neurophysiol.* **40**, 480–488 (1976)

Hodgkin, A.L.: The subthreshold potentials in a crustacean nerve fibre. *Proc. R. Soc. London, Ser. B* **126**, 87–121 (1938)

Hodgkin, A.L.: *The conduction of the nervous impulse.* Liverpool: University Liverpool Press (1964)

Katz, B.: *Electric excitation of nerve.* Oxford: Oxford University Press (1940)

Rinzel, J.: On repetitive activity in nerve. *Fed. Proc.* **37**, 2793–2802 (1978)

Rushton, W.A.H.: Initiation of the propagated disturbance. *Proc. R. Soc. London, Ser. B* **124**, 210–243 (1937)

Stein, R.B.: The frequency of nerve action potentials generated by applied currents. *Proc. R. Soc. London, Ser. B* **167**, 64–86 (1967)

Sunday, D.M., Freeman, W.J.: Nonlinear coupling of wave and pulse activity in a neural mass. *Physiologist* **18**, 42 (1975)

11. The Use of Euclidean Distance in 64-space and Behavioral Correlates to Optimize Filters for Gamma AM Pattern Classification

Experiments with untrained animals showed that bulbar responses to novel odorants that were presented in sessions without reinforcement failed to develop novel AM patterns. These stimuli initially elicited an orienting response in company with replacement of gamma bursts by broad spectrum fluctuation resembling background activity. This suppression of bursts disappeared after 3 or 4 trials. If the novel odorant was followed by reinforcement, burst suppression gave way to renewed formation of gamma bursts also within 3 or 4 trials. My search for event-related content in bulbar gamma activity was guided by two hypotheses. First, when unconditioned stimuli were paired with one of two discriminative conditioned stimuli (the CS+) but not the other (CS–), resulting in a conditioned response (CR) to the CS+, and not to the CS–, randomly alternated in the same training session, odorant-related content must exist in the form of a pattern of neural activity in the olfactory bulb between the arrival of a CS at the nose and the performance of a correct CR. Second, this pattern elaborated by synaptic currents of the granule cells would be detected in field potentials in the bulbar AM patterns of gamma activity.

Detection of the gamma AM patterns required adaptation of techniques in small animal surgery, olfactory conditioning, computer file management, signal identification, digital image processing, multivariate statistics, and nonlinear dynamics to achieve the requisite degree of precision in EEG measurement (Freeman 1980; Davis and Freeman 1982; Freeman *et al.* 1983; Viana Di Prisco and Freeman 1985). The behavioral paradigm had to involve a discrimination between two odorants, because if the animals were conditioned to a single odor, they gave a response to any odorant that was different from the background. Two reliable response measures were necessary, one to verify perception of the reinforced conditioned stimulus (CS+) and the other for the unreinforced conditioned stimulus (CS–). The reinforcing UCS had to be a reward, because the rabbits tended to conflate multiple conditioned stimuli into a single warning category during aversive conditioning. Licking for water by thirsty rabbits was chosen as the CR to a CS+ with reinforcement, and sniffing without licking was chosen as the CR a CS– that was not rewarded. The sniff response was chosen because of the rapidity and ease with which it was acquired and with which it could be measured (Davis *et al.* 1982; Freeman *et al.* 1983), in comparison to other responses that were more commonly used such as eyelid closure, nictitating membrane retraction, paw flexion, or modification of heart rate. This autoshaped response was particularly well adapted to showing when a rabbit had learned to detect and discriminate an odorant (Viana Di Prisco and Freeman 1985). It showed that they did so in 3 or 4 trials in a single session, as compared with the other measures that required 70 or more trials for each of 4 or more sessions for the subjects to reach criterion.

Measurement of the waveforms of gamma bursts required the development of novel methods for curve fitting, using as a nonlinear basis function a cosine with both amplitude and frequency modulation, the better to adapt to the waxing and waning shapes of gamma bursts (Figure 1 in Chapter 11). A signal:noise criterion for goodness of fit of the sum of basis functions to each gamma segment was devised, which was the sum of squares of the residuals divided by the sum of squares of the fitted curve. The assay of standard errors of measurement of amplitude and of phase was calibrated by fitting the basis functions to a known cosine signal embedded in "noise" from a random number generator. This analysis showed that the technique of fitting a single cosine to an EEG segment gave standard errors approaching a standard deviation of EEG amplitude and a full radian (57.2°) of phase at the median signal:noise ratio (~1:3). Several years of development of adaptive filters to measure the "clutter" (non-white noise) simultaneously with the signal reduced the errors of measurement by a full order of magnitude and increased the median signal:noise ratio of measurement to levels approaching 4:1 (Figure XI, A). Analysis of the spatial patterns of amplitude and phase in gamma bursts could not have been achieved without this precision of measurement.

The output of measurement of each burst had the form of multiple matrices of parameter values, each matrix specifying a 64×1 column vector for the AM or phase pattern. The vectors decomposed each gamma event, including the residuals, so that each part could be tested for its behavioral correlation, just as a chemical mixture can be separated into fractions, each of which can be assayed for its concentration of an active substance. The criterion for success was the maximal separation of the CS+ bursts from the CS– bursts (Figure 8 in Chapter 11). The matrix for each burst fixed a point in 64-space, so a set of bursts from the same condition formed a normally distributed multivariate cloud in that space. The CS+ and CS– bursts together formed an elliptical cloud that was distinctly separated from the spherical cloud of points for the control bursts. The CS bursts proved to be the most difficult to separate into two clouds (Figure 9 in Chapter 11).

I devised an assay of CS burst separation that was fast and easy to apply in succession to each of the matrices (including the residuals) for all animals to optimize the degree of separation. The assay directed the choice of the basis functions and the temporal and spatial filters, including spatial deconvolution with the point spread function and range normalization (Figure 12 in Chapter 11). An important step was to equalize the variance by recognizing that about 20% of the bursts were outliers, in the sense that their points were widely scattered in 64-space and did not conform to any cloud. In retrospect these stray bursts may have manifested failure of the olfactory mechanism to converge to an answer in the form of a pre-existing basin of

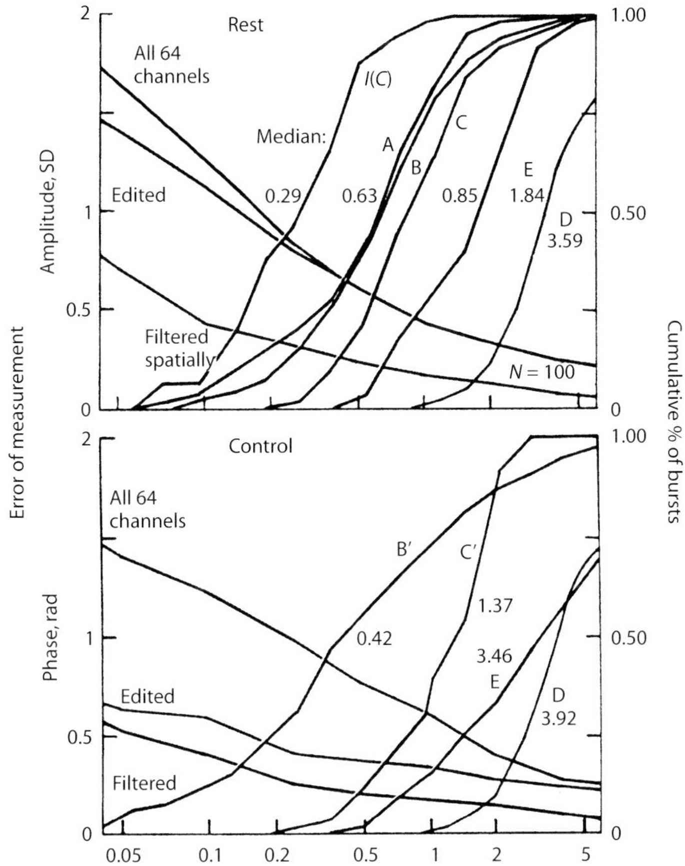

Figure XI, A. The method for estimating the standard error of measurement of gamma bursts is illustrated. A set of 64 cosine waves was generated at 40 Hz at zero phase and with an amplitude distribution that simulated the AM pattern of a typical burst. The "noise" was added from a random number generator independently to each trace, and the set was "digitized" (truncated at 12 bits) and filtered in the same way as EEGs. The peak frequency was estimated from the fast Fourier transform (FFT) of the ensemble average, and a sine wave at that frequency was fitted to all 64 traces. The signal:noise ratio was calculated from the sum of squares of the fitted curves divided by the sum of squares of the residuals. The curves sloping downwardly to the right show the standard deviations of the estimates of amplitude and of phase from the known values as a function of signal:noise ratio. The uppermost curve resulted without preprocessing, and the curves beneath it show successive refinements in preprocessing. The abscissa shows the signal:noise ratio on a log10 scale. The ordinates at the left show the standard deviation of amplitude and phase measurements of simulated standard data from their known 64 values. "Edited" refers to deleting amplitude or phase values for which the correlation coefficient between the trace and the ensemble average was less than 0.2. "Filtered" refers to a spatial low pass filter (Freeman and Baird 1987). The ordinates at the right show percentage of the total number of rest or control bursts (100 bursts total in the upper section comprising 20 from each of 5 rabbits at rest and 300 bursts total in the lower section comprising 60 control bursts from each of 5 rabbits under appetitive conditioning, having at least the signal:noise ratio on the abscissa. Rad = radians for phase. This approach showed that fitting a cosine to a gamma burst gave a median signal:noise ratio of 0.29:1 and an error in phase of about a radian (57 degrees). Improvements raised the median signal:noise ratio by an order of magnitude and reduced the errors or measurement correspondingly. All understanding of the spatial patterns of gamma bursts has come by virtue of these improvements in measurement. From Freeman and Viana Di Prisco (1986, Figure 5, p. 758).

attraction (Chapter 14). From my understanding of bulbar dynamics through my models, I interpreted these as failures of the olfactory pattern recognition mechanism to converge to a learned attractor, and I devised criteria by means of which to identify them (Figure 11 in Chapter 11) independently of their spatial patterns specifying locations of their points in hyperspace.

The optimized filters enabled classification of gamma bursts in each of four rabbits. A fifth rabbit served as a control, because it never learned to discriminate the CS+ and CS– odorants, and its cloud of CS points from the EEG bursts stubbornly failed to resolve into two separate clouds. But this was not enough. It was also necessary to show that the clouds of points for each rabbit stayed relatively fixed in measurement space as long as the odorants and reinforcements were unchanged, but that a distinct jump in the locations of the clouds of points, reflecting a global change in spatial patterns of the EEG bursts, took place whenever an odorant or a reinforcement contingency was changed (Figure XI, B), because a gradual and continuous change in patterns could be ascribed to extraneous factors such as aging

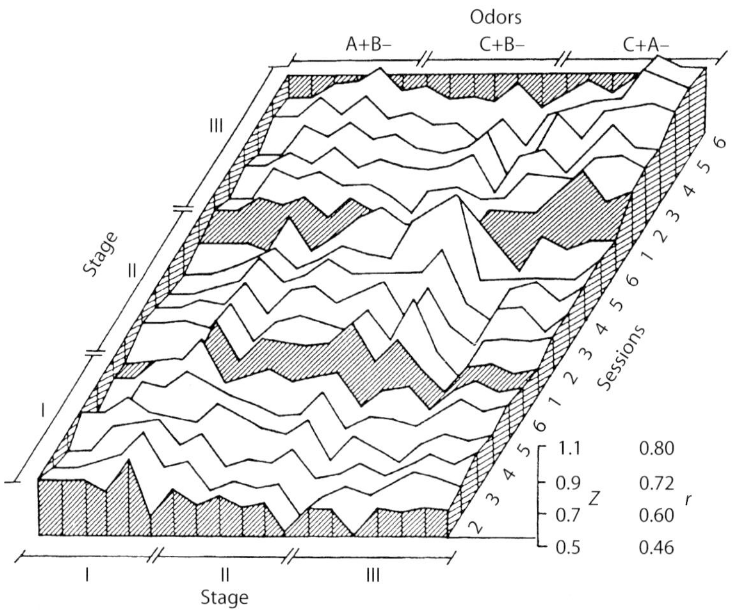

Figure XI, B. Derivation of spatial AM patterns of gamma activity over periods of months and years have shown that gradual changes may take place whether or not the animals are trained to respond to selected odorants. The correlation of AM pattern change with learning requires that a change in behavior with new learning be accompanied by a change in AM pattern that exceeds the background fluctuation and persists. This evidence was derived in a later report (Freeman and Grajski 1987). The bursts were analyzed from 6 sessions in each of 3 Groups. The CS+/CS– configuration was kept constant in each Group. Discontinuities were sought when the configuration was changed between Groups I and II, and between Groups II and III. The gamma bursts were decomposed by factor analysis, and the factor loadings giving the spatial modes from each session were compared with those from the other 17 sessions. Each point specified the average correlation coefficient expressed by Fisher's z-transform over the first six factors from all subjects. Each trace showed the mean correlation of the session designated at the left with the session shown on the abscissa. The sustained jumps are shown by the three successive plateaus. From Freeman and Grajski (1987, Figure 3, p. 771).

of the animal, deterioration of the neural tissue under the array, or growth of the skull with displacement of the array. Subsequent work showed that the jumps in AM patterns of bulbar gamma activity were facilitated by local injection of norepinephrine and were blocked by local injection of propanolol (Gray *et al.* 1989).

EEG Spatial Pattern Differences with Discriminated Odors Manifest Chaotic and Limit Cycle Attractors in Olfactory Bulb of Rabbits

Introduction

The immediate aim of this study was to find reproducible but differing patterns in the EEG of the olfactory bulb that recur with presentation of different odors. The underlying purpose was to use these patterns to explain how the bulb operates on its receptor input in the process of olfactory discrimination.

Rabbits that are trained to respond by licking to one odor with reinforcement (CS+) but not to another odor without reinforcement (CS–) tend to sniff each odor before making a discriminative response. By inference, between the onsets of the sniff and the lick response the olfactory bulb has already received the input from the receptors activated by the CS+ or CS– odor, and it is generating neural activity for transmission to the olfactory cortex that subserves discrimination between the two odors as the basis for response selection. Even if the odor discrimination is not completed in the bulb, the requisite information must be contained in differing bulbar neural activity patterns. If the EEG adequately manifests those activity patterns, then a distinctive and reproducible difference should exist in some aspect of the EEG during presentation of each discriminated odor (Viana Di Prisco and Freeman, 1985).

In this study the EEG is decomposed by filters into various components, and each is tested for its efficacy to classify bursts correctly to preceding control, CS+ odor, and CS– odor periods. The procedure is analogous to a process of chemical purification, with the fractions tested for efficacy after each separation. Eventually a fraction is found, which serves optimally to classify EEG records correctly as belonging to CS+ or CS– trials. This report describes briefly the separation process, the test of efficacy, the distinguishing characteristics of the EEG, and the implications for the neural function of the bulb in sensory processing.

In regard to mechanism the bulbar neurons form an interactive neural mass that is distributed in the two dimensions of the bulbar surface. Interactions include mutual excitation, mutual inhibition and negative feedback between mitral (excitatory) and granule (inhibitory) cells (Freeman 1975). Owing to the lengths of basal dendrites and axon collaterals running parallel to the surface the interactions are diffuse and widespread. To a good approximation this distributed system can be described as a set of coupled oscillators (Freeman 1979). Owing to nonlinearities in the dynamics of the component neurons, we can predict that the bulb has multiple stable states. Three classes of these states are identified for this type of system (Garfinkel 1983): equilibrial, limit cycle and chaotic. The types of output characteristic of these three states are respectively tendencies to go to rest, to enter periodic oscillation or to sustain aperiodic, pseudorandom activity. The time-series output in an equilibrial state under perturbation from noise may closely resemble chaotic activity but must be distinguished by its degrees of freedom. The further aim here is to classify the EEG patterns observed before and during odor stimulation into these three types. Their occurrence or absence indicates the presence or absence of equilibrium points, limit cycles and chaotic domains, that constitute the attractors (Garfinkel 1983, Abraham and Shaw 1982) in the state space of the bulbar neural mechanism for odor identification.

This theory provides a necessary basis on which to analyze the EEG data and to explain how the neural mechanism of the bulb discriminates odors.

2 Methods

2.1 Establishing the Data Base

In a previous study (Viana Di Prisco and Freeman 1985) thirsted rabbits were trained to respond with licking to one odor (the CS+) paired with intraoral water delivery, and not to lick on presentation of another odor (the CS–). The rabbits tended to sniff to both odors but more frequently to the CS–. In a set of six sessions four of five rabbits acquired strongly the CR+ (licking with or without sniffing) and half as strongly the CR- (sniffing without licking) in the first two sessions and maintained these levels of responding thereafter. The fifth rabbit often sniffed but seldom licked and then indiscriminately to both the CS+ and CS–.

The EEG was recorded from the olfactory bulb by a 64-channel array of electrodes (8×8, 4×4 mm) surgically placed on the lateral surface. When the behavioral pattern was stabilized by training, the bulbar EEG underwent a reproducible change on each trial after delivery of an odor and after onset of the sniff response but before onset of the lick response (if it occurred) and before the UCS presentation (if any). In this change the EEG amplitude decreased by 35% on the average over subjects, the frequency on the average decreased, and the spatial patterns of root mean square (rms) amplitude changed slightly but significantly.

The bulbar EEG consisted of a recurrent burst oscillating at a frequency between 20–100 Hz for about 0.1s with each inhalation and riding on the crest of a slow wave synchronous with respiration, usually at 3–5 s^{-1}. On each trial a 3 s control record was immediately followed by a 3 s test record during which the CS was delivered and the CR+ and CR- were sought. After each trial the EEG records were displayed on an oscilloscope; three bursts were selected from the control period (labeled C1, C2, and C3 starting from the end of the control period), and three bursts were taken from the test period (labeled T1 between odor and sniff onsets and T2, T3 between sniff and lick onsets). Each stored segment contained 64 EEG traces digitized at 2ms intervals for a duration of 76 ms.

Each session contained 10 CS+ and 10 CS– trials randomly interspersed. The data summarized here were taken from sessions four to six when response patterns were stable. Each rabbit yielded 360 bursts: 180 control bursts and 90 CS bursts with each odor. Further details have already been published (Viana Di Prisco and Freeman 1985).

Each EEG was amplified (10 K), filtered (3db fall-off at 10 Hz and 300 Hz), digitized (8 bits/read, 10 ms read time, and 2 ms interval between time frames), and stored in a 64×38 matrix. The data on each channel were corrected for small differences in preamplifier gain, digitally smoothed by averaging over adjacent time samples, and detrended to remove the remaining respiratory wave. Bad channels (usually 2 to 4 and not more than 10) were replaced by the average of adjacent pairs. The correlation matrix of each trace with the burst ensemble average was computed; traces with correlation coefficients less than 0.20 were similarly replaced (an average of 2.2/burst), because by inspection of records these non-correlated waves were either from edge electrodes contaminated with nonbulbar EEG or were obscured by transient noise. The ensemble average $E(t)$ was then reformulated as the best available estimate of the EEG time course of each burst.

2.2 Decomposing the EEG Burst

Each burst was decomposed by the combined use of spectral analysis and curve-fitting. The most effective of several procedures was as follows (Freeman 1985a). The Fourier transform was taken of the ensemble average $E(t)$. The frequency and phase were identified for the highest gain; these values were used as initial guesses to fit a cosine wave to $E(t)$ in order to determine the amplitude V, frequency f, and phase p coefficients with a criterion of least squares deviation. The equation was

$$v(t) = \sum^{i} [1 + AM_i (t - t_m)] V_i \cos[FM_i (t - t_m) t + 2\pi f_i t + p_i],$$

where t_m was the midpoint of each trace, AM_i was the fixed amplitude modulation coefficient, FM_i was the fixed frequency modulation coefficient (both modulations linear with time) and $i = 1$ for the 1st component to $i = 5$ for the 5th. This first curve was subtracted from $E(t)$. The procedure was repeated on the interim residue 4 times, leaving a final residue of less than 3% of the energy of $E(t)$. Then the 5 frequency and 5 phase values were fixed, and each of the 64 waves was fitted by linear regression with the sum of the 5 cosine waves. This yielded 5 8×8 matrices of amplitude coefficients that represented the spatial patterns of the 5 components, and an 8×8 matrix of the rms residuals, estimated to contain an average of 19% of the total energy of control bursts and 25–27% of odor bursts, that had lower amplitude.

Each 8×8 amplitude matrix was subjected to a two-dimensional spatial Fourier transform. A low-pass 4th-order exponential filter (Gonzalez and Wintz 1977) was applied to the spatial frequency matrix with its 3 db fall-off set at 0.5c mm^{-1}. This operation served to remove activity that was generated at spatial frequencies above the spectral range accessible to the bulbar granule cell generator (Freeman 1978, 1985b). The amount of energy removed from the total energy in the components was estimated to be on the average 16 ± 2% in control bursts and 23 ± 3% of odor bursts. The inverse Fourier transform was applied to take the filtered data back to the spatial domain.

The next operation was spatial deconvolution by the "software lens," which served to compensate for the distortion of the granule cell field potential by the volume conductor properties of the bulb (Freeman 1980). The depth of focus was set at 0.49 mm by systematic search to find the optimal value, a few microns less than the average depth of the granule cell generator in these rabbits estimated post-mortem from the depth of the mitral cell layer. The procedure was applied to each of the 5 amplitude matrices in turn.

In an alternative procedure the 64 values of phase were determined from the Fourier transform of each waveform or by curve-fitting using nonlinear regression. The phase matrices were not useful for burst classification and are not further considered here.

The final steps were to correct the amplitude and phase values at each frequency for the attenuation and phase distortion caused by bandpass filtering, smoothing, and the digitizing time lag over the 64 channels, and to normalize the 360×64×5 amplitude values for each subject across sessions by channel to zero grand mean and unit standard deviation for each channel prior to statistical analysis (Freeman 1985c).

3 Results

3.1 The Distributions of Spectral Energy

Some representative examples of the ensemble average are shown in Fig. 1 together with the Fourier transform, at the stage of identification of the first component and calculation of the interim residue, illustrating the multiple subsidiary peaks found in the spectrum. The mean values for the relative energy (the sum of 64×38 squares

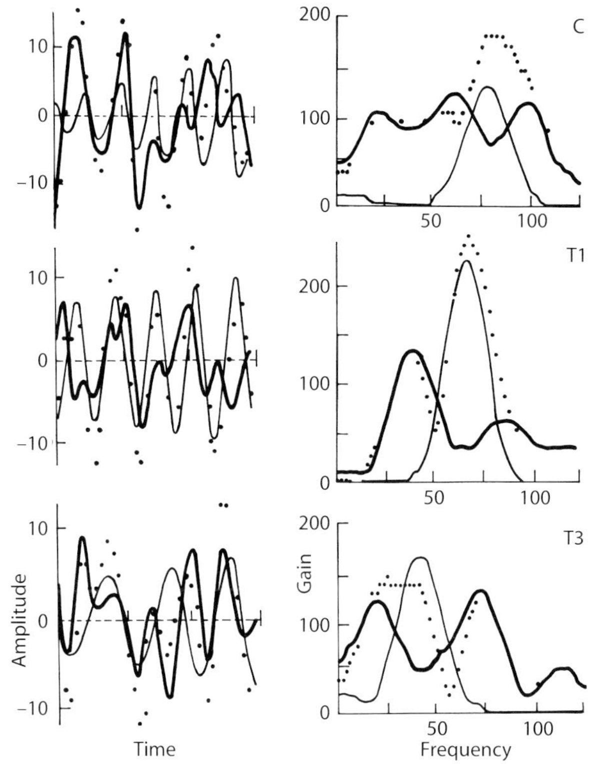

Fig. 1. Three examples are shown of the decomposition of bursts from one subject and trial. Left the dots show the digitized ensemble average; the light curves show the fitted dominant components; the dark curves show the residuals after the first subtraction. Right the Fourier transform gives the spectral gains of the data, the dominant component and the residuals. C control bursts; T1 first odor burst; T3 third odor burst. Time in ms, frequency in Hz. Freeman (1985a)

of each component divided by the total sum of 64×38 squares of the EEG) in each of the 5 components and the residuals for the 64 traces in each burst are shown in Fig. 2.

Figure 3 shows the variation over bursts in the energy distribution among components. The left-most curve is for the first component; the second is for the sum of the first two, and so on. The abscissa shows the relative energy content; the ordinate shows the proportion of 900 bursts having the energy content at each level. The compilation was done separately for control and odor bursts to show the tendency for relative energy to be more broadly distributed beyond the first component in the odor bursts.

The distribution of frequencies is shown in Fig. 4 for both control and odor bursts having dominant frequencies (f_1 in equation in Methods) > 55 Hz for all subjects (upper frame) and one exemplary subject (lower frame). The 5 values from each burst tended to spread across the frequency range from 10 to 125 Hz; the mean interval was 21.1 ± 5.3 Hz (standard deviation). Not every frequency subrange of 20 was represented in each burst; the interval was then greater by a factor of about 2 about the missing frequency value. The weighted mean frequency was 68.7 ± 7.6 Hz

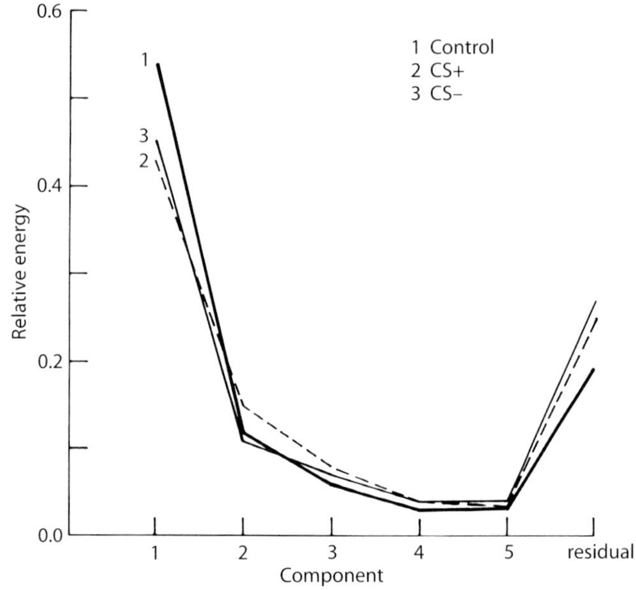

Fig. 2. The curves show the average relative energy in the five components ranked in order of size and in the residuals (the sum of squares in each component divided by the total sum of squares of the 64×38 data values in each burst). Separate tallies are shown for the control, CS+ and CS− bursts. Decomposition by curve-fitting is an artifice for measurement and further processing; it is not intended to suggest that all the wave-forms manifest independent oscillators; the aim is incorporation of the shapes of the EEG into numbers so that covariance with behavioral variables can be sought

for the control bursts and 61.5 ± 5.3 Hz for the odor bursts, reflecting a small but unequivocal decrease in average burst frequency with odor presentation.

The spectral distributions of the energy are shown in Fig. 5. The dominant frequency of most bursts shifted randomly within the spectral zone of 55 to 75 Hz. Comparing the control and odor bursts, upon odor arrival there was a substantial decrease in the energy in the 55–75 Hz region. On the average there was an increase in the energy in the 20–50 Hz region, both in the relative fraction of total energy (averaging 2 fold) and in the absolute level of energy (averaging +12%), partially off-setting the average 35% decline in burst rms amplitude upon odor presentation. Despite this spectral shift the slight majority of odor bursts (60%) had dominant components in the 55–75 Hz range. This average shift was localized to a subset of odor bursts, which had components in the frequency ranges > 55 Hz and < 55 Hz, but in which the energy in a lower frequency component was greater than in a higher frequency component (shaded histogram in Fig. 5), as opposed to the more common ratio (clear histogram). This type of burst was found in approximately 40% of odor bursts of both kinds, and also in about 10% of control bursts.

Bursts in the upper frequency range were more coherent by several measures. Two of these are shown in Fig. 6. The upper frame shows a histogram of dominant component frequencies (denoted f_1) over the set of odor and control bursts from only those trials on which a correct CR occurred. The mean and standard deviation of the fraction of power in the dominant component of each burst are shown with its

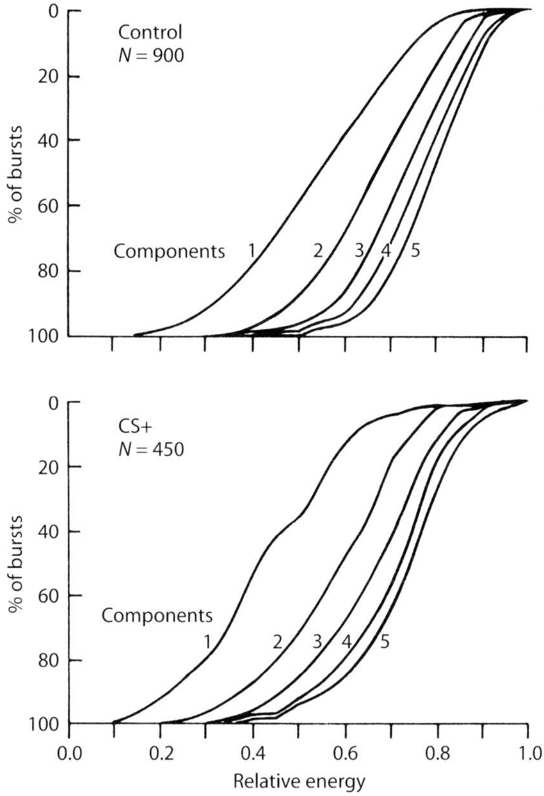

Fig. 3. Variation in energy distribution is shown for 900 control bursts and for 450 CS+ odor bursts. The abscissa shows the energy of the dominant component (1) and the increased amount with the addition of other components (2–5). The ordinate shows the percentage of all bursts having at least the designated amount of energy included in the specified number of components. Left frame: control bursts from five rabbits over three sessions; right frame odor bursts for CS+. (Freeman 1985a)

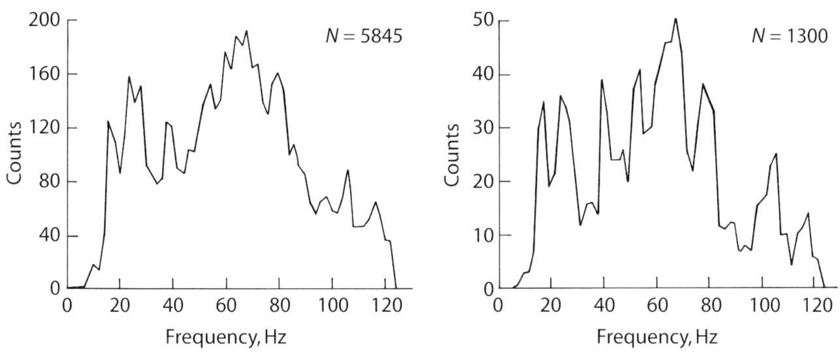

Fig. 4. Left frame shows an example of the distribution of the frequencies of the five components of 1169 bursts having $f_1 > 55$ hz from five subjects. Right frame is an example from one subject. The limits of the spectral distributions near l0 Hz and 120Hz were determined by filtering. Within these limits there was no evidence from any rabbit for fixed frequencies or integer multiples to suggest harmonics. The mode of the intervals between components was 19 Hz

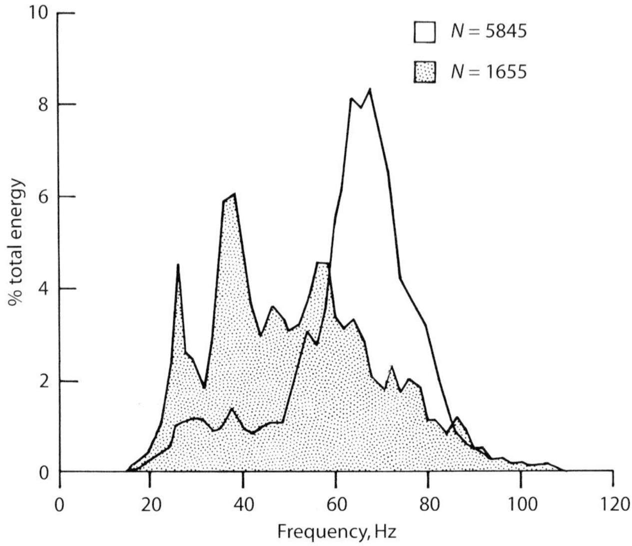

Fig. 5. Weighting each frequency count by the relative energy of the component gave an energy distribution that approximated the average power spectrum without the residuals. The values are shown separately for bursts with $f_1 > 55$ (clear) and $f_1 < 55$ Hz (shaded). The latter type of burst occurred four times more often in test periods than in odor periods. (Freeman 1985a)

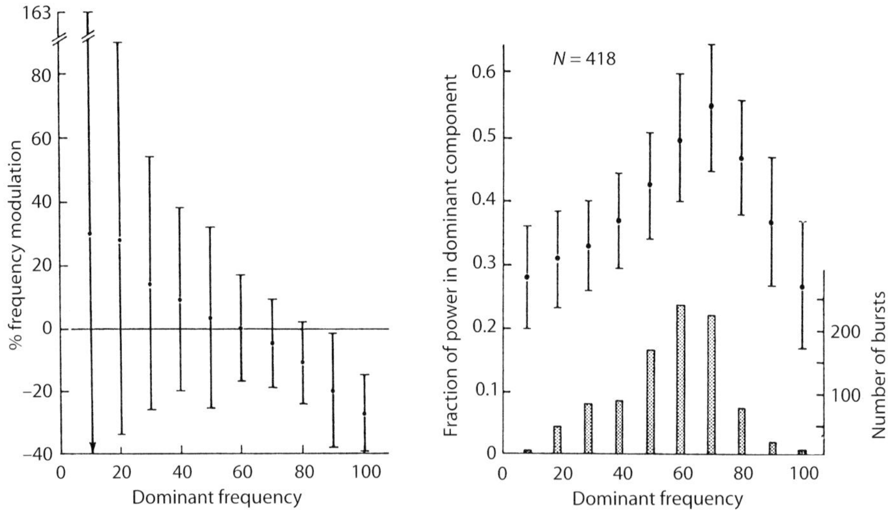

Fig. 6. *Right frame:* the shaded bars show the number of bursts having dominant component frequencies f_1 in the indicated frequency intervals. The points and vertical lines show the mean and standard deviation of the fraction of total burst energy contained by the dominant component in each designated frequency range. Maximal fractions were found in the range 55–75 Hz. *Left frame* the means and standard deviations are shown for the parameter FM_1 (equation in Methods). This is expressed here as % frequency modulation of mean frequency f_1 of the dominant component. Negative values represent a constant rate of decrease in frequency with time. The variation in FM_1 between bursts was minimal for frequencies in the range of 55–75 Hz. (Freeman 1985a)

frequency f_1 in the ranges indicated. The maximal concentration of energy was in the upper frequency range. The lower frame shows the means and standard deviations of the frequency modulation parameter FM_1 (see Methods) for f_1 similarly grouped. There was a tendency for frequency slowing ($FM_1 < 0$) in bursts with $f_1 >$ 60Hz, and for acceleration ($FM_1 > 0$) in bursts with $f_1 < 60$ Hz. The variance in FM_1 increased exponentially with decreasing f_1 frequency.

3.2 Classification of Spatial Patterns

The mean density plots of sets of the five components were similar on visual inspection to the mean density plots of rms bursts amplitude for each rabbit. As previously described (Freeman and Schneider 1982, Viana Di Prisco and Freeman 1985) each rabbit had its unique spatial pattern of EEG amplitude, and the spatial patterns of the five components tended to conform to this pattern, although with greater variance for low energy components of individual bursts. An example of the patterns of the means of five components is shown in Fig. 7 for one rabbit; the control bursts and the CS+ and CS− T2, T3 bursts were averaged after deletion of those bursts having $f_1 < 55$ Hz.

The degree of similarity of the amplitude V_1 of dominant components within each subject and subgroup was evaluated by computing the product moment correlation coefficient r of each 1×64 matrix V_1 with other matrices V_1 and averaging by

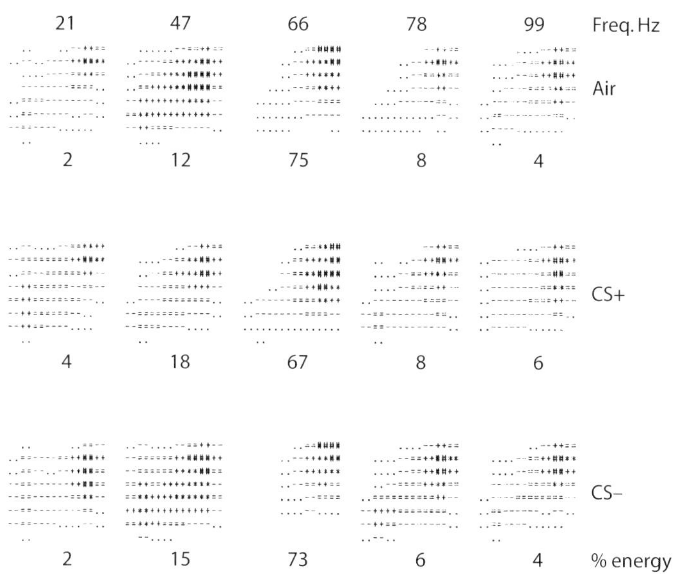

Fig. 7. Digital density plots were used to display 8×8 maps of the averages of the five components of control CS+, and CS− odor bursts (respectively upper, middle and lower rows for one subject) for those bursts having f_1 < 55 Hz. The mean of the frequencies is shown above each column; the proportion of relative energy expressed as % is shown under each frame. The spatial patterns of energy conformed in the main to the pattern of rms amplitude for each animal

z-transform over subjects and burst pairs to find r. As expected, the average value r decreased monotonically with decreasing energy content of the components. Of greatest interest were the values for the dominant components within the subsets of bursts with $f_1 > 55$ Hz. The dominant components of control bursts (based on sets of three bursts within trials) gave $\bar{r} = 0.97$, whereas this component for both CS+ and CS– bursts gave $\bar{r}_+ = 0.84$ and $\bar{r}_- = 0.88$ respectively, reflecting the lower amplitude of the odor bursts. The cross-correlation of control burst pairs C+ and C- preceding CS+ and CS– trials gave $\bar{r} = 0.85$, reflecting the introduction of between-trial variability. The comparison of the dominant components of CS+ versus CS– bursts gave $\bar{r}_{+-} = 0.58$, indicating that these spatial patterns across trials were substantially less similar to each other than were the control bursts across trials (Freeman 1985a).

The within-trial correlation of dominant components from bursts with $f_1 < 55$ Hz (those in which the lower frequencies tended to dominate) gave $\bar{r} = 0.12$, and similarly for the secondary and yet smaller components, and for dominant versus secondary, indicating that the components from these bursts did not form coherent classes within or over trials.

A Euclidean distance measure was used to evaluate differences between bursts with $f_1 > 55$ Hz. Each burst was treated as a point in a 64-dimensional space defined by its dominant component. The centroid was computed for each subject for the two types of odor bursts (CS+ and CS–) and for the sets of preceding control bursts (C+ and C–). The sets of distances from points to each centroid (their own or the centroids of other sets of bursts) were normally distributed. The standard error (SE) was computed for each of the four sets. The distance between each pair of centroids was divided by the pooled standard error of the two distributions to give a t-distance. Examples are given in Table I for the five rabbits ranked in the order of

Table 1. *Upper left:* the numbers of *CR+* (lick) and *CR–* (sniff only) responses are shown for 30 *CS+* and 30 *CS–* trials. The conditioning procedure used established a lower rate for *CR–* than for *CR+*. *Upper right:* the Euclidean distance measures are shown between control (*C+*, *C–*) and odor (*CS+*, *CS–*) centroids, f_1>55 Hz. *Lower left:* the proportions are shown of correct classification of bursts by the Euclidean distance measure. Chance level is 50%. *Lower right:* The results are summarized for the *t*-distance measure based on nonlinear mapping (see Fig. 9)

Number in 30 trials							
CR+		*CR–*		*t*-distance in 64-space			
CS+	*CS–*	*CS–*	*CS+*	air.air	air. *CS+*	air.*CS–*	*CS+*.*CS–*
25	2	18	5	2.58	5.78	3.87	4.97
25	2	10	3	2.98	7.69	12.55	4.87
20	9	13	5	1.28	11.27	4.27	5.08
19	3	10	9	2.87	3.21	4.74	4.20
4	4	7	23	2.01	10.41	14.60	1.22
Classification % correct				*t*-distance in nonlinear map			
Air.air	air.*CS+*	Air.*CS–*	*CS+*.*CS–*	Air.air	Air.*CS+*	Air.*CS–*	*CS+*.*CS–*
49	83	94	88	0.08	7.03	1.63	1.40
62	82	83	84	0.31	0.63	1.20	0.75
59	84	75	78	0.12	1.59	0.79	0.97
52	76	78	72	0.51	2.44	8.31	1.19
67	83	86	52	2.68	1.87	7.27	0.12

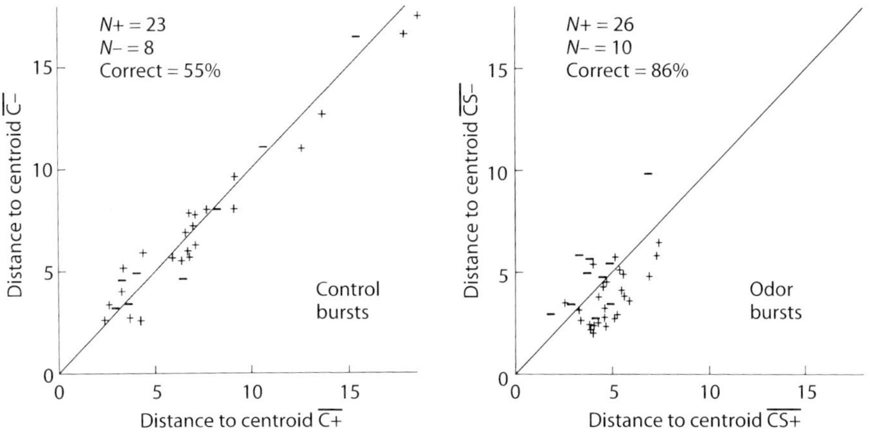

Fig. 8. A centroid was found in 64-space for each subgroup: CS+, CS−, and the two preceding controls C+ and C− by calculating the mean for each channel over the subgroup. The Euclidean distance was calculated for each control burst of both types to the C+ and C− centroids (upper frame), and for each odor burst of both types to the CS+ and CS− (lower frame). If the burst point lay closest to its own centroid it was labeled "correct", otherwise "incorrect". The aim of measurement was to separate the odor bursts from each other (CS+ from CS−) without separating the control bursts (C+ from C−). The efficacy of various procedures was expressed by the difference between % correct odor bursts less % correct control bursts. (Freeman 1985a)

reliability of responding correctly according to the behavioral criteria. The *t*-distances between the two control centroids (C+, C–) were relatively small for all rabbits in comparison to the t-distances between control and CS+ or CS– centroids. The *t*-distances were relatively large also between the CS+ and CS– centroids, but only for those four rabbits showing behavioral evidence for discrimination. The (C+, C–) and (CS+, CS–) *t*-distances were greater for the subsets of trials on which correct responses occurred (as shown in Table 1) than for the entire sets of trials irrespective of the nature of the CR's.

Each burst was classified as "correct" if its *t*-distance was shortest to the centroid to which it contributed, or as "incorrect" if the distance was shorter to an apposing centroid (Fig. 8). As shown in Table I the proportion of correct classification of control bursts between the two control centroids averaged 58%, which was near the chance level of 50% . The average % correct classification of odor versus control bursts was 82% for five subjects; that between CS+ and CS– odor bursts averaged 81% for the four trained subjects.

Each of the sets of 5 components was evaluated by the Euclidean distance measure and also by cluster analysis using the nonlinear mapping procedure of Sammon (1969). The only effective classification of bursts into clusters relating to control, CS+, and CS– conditions was provided by the 55–75Hz dominant component with $f_1 > 55$ Hz. Bursts with $f_1 < 55$ Hz did not cluster in accordance with behavioral criteria by any components. Figure 9 shows an example of the controls and standard errors of the clusters for one subject, in which the odor bursts were widely separated from control bursts and less widely from each other. An empirical *t*-distance was computed for each pair of clusters by dividing the distance between centroids by the pooled standard error. These values for control-control, control-

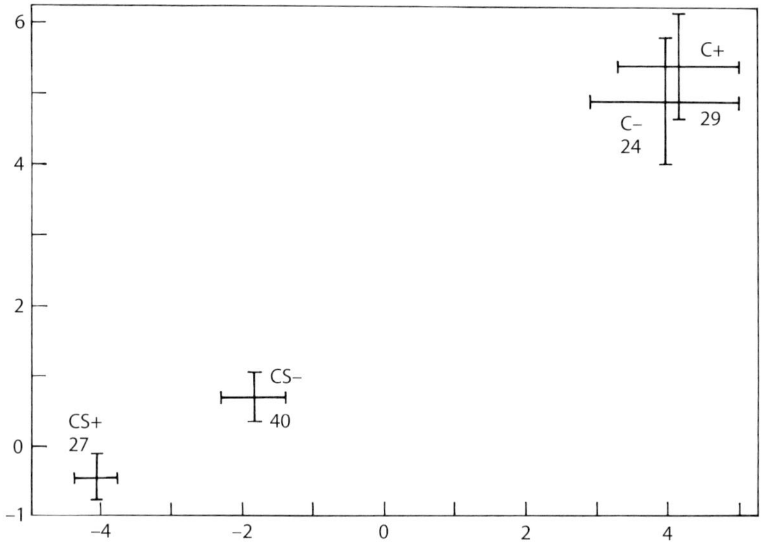

Fig. 9. An example of the results of nonlinear mapping is shown for bursts with $f_1 > 55$ Hz taken from four classes: control bursts prior to CS+ odor delivery (C+), control bursts prior to CS– odor delivery (C–), CS+ odor bursts and CS– odor bursts. This procedure of cluster analysis preserves the relative distances among bursts but presents the two-dimensional display in a mode that minimizes the apparent scatter of those that are most similar to each other in spatial pattern. The results of further statistical analysis of the means of the centroids and their standard errors (crosses) are shown in Table 1. In all rabbits the means for controls C+ and C– were not separated, whereas those for control and odor bursts were widely separated. In four of five rabbits the CS+ and CS– clusters were separated; these rabbits also showed behavioral evidence for odor discrimination

odor and CS+ versus CS– are given in Table 1. The four rabbits showing behavioral evidence of odor discrimination showed evidence by this criterion as well for spatial differences between CS+ and CS– burst patterns, whereas the rabbit that did not discriminate also did not give bursts that differed in spatial pattern between CS+ and CS–. However, in contrast to the distance from the Euclidean measure that were normally distributed, these distances from nonlinear mapping did not appear to conform to the normal distribution; because they were too few in number to form an empirical distribution, their statistical significance was not evaluated (Freeman 1985a).

The optimal cut-off level for f_1 was evaluated by repeatedly calculating the t-distances while raising the level in steps of 5 Hz beginning at 30 Hz and deleting bursts with f_1 below the level. Also deleted were bursts with $-50\% < FM_1 < 50\%$ and with $V_1 < 0.15$, both levels found by trial and error and accounting for 6% of bursts in this set. The procedure was done in two ways. One included only the C2, C3, T2, and T3 "discrimination" bursts on the premise that these last two bursts were more likely to have occurred after discrimination had been achieved and thereby to give a clearer separation (Fig. 10, left frame). The other included all three test bursts and their control bursts to give a larger total number (right frame). Both sets included only trials on which a correct CR+ or CR– occurred, and only the 4 subjects that learned the CR+. Optimal separation of odor bursts with minimal separation of control bursts occurred in both ways with the cut-off at 55 Hz. A more sensitive

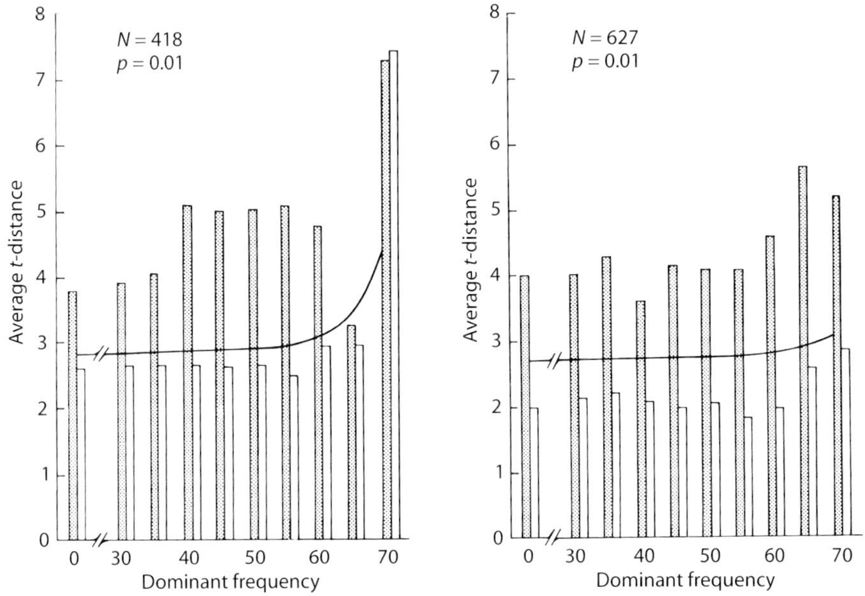

Fig. 10. Evaluation of the cut-off level for low frequency bursts was done by repeating the Euclidean distance measure after deleting the bursts below a cut-off level first at 30 Hz and then up to 70 Hz at increments of 5 Hz. The effects on the *t*-distances between CS+ and CS– centroids (shaded bars) and between C+ and C– centroids (open bars) are summarized for trials on which a correct CR occurred in four rabbits. The left frame includes only C2, C3, T2 and T3 bursts; the right frame includes all six bursts from each trial. The assignment of the *p*-value is based on the number of burst pairs remaining, on the normality of the distributions of the *t*-values, and on the assumption of independence among bursts. Optimal separation of odor bursts with minimal separation of control bursts was found at a cut-off of 55 Hz. (Freeman 1985a)

measure was the % correct classification of odor bursts minus the % correct classification for control bursts. Figure 11 shows that this measure peaked with the cut-off at 55 Hz for both groups of data. The peak resulted on the one hand from effective separation of odor bursts and on the other hand from minimal separation of the control bursts (Fig. 11, upper frames). On the premises that control bursts preceding CS+ and CS– odor bursts were from a common control state, the minimal separation was the "correct" outcome for the control bursts. In contrast, the classifications of bursts from trials for the subject with no evidence for a CR+ to the CS+ were near chance levels.

The % difference measure was adopted as a standard "touchstone" to evaluate the effectiveness of the various procedures for data processing. This included re-evaluation of the depth of focus for spatial deconvolution (Fig. 12, lower right), leading to adoption of 0.49 mm as the optimal value (upper right) for the four subjects, and re-evaluation of the cut-off frequencies for spatial filters (lower left). The measure also was repeated on each of the three test bursts, T1, T2 and T3 separately, with the result that the % difference was found to increase progressively over the 3 sequential bursts (Fig. 12, upper right). Among those trials on which a correct CR occurred at least one EEG burst was labeled as correct on 76% of them. The first "correct" burst

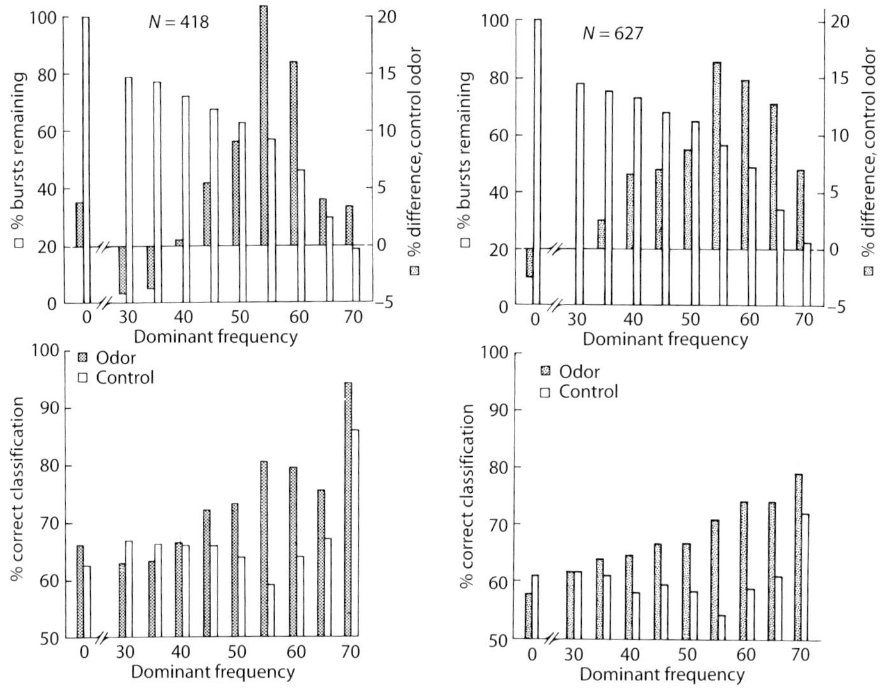

Fig. 11. The difference in % correct classification between odor and control bursts was a more sensitive indicator of the optimal value for the cut-off frequency. The groups are described in Fig. 10. The % difference was adopted as simple and versatile "touchstone" correction feedback on techniques for data processing, with the proviso that separation should be maximal for odor bursts and minimal for control bursts

occurred on 59% of these trials as T1, on 28% as T2, and on 13% as T3. This indicated that a "correct" burst was more likely than not to occur after odor onset but before the sniff response, and that if not, convergence to a "correct" EEG pattern occurred thereafter. This set of three test bursts was selected from a total of two to four times that number in each test period; from CR latency measurements only one "correct" burst sufficed to initiate a correct CR, and the sample taken may not have included a "correct" burst.

The question was asked, whether the pattern separability shown by Fig. 11 (the upper frames) was owing to any number of channels less than 64, on the premise that selected channels might reflect focal neural activity that existed only for each

---

Fig. 12. Four further uses are shown of the % difference in classification as a "touchstone" for procedural evaluation.

At lower right is the efficacy of separation as affected by the parameter in spatial deconvolution (the "software lens") that determines focal depth. The data are averaged from four rabbits; for 3 the optimal depth was at 0.49 mm and for the other at 0.54 mm. This distance was slightly less than the mean depth of the mitral cell layer (0.58 mm), which is the usual site of reversal for the granule cell dipole field of potential (Freeman 1975, 1980). From Freeman (1985b).

At lower left the "touchstone" is used to evaluate the effects of n-th order exponential spatial filters on the EEG (Gonzalez and Wintz 1977). The upper curve shows that maximal separation of odor bursts occurs with a

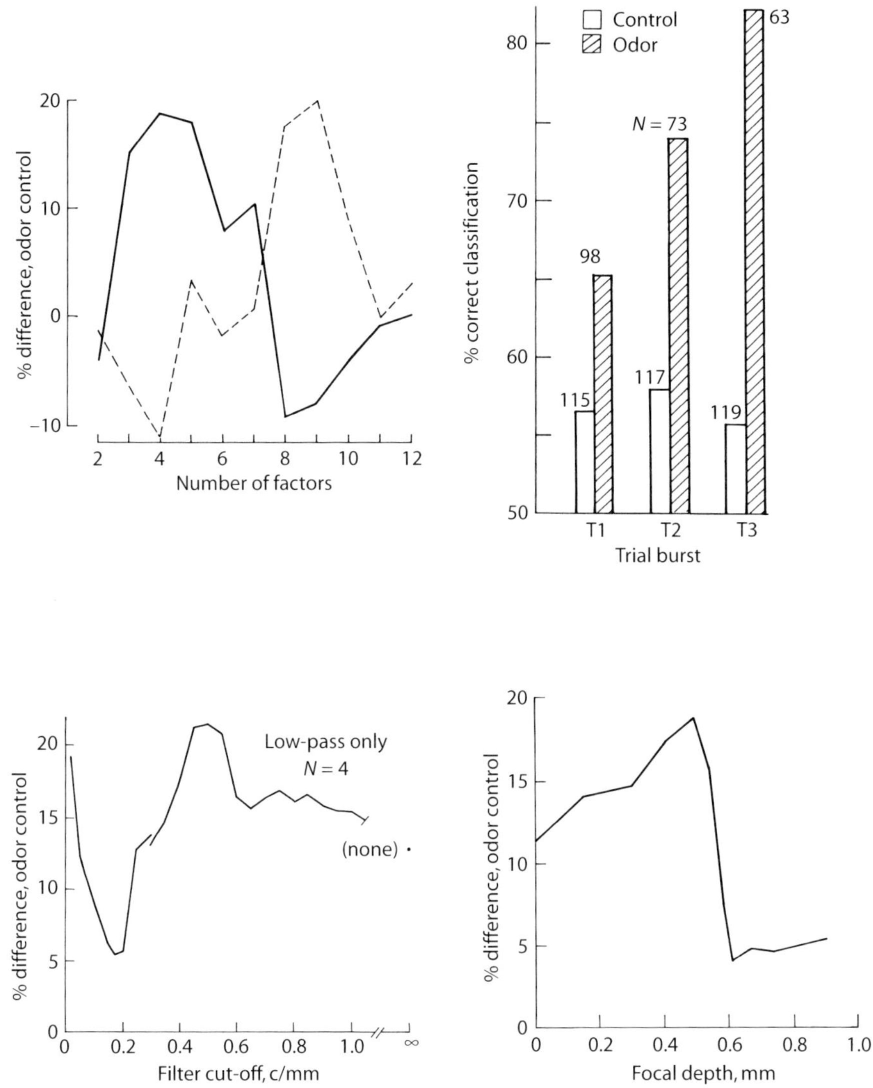

4th-order low-pass filter cut-off frequency (down 3 db) at 0.5 $c\,\text{mm}^{-1}$, as predicted by spatial spectral analysis of the bulbar EEG. The lower curve shows that minimal separation occurs with a 2nd order high-pass filter set at 0.17 $c\,\text{mm}^{-1}$, as predicted by the average of measurements of the widths of active EEG foci, and near 0.1 $c\,\text{mm}^{-1}$ as predicted by the width of the electrode array. The order parameter (n) is optimized in the same way. These results prove that the odor-specific information exists in the spatial pass-band that is characteristic of the bulbar granule cell EEG generator (Freeman 1975, 1980).

The upper right frame shows the increase in separation of odor bursts progressively with time into each trial; the bursts were taken about 0.5 s apart. N denotes the number of bursts in each group. The numbers decrease with time owing to an increase in the proportion of bursts with $f_1 < 55$ Hz during odor presentation.

The upper left frame illustrates the use of the % difference measure to evaluate the optimal number of factors to be used in the Euclidean distance algorithm after factor analysis of the data. Two examples are shown from the four rabbits, one with maximal separation using 4 factors, the other with 9 factors. When all factors with nonzero eigenvalues are used, the subgroups all approach 100% correct classification, and the % difference measure goes to zero. (From Freeman 1985c) odor. This was tested by repeating the % difference measure for each subject while reducing the number of channels used for the Euclidean distance. Those channels deleted either had means below a cut-off level or had variances above or below a cut-off level, with progressive change in the levels of cut-off. In all of these tests the separation of bursts was progressively impaired with deletion of channels, indicating that all 64 of the EEG traces contributed to the burst identification.

The dimensionality was evaluated by factor analysis of the data following normalization by channel. The hypothesis was that the signature pattern of each subject and its three odor conditions (including the background) were represented by four spatial patterns of amplitude of the dominant component, and that these patterns might be accessed through the sets of eigenvectors from the correlation matrix. The data set for each subject was re-expressed in this new coordinate system of the factor space; the channel amplitudes were replaced by factor scores. These scores were used in the Euclidean distance measure to determine the optimal number of factors to estimate the degree of separation. Two examples are shown in Fig. 12 (upper left) of the change in the % difference measure as the number of factors was increased from 2 to 12. In two subjects the optimal number of factors was 4; in two others it was 8 or 9; these subsets of factors incorporated 92% of the variance. The average % difference of 24.2% (80.0%–55.8%) did not differ from the value of 20.8% (80.2%–59.4%) from analysis of channels (Fig. 11, left frame). Under varimax rotation the optimal number factors ranged from 3 to 5 over the four subjects with the % difference at 24.8% (78.4%–53.6%); under oblique rotations the % difference deteriorated, indicating that the factors indeed were orthogonal. In this way the number of variables used for classification of bursts (the 3 to 5 factors) was less than the number of bursts (ranging from 8 to 42 and averaging 17) in every classification subgroup (C+, C–, CS+, and CS–), on the average about four fold.

Three tests were made for factorial invariance. In the first test factor analysis was repeated n times over n cases for each subject, with deletion of a different burst on each run ("jack-knifing"). The grand mean for the % difference was 20.6% (73.9%–53.3%). The standard deviations about the subject means (ranging from –1.4% to 38.2%) ranged from 2.98% to 5.63% with an average value of 3.78%. The distribution of % difference values combined after normalization by subject was not significantly different from normal by the chi square test, indicating that the 99% confidence interval for the % difference was +/–8.80% . By this criterion the % difference values reflected differences between CS+ and CS– bursts that were significant for each subject at $p < 0.01$.

In the second test factor analysis was repeated on subsets of the data, and the similarity of the factor patterns was determined from correlation matrices of the factors. Table 2A shows the matrix resulting from factor analyses of the even- and odd-numbered bursts for each subject. The correlation coefficients were averaged over subjects by Fisher's z-transform, after reranking factors in three instances when their order by size of the eigenvalues had changed. The diagonal entries for the first four factors averaged 0.904, showing that both subsets of the data supported convergence to a common set of spatial patterns. Table 2B shows the results from performing factor analysis on three sets of bursts, one from each of the three sessions at weekly intervals that provided the whole data set, and averaging over

Table 2. The results are shown of using correlation over factors derived from analysis of subsets of the data to test for factorial invariance. In Table 2A the data were divided into two subsets by partitioning the entire data set into two parts, one of odd-numbered bursts and one of even-numbered bursts. In Table 2B the factor analysis was done on the data from each of the consecutive three behavioral sessions yielding the bursts. The asterisks denote the correlation coefficients of the first four factors (re-ranked in the few instances where the order by rank of the eigenvalues differed over sessions) averaged (by Fisher's *z*-transform) over subjects and (in Table 2B) inter-session comparison. These high coefficients reflect the stability of the factor patterns over sessions. From Freeman (1985c)

A

Factors	I	II	III	IV	V	VI
I′	0.94*	−0.23	−0.31	0.02	−0.05	−0.07
II′	−0.08	0.95*	−0.29	−0.32	−0.14	0.10
III′	−0.31	−0.21	0.86*	0.04	−0.11	0.02
IV′	−0.10	−0.37	−0.18	0.80*	−0.09	0.02
V′	−0.21	−0.11	0.01	0.09	0.41	0.21
VI′	0.06	−0.13	−0.03	0.04	0.21	0.33

B

Factors	I	II	III	IV	V	VI
I′	0.86*	−0.09	−0.28	−0.02	0.00	0.01
II′	−0.19	0.94*	−0.35	−0.38	−0.19	−0.13
III′	−0.34	−0.20	0.79*	−0.15	−0.17	0.04
IV′	0.06	−0.49	0.07	0.83*	−0.08	−0.00
V′	−0.17	−0.13	0.06	0.14	0.46	0.22
VI′	0.07	−0.14	−0.06	0.04	0.17	0.39

subjects and session pairs. The high values on the diagonal again reflected the stability of the factor patterns extracted from the EEG over a period of weeks.

In the third test factor analyses were done independently on the odd- and even-numbered bursts respectively as "training" and "validation" sets. In this test the entire set of trials was used, including those with and without correct CR's, in order to increase the number of cases; this reduced the average % difference to 17.9% (70.5% – 52.6%). Factor scores were determined for all bursts with respect to both sets of factors. The lower left half of Table 3 shows the % difference measure of classification with factor scores of the same subsets of bursts as used to derive the factors (the "training" sets); the lower right half shows the % difference measure with factor scores of the omitted subsets (the "validation" sets). The results showed that the factors derived from one half of the data were no less effective in classifying the other half, the differences being within the error of measurement as determined by "jack-knifing".

Discriminant analysis was done on the factor scores after varimax rotation. The two sets of C+ and C- control bursts were pooled, and the question was asked whether by two discriminant functions the three groups of control, CS+, and CS– bursts were from a common population. This case was rejected by Wilks lambda criterion at $p = 0.01$ for all four rabbits, with F values ranging from 2.69 (d.f. 8,108) to 15.79 (d.f. 8,202). The over-all level of correct classification of the individual bursts with respect to the three centroids by a Euclidean distance measure (Cooley and Lohnes 1962) was over 80% for the four rabbits.

Table 3. The bursts from all trials (whether or not a *CR* occurred) are divided into those odd- and even-numbered. Classification by the % difference measure on four factors is as effective (within the precision of the measure determined by "jack-knifing") for "validation" subsets (lower right) as for "training" subsets (lower left). Subject 5 showed no behavioral evidence for odor discrimination. From Freeman (1985c)

Numbers of bursts in separated groups

		Odd				Even			
Subject	Total	*C+*	*CS+*	*C–*	*CS–*	*C+*	*CS+*	*C–*	*CS–*
1	86	13	7	10	13	9	11	13	10
2	80	6	19	5	10	9	16	6	9
3	120	9	19	12	20	19	10	14	17
4	92	13	10	11	12	7	14	12	13
5	120	17	13	16	14	8	22	10	20

% difference classification measure

	Odd*odd	Even*even	Odd*even	Even*odd
	$X_1 \times X_1$	$X_2 \times X_2$	$X_1 \times X_2$	$X_2 \times X_1$
1	26.2	18.6	26.2	28.1
2	–5.2	18.2	–.52	26.6
3	9.2	16.1	9.2	18.1
4	15.2	4.3	19.0	12.3
Avg.	11.4	14.3	12.3	21.3
5	–4.3	–8.8	–1.3	–.27

4 Discussion

These results show the EEG bursts (optimally about 80%) can be sorted into classes relating to the control state and the two odor states on the basis of the spatial pattern of amplitude of the dominant cosine component. The separation holds for bursts from trials with a correct CR+ or CR–; on incorrect and nonresponding trials the separation approaches chance levels. On the majority of correct trials the separation begins with the first burst after CS onset, and it improves progressively through the third burst. The separation of CS+ and CS– odor bursts holds only for the four of five rabbits that learned the CR+, although the fifth rabbit that did sniff to both odors shows a significant difference between control pattern and the common odor spatial pattern. On these grounds we conclude that the EEG manifests odor-specific information, and that this information takes the form of stable and reproducible spatial patterns embedded in the neural activity accompanying the odor and the control states both within and across the sessions having stable behavior.

These results are based on the exclusion from factor and cluster analysis of bursts having a dominant frequency $f_1 < 55$ Hz. The validity of our interpretation hinges on a physiological explanation of the origin and significance of these as aberrant bursts. We characterize them as having frequencies spread relatively widely but usually in the 20–50 Hz range; a tendency to augmenting frequency within each burst; a large variation in the amount of frequency modulation from each burst to the next; a relatively low fraction of total burst energy in the dominant component; and a spatial amplitude pattern that varies unpredictably, in contrast to the reproducible spatial patterns of bursts with $f_1 > 55$ Hz. All of these features indicate that bursts with $f_1 < 55$ Hz are less coherent than those with $f_1 > 55$ Hz.

In these respects they resemble the background or interburst activity. This so-called "spontaneous" activity is most clearly seen in waking, unmotivated animals in which bursts seldom occur. It has the appearance of random activity in several respects; its time series is irregular and unpredictable; its spectrum has multiple low peaks that vary from one segment to the next; its amplitude histogram is Gaussian; and the envelope of the time series conforms to the Rayleigh distribution (Freeman 1975). Remarkably, however, in every time segment the same seemingly random time series appears on all electrodes in the array, and the ensemble average of the spatial rms amplitude over multiple interburst segments has the same form as the unique signature pattern for each rabbit, which we derive from ensemble averages of dominant components from the coherent and the incoherent bursts.

We conclude that this activity is not random. On the one hand, there is no known random generator located in the bulb that can serve as a noise source to produce this temporally widespread incoherent activity with a common instantaneous frequency at all times. Yet the source must lie within the bulb, because (a) the receptors do not have the physiological basis for the requisite coordination of their activity, other than that imposed at low frequency (3–5 Hz) by inhalation, and (b) the activity persists after transection of the connections of the bulb to the brain. On the other hand, we know that the bulb can be regarded as a distributed system of coupled oscillators. From simulation studies we have the capability of generating chaotic activity with such a system (Freeman 1979), as well as both steady state and limit cycle activity patterns. We conclude that the background state is the manifestation of a chaotic attractor of the bulbar mechanism and not of an equilibrium or point attractor under perturbation. We infer that the low-level basal and interburst activities are manifestations of a bulbar chaotic attractor, and that the great majority if not all of bursts with $f_1 < 55$ Hz are manifestations of one or more chaotic attractors, not that of the basal state (its peak frequency exceeds 55 Hz) but another one or more with properties similar to it.

A stable and reproducible spatial pattern of coherent activity repeatedly emerges from chaos and is manifested in the dominant component of the control bursts in each rabbit with $f_1 > 55$ Hz; this indicates that a limit cycle attractor exists in the mechanism, to which the output tends to converge upon most inhalations. The discovery of differing stable spatial patterns during inhalation of CS+ and CS– odors implies that a limit cycle attractor may exist for each discriminable odor.

We suggest that the changes in spatial EEG pattern that occur during learning to respond to an odor (Freeman and Schneider 1982, Viana Di Prisco and Freeman 1985, Gray *et al.* 1985) reflect a process of synaptic modification and nerve cell assembly formation (Freeman 1979, 1981) by which a new limit cycle attractor is formed. Looking toward the receptors, a basin is formed such that, under the initial conditions set by the input during learning, the bulbar mechanism moves toward one or another differing limit cycle state; looking toward the olfactory cortex, the output of the bulb is a reproducible spatial pattern of activity that is determined by the attractor of the basin selected by the input. We suggest that the control burst pattern is determined by learning in respect to the collection of background odors, and that the background, CS+ and CS– spatial patterns represent three of a repertoire which includes other undefined limit cycle attractors.

By this interpretation each inhalation constitutes a test of the environment by the rabbit, that may result in convergence of the bulbar mechanism to a limit cycle attractor, but that may result on some inhalations in failure of convergence with an indeterminate outcome. We speculate that this event may be manifested in those bursts in which the energy is shifted to a lower frequency manifesting a chaotic attractor. This might provide the basis for a clear signal to the rabbit of indeterminacy, leading to re-testing or to initiation of an orienting response.

The spatial pattern relating to each CS could not be localized to any smaller number of channels than the set of 64. When channels were deleted by any of several criteria relating to amplitude, phase, variance, or change from control, the efficacy of separation by the % difference measure and by nonlinear mapping diminished. This did not result simply from reducing the dimensionality of the data sets, as shown by factor analysis. This result indicates that the spatial pattern of each attractor exists widely over the bulb, perhaps the whole of it (Bressler 1984), that our electrode array gives us only a spatial sample, and that each part of the sample contributes to the resolution of pattern differences. In the manner of analysis of a large scene, parts may be deleted without loss of the global image but with diminished resolution from other patterns.

These results leave questions unanswered regarding the size, internal structure, and fineness of spatial grain of the pattern of a limit cycle attractor. The array window covers less than 20% of the bulbar surface, and the interelectrode distance (0.5 mm) is twice the mean interglomerular distance. Four to 8 times the present number of electrodes would improve spatial resolution by increasing the window size and decreasing the interelectrode distance, but at great experimental cost if temporal resolution were not to be sacrificed. At present the centroids of clusters are clearly separable, but the percentage of correct classification of individual CS+ and CS– bursts based on the results of the distance measure averages 80% . This is near the average 74% of correct behavioral responses by discriminating rabbits for the CS+, but there is not a one-to-one correspondence between correctness of classification by the behavioral and EEG measures over the set of 180 odor bursts from each rabbit. Improved pattern resolution is needed, including more extensive sampling of the EEG throughout the interval between CS and CR. The manner of read-out of limit cycle and chaotic attractor patterns is undetermined. This should depend on spatial as well as temporal integration by the cortex over the bulbar output, but the kernels of the integrals and the cortical operations on bulbar output are as yet unknown.

The KII model (Freeman 1979) that has been used to simulate the temporal and spatial patterns of EEG waves has not been adequately explored for the properties of compound or multiply co-active templates to simulate nerve cell assemblies. Regarding chaos these new findings should be simulated with coupled nonlinear differential equations, in the manner that we have simulated the onset and termination of bursts in the form of limit cycle activity. Our qualitative analysis of chaotic activity must be verified by quantitative analysis, including estimation of the fractal dimensionality of the putative chaotic attractors (Freeman 1985d).

The results explain why the analysis of burst root mean square (rms) amplitude patterns (Viana Di Prisco and Freeman 1985) failed to detect reliable differences between CS+ and CS− bursts. First, the differences are not clear-cut unless the patterns are spatially deconvolved by the "software lens" (Freeman 1980). This operation is very sensitive to noise and cannot be used effectively unless the noise is attenuated with both spatial and temporal filters. Further, spatial filtering cannot at present be done prior to Fourier decomposition of the time series at each channel. Second, the bursts with $f_1 < 55$ Hz, which we propose to call chaotic, constitute a massive source of variance. When, all bursts are pooled, an essential condition for factor analysis is violated, and the results lead to excessive scatter of the classes of CS burst patterns. The results also explain how the 64 channel array can suffice to detect burst pattern differences. If the spatial pattern of the limit cycle attractor is widely distributed in the bulb, then the precision of anatomical placement of electrodes is secondary, and the degree of resolution depends mainly on the number of available electrodes.

We conclude that the bulb operates as a global dynamic system in odor discrimination (Freeman 1985c, Freeman and Skarda 1985). Its background activity is determined by a chaotic attractor (Freeman 1985d) that allows for easy transition to other attractors. Input during inhalation determines the basin for another attractor and forces a bifurcation by an attendant parametric increase in feedback gain (Freeman 1979). For each odor discriminated there is a learned, limit cycle attractor, that determines a global spatial burst pattern signifying the presence of that odor. If convergence fails, or if there is no attractor corresponding to the input odor, then a chaotic burst occurs, typically at a lower frequency, signifying the need for re-testing, orienting, habituation or learning depending on reinforcement (Gray *et al.* 1986). This allows for updating of the mechanism in an ever changing environment.

Acknowledgements

Reprinted with permission by Springer-Verlag, from: "EEG Spatial Pattern Differences with Discriminated Odors Manifest Chaotic and Limit Cycle Attractors in Olfactory Bulb of Rabbits", *Brain Theory*, pp. 97–119, 1986.

Supported by a grant MH06686 from the National Institute of Mental Health and by a grant NS16559 from the National Institute of Health

References

Abraham RH, Shaw CD (1983) *Dynamics – the geometry of behavior.* Aerial Press, Santa Cruz

Bressler SL (1984) Spatial organization of EEGs from olfactory bulb and cortex. *Electroencephalographic Clin Neurophysiol* 57: 270–276

Cooley WW, Lohnes PR (1962) *Multivariate procedures for the behavioral sciences.* Wiley, New York

Freeman WJ (1975) *Mass action in the nervous system.* Academic Press, London New York

Freeman WJ (1978) Spatial frequency analysis of an EEG event in the olfactory bulb. In: Otto DA (ed) *Multidisciplinary perspectives in event-related brain potential research*, pp 531–546. US Govt Printing Office, EPA-600/9-77-043

Freeman WJ (1979) EEG analysis gives model of neuronal template-matching mechanism for sensory search with olfactory bulb. *Biol Cybern* **35**: 221–234

Freeman WJ (1980) Use of spatial deconvolution to compensate for distortion of EEG by volume conduction. *IEEE Trans Biomed Eng* **27**: 421–429

Freeman WJ (1981) A physiological hypothesis of perception. *Perspect Biol Med* **24**: 561–592

Freeman WJ, Viana Di Prisco, G. (1986) Relation of olfactory EEG to behavior: Time series analysis. *Behavioral Neuroscience* **100**: 753–763.

Freeman WJ, Baird B (1987) Relation of olfactory EEG to behavior: Spatial analysis: *Behavioral Neuroscience.* **101**: 393–408

Freeman WJ, Grajski, K.A. (1987) Relation of olfactory EEG to behavior: Factor analysis; *Behavioral Neuroscience*, **101**: 766–777.

Freeman WJ (1987) Analytic techniques used in the search for the physiological basis of the EEG. In: Gevins A, Remond A (eds). *Handbook of Electroencephalography & Clinical Neurophysiology* Vol 3A, Part 2, Ch. 18. Amsterdam, Elsevier. pp. 583–664.

Freeman WJ, Schneider W (1982) Changes in spatial patterns of rabbit olfactory EEG with conditioning to odors. *Psychophysiology* **19**: 44–56

Freeman WJ, Skarda CA (1985) Nonlinear dynamics, perception, and the EEG; the neo-Sherringtonian view. *Brain Research Reviews* **10**: 147–175.

Garfinkel A (1983) A mathematics for physiology. *Am J Physiol* **14**: R455-R466

Gonzalez RC, Wintz P (1977) Digital image processing. *Addison-Wesley*, Reading

Grassberger P, Procaccia I (1983) Measuring the strangeness of strange attractors. *Physica* **9D**: 189–208

Gray CM, Freeman WJ, Skinner JE (1985) Associative changes in the spatial amplitude patterns of rabbit olfactory EEG are norepinephrine dependent. *Behavioral Neuroscience* **100**: 585–596.

Sammon JW (1969) A nonlinear mapping for data structure analysis. *IEEE Trans Comput* **18**: 401–409

Viana Di Prisco G, Freeman WJ (1985) Odor-related bulbar EEG spatial pattern analysis during appetitive conditioning in rabbits. *Behav. Neurosci* **99**: 964–978

12. *Simulating Gamma Waveforms, AM Patterns and $1/f^{\alpha}$ Spectra by Means of Mesoscopic Chaotic Neurodynamics*

Piecewise linear analysis modeled gamma activity as the output of an adaptive filter with white noise input (Elul 1972). Under this model a point attractor was postulated to govern the dynamics of the olfactory system under noise, which accounted for the broad spectrum of its background activity in the gamma range, the Gaussian amplitude density distribution, and the Rayleigh distribution of the peak amplitudes of the oscillations (Figure 3.13, c, p. 148 in Freeman 1975), but which failed to account for the $1/f^{\alpha}$ power spectral density and the spatial patterns of phase cones (Figure I in Prologue, Sections B and C). The gamma burst on inhalation was thought to be governed by a Hopf bifurcation to a limit cycle attractor. The amplitude and frequency modulations of bursts were ascribed to the brief time of access during inhalation, which allowed for an approach to the attractor, but with the next bifurcation coming too quickly to allow the state to settle into a stable orbit.

This model was proven to be wrong by data from the algorithm that was designed to evaluate the standard errors of measurement of the phase and amplitude of gamma bursts (Figure XI, A in the comment on Chapter 11). The method was to generate a cosine wave at fixed frequency, amplitude and phase on 64 channels, give the set of cosines an amplitude distribution having the shape of a bivariate normal surface, which approximates a typical form of an AM pattern (Freeman 1978a,b; Freeman and Baird 1987), and then add noise from a random number generator. In accordance with the hypothesis of noise input, independent random numbers were assigned to each channel. The cosine of the signal+noise on each channel was convolved with the other 63 sets of values in accordance with the known spatial point spread function of the granule cell generator (Freeman 1972d, 1980). This replicated the experimental finding that the correlation between traces of gamma activity decreased with increasing distance between channels of recording the EEG (Freeman 1979c). This algorithm sufficed to give the required standard errors of measurement of phase and amplitude, but it also demonstrated a striking disparity. The calculated signal:noise ratios fell far more rapidly with distance than did the observed ratios. Contrary to the impression from visual inspection of the plots of the 64 time series of the gamma bursts, the "noise" in the residuals from curve-fitting the EEGs was not independent across the 64 channels. The traces merely appeared to be uncorrelated, owing to irregularities contributed by the conic phase gradients (Figure 1 in Prologue). The same waveform of "noise" was present on every channel of the EEG bursts, although at differing amplitudes and latencies or phases. The hypothesis of a limit cycle under noise was untenable.

The broad-spectrum aperiodic gamma activity could only arise in common over the whole bulb if it was driven by an external "pacemaker", such as that suggested in the thalamus (Andersen and Andersson 1968), or if it was an emergent property of the bulbar populations. Review of previous data from isolation of the bulb by surgery or cryogenic blockade (Gray and Skinner 1988) showed that continual broad-spectrum bulbar activity was not totally abolished by isolation. It was severely attenuated between inhalations, and it tended to very narrow band oscillation that suggested an approach to limit cycle activity during inhalations. Measurements of axonal conduction velocities (Freeman 1962, 1974a, 1975) had shown that the spatial divergences and temporal dispersions on transmission by the primary olfactory nerve and the medial olfactory tract acted as powerful low pass filters, which would prevent any peripheral or centrifugal pacemaker from globally driving the bulb in the upper part of the gamma range. I concluded that the broad-spectrum background gamma activity must be regulated by a chaotic attractor (Freeman 1988b) within the olfactory system, not just the bulb.

To test whether this hypothesis was admissible, I constructed a model of the olfactory system with the bulb, anterior olfactory nucleus (AON) and prepyriform as its main parts (Figure 1 in Chapter 12), each with its characteristic frequency incommensurate with the others (Figure 2 in Chapter 12), as indicated by evoked potential studies. There were four long feedback pathways, in which the conduction delays and temporal dispersions were evaluated from prior physiological measurements of axonal propagation on the medial olfactory tract. The most critical step was to introduce an excitatory feedback pathway from the cells of the anterior olfactory nucleus to the periglomerular cells in the bulb, which had been described anatomically by Luskin and Price (1983), but for which no physiology had yet been done. The gain of this feedback pathway (L4 in Figure 1 in Chapter 12) in the model proved to be the most sensitive parameter for the control of the spectrum and the stability of the aperiodic activity.

The model sufficed to replicate the waveforms (Figure 3 in Chapter 12) and statistical properties (Figure 4 in Chapter 12) of the normal background activity and the waveforms of induced bursts (Figures 5 to 7 in Chapter 12), showing that both background and induced gamma activity manifested a chaotic attractor. The validity of the model was supported by the demonstration that it also sufficed to generate abnormal activity in the form of partial complex seizures (Figures 8 to 12 in Chapter 12) with parameter values that replicated the experimental conditions under which these epileptic seizures are elicited (Freeman 1986).

Simulation of Chaotic EEG Patterns with a Dynamic Model of the Olfactory System

Abstract

The main parts of the central olfactory system are the bulb (*OB*), anterior nucleus (*AON*), and prepyriform cortex (*PC*). Each part consists of a mass of excitatory or inhibitory neurons that is modeled in its noninteractive state by a 2nd order ordinary differential equation (*ODE*) having a static nonlinearity. The model is called a KO_e, or a KO_i set respectively; it is evaluated in the "open loop" state under deep anesthesia. Interactions in waking states are represented by coupled KO sets, respectively KI_e, (mutual excitation) and KI_i (mutual inhibition). The coupled KI_e, and KI_i sets form a *KII* set, which suffices to represent the dynamics of the *OB*, *AON*, and *PC* separately. The coupling of these three structures by both excitatory and inhibitory feedback loops forms a *KIII* set. The solutions to this high-dimensional system of *ODE*s suffice to simulate the chaotic patterns of the EEG, including the normal low-level background activity, the high-level relatively coherent "bursts" of oscillation that accompany reception of input to the bulb, and a degenerate state of an epileptic seizure determined by a toroidal chaotic attractor. An example is given of the Ruelle–Takens–Newhouse route to chaos in the olfactory system. Due to the simplicity and generality of the elements of the model and their interconnections, the model can serve as the starting point for other neural systems that generate deterministic chaotic activity.

1 Introduction

An obvious experimental fact about the cerebral cortex, including the various parts of the olfactory system, is its continual background or so-called "spontaneous" activity revealed in both single cell unit and electroencephalographic (EEG) "brain wave" recordings. It is so robust that it endures all but the most drastic measures such as near-lethal levels of anesthesia, several minutes of asphyxia, or complete surgical isolation of a slab of cortex. It is often called "noise" because of its unpredictable and seemingly random appearance. In particular, the olfactory EEG has a Gaussian amplitude histogram, a rapidly attenuating autocorrelation function, and a broad spectrum that resembles "1/*f* noise" (Freeman 1975). Similarly, the pulse trains of single neurons in various parts of the olfactory system yield Poisson distributions in their interval histograms and aperiodic auto- and cross-correlation functions tending to steady means after refractory periods.

In earlier studies we assumed that the basal EEG at the low amplitude characteristic of waking but nonmotivated animals was akin to "shot noise" from passing "white noise" through a band-pass filter, and we showed that the distribution of values of the envelope of the EEG conformed to the Rayleigh function (Freeman 1975). This model, treating the EEG as analogous to Brownian motion, led to the prediction that the time-dependent correlation between pairs of EEG traces from 2 points in the same area of cortex should decrease with increasing distance between the points.

Systematic studies of the olfactory EEG confirmed this prediction (Freeman and Baird 1987). However, the correlation gradient was not due to independence of activity in different parts of the same structure; it was due to a phase gradient of a common signal. Within each part (the olfactory bulb, *OB*, anterior olfactory nucleus, *AON*, and prepyriform cortex, *PC*, the same waveform was found at all points, albeit with differing amplitudes and small variations in time lag (Freeman and Skarda 1985) with respect to the ensemble average of a set of EEG traces taken simultaneously from multiple points on the same structure. The results of various surgical manipulations showed that the common pattern of activity of the *OB* in particular was not imposed from outside (Freeman 1986). This led us to postulate that the random-seeming basal activities of the *OB*, *AON*, and *PC* were emergent properties of the interactive neurons of the entire olfactory system (Freeman and Viana Di Prisco 1986). In order to test this postulate, we assembled a system of ordinary differential equations (*ODE*s) in conformance with the anatomical and physiological properties of the olfactory system, and we determined some of the parameter ranges in which this model generated chaotic activity resembling the basal EEG. By serendipity, we found that this model sufficed to simulate an informative degenerate type of activity seen in epileptic seizures of the olfactory system.

We found support for this endeavor in several introductory works, including those by Schuster (1984), Abraham and Shaw (1985), Shaw (1984), Garfinkel (1983), Holden (1986), and a fascinating essay by Rössler (1983) on a hierarchy of chaos, which introduced us to the unavoidability of chaos in the dynamics of large scale interactive systems (Shaw 1984), and the possibility of describing the geometric structures of strange attractors in the nervous system.

2 Construction of the Model

The basic module is called a *KO* subset. It represents the dynamics of a local collection of non-interactive neurons in parallel, being either all excitatory or all inhibitory. The neurons simultaneously perform 4 serial operations: nonlinear conversion of afferent axonal impulses to dendritic currents; linear spatiotemporal integration; nonlinear conversion of summed dendritic current to a pulse density; and linear axonal delay, temporal dispersion, translation, and spatial divergence. The wave to pulse function is static and sigmoidal with bilateral saturation. In feedback loops, the saturation restricts the pulse to wave conversion to its small-signal linear range. By proper choice of a dimensionless variable the dynamics of each *KO* subset is described by a linear *ODE* cascaded with the static sigmoid function. The two fixed time constants a and b and the fixed scaling parameter of the sigmoid function Q_m have been evaluated experimentally (Freeman 1975, 1979a–c).

The operation of the basic integrator, the *KO* subset, is split into two parts, a linear time-dependent operator $F(t)$ and a nonlinear time-invariant part $G(v)$. The equation for the linear part in voltage v with time t is

$$abF(v_n) = \ddot{v}_n + (a+b)\dot{v}_n + abv_n \,, \tag{1}$$

where $a = 220/\text{s}$ and $b = 720/\text{s}$ are fixed rate constants for all *KO* subsets that determine the carrier frequency for the *KII* set. The equation for the nonlinear part in the input variable v and the output variable p are

$$Q = Q_m\{1 - \exp[-(e^v - 1)/Q_m]\}, \quad v > -u_0 \tag{2a}$$

$$Q = -1, \quad v \le -u_0 \,, \tag{2b}$$

$$u_0 = -\ln[1 - Q_m \ln(1 + 1/Q_m)], \tag{2c}$$

$$p = u_0(Q+1) \tag{2d}$$

where $Q_m = 5.0$ is the asymptotic maximum of the sigmoid curve. For stability over all values of Q_m, the synaptic conversion is weighted by

$$u'_0 = u_0 d_1 \exp(-d_2 u_0), \tag{3}$$

where $d_1 = 1.64$ and $d_2 = 0.53$ are determined empirically (Freeman 1979b).

A KI_e subset is formed with two KO_e subsets in feedback relation. A KI_i subset is formed with two KO_i sets in feedback and with inversion of the output voltage of each before input to the other. A *KII* subset is formed by feedback between a KI_e subset and a KI_i subset.

Each output is multiplied by a gain coefficient k_{ij} where $v_j = k_{ij}p_i$. The k_{ij}'s are fixed at appropriate values $0.2 < k_{ij} < 2.0$ to optimize stability and sensitivity. Four types of internal connection are identified: k_{ee}, k_{ei}, k_{ii}, and k_{ie}. The equations for the full *KII* set with N subsets are:

$$F(v_{e1,j}) = k_{ee}\, p_{e2,j} - k_{ie}(p_{i1,j} + p_{i2,j}) + \sum_{k \ne j}^{N} k_{ee}\, p_{e1,k} + I_j, \tag{4a}$$

$$F(v_{e2,j}) = k_{ee}\, p_{e1,j} - k_{ie}\, p_{i1,j} \,, \tag{4b}$$

$$F(v_{i2,j}) = k_{ei}\, p_{e1,j} - k_{ii}\, p_{i1,j} \,, \tag{4c}$$

$$F(v_{i1,j}) = k_{ei}(p_{e1,j} + p_{e2,j}) - k_{ii}\, p_{i2,j} + k_{ii} \sum_{k \ne j}^{N} p_{i1,k}, \tag{4d}$$

where I_j is the input to the j-th *KII* subset. Summation is over all other *KII* subsets. A worked example is presented in the Appendix.

The interaction within a local population of like neurons, either excitatory or inhibitory, is modelled by positive feedback between two KO_e subsets of two KO_i subsets, giving respectively a KI_e or KI_i subset. The negative feedback interaction between excitatory and inhibitory neurons is represented by two KI subsets forming a KII_{ei} subset. An array of subsets forming a full KII set suffices to represent the main dynamics of each of the 3 main parts of the olfactory system (Fig. 1).

A detailed description of lumped and distributed KII dynamics has already been given (Ahn and Freeman 1974a,b, 1975; Freeman 1979a–c). The KII_{OB} set receives excitatory input from a KO_R set representing the sensory receptors, and it transmits by excitation through the lateral olfactory tract (LOT) to both the KII_{AON} and KII_{PC} sets. The KII_{AON} set transmits forward to the KII_{PC} set by excitation, back by excitation to the inhibitory interneurons of the KII_{OB} set, and also by excitation back to a KI_e set at the input of the KII_{OB} set (Luskin and Price 1983). This last link constitutes excitatory positive feedback. The KII_{PC} set gives negative feedback both to the KII_{AON} set and by a longer loop simulating the PC projection neurons (C) to the

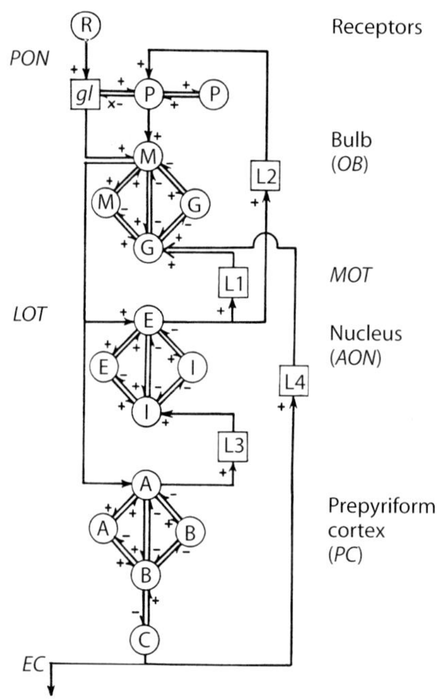

Fig. 1. Flow diagram for the equations of the olfactory system. Each *circle* (except R) represents a second order nonlinear differential equation (Appendix). Input from receptors (R) by the primary olfactory nerve (PON) is to periglomerular (P) and mitral (M) cells through the glomeruli (gl) subject to attenuation (x–), with connections to granule cells (G). Output by the lateral olfactory tract (LOT) is to the superficial pyramidal cells of the *AON* (E) and *PC* (A), each with inhibitory neurons respectively (I) and (B). Output of the *PC* is by deep pyramidal cells (C) into the external capsule (EC) and centrifugally to the *AON* and *OB* in the medial olfactory tract (MOT). The *AON* also feeds back to the granule cells (G) and the glomerular layer (P). Excitation is (+); inhibition is (–). Latencies (L1 to L4) are calculated from measurements of the conduction velocities and distances between structures. Each part is treated as a lumped system in this approximation. Each path is assigned a gain, for example, $k_{MG} = k_{ei}$ in the *OB*, k_{ME} from the *OB* to the *AON*, k_{EG} from the *AON* to the *OB*, etc.

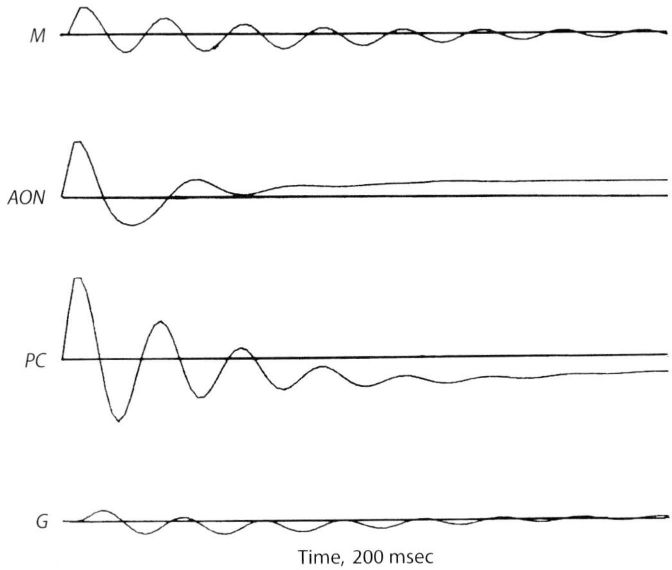

Fig. 2. Impulse responses of the three *KII* sets simulated for M (mitral unit activity), A of the *AON* (EEG), E of the *PC* (EEG) and G of the *OB* (EEG activity). The internal gains, k_{ee}, k_{ei}, k_{ie}, k_{ii}, are: *OB* (0.25, 1.50, 1.50, 1.80); *AON* (1.50, 1.50, 1.50, 1.80); *PC* (0.25, 1.40, 1.40, 1.80). The non-zero equilibria are not directly detectable with electrophysiological recordings of impulse responses. The negative value for the *PC* is consistent with the silence of the *PC* after section of the LOT through the *AON* (Freeman 1975)

KII_{OB} set. These feedback connections between the 3 *KII* sets constitute a *KIII* set. Thus far it has been studied only in its lumped form.

In order to solve the equations (by numerical integration with the forward Euler method), a gain coefficient k_{jk} must be evaluated for each connection shown by an arrow in Fig. 1 from the j-th to the k-th lumped *KO* set. The k_{jk}'s within the three *KII* sets are fixed at values listed in the legend of Fig. 2, so as, when isolated from each other, to yield impulse responses that fit the forms of averaged evoked potentials derived from these three structures in waking animals. Included also are a KI_e set at the input to the *OB*, representing the dynamics of the periglomerular interneurons (P, $k_{ee} = 1.0$) and a KO_e set representing the output neurons of the *PC* (C, $k_{BC} = 1.0$). (The values cited are typical, not essential; the model is quite robust in respect to parameter variation.)

The connection strength between *KII* sets are all set to unity, as for example $k_{ME} =$ 1.0 denoting the gain of the *LOT* axons from excitatory (mitral, *M*) cells in the *OB* to excitatory (superficial pyramidal cells, *E*) in the *PC*, except for $k_{pm} = 0.1$ to 0.3 from the excitatory periglomerular cells (*P*) to the mitral cells (*M*). These values suffice to generate interesting simulations of the EEG. Variations are achieved by modifying one or more within a range of ± 1.0.

The latencies L1 to L4 are determined as follows. The conduction velocity of the *LOT* is 5–10 m/s in its main portion; this gives a short latency so that its conduction delay is negligible. The medial olfactory tract (*MOT*) has smaller axons with velocities of 1.0–1.3 m/s (Freeman 1975). The distances in the rabbit vary between 1–10 mm from the *AON* to the *OB* (L1 and L2) and the *PC* to the *AON* (L3) and from 6 to 12 mm from the *PC* to the *OB* (L4). For each path there is a distribution of both velocities and distances (Scott *et al.* 1985), so that each delay in the model is represented by a mean and a range (L1 = 5.5 ± 4.0 m/s; L2 = 6.5 ± 5 m/s; L3 = 5.5 ± 4.0 m/s; L4 = 15 ± 7 m/s). Delay combined with dispersion is an operation of temporal convolution that serves as a low-pass filter.

3 Solutions of the Equations

When given an arbitrarily small initiating pulse at *R* (receptor input) and no further input, the system generates continuing activity (Fig. 3) that has the statistical

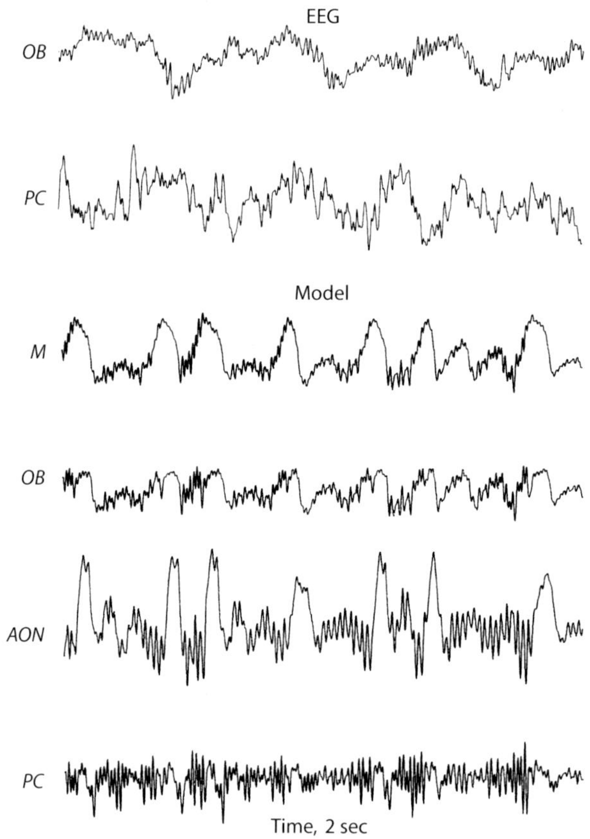

Fig. 3. Examples of chaotic background activity generated by the model, simulating bulbar unit activity (M) and the EEGs of the *OB*, *AON* and *PC*. Q_m=5.0, k_{ME}=1.5, k_{EG}=0.67, k_{EP}=1.0, k_{PM} = 0.1, k_{MA} = 1, k_{EA} = 1.5, k_{A1}, = 1.0, k_{AP}= 1. The top two traces are representative records of the *OB* and *PC* EEGs from a rat at rest breathing through the nose

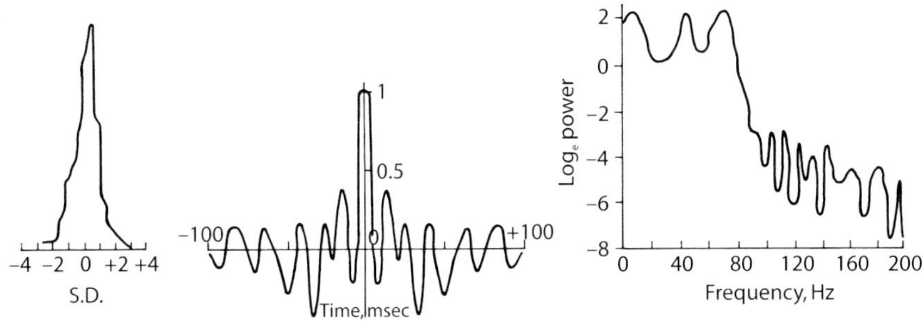

Fig. 4. Examples of the amplitude histogram (left) autocorrelation function (center) and power spectrum (right) of the simulated *PC* EEG, which are comparable with experimental results (Freeman 1975, 1986)

properties (Fig. 4) of the background EEG of resting animals. Each time series is identical to itself if the system is re-started with the same initial conditions, but if the input is changed by any amount (e.g. ± 10%) the outputs diverge progressively over time. No segment ever repeats itself in time. Plots of each variable against another or against itself displaced in time reveal thick tangles that resemble similarly derived plots of the EEG (Fig. 5). Close visual inspection of resting EEGs and their simulations shows that they are not indistinguishable, but statistical procedures by which to describe their differences have not yet been devised. Both appear to be chaotic on the basis of the properties listed, but the forms of their chaotic attractors and their dimensions have not yet been derived.

When the *KIII* set is placed into a chaotic state and input is given through R to simulate a surge of receptor input during inhalation, a simulated EEG burst appears (Fig. 6, left). After cessation of the input, the *KIII* set lapses into its chaotic basal state. If the input is sustained, then a near-periodic oscillation results (Fig. 6, right) with a narrow spectral peak for each variable. Plots of each variable against another reveal seemingly random variation in phase or equivalently in frequency (Fig. 7), indicating that despite its near periodicity this "burst" activity is also chaotic.

The *KIII* model is robust in respect to generating chaotic activity with a variety of parameter settings, but certain aspects appear to be critical. In each *KII* set, the strength of mutual inhibition has to exceed that of mutual excitation ($k_{ii} > k_{ee}$). The means of the distributed feedback delays should exceed a half cycle duration of the characteristic frequency of the *KII* set receiving the delayed feedback, and their ranges must be a substantial fraction of twice their means. The excitatory output of each *KII* set should be given to the KI_i set of the *KII* set to which it feeds back. By itself the KII_{OB} set is capable only of steady state activity ("d.c.") or oscillation at one frequency, not chaotic activity. Coupling of the KII_{OB} set with the KII_{AON} set (k_{EG}) results in multi-frequency oscillation that exactly repeats a complex pattern about every 0.3 s. Feedback coupling of both *KII* sets through the KI_p set (k_{AP} and k_{PM}) results in chaotic segments that repeat at about 3/s. Introduction of the KII_{PC} set (Fig. 1) to attain the *KIII* level deregularizes the 3/s activity.

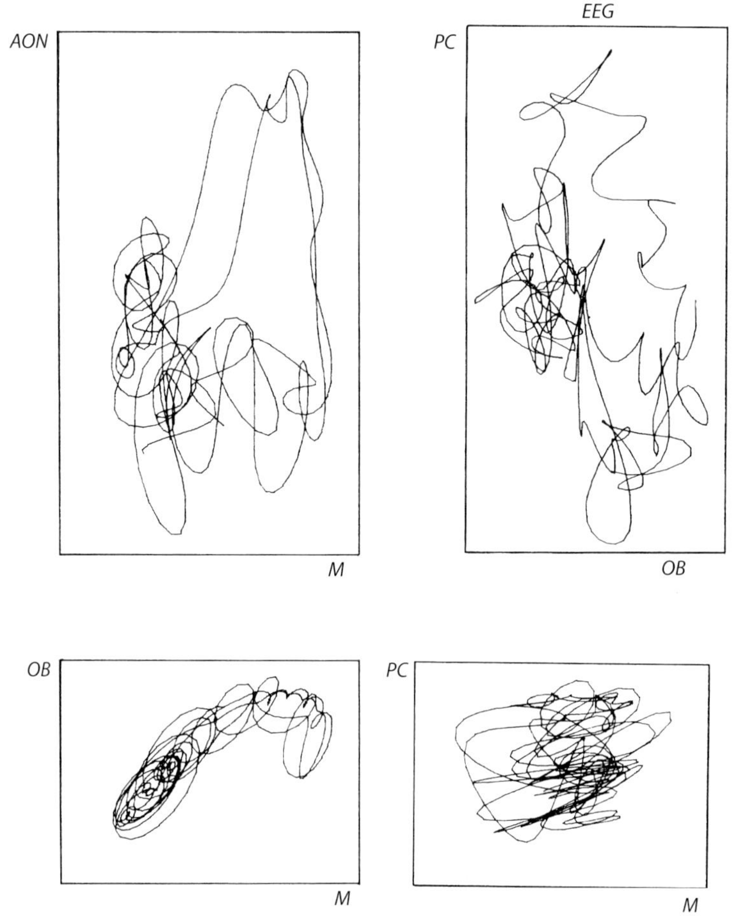

Fig. 5. Tangles showing the relations between pairs of simulated EEGs (0.5 s) of the model, and comparing these with the plot of the *OB* EEG against the *PC* EEG. Rotation is counterclockwise, because increases in mitral (M) activity precede increases in the others

Most important for operation of the *KIII* set is an internal source of sustained excitation. The three *KII* sets are stable in themselves and are given only point attractors by their fixed internal gain settings (Fig. 2). The feedback loops between them create the possibilities for instability, but fluctuation or oscillation emerges only under maintained excitatory action. The KI_P set provides such action. If its action is steady, the resulting oscillation appears to be complex and high dimensional, but it repeats itself and is not chaotic. The essential chaoticizing process in this model is the delayed feedback, from the KII_{AON} set to the KI_P set (k_{ep}). Any similar excitatory feedback, in parallel with inhibitory feedback but with differing distributed delays, might suffice for this action and may co-exist in these or yet other parts of the olfactory system, this KI_P action being at present merely the best understood.

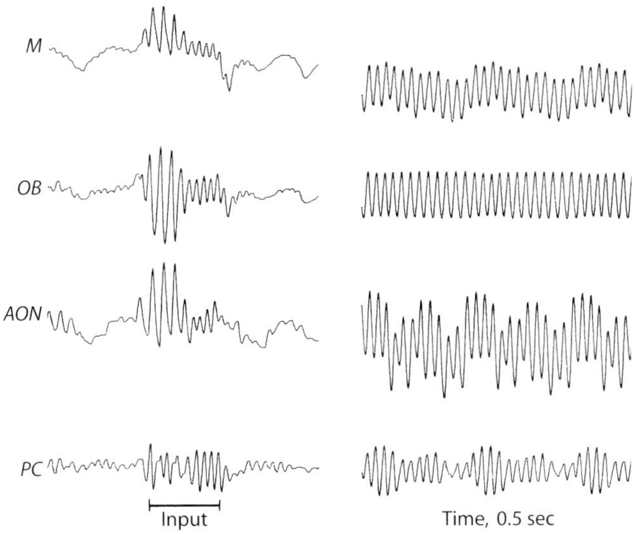

Fig. 6. *Left*: a simulated burst induced by giving a surge of input at *R* similar to receptor input density during inhalation and exhalation lasting 0.2 s. *Right*: sustained input onto pre-existing chaotic activity

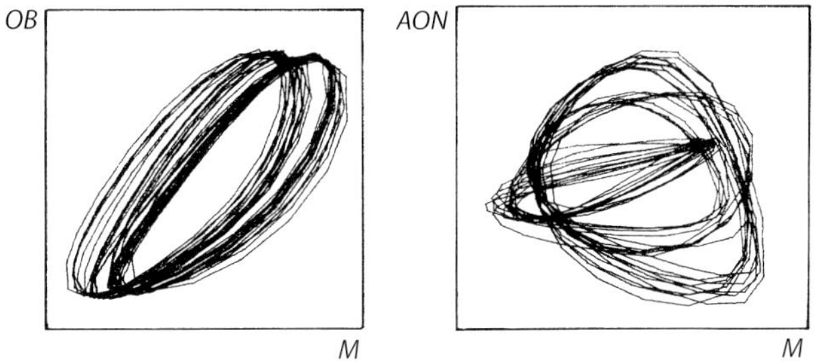

Fig. 7. Representative plots of pairs of outputs of the model during steady state input at *R*, revealing the pseudorandom variation in phase. Rotation is counterclockwise

4 Modification for Seizure Activity

Three changes in the *KIII* model together suffice to alter the pattern of activity from that resembling the background EEG to a 3/s spike train resembling a seizure (Fig. 8). One is reduction of the LOT input to the *PC* to 20% of its normal value (e.g. reducing k_{MA} and k_{EA} from 1.0 to 0.2). The 2nd change is decreasing feedback from the KII_{PC} to KII_{AON} (reducing k_{AI} from 1.0 to 0.1). The 3rd change is increasing the feedback of the *AON* to the *OB* (increasing k_{EG} from 0.67 to 1.8). The seizure spike train is initiated with a very small pulse at time zero; it grows over 0.6 s and thereafter repeats indefinitely. The similarity of the recorded and simulated traces is

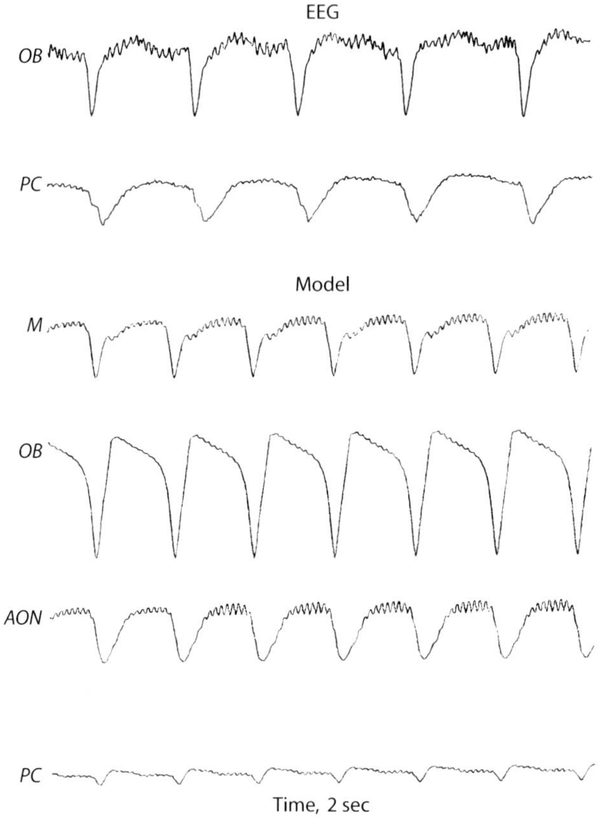

Fig. 8. Examples of 2-s time segments of EEGs recorded from a rat during a seizure, comparing these with the outputs of the model (see Fig. 3)

further shown by plotting each against itself displaced in time. An example in Fig. 9 compares the plots for the *OB* recording and the simulated output of the granule cells, both with time displacements of 30 ms.

These three changes in this example are restricted to the gain coefficients between the *KII* sets. Other modifications suffice, that are made in the gain coefficients within *KII* set and the height (Q_m) of the sigmoid curves of component *KO* sets. The directions of change that are effective serve to reduce the activity of excitatory elements and to increase that of inhibitory elements. Systematic study of the varieties of simulated seizure patterns and their relations to the varieties of recordings from animals has not yet been undertaken.

The variants in seizure forms appear thus far to have a common basic structure. This is shown by plotting the output of KII_{AON} (KO_1) against the output of KII_{OB} from granule cells (KO_G) at different levels of input from the KI_P set to the KII_{OB} set. At low simulated periglomerular input ($k_{PM} = 0.1$), the spikes occur regularly with no oscillations between spikes. The output is a limit cycle (Fig. 10, upper left) in a plane. At intermediate KI_P input ($k_{PM} = 0.2$) oscillations appear between spikes that repeated regularly. This is a multi-frequency limit cycle that is projected into the

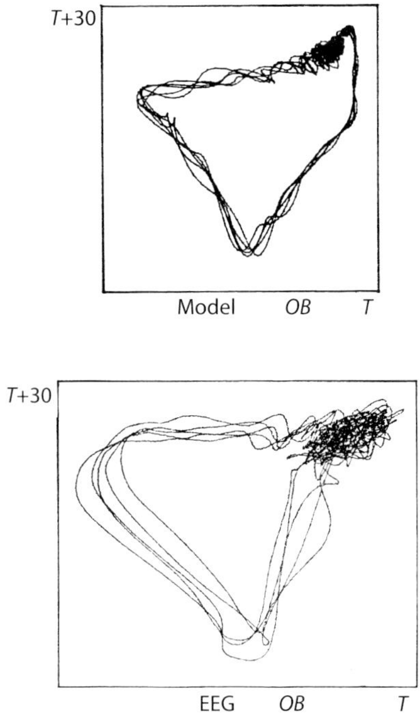

Fig. 9. Comparison of the output of the trace from granule cells (*G*) in the model (above) with the *OB* seizure from a rat (below),each plotted against itself lagged 30 ms in time (Ruelle plots). Duration is 1.0 s; rotation is counterclockwise.

plane of display. At higher KI_P input both the spike and the intervening fluctuations became unpredictable and non-reproducing, indicating that the *KIII* system is now chaotic. An alternate perspective is shown by plotting the simulated output of the mitral cells (KO_M) against that of the granule cells (KO_G) (upper right). The plot of the simulated EEGs (*OB* vs. *PC*) resembles the plot of the *OB* seizure trace against the *PC* seizure trace (lower right).

5 Interpretation of the Model

The olfactory forebrain is no more or less complex than other parts of the central nervous system, though better known than most. The *KIII* model of it is a simplification serving to emphasize certain features, in particular the capacity for generating chaotic background activity. The flow diagram in Fig. 1 is abstracted from anatomical studies, thereby deleting all but the most crucial details. The influences are neglected of the many other parts of the brain with which the olfactory system is inextricably interconnected and interactive. The numerous distinguishable active states relating to waking, sleep, etc. are reduced to four main classes of state: the

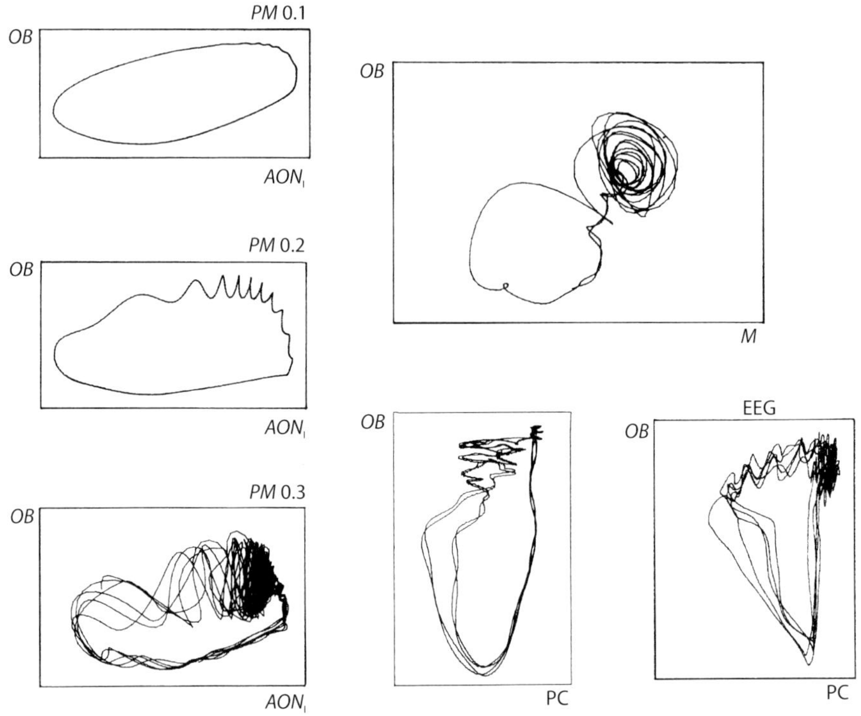

Fig. 10. The traces at left derived from a time segment of the spike train output (1.0 s duration) of the model for 3 values of K_{PM}, showing a low dimension limit cycle (above), a high dimension limit cycle, and a chaotic attractor. (In one sense, the limit cycle has only one dimension, that along its trajectory, but in another sense, it exists in multiple dimensions such that it never crosses itself). Reconstruction of the chaotic attractor in 3 dimensions shows that it is a 2-torus without detectable orifices or folds. The upper right frame shows a short segment (0.25 s) from a different perspective. The lower right frames compare the accessible *OB* and *PC* EEG traces during a seizure in a rat with the comparable output variables of the model. While related, they are not identical.

"open loop" state of deep anesthesia, resting chaos, the oscillatory burst, and the seizure state.

Examples of *OB* EEG traces in these four states (Fig. 11) show that they occur in a hierarchy, with the least energy expenditure and neural activity under anesthesia and the most during seizures. Based on these activity patterns and their simulations a global phase portrait is attempted in Fig. 12, with accompanying bifurcation diagram in Fig. 13. Each "surface" in Fig. 12 should be viewed as the "potential well" in the two representative surface dimensions of the state variables of a KO_e set and a KO_i set in any one *KII* set with activity amplitude on the vertical axis.

Under deep anesthesia there is no "background" activity, and if the system is perturbed with an excitatory pulse as by electrical stimulation, it returns quickly to rest without oscillation. This implies that each *KO* set is governed by a point attractor, and that these points may all coincide at a zero activity level. There is no detectable response to an inhibitory input.

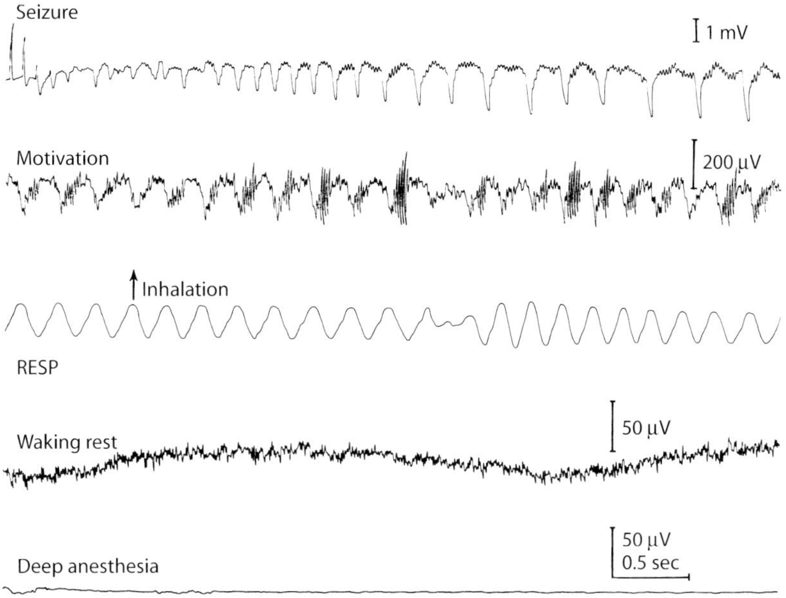

Fig. 11. Four classes of states are identified for the olfactory system from EEG traces. Fluctuations are suppressed under deep anesthesia (lowest trace). In waking but unmotivated animals the amplitude is low and the trace is irregular and unpredictable. Under motivation the irregular activity is interrupted by brief oscillatory bursts following activation of the *OB* by receptors on inhalation. Under several seconds of intense electrical stimulation of the LOT (top trace) an epileptic seizure is released. It is initiated after the failure of excitatory input transmission as shown by the decreasing responses at left to the last 5 pulses of the stimulus train. The seizure spike train then progressively emerges from a relatively quiet poststimulus state.

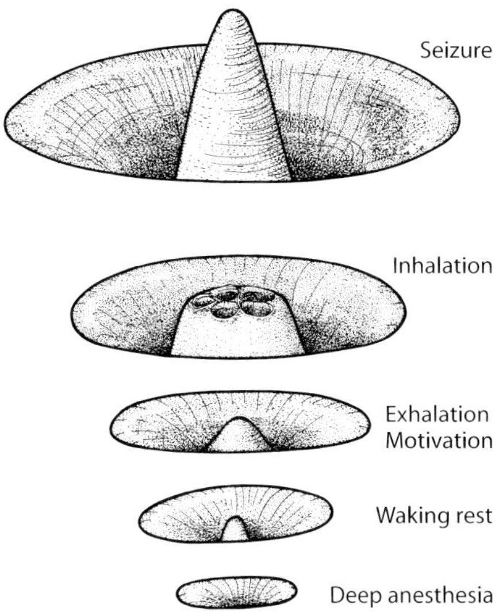

Fig. 12. A phase portrait of the high-dimensional olfactory dynamics is shown in two representative dimensions of the output of a KO_e set and a KO_i set within one *KII* set. Elevation is proportional to amplitude.

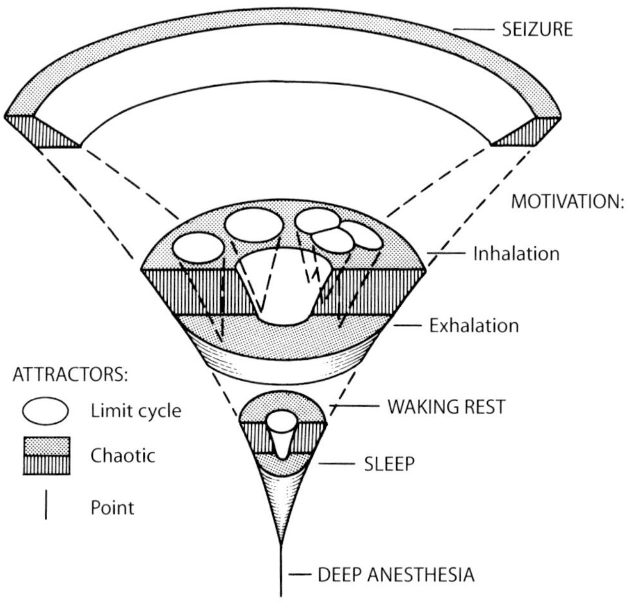

Fig. 13. Bifurcation diagram corresponding to the phase portrait in Fig. 12.

In the state of waking rest the low-amplitude chaotic activity implies that the point attractor is replaced by one or more point repellors, shown by the central peak in the potential well in Fig. 12, and that the dynamics is governed by a chaotic attractor indicated by the flat-bottomed ring of the potential well. Impulse driving whether excitatory (outwardly onto the plate) or inhibitory (inwardly onto the central peak) results in an exponentially decaying oscillation (Fig. 2) appearing as a counter-clockwise spiral on the surface.

This dynamic portrait persists under increasing motivation, mediated by centrifugal controls on the olfactory system, but mainly during exhalation. During inhalation, the third state appears as manifested by the oscillatory burst. In previous studies (Freeman 1975, 1986) this burst has been described as an abortive convergence to a limit cycle attractor. Simulations now suggest that an attractor for the burst does exist, that it is easily differentiated form the basal chaotic attractor, that it is also chaotic, but that it confers a higher degree of spatial and temporal order on *OB* activity than does the basal chaotic attractor. The term "limit cycle" attractor serves to connote this augmented order, but a more appropriate term may be "patterned" or "spatially patterned" attractor, in recognition of the experimental finding that EEG bursts can be classified, by use of the spatial pattern of amplitude modulation of each burst, in respect to stimulus odors that animals are trained to respond to (Freeman and Viana Di Prisco 1986). The results suggest that a "learned" patterned attractor is formed by synaptic modification in the *OB* for each discriminable odor (Freeman and Skarda 1985). These are latent during exhalation and come into existence with each inhalation in the form of a mosaic of basins. We conceive that the receptor input to the *OB* as determined by odor stimulation selects the basin for the appropriate attractor. Convergence is sustained only during each inhalation and its aftermath; the patterned attractors all vanish during exhalation, with return to the basal chaotic state.

This description suggests four functions for the ubiquitous chaotic activity. First, it provides rapid and unbiased access to all of the collection of latent attractors, any of which may be selected unpredictably on any inhalation, depending on the olfactory environment. The low-dimensional "noise" is "turned off" at the moment of bifurcation to a patterned attractor, and it is "turned on" again on reverse bifurcation as the patterned attractor vanishes. Second, the chaotic attractor provides for global and continuous spatiotemporally unstructured neural activity, which is essential to maintain neurons in optimal condition by preventing atrophy of disuse in periods of low olfactory demand. Third, one of the patterned attractors provides for responding to the background or contextual odor complex. It appears that a novel odor interferes with the background and leads to failure of convergence to any patterned attractor, and that chaotic *OB* output may then serve by default to signal to the *PC the presence of a significant but unidentified departure from the environmental status quo detected by the receptors R* (Freeman and Viana Di Prisco 1986). The correct classification of a novel odor by this scheme can occur as rapidly and reliably as the classification of any known odor, without requiring an exhaustive search through an ensemble of classifiable patterns that is stored in the brain. Fourth, the chaotic activity evoked by a novel odor provides unstructured activity that can drive the formation of a new nerve cell assembly by strengthening of synapses between pairs of neurons having highly correlated activity (the Hebb rule in its differential form). Thereby chaos allows the system to escape from its established repertoire of responses in order to add a new response to a novel stimulus under reinforcement.

These speculations are perhaps too narrowly focused onto the dynamics of the olfactory system. An informative essay by Conrad (1986) presents a larger picture of the "uses" of chaos in the processes of adaptive behavior, the generation of diversity in formation of a repertoire of responses, and the selection of optimal responses in an unpredictable and evolving environment.

Whatever its roles chaotic activity pervades normal cortical function. The experimental and theoretical results converge to show that the separated parts tend to be governed by equilibrium attractors, and that one or more chaotic attractors emerges under coupling. Intuitively, it is clear that the smoothed and delayed output of the *AON* is fed back into the *OB* as an excitatory bias, leading to an expansion of *OB* and then *AON* activity, and also concomitantly as an inhibitory bias, which, combined with the saturation effects of the sigmoid nonlinearity, continually compresses, curbs and folds back the exuberant outgrowth of activity under excitatory bias.

Epileptic seizure in the animal is initiated by a few seconds of repetitive electrical stimulation of the *LOT* at high intensity, which culminates in a catastrophic collapse of the transmission from the *LOT* to the *AON* to the *PC*. In the example of Fig. 11, there are 5 electrically evoked compound *EPSP* responses in the *PC* to *LOT* stimulation, the first shown being the last of a 5 s train of responses at that amplitude. The collapse of the *EPSPs* manifests the necessary and sufficient condition for the emergence of the seizure spike train (Freeman 1962). The simulation of this seizure activity requires an equivalent structural change in the model. These findings

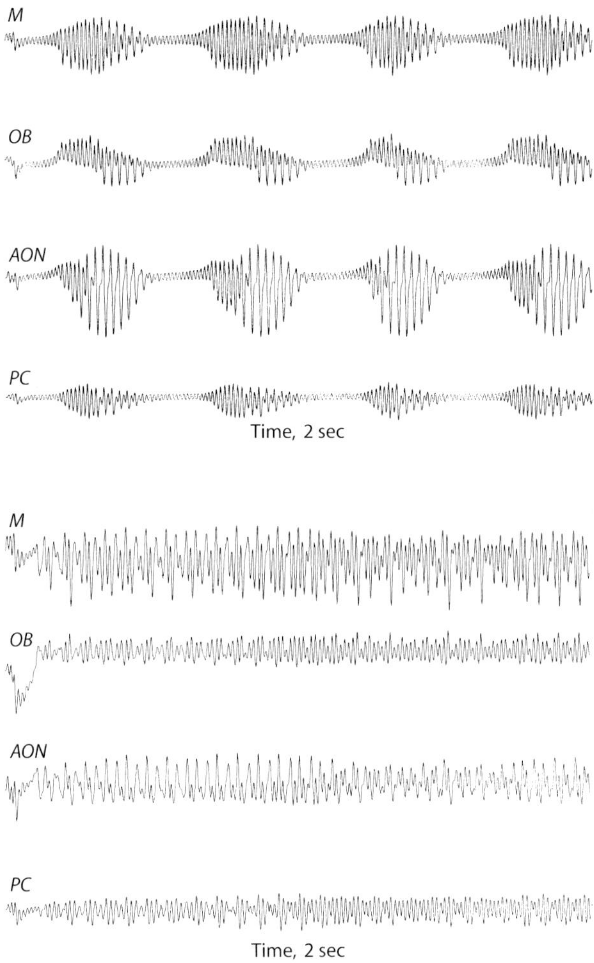

Fig. 14. A worked example of the first 2 s of activity following impulse input to the system at $t = 0$ s. See Appendix

indicate that the seizure is the result of a degenerate phase transition in the feedback mechanism that normally provides for chaotic background activity. The transition is due to an asymmetry of excess inhibitory activity over excitatory activity, which violates one of the requirements for stability of the *KII* set (Freeman 1979b).

This reduced system allows us to explore possible neural mechanisms for chaotic activity, one of which is described by Schuster (1984) as the Ruelle–Takens–Newhouse route to chaos. This involves manipulation of a parameter in the KII_{AON} to KII_{OB}, excitatory feedback pathway, such as k_{PM}. For $k_{PM} = 0$ the *KIII* set has a point attractor. With increasing strength of k_{PM}, there are two Hopf bifurcations, first to a limit cycle at a low frequency corresponding to the *KIII* feedback delay time, and second to a 2-torus at a higher frequency as well corresponding to feedback delay time within a *KII* set. With further increase in k_{PM} the *KII* system bifurcates to a chaotic attractor. Reconstruction of a model of the attractor in clay from serial thick "Poincaré" sections show that it is a 2-torus with a thick spiral region corresponding to the *KII* oscillations and a thin loop corresponding to the spikes (Babloyantz and Destexhe 1986). With still further increase in k_{PM} the torus is destroyed as the *KII* system bifurcates into the high-dimensional space characteristic of hyperchaos (Rössler 1983) and the normal EEG.

Other "routes" to chaos may exist and be found in conjunction with manipulation of other parameters. The experimental study of the return of chaotic EEG patterns following suppression by an anesthetic would provide useful and important perspectives on the model. It is unlikely that any one parameter or pathway will be found invariably to hold in a system of this high dimension. Each may have its value that provides a certain insight into the global mechanism. We emphasize that the most crucial condition for this analysis is the experimental finding of the commonality of waveform manifesting phase locking of the chaotic activity within each main part, which directs us to the lumped systems and away from the two surface dimensions in which the neurons of each part are distributed. It is presently our view that those dimensions are reserved for the expression in neural activity of information that is behaviorally significant, and that partial differential equations are not yet called for to describe the dynamics of chaotic neural activity.

Appendix

Worked examples are presented using the flow diagram in Fig. 1. Each circle signifies a *KO* neural subset governed by a linear second order equation from (1):

$$F(v_j) = \ddot{v}_j + 0.94\dot{v}_j + 0.158v_j,$$

where the two coefficients derive from the short and long rate constants $a = 0.22/\text{ms}$ and $b = 0.72/\text{ms}$. The time interval for numerical integration by the forward Euler method is 0.5 ms. The state variable v_j represents the dendritic current density in A/sq cm cortical surface area, and p_j represents axonal pulse density in pulse/s/sq cm. The conversion at synapses is fixed by the constant k_{jk} between the j-th and k-th subset. The conversion at trigger zones is governed by a static nonlinearity (see text).

For convenience it is made equal for all subsets by $Q_m = 5.0$, so that by (2) and (3)

$$p_j = 6.6 - 5.5\exp(-0.2e^{v_j} - 0.2). \tag{A1}$$

The input is $I_R = 1$ at $t = 0$ and zero for all other times. Then for the traces in Fig. 14

PG:

$$F(v_{P1}) = 0.25p_{P2} + (0.25/11)\sum_{J=15}^{25} p_{E1}(J) + 0.1I_R \tag{A2}$$

$$F(v_{P2}) = 0.25p_{P1}$$

OB:

$$
\begin{aligned}
F(v_{M1}) &= 0.25p_{M2} - 1.5p_{G1} - 1.5p_{G2} + 0.5(p_{P1} + p_{P2})/2 + I_R \\
F(v_{M2}) &= 0.25p_{M1} - 1.5p_{G1} \\
F(v_{G2}) &= 1.5p_{M1} - 1.8p_{G1} \\
F(v_{G1}) &= 1.5p_{M1} + 1.5p_{M2} - 1.8p_{G2} \\
&\quad + (1/9)\sum_{J=14}^{22} p_{E1}(J) + (k_{AG}/17)\sum_{J=22}^{38} p_C(J)
\end{aligned}
\tag{A3}
$$

AON:

$$
\begin{aligned}
F(v_{E1}) &= 1.5p_{E2} - 1.5p_{I1} - 1.5p_{I2} + 1.35p_{M1} \\
F(v_{E2}) &= 1.5p_{E1} - 1.5p_{I1} \\
F(v_{I2}) &= 1.5p_{E1} - 1.5p_{I1} \\
F(v_{I1}) &= 1.5p_{E1} + 1.5p_{E2} - 1.8p_{I2} + (k_{A1}/15)\sum_{J=17}^{31} p_{A1}(J)
\end{aligned}
\tag{A4}
$$

PC:

$$
\begin{aligned}
F(v_{A1}) &= 0.25p_{A2} - 1.4p_{B1} - 1.4p_{B2} + p_{M1} \\
F(v_{A2}) &= 0.25p_{A1} - 1.4p_{B1} \\
F(v_{B2}) &= 1.4p_{A1} - 1.8p_{B1} \\
F(v_{B1}) &= 1.4p_{A1} + 1.4p_{A2} - 1.8p_{B2} \\
F(v_C) &= -p_{B1}.
\end{aligned}
$$

The integer J counts backward in time in steps of 0.5 ms. The four representative outputs for the first two seconds after initiation by a unit impulse at $t = 0$ are from $M1$, $G1$, $E1$, and $A1$. In the upper frame the feedback from PC to AON and PC is blocked by setting $k_{AI} = 0$. This example shows an intermediate step yielding low level chaotic output that illustrates the interaction between OB and AON. In the lower frame $k_{AI} = 1$ and $k_{AG} = 1$. This illustrates the hyperchaos that characterizes the EEG. These conditions constitute a convenient starting point for exploring the capability of the model to generate a variety of signals ranging from the respiratory control activity to microsaccades in the oculomotor control system. It would appear that the most sensitive and least robust aspect of the model lies in the values chosen for the ranges of the four delay and dispersion elements L1–L4. Systematic development will require an evaluation of the basins that correspond to various combinations of values. The further questions arise, how does the nervous system maintain desired regimes in the face of changes in brain size and the maturation of axons affecting delays, and do the well-known changes in EEG with development correspond to these parametric perturbations at least in part?

Acknowledgements

Reprinted with permission by Springer-Verlag from: "Simulation of Chaotic EEG Patterns with a Dynamic Model of the Olfactory System", *Biological Cybernetics* **56**, 139–150 (1987)

Supported by a grant MH06686 from the National Institute of Mental Health

References

Abraham RJ, Shaw CD (1983, 1985) *Dynamics, the geometry of behavior*. Ariel Press, Santa Cruz, CA, pp 220 (part 1), pp 137 (part 2), pp 121 (part 3)

Ahn SM, Freeman WJ (1974) Steady-state and limit cycle activity of mass of neurons forming simple feedback loops (1): Lumped circuit model. *Kybernetik* **16**: 87–91

Ahn SM, Freeman WJ (1974) Steady-state and limit cycle activity of mass of neurons forming simple feedback loops (II): Distributed parameter model. *Kybernetik* **16**: 127–132

Ahn SM, Freeman WJ (1975) Neural dynamics under noise in the olfactory system. *Biol Cybern* **17**: 165–168

Babloyantz A, Destexhe A (1986) Low dimensional chaos in epilepsy. *Proc Nat l Acad Sci USA* **83**: 3513

Conrad M (1986) What is the use of chaos? In: Holden AV (ed) *Chaos*. Manchester University Press, Manchester UK, pp 3–14

Freeman WJ (1962a) Alterations in prepyriform evoked potential in relation to stimulus intensity. *Exp Neurol* **6**: 70–84

Freeman WJ (1962b) Comparison of thresholds for behavioral and electrical responses to cortical electrical stimulation in cats. *Exp Neurol* **6**: 315–331

Freeman WJ, (1975) *Mass action in the nervous system*. Chap. 7. Academic Press, New York, p 489

Freeman WJ (1979a) Nonlinear gain mediating cortical stimulus-response relations. *Biol Cybern* **33**: 237–247

Freeman WJ (1979b) Nonlinear dynamics of paleocortex manifested in the olfactory *EEG*. *Biol Cybern* **35**: 21–37

Freeman WJ (1979c) EEG analysis gives model of neuronal template-matching mechanism for sensory search with olfactory bulb. *Biol Cybern* **35**: 221–234

Freeman WJ (1983) Dynamics of image formation by nerve cell assemblies. In: Basar E, Flor H, Haken H, Mandell AJ (eds) *Synergetics of the brain*. Springer, Berlin Heidelberg New York, pp 102–121

Freeman WJ (1985) Techniques used in the search for physiological basis for the EEG. In: Gevins A, Remond A (eds) *Handbook of electroencephalography and clinical neurophysiology*, vol 3A, part 2, chap 18. Elsevier, Amsterdam

Freeman WJ (1986) Petit mal seizure spikes in olfactory bulb and cortex caused by runaway inhibition after exhaustion of excitation. *Brain Res Rev* **11**: 259–284

Freeman WJ, Baird B (1987) Relation of olfactory EEG to behavior: Spatial analysis. *Behavioral Neuroscience*. **101**: 393–408.

Freeman WJ, Skarda CA (1985) Spatial EEG patterns, nonlinear dynamics and perception: the neo-Sherringtonian view. *Brain Res Rev* **10**: 147–175

Freeman WJ, Viana Di Prisco G (1986) EEG spatial pattern: Palm G, Aertsen A (eds) *Brain theory*. Springer, Berlin Heidelberg New York

Garfinkel A (1983) A mathematics for physiology. *Am J Physiol* **245**: (*Regul. Integr. Comp. Physiol.*, **14**, R445-R446)

Holden AV (ed) (1986) *Chaos*. Manchester University Press, Manchester UK, p 324

Luskin MB, Price JL (1983) The topographic organization of associated fibers of the olfactory system in the rat, including centrifugal fibers to the olfactory bulb. *J Comp. Neurol* **216**: 264–291

Rössler OE (1983) The chaotic hierarchy. *Z Naturforsch* **38a**: 788–801

Schuster HG (1984) *Deterministic chaos*. Physik, Weinheim, p 220

Scott JW, Ranier EC, Pemberton JL, Orona E, Mouradian LE (1985) Patterns of rat olfactory bulb mitral and tufted cell connections to the anterior olfactory nucleus pars extrema, *J Comp Neurol* **242**: 415–424

Shaw R (1984) *The dripping faucet as a model chaotic system.* Aerial Press, Santa Cruz CA, p 113

Shepherd GM (1972) Synaptic organization of the mammalian olfactory bulb. *Physiol Rev* **52**: 864–917

13. *Tuning Curves to Optimize Temporal Segmentation and Parameter Evaluation of Adaptive Filters for Neocortical EEG*

The neocortex is the motherlode of brain function. This mammalian organ enables bats to outfly birds, dolphins to outswim sharks, and humans to outwit just about everybody but Mother Nature. The application of mesoscopic dynamics to neocortex has been slow due to lack of the most basic knowledge. Which populations in what layers of cortical neurons generate the major components of the dendritic currents underlying the EEG, particularly the gamma band? What are the depths of their zero isopotential surfaces? Are the dipole fields of pyramidal cells asymmetric, and are they coupled with closed fields of interneuronal stellate cells? What are the widths of the bell-shaped distributions of dendritic potentials at the pial surface of single neurons or small dendritic bundles (their "point spread functions" in the language of optics)? What are the open loop rate constants of the neocortical dendritic generators? What other factors contribute to the temporal spectra (Figure 5 in Chapter 13) and spatial spectra (Figure 11 in Chapter 13) of the subdural electrocorticograms (see comment before Chapter 9)? What are the gains of the three types of feedback in cortex among the excitatory and inhibitory populations (Chapter 4)? What are the characteristic closed loop frequencies in the resting states, which reflect the set points that are determined by these diverse functional connection strengths (Figure 2 in Chapter 12)? Numbers of connections alone cannot tell the story; whereas in neocortex the excitatory neurons outnumber the inhibitory neurons by over 5 to 1 and in hippocampus by 20 to 1, in the prepyriform the ratio is closer to 1 to 1, and in the olfactory bulb it exceeds 1 to 100 (Freeman 1972e). What are the transfer functions in the small-signal near-linear range of neocortical dynamics (Chapter 3)? What are the parameters of the static asymmetric sigmoid function, which is to be revealed by calculating the pulse probability of single neurons conditional on the neocortical gamma EEG (Figures 4 and 5 in Chapter 10)? What are the kernels of the spatial integrations performed by the convergent-divergent pathways that carry cortical output (Freeman 1992a)? Sixty years of neglect of the basic biophysics of neocortical EEG, as compared with the paleocortical EEG (Freeman 1987a), will not be quickly remedied.

The cytoarchitectures of six-layered neocortices are more varied and complex than those of three-layered paleocortices (Sholl 1956; Braitenberg and Schüz 1991), and so also are the characteristics of neocortical EEG (Lopes da Silva 1991). Prominent peaks are often seen in the theta (3–7 Hz), alpha (8–12 Hz), and gamma (20–80 Hz including "40 Hz") ranges, but they tend to vanish when the cortices go into action, as indicated by the onset of intentional behavior, suggesting that narrowband oscillations may indicate a form of idling (Walter 1953) when mesoscopic dynamics are on hold. In the active state that for years was referred to by clinicians as "desynchronized" (Luchins 1990; Barlow 1993), the temporal spectra of neocortical EEG in small mammals show very little in the way of peaks (Figure 5 in Chapter 13). There is no clear evidence of "forcing functions" comparable to the receptor input with respiration in the olfactory system (Figures 2 and 3 in Chapter 13), and while there are bursts of gamma, their relations to perceptual and motor events in behavior are difficult to measure and verify (Bressler *et al.* 1993; Tallon-Baudry *et al.* 1998; Rodriguez *et al.* 1999), especially the EEG segments with aperiodic oscillations and broad temporal spectral power distributions.

Despite differences between the waveforms of paleocortical and neocortical EEG, there are good reasons to believe that the basic form of the spatial "code" in mesoscopic dynamics of paleocortex, the AM pattern, should also be found in neocortical EEG, because of the similarities of neuron types, connection densities, and chemistries, and their biophysical functions (Freeman 1992b), and because, most importantly, at some combinatorial level every sensory systems must have a "code" that is compatible with those of all the others, in order that their outputs be combined easily and effectively into Gestalts. The state transitions between wings of global attractors, which Kaneko (1997, 1998) identifies with Milnor ("fragile") attractors, provide in the sensory cortices the essential link between the microscopic sensory-driven activity and the mesoscopic constructs of the cortices. Thereby the sensory cortices constitute a unique interface between the arrays of sensory neurons and the limbic system of the brain (Maclean 1969; Eichenbaum *et al.* 1992; Miyashita 1995; Buzsáki 1996). The neocortices screen and accept environmental input, compact it, abstract, generalize, and create meaning for transmission to sites of multisensory convergence, with laundering (Freeman 1991a, 1991c) along the way through spatiotemporal integration that removes the irrelevant details peculiar to each sensory mode. The limbic system sends preparatory 'preafferent' messages (Kay and Freeman 1998, after von Holst and Mittelstädt 1950), also termed 'corollary discharges' (Sperry 1950), which are shaped by prior Gestalts in the ongoing stream, and which prepare the sensory cortices for the succeeding inputs that have been solicited by intentional behavior (Freeman 1995; 1999a).

Evidence has now been obtained from trained rabbits that the visual, auditory, and somesthetic neocortices (Figure 1 in Chapter 13) generate AM patterns (Figure 7 in Chapter 13), and that they occur episodically between the CS and CR onsets (Figure 12 in Chapter 13). Their forms are comparable to those in the olfactory bulb, but not to those in the prepyriform cortex. They differ not in their AM patterns (Bressler 1988) but in their phase patterns. The spatial pattern of phase of the prepyriform EEG conforms in its directions and slopes to directions and velocities of the transmission pathway by which the bulb drives the cortex. The phase patterns of the sensory neocortices, like those of the olfactory bulb, do not conform to the properties of the afferent pathways. Instead they also have the form of a cone, and the location and sign (maximal lead or lag) of the apex vary at random (Figure I in Prologue, Sections B and C), indicating that the apex of the phase cone marks the site of nucleation for the last preceding cortical state transition. The difference in phase patterns suggests that the primary receiving areas of the sensory systems are capable of autonomous organization of perceptual messages on impact from sensory receptors, and that at least some of the targets of cortical transmission are driven by their integrated input and perform further as yet undefined operations, including feedback control of the bulb, without invoking the capacity for self-organization.

Spatiotemporal Analysis of Prepyriform, Visual, Auditory, and Somesthetic Surface EEGs in Trained Rabbits

Summary and Conclusions

1. Spatial ensemble averages were computed of 64 traces of electroencephalograms (EEGs) simultaneously recorded from 8×8 arrays over the epidural surfaces of the prepyriform (PPC), and visual, somatic, and auditory cortices. They revealed a common waveform across each array. Examination of the spatial amplitude modulation (AM) of the waveform revealed classifiable spatial patterns in short time segments. The AM patterns varied within trials following presentation of identical conditioned stimuli, and also between trials with differing stimuli.
2. Prepyriform cortical EEGs revealed strong correlates with the respiratory rhythm; neocortical EEGs did not.
3. Time ensemble averaging of the PPC EEG attenuated the oscillatory bursts, indicating that olfactory gamma oscillations (20–80 Hz) were not phase-locked to the times of stimulus delivery but instead to inhalations. Time ensemble averages of neocortical recordings across trials revealed average evoked potentials starting 30–50 ms after the arrival of the stimulus.
4. Average temporal fast Fourier transform (FFT) power spectral densities (PSD) from pre- and post-stimulus PPC EEG segments revealed a peak of gamma activity in olfactory bursts.
5. The logarithm of the average temporal FFT PSDs from pre- and post-stimulus neocortical EEG segments, when plotted against log frequency, revealed a $1/f$-type spectra in both pre- and post-stimulus segments for CS– and CS+ stimuli. The α'- and β'-coefficients from the regression of Eq. [2] onto the average PSDs were significantly different between pre- and post-stimulus segments, owing to the evoked potentials, but not between CS– and CS+ stimulus segments.
6. Spatiotemporal patterns were invariant over all frequency bins from 20–100 Hz in the $1/f$ domain. Spatiotemporal patterns in the 2–20 Hz domain progressively differed from the invariant patterns with decreasing frequency.

7. In the spatial frequency domain, the logarithm of the average spatial FFT power spectra from pre- and post-stimulus neocortical EEG segments, when plotted against the log spatial frequency, fell monotonically from the maximum at the lowest spatial frequency, downwardly curving to a linear $1/f$ spectral domain. This curve in the $1/f$ spectral domain extended from 0.133–0.880 cycles/mm in the PPC and from 0.095–0.624 cycles/mm in the neocortices.
8. Methods of FFT and PCA EEG decomposition were used to extract the broad-spectrum waveform common to all 64 EEGs from an array. AM patterns for the FFT and PCA components were derived by regression. They were shown by cross-correlation to yield spatial patterns which were equivalent to each other and to AM patterns from calculation of the 64 root mean square (RMS) amplitudes of the segments.
9. Each spatial AM pattern was expressed by a 1×64 column vector and a point in 64-space. Similar patterns formed clusters, and dissimilar patterns gave multiple clusters. A statistical test was devised to evaluate dissimilarity by a Euclidean distance metric in 64-space.
10. Significant spatial pattern classification of CS– versus CS+ trials (below the 1% confidence limit for 20 of each) was found in discrete temporal segments of post-stimulus data after digital temporal and spatial filter optimization.
11. Varying the analysis window duration from 10 ms to 600 ms yielded a window length of 120 ms as optimal for pattern classification. A 120 ms window was subsequently stepped across each record in overlapping intervals of 20 ms. Windows in which episodic, significant CS+/CS– differences occurred lasted 50–200 ms and were separated by 100–200 ms in the post-stimulus period.
12. Neocortical spatial patterns changed under reinforcement contingency reversal, showing a lack of invariance in respect to stimuli and a dependence on context and learning, as previously found for the olfactory bulb and PPC.
13. The EEG data contributing to classification were homogeneously distributed across wide temporal and spatial spectral bands and across the spatial array of electrodes. Patterns could be resolved with as few as 16 channels. No channel was more or less contributory to classification than any other.
14. The aperiodic waveforms, the rapid global state changes, the context-dependence of the AM patterns, and the homogeneous distribution of neural activity suggest that the neural events formed during perception are constructed by cooperative population dynamics in both paleocortex and neocortex. These characteristics so far provide the most powerful evidence for spatially coherent, aperiodic oscillations manifesting macroscopic cortical states that are spatially continuous over areas more than 5 mm in diameter and last less than 200 ms.

Introduction

Correlation of spatial patterns of neural activity in the olfactory receptor layer and olfactory bulb with classes of odors was first postulated by Adrian (Adrian 1950). Subsequent studies of the spatial patterns of activity among receptors (Moulton 1976) and periglomerular cells with 2-deoxyglucose (Lancet *et al.* 1982) or optical dye recording (Kauer 1987), and of the requisite degree of topographic mapping in the primary olfactory nerve between them (Freeman 1974), have supported this hypothesis of sensory coding. However, unit studies of mitral cells have thus far failed to yield an unequivocal central neurosensory code for odors (Thommeson 1978; Wells *et al.* 1989).

Studies of electroencephalograms (EEGs) in the olfactory bulb revealed close statistical correlations in both time, space, and EEG amplitude with nearby single and multiple unit activity (Freeman 1975; Eeckman and Freeman 1990, 1991). Large numbers of neurons were active in response to odorant stimuli, so that the failure of a spatial sensory code to previously emerge was possibly due to inadequate sampling by microelectrode recording. On the premise that dendritic potentials gave access to neuron population activity, spatial analysis of EEG patterns from arrays of 64 epidural electrodes was undertaken. Evidence was sought for spatial coding at a macroscopic, or neural population level. Within training experiments, spatial patterns were found in respect to odors that rabbits had learned to discriminate (Freeman and Viana Di Prisco 1986). The patterns took the form of amplitude modulation (AM) of a spatially coherent oscillation in the gamma frequency (20–80 Hz) range (Bressler and Freeman 1980).

However, the spatial patterns lacked invariance with respect to odorant conditioned stimuli, showing instead a dependence on brain state, behavioral context, and training history (Freeman 1991c). This context dependence referred to the meaning of the odors for the subjects, including both positive (i.e. food or drink reward) and negative (i.e. a mild shock or other aversive stimuli) reinforcement. The patterns did not reflect a strictly sensory code. A nonlinear dynamic model was constructed to show how a chaotic carrier wave could be generated (Freeman 1987), and how it might be modulated to give context-dependent spatial patterns in response to learned input (Yao and Freeman 1990).

The aim of the present report is to extend the approach of multichannel surface EEG recording and analysis to the visual, auditory, and somatic neocortices. Analyses of neocortical recordings from multiple electrodes on the human scalp (Walter 1953; Livanov 1977; Barlow 1993) and from arrays placed onto the exposed cortex (25 channels by Lilly (1949, 1954); up to 100 channels by Livanov (1977); and 400 channels by DeMott (1966) showed the existence of complex patterns of aperiodic activity (Mickle and Ades 1954). DeMott (1966) concluded that sensory input "is presented to the cortex not as a map, but as a very complex spatial-temporal sequence, in which every part of the cortex participates in displaying information from every part of the [sensory] field" (p. 29). Those efforts failed to yield a theory of sensory or perceptual coding. Previous results from the study of olfaction (Freeman 1987) suggested that the failure was due to inadequate sampling of EEG data from trained animals. Here we report on the analysis of a massive data base of 50+ gigabytes, representing an iterative analysis of two years of 64-channel EEG recordings from 14 subjects and four cortical areas, enabled by recent developments in high speed computing. The results show that the fundamental characteristics of EEGs from the three sensory neocortices resemble olfactory EEGs in displaying spatial AM patterns of a spatially coherent, aperiodic carrier wave, which serve to classify EEG segments with respect to discriminated sensory stimuli. These AM patterns reflect the context and history of training, not merely stimulus coding and classification.

Methods

Animal Preparation

Square arrays of 64 (8×8), 0.25 mm diameter stainless-steel, epoxy-insulated wire electrodes were prefabricated with connectors (Eastman 1975) prior to surgery. Interelectrode distances for paleocortical and neocortical arrays were 0.56 mm and 0.79 mm respectively. In order to make the array as large as possible without spatial aliasing, the maximum allowable interelectrode spacing was found which provided the optimal spatial sampling frequency as determined by spatial spectral analysis (Freeman and Baird 1987).

Each of 14 female New Zealand White (NZW) rabbits (2.5–4.5 kg) was chronically implanted with an epidural array on the prepyriform cortex (PPC), the visual (Thompson *et al.* 1950; Hollander and Halbig 1980; Hughes and Vaney 1982), auditory (Galli *et al.* 1971; McMullen and Glaser 1982), or somatic (Gould 1986) cortex. There were four PPC, two visual, five auditory, and three somatic implants; all of the arrays were placed onto the left cortical hemisphere (Figure 1). During array implantation a full surgical plane of anesthesia (4% isoflurane/O_2 mixture) was maintained via endotrachial intubation; standard aseptic techniques were observed. After rectangular trephination, an array was placed directly onto the intact dura. K-Y jelly was applied to protect exposed dura at the margins of the array, and the opening was sealed with acrylic cement used to fix the array to the

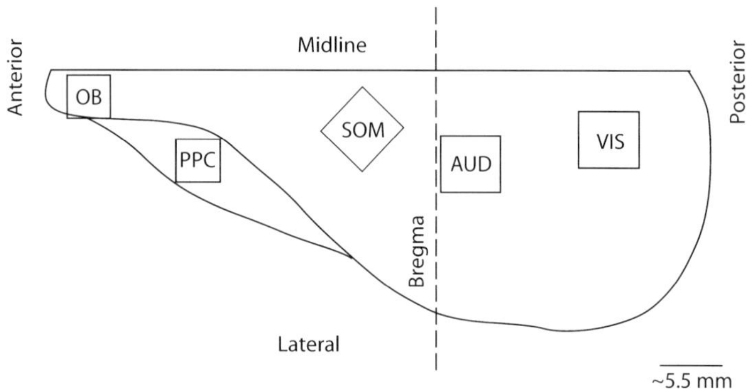

Fig. 1. Schematic of electrode array implant sites. Each array was chronically implanted onto the epidural surface of the left cortical hemisphere of the NZW rabbit. This figure represents the left, lateral surface of the cortex and demarcates the general location of each array placement. Note that the OB implant was located on the dorsal aspect of the bulb and has been included in this schematic as a reference to previous work. PPC, prepyriform cortex; SOM, somatosensory cortex; AUD, auditory cortex; VIS, visual cortex.

calvarium. Structural ground and reference screws were inserted at six sites into the skull (two anterolateral to the coronal suture, two on the medial aspect of the supraorbital ridge, and two 1.0 cm posterolateral to the bregma). Drying and infection of the exposed soft tissues and corneas was prevented with periodic application of 5% baytril/saline solution. Buprenorphine (0.00875 mg/kg SC) was postoperatively administered. All procedures were conducted according to protocol approved by the U.C. Berkeley Animal Care and Use Committee with veterinary supervision by the Office of Laboratory Animal Care.

Post-Mortem Animal Disposition

After all experiments were complete, each animal was euthanized by pentobarbital overdose (120 mg/kg Nembutal IV). Each rabbit was perfused with 10% formalin and an autopsy was performed to verify placement of the electrode array, to check for damage of the underlying cortical tissue, and to check for regrowth of bone over the cortical window.

Recording

The EEG was recorded monopolarly with respect to that cranial reference electrode nearest the array and amplified by fixed-gain (10K) ISO 4/8 differential amplifiers manufactured by World Precision Instruments. Each channel was filtered with single pole, first order analog RC filters (6 dB/octave falloff) set at 100 Hz (3 dB point) and 0.1 Hz. Records of sixty-four 12-bit samples multiplexed at 10 μs were recorded at a 2 ms digitizing interval (500 Hz) for 6 seconds and stored as signed 16-bit integers. The incremental time delay caused by multiplexing of the EEG was corrected off-line. Bad channels associated with movement artifact or EMG were identified by visual editing and replaced off-line by averaging the signals of two vertically or horizontally adjacent channels. For each experiment there were no more than ten bad channels.

Conditioning

After a week for post-operative recovery each rabbit was familiarized with the experimental setup (Viana Di Prisco and Freeman 1985), while placed in a restraining box in order to decrease movement artifact in the EEG recordings. A pneumograph belt was attached around the chest for a digital record of respiration; surgical clips were placed onto the posterolateral aspect of the left cheek for unconditioned stimulus delivery; and the restrained animal was placed into an electrically shielded chamber with adequate ventilation and a source of white noise at 72 dB. The PPC rabbits were fitted with a nose cone for odor delivery and the auditory rabbits were fitted with earphones for delivery of auditory stimuli. Each rabbit was given a stimulus modality-specific to the cortical area of implantation (e.g. rabbits implanted with arrays over the visual cortical areas received visual stimuli, etc.). The EEG was analyzed for movement artifacts and for the presence of stimulus-specific activity, including oscillations in the PPC and evoked potentials in the neocortical sites, to confirm usable EEG recordings.

After familiarization there was one, basic experimental paradigm. Each rabbit was classically conditioned to discriminate between two different modality-specific stimuli. Each stimulus was delivered on 20 trials randomly sequenced in intertrial interval and stimulus order, thereby yielding a 40 record experiment. The 6 second recording period was divided into 3 seconds of pre-stimulus EEG recording and 3 seconds of post-stimulus EEG with the stimulus arriving at 3000 ms (3500 ms for olfactory stimuli). During the recording experiment the rabbit was trained to discriminate between an unreinforced stimulus (CS–) and a reinforced stimulus (CS+) paired to a mild electric shock (unconditioned stimulus, US), which had sufficient intensity to elicit a skin twitch (unconditioned response, UR). The US was delivered via a pair of skin clips placed superficially at the base of the left ear or onto the posterolateral aspect of the left cheek. The US arrived at the end of the 6-second trial period and consisted of four to five electrical pulses (1–5 mA) delivered in a window of 10 ms. Previous work (Davis and Freeman 1982) demonstrated acquisition of a conditioned response (CR) during training to reinforced olfactory stimuli within 3–4 trials. Presence of a CR was measured as a change in the post-stimulus respiration (Freeman *et al.* 1983). This basic paradigm was replicated once during the second and third weeks. The CS–/CS+ contingencies were reversed during the third week with the same animal; the CS+ that was initially paired with the US became the CS– and vice versa. Each rabbit was trained to the basic paradigm once per week for a total of three experiments per animal at the conclusion of this project. In the case of the PPC and auditory implants, one of each animal did not survive through the last experiment.

Stimuli

Four pairs of discriminable stimuli were used throughout these experiments. Olfactory stimuli typically included food flavorings (e.g. peppermint and coconut extracts) delivered by an olfactometer which injected each test odor into a constant stream of charcoal-filtered air flowing into a mask attached over the rabbit's muzzle (Freeman and Schneider 1982). Arrival time of each odor occurred at 3500 ms (the measured time lag through the olfactometer). Odorants were removed from the mask by a mild vacuum sufficient to prevent the spread of the odorants into the recording chamber. Duration of the olfactory stimuli was 2.5 seconds.

Visual stimuli consisted of two full-field flashes varying only in intensity lumens (3.6 s. 2.8 ft-cd). During presentation of visual stimuli each animal was placed into a dark chamber with no background light. Duration of the visual stimuli was 10 ms. Auditory stimuli consisted of two amplitude-invariant sinusoids differing only in frequency (500 Hz vs. 5000 Hz), delivered binaurally via a pair of Sony MX-2 stereo earphones at 72–84 dB above 72 dB white noise within the recording chamber. Duration of auditory stimuli was 100 ms. Somatic stimuli consisted of an air puff onto the contralateral (to the implant site) cheek or onto the contralateral hindquarters. Duration of somatic stimuli was 3.0 seconds.

Data Analysis

Averages of EEG Segments

Display of paleocortical and neocortical EEGs was by the spatial ensemble average (SEA) and standard deviation (SD) from single trials across 64-channels. Temporal ensemble averages (TEA) were computed from SEA's across 40 experimental trials (20 CS–/CS+) to give the average evoked potential (AEP) and its SD for each experiment. Temporal spectra of the EEGs were calculated by applying the fast Fourier transform (FFT) (Press *et al.* 1988) to contiguous 500 ms pre- and post-stimulus segments of SEA's. Each time series was padded with zeros until the number of points reached a power of two, and each time series was smoothed with a Hamming window prior to FFT decomposition. Power spectral densities (PSDs) derived from the FFT were averaged across the 20 CS– and 20 CS+ records for each time window, thereby yielding the average PSD and corresponding standard errors (SE) for all frequencies below the 100 Hz low-pass analog filter cutoff. In order to test how much the average PSD deviated from a $1/f$-type spectra in the gamma range, a line described by Eq. [1] was linearly regressed onto the log-log PSD plot in the 20–80 Hz range,

$$\log_{10} P = \alpha - \beta \log_{10} f \qquad [1]$$

Equation [1] can be rewritten as

$$P_k(f) = 10^{\alpha'_k} / f^{\beta'} \forall k \in \{1, 2, \ldots, 64\} \qquad [2]$$

where P is the PSD of each channel (k) at every frequency (f); α is the y-intercept of the PSD regression line and $-\beta$ is the slope of the regression line from Eq. [1]. Since regression of a line onto the log-log frequency spectra naturally biases the high frequency portion of the average log-log PSD, a curve described by Eq. [2] was regressed onto the temporal spectra. Calculation of the nonlinear coefficients, α' and β', was determined by using α and β, respectively, as initial guesses for the Levenberg–Marquardt method of nonlinear regression (Press *et al.* 1988) of Eq.[2] onto the average 20–80 Hz PSD. The average PSD ±SE expressed to the 95% confidence level was plotted against the curve described in Eq.[2]. Regression onto the 20–80 Hz domain of the neocortical PSD as opposed to the 20–100 Hz domain was chosen because these frequencies were minimally affected by the 100 Hz RC filter. Deviations from a $1/f$ spectra were indicated by those areas of the regression curve lying outside of the SE curves. Similar regression analyses for the PPC data sets were conducted using the 10–50 Hz portion of the frequency spectra. This frequency range was expected to reveal excess power in the 50–80 Hz band, which could force a poor fit of Eq. [2] to the PPC PSD.

Derivation of Spatiotemporal Pattern from 64 Traces

Spatial patterns of EEG activity were calculated by decomposing or representing EEG time segments of specified length from the 64 channel recordings. Four methods were used to derive spatial patterns: FFT (Press *et al.* 1988) and modified FFT decomposition (Freeman and Viana Di Prisco 1986; Freeman 1987); principle component analysis (PCA) (Freeman and Van Dijk 1987; Maier *et al.* 1987); and root mean square (RMS) (Freeman 1978) computation of amplitude activity. FFT decomposition yielded the spatiotemporal pattern of PSD activity at a certain frequency. The modified FFT returned the PSD at the dominant frequency varied by linear amplitude and frequency modulation across the EEG segment. Modified FFT and PCA algorithms yielded the waveform common to all 64 channels. Use of the RMS was justified by the small value of the residuals, after removal of the dominant component by PCA. The spatial pattern of each common waveform was expressed by the 8×8 PSD matrix of the largest modified FFT component, the 8×8 matrix of the factor loadings of the dominant factor of PCA, or the 8×8 matrix of RMS amplitudes. Each matrix was represented as a 1×64 column vector $\overline{V}$,

$$\overline{V} = R_k \, \forall k \in \{1, 2, \ldots, 64\} \qquad [3]$$

where R is the spatial pattern matrix for each channel (k). The degree of conformance between the modified FFT, PCA, and RMS analysis techniques was measured by calculating the nonparametric correlation coefficients (r) between sets of patterns. Each correlation coefficient was normalized to a z-score prior to averaging and re-transformed to its r-value with its asymmetric SDs.

Relating Spatiotemporal Pattern to Temporal Frequency Spectrum

To determine whether the spatiotemporal patterns of EEG PSDs differed across the temporal frequency spectra, 128 ms and 256 ms pre- and post-stimulus windows of experimental EEG data taken from 20 CS– and 20 CS+ records were transformed into 1×64 vectors with the RMS and FFT algorithms. The normalized RMS pattern was employed as a reference pattern. The Euclidean distance from the reference pattern to each pattern reconstructed from the normalized PSD from the first frequency (2 Hz) of the FFT decomposition was calculated and averaged across 40 records. This process was repeated for each normalized PSD bin, thereby yielding a pre- and post-stimulus frequency spectrum of distances from the reference pattern to each frequency pattern,

$$D_f(\overline{V}', \overline{V}''^{(f)}) = \sqrt{\sum_{k=1}^{64} (V'_k - V''^{(f)}_k)^2} \, \forall f \in \{2, 4, \ldots, \tilde{n}\} \qquad [4]$$

where $\overline{V}'$ is the 1×64 RMS reference vector; $\overline{V}''^{(f)}$ is the 1×64 FFT PSD vector for each frequency bin (f) ranging from 2 Hz to the last frequency bin ($\tilde{n}$) under 100 Hz; and D_f is the Euclidean distance from the reference pattern to the FFT PSD pattern reconstructed at each frequency bin.

Spatial Pattern Classification

The search for significantly different, stimulus-induced, spatial CS–/CS+ patterns was conducted by stepping a T ms window along individual records at over-lapping 20 ms time intervals, where T was varied from 10 to 500 ms in search of an optimal value for classification. Each window, having 64 EEG traces, yielded a 1×64 column vector in amplitude $\overline{V}$,

$$\overline{V}^{(i,j)} = (R_k^{(i,j)}) \forall k \in \{1, 2, \ldots, 64\}, \quad j \in \{1, 2, \ldots, n\}, \; i \in \{1, 2, \ldots, 40\} \qquad [5]$$

where $\overline{V}^{(i,j)}$ is the 1×64 spatial pattern vector at time window (j), record (i); n is the number of time windows, and R is the component amplitude for each channel (k). This analysis yielded 20 CS– patterns and 20 CS+ patterns for each time step. These patterns were further divided into two subgroups of 10 CS– patterns and two subgroups of 10 CS+ patterns. Each pattern defined a point in 64-space and was normalized to zero mean and unit standard deviation to remove stimulus specific amplitude differences between patterns. For each step (j) in a subgroup (m) a series of centroids $\overline{C}^{(m,j)}$ was calculated in 64-space,

$$\overline{C}^{(m,j)} = 1/10 \sum_{i=10(m-1)+1}^{10m} \overline{V}^{(i,j)} \forall m \in \{1, 2, 3, 4\}, \; j \in \{1, 2, \ldots, n\} \qquad [6]$$

where m = 1, 3 represents the CS– centroids and m = 2, 4 represents the CS+ centroids for each time window. Classification of patterns as CS– or CS+ was determined by calculating the Euclidean distances of all patterns in subgroups 1, 2 to centroids 3, 4 within each of n windows,

$$D(\overline{C}^{(m,j)}, \overline{V}^{(i,j)}) = \sqrt{\sum_{k=1}^{64} (C_k^{(m,j)} - V_k^{(i,j)})^2} \qquad [7]$$

This process was repeated by cross-classifying all patterns in subgroups 3, 4 by centroids 1, 2. A spatial pattern in a test subgroup was classified correctly if the distance between that pattern and the centroid of its training subgroup was less than the distance between the same pattern and the centroid of the opposite training subgroup,

$$D(\overline{C}^{(2\tilde{m}-1,j)}, \overline{R}^{(i,j)}) < D(\overline{C}^{(2\tilde{m},j)}, \overline{R}^{(i,j)}) \forall \tilde{m} \in \{1, 2\} \quad i \in \{20(2-\tilde{m})+1, \ldots, 20(3-\tilde{m})\} \qquad [8]$$

where $\tilde{m}$ = 1 represents centroids 1, 2 and indexes i to patterns 21–40; and where $\tilde{m}$ = 2 represents centroids 3, 4 and indexes i to patterns 1–20. What emerged was a temporal sequence of ratios (x:40) where x is the number of correctly discriminated spatial patterns within each window (j). The probability (p) that the number of patterns that were correctly cross-classified was significant was determined with a cumulative binomial probability distribution scale,

$$p_j = \sum_{r=x}^{q} (q!/r!(q-r)!) \tilde{p}^r (1-\tilde{p})^{q-r} \forall j \in \{1, 2, \ldots, n\} \qquad [9]$$

where q is the number of trials (40) (Press *et al.* 1988; Barrie and Freeman 1994). Since the calculation of Eq. [9] assumed that there was an equal probability that any spatial pattern could be classified as a CS– pattern or as a CS+ pattern, $\tilde{p}$ = 0.5. This calculation was repeated for each time window giving n probability values in a time series: p_j, where $j = 1,... n$. The average change for probability values across a series of analysis windows was defined as

$$\overline{\Delta p} = 1/(n-1)\sum_{j=2}^{n}|\log_{10}(p_j) - \log_{10}(p_{j-1})| \quad [10]$$

1-D and 2-D Temporal and Spatial Digital Filtering

Previous work demonstrated significantly improved classification of paleocortical EEG spatiotemporal patterns after temporal (Freeman and Viana Di Prisco 1986) and spatial (Freeman and Baird 1987) digital filtering. Optimizing temporal filters involved searching a set of band-pass filter parameter spaces in order to maximize the number of correctly classified post-stimulus patterns while minimizing the number of falsely classified pre-stimulus patterns. This band-pass temporal filter space was defined by a variable low-cut setting beginning at 5 Hz and stepped every 5 Hz until the constant high-cut setting of 100 Hz (the high-cut analog filter setting defining the usable experimental frequency space) was reached. For each band-pass filter setting the experimental data was digitally 1D-filtered, transformed into a series of RMS spatial patterns, normalized, and re-classified. This analysis yielded a band-pass temporal filter tuning curve for each individual data set. From the tuning curve an optimal band-pass temporal filter setting was employed throughout the remainder of the analysis of that data set.

Spatial spectral analysis of the 8×8 RMS amplitude matrix from a 120 ms segment of EEG was done by imbedding it in a 32×32 matrix of zeros, translating the data to bring the zero spatial frequency values to the center of the 2D plot of power, and applying a two-dimensional FFT (Freeman and Baird 1987; Press *et al.* 1988). Spatial PSDs derived from the 2D-FFT were radially integrated in order to obtain the PSD in cycles/mm,

$$S_i = 1/c\sum_{h=0}^{D_i'}\sum_{v=0}^{D_i'} P_{h,v}' \quad \text{where } D_{i-1}' \le \sqrt{h^2+v^2} \le D_i' \;\; \forall i \in \{1, 2, \ldots, 15\} \quad [11]$$

where P' is the 2D PSD; h is the horizontal distance of P' from the center of the array; v is the vertical distance of P' from the center of the array; D' is the radius (in mm) of a concentric circle mapped onto the 32×32 matrix; i is the index of one concentric circle (15 were mapped onto the matrix of PSDs); c is the number of PSDs satisfying the condition $D_{i-1}' \le \sqrt{h^2+v^2} \le D_i'$; and S is the average 2D PSD for those values within a circle containing the center four electrodes for the first integration or an annulus containing 4–6 electrodes for subsequent integrations. Radially integrated spatial PSDs were subsequently averaged across the 20 CS– and 20 CS+ records for contiguous pre- and post-stimulus portions of an experiment analyzed with the RMS algorithm, thereby yielding the average spatial PSD in cycles/mm and ±SE expressed to the 95% confidence level. Curves were regressed onto the average 2D-PSD using Eq. [2] for only the high frequency portion of the PSD defined as that above 0.09 cycles/mm and that below the spatial Nyquist frequency (0.9 cycles/mm for the PPC 2D-FFT and 0.62 cycles/mm for the neocortical 2D-FFT).

Spatial filters were optimized (after the data set had been filtered at an optimal band-pass temporal filter setting) by imbedding the 8×8 RMS amplitude matrix in a 32×32 matrix of zeros, translating the data, calculating the 2D-FFT, and applying an exponential band-pass digital spatial filter defined by a variable low-cut setting beginning at the minimal resolution of the 2D-FFT and stepped at every subsequent resolvable spatial frequency until the constant high-cut setting was reached. The constant high-cut setting was defined as that immediately below the spatial frequency artifact (the spatial Nyquist frequency) induced by the inter-electrode spacing of the array (Freeman and Baird 1987). The subsequent 32×32 matrix of 2D-FFT values was inverse transformed back into an 8×8 matrix of spatially filtered RMS values. Optimization of spatial filters involved searching the band-pass spatial filter parameter space in order to maximize the number of correctly classified post-stimulus patterns while minimizing the number of falsely classified pre-stimulus patterns. This analysis yielded a band-pass spatial filter tuning curve for each individual data set. Each data set was ultimately classified using an optimized temporal and spatial filter setting.

Quantifying the Spatial EEG Distribution of Neural Activity Across Channels
The contribution in EEG activity from any electrode as part of a subset of electrodes from the 64-channel array was assayed by repeating the cross-classification measure after groups (0, 2, 4, 8, 16, 24, 32, 48, 56, 63) of randomly selected channels were deleted. Groups of random channels were selected by a multiplicative congruential random-number generator with a period of 2^{32}. This analysis was iterated 50 times for each group of deleted channels. The average number of patterns classifying below chance levels ($p < 0.50$) across analysis windows was calculated for each group. Standard errors were expressed to the 99% confidence level. The distribution of neural activity across the array was assayed by keeping a running total of how many times each channel was deleted across groups, and how many cases classified correctly when that channel (as a subset of other randomly deleted channels) was absent. This yielded an average number of spatial patterns which classified correctly as a function of individual channel deletion. The questions were asked, which channels (if any) were more or less important than average for classification, and how did the goodness of classification decline with the number of channels deleted?

Results

Temporal Analysis

Inspection of raw EEGs from the PPC conformed to previous results (Freeman and Schneider 1982; Bressler 1987). A spatial ensemble average (SEA) of the PPC recordings across the 64-channel array retained the series of high frequency bursts

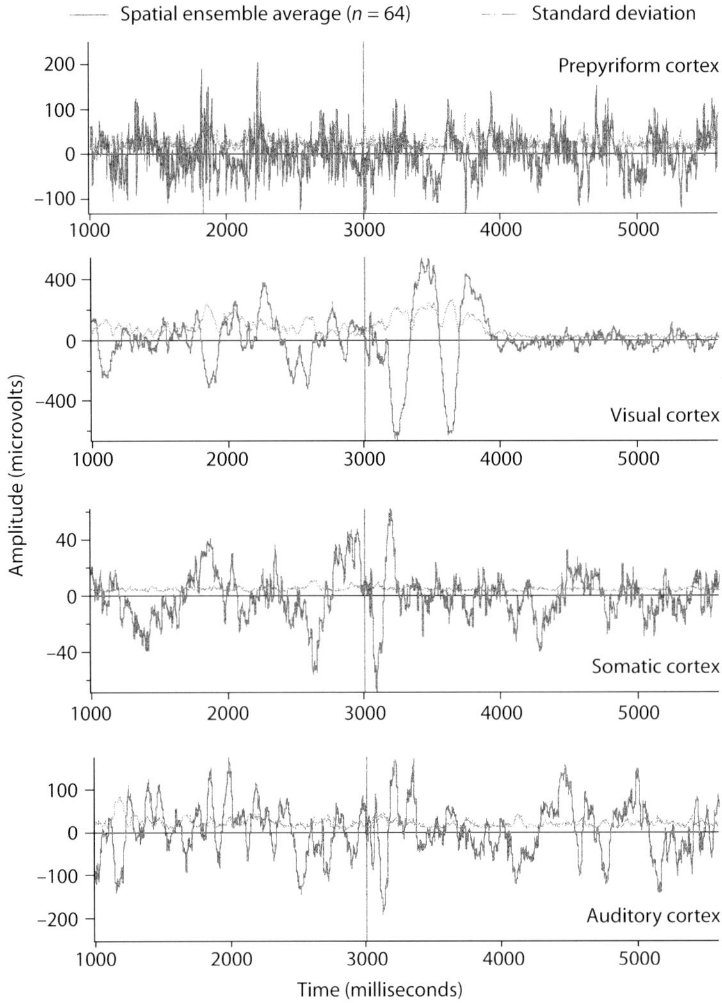

Fig. 2. Examples of spatial ensemble averages (*solid curves*) and standard deviations (*dashed curves*) from four cortical areas, calculated by averaging 64 EEGs simultaneously recorded during one experimental trial. In all recordings the stimulus was initiated at $t = 3000$ ms. In the case of olfactory stimuli there was a 500 ms delay, so that the odor arrived at the rabbit's nose cone at approximately $t = 3500$ ms.

separated by broad-band interbursts all riding on a low frequency wave highly correlated with the respiration (Figure 2, *upper trace*). The same procedures applied to the neocortical data sets yielded the SEA time series with no bursts but with an evoked potential marking the arrival at the sensory cortex of the afferent volley of action potentials relayed through the thalamus (Chang 1959) from the periphery (Figure 2, *lower three traces*). The low standard deviation (SD) time series for each SEA indicated high spatial coherence across the neocortical arrays. When the raw PPC EEG recording from a single channel was plotted against the respiratory wave, each olfactory burst began one-quarter cycle into the inspiratory phase and lasted

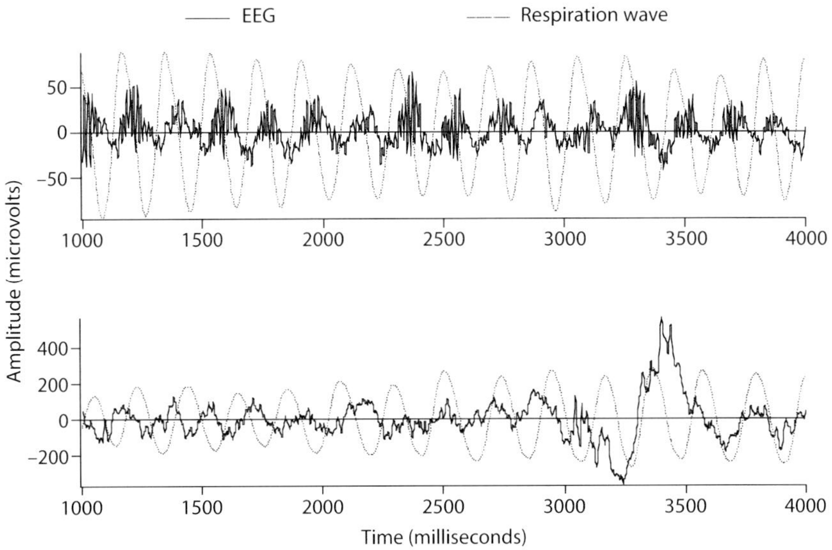

Fig. 3. Respiration, as measured with a pneumograph (*dashed curves*), and EEGs (*solid curves*) were correlated in the prepyriform cortex (*upper traces*) but not in the visual cortex (*lower traces*)

on the order of 80–200 ms (Figure 3, *upper trace*). Raw neocortical EEG recordings plotted against the respiratory wave showed no such correlation (Figure 3, *lower trace*), indicating that respiratory activity was not a forcing function for neocortical events.

Averaged 20 CS– and 20 CS+ SEA's from a single experiment yielded the temporal ensemble average (TEA), including the average evoked potential. The PPC TEA showed that temporal averaging of olfactory EEGs blurred burst patterns, because oscillations within bursts were not phase-locked to the onset of the olfactory stimulus (Figure 4, *upper trace*). Neocortical TEA's revealed activity time-locked to the stimulus arrival. Background EEG activity was smoothed, increasing the signal to noise ratio of the evoked potential above the neocortical SEA's (Figure 4, *lower three traces*). Within all four sensory modalities the SD's for the TEA's were greater than the SD's from the SEA's due to high variance in the spatial EEG patterns across trials.

Temporal spectra were used to locate peaks, particularly in the gamma band (Bressler *et al.* 1980, 1984, 1987, 1993; Gray *et al.* 1989; Freeman 1991b; Engel *et al.* 1992; Koch 1993; Freeman and Barrie 1994; Gray 1994). Averaged spectra from 500 ms windows of post-stimulus (3000–3500 ms) EEG were computed over 40 records from the prepyriform, visual, somatic, and auditory cortices and plotted against log frequency. Power spectra were calculated from the SEA, since the average PSD across each of the 64-channels was statistically indistinguishable from the PSD of the SEA. Curves described by Eq.[2] were regressed onto the 10–50 Hz frequency range for the PPC data and onto the 20–80 Hz frequency range for the neocortical data and plotted against each average PSD and ±SE (95% confidence level). Because of an expected excess of power in the 50–80 Hz domain (the range for olfactory bursting), regression onto the PPC PSD utilized only those frequencies in

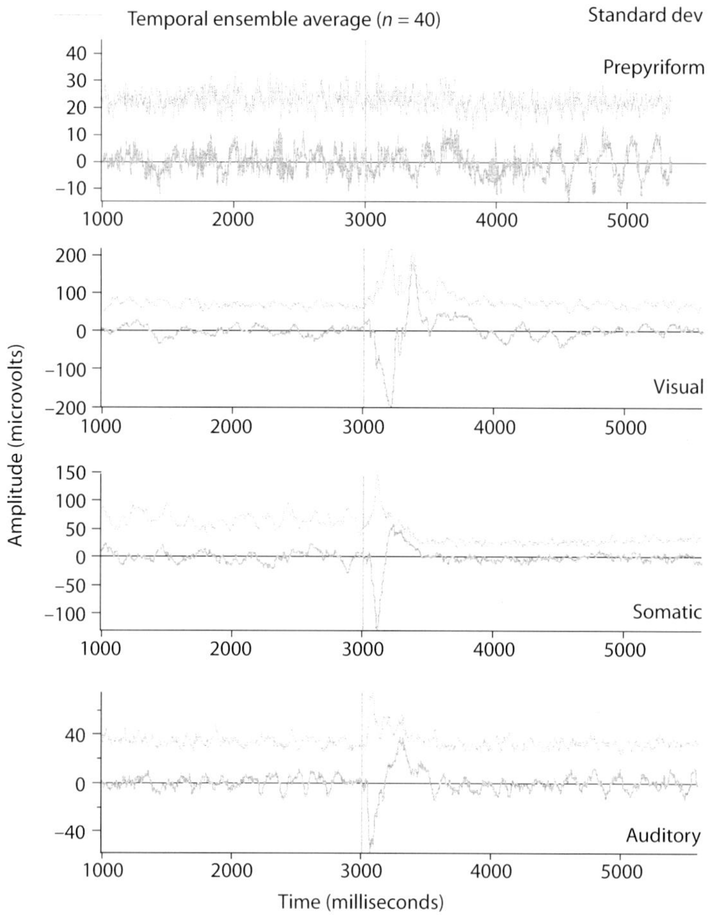

Fig. 4. Temporal ensemble averages (TEAs) (*solid curves*) and standard deviations (*dashed curves*) were calculated for the prepyriform, visual, somatic, and auditory cortices by averaging the spatial ensemble averages (SEAs) of 40 records from one experiment. Temporal averaging of the PPC SEAs blurred the pattern of olfactory bursts as shown in Fig. 3 (*upper traces*). Temporal averaging of the neocortical SEAs revealed that there was EEG activity time-locked to the arrival of the stimulus as demonstrated by the increased signal-to-noise ratio of the evoked potential above the neocortical SEAs and by the smoothing of the control-period TEA. There was no analogous time-locked PPC TEA activity.

the most linear portion of the PSD (10–50 Hz). The criterion was adopted that a spectrum was $1/f$ if its regression curve lay completely within the standard errors. Deviant peaks in the gamma frequency range were sought, because gamma activity has been proposed as a vehicle for feature binding (Gray *et al.* 1989; Freeman 1991a; Eckhorn *et al.* 1992; Engel *et al.* 1992). The PPC spectra closely matched previous results (Boudreau and Freeman 1963; Freeman 1975; Bressler 1987) with peaks in power spectral density at frequencies between 50–80 Hz (Figure 5, *upper*). Neocortical results (Figure 5, *lower three*) illustrated that there were no statistically significant spectral deviations within the gamma frequency range for the neocortical EEG. Occasional deviations from $1/f$ within the gamma range were not

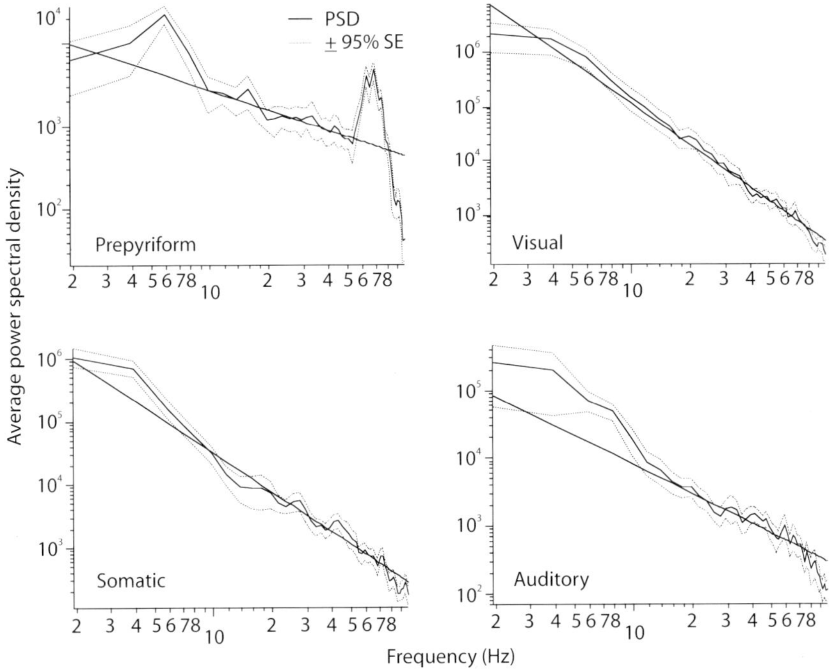

Fig. 5. Average temporal power spectral densities (PSDs, *solid curves*) and ±95% confidence intervals (*dashed curves*) were calculated from the prepyriform, visual, somatic, and auditory EEGs. Each PSD was averaged across 40 records from a Hamming smoothed, 500 ms, post-stimulus EEG window. Regression curves calculated by using Eq.[2] were fit to the 10–50 Hz domain of the PPC PSD and the 20–80 Hz gamma domain of the neocortical PSDs.

reproducible across other neocortical data sets. The 80–100 Hz PSD segment of the regression curve for the neocortical recordings lay above the SE curves, indicating that EEG activity was attenuated by the low-pass analog RC filter.

Differences between the pre- and post-stimulus PSD across all cortical areas were sought by calculating the average, normalized α'-coefficients and β'-coefficients from the regression of Eq. [2] onto the average CS–/CS+ PSD within the 20–80 Hz range for the neocortical data sets. The average α'- and β'-coefficients were plotted with the corresponding ± SE (95% confidence level) for the visual and somatic cortices (Figure 6). For all neocortical recordings, the average, normalized α'- and β'-coefficients of the average PSD defined baseline pre-stimulus levels. After stimulus onset and within the first 128 ms post-stimulus period, the α' coefficient for the gamma band visual cortical recordings increased for approximately 750 ms. Somatic and auditory records showed an immediate post-stimulus increase in α' for the gamma band, and a subsequent and sustained decrease in α' for the duration of the post-stimulus period (Figure 6). Every change in α' was associated with an inverse change in the steepness of slope for the $1/f$ band PSD, as indicated by β'. The values for pre- and post-stimulus α'- and β'-coefficients for the average CS– and CS+ PSDs lay within the SE curves, indicating that spectra between the discriminanda did not differ. These findings held for all sensory cortices and

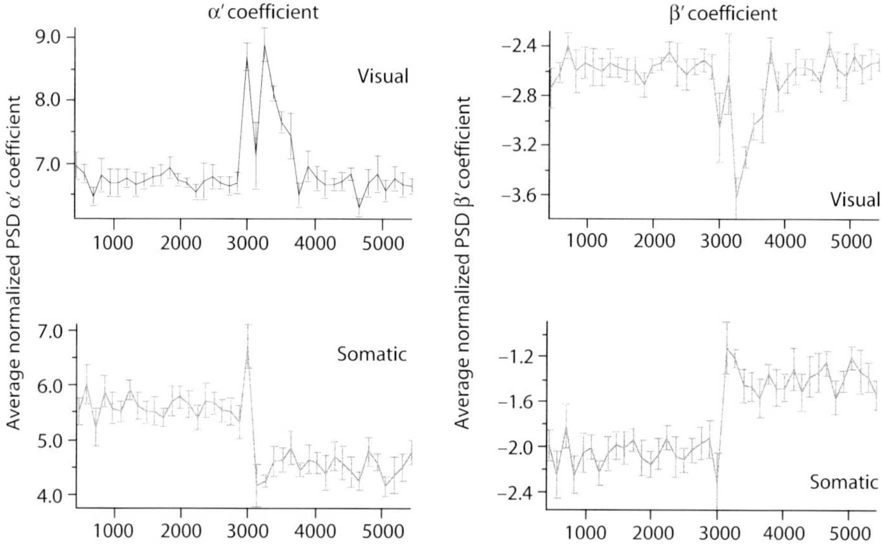

Fig. 6. The α'-coefficients (*left frames*) and β'-coefficients (*right frames*) were calculated by regression of Eq. [2] onto the 20–80 Hz band of the PSD, averaged across 40 CS–/CS+ EEG records from each experiment. The average PSDs from individual experiments were normalized and averaged across visual (n=6) and somatosensory (n=8) experiments. Coefficients from the auditory EEGs conformed to the pattern for somatic cortex.

animals. The importance of changes in the gamma band α'- and β'-coefficients could reflect changes in the fractal dimension of the dynamics of the EEG generated from the neocortex (Higuchi 1990).

Spatiotemporal Pattern Representation

In order to determine the spatial amplitude pattern of the common waveform, each 64-channel segment of EEG data (Figure 7, *left*) was transformed into a spatial pattern (Figure 7, *right*) by the methods of FFT, modified FFT, PCA and RMS. With the modified Fourier method a sum of five cosines was fit to the data (Freeman and Viana Di Prisco 1986). The dominant frequency component comprised, on average, 60–70% of the EEG variance. Components 2–5 had the same spatial pattern as the dominant component. It also gave the spatial pattern of phase for each frequency; its use is discussed elsewhere (Freeman and Baird 1987; Freeman 1991a) and was previously thought to be the ideal method for resolving EEG spatial patterns. This method is computationally intensive, and it does not work well with EEG segments with a $1/f$ spectra instead of narrow-band bursts, but its use was included with this analysis in order to quantitatively contrast modified FFT spatiotemporal patterns resolved in the neocortex with other methods of EEG spatial pattern resolution utilized throughout the remainder of this study. The first PCA component represented 90–99% of the EEG variance. Its factor loadings defined the spatial pattern or mode. Figure 8 (*upper*) illustrates representative modified FFT and PCA decompositions for the average waveform from Figure 7 (*left*). Since the first PCA component

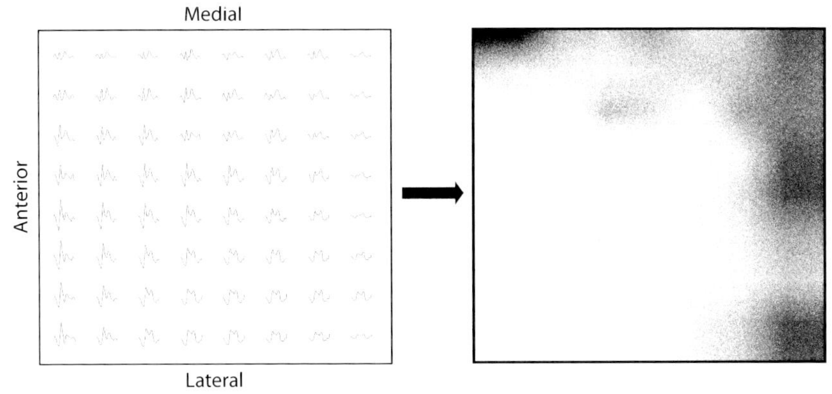

Fig. 7. A representative 120 ms segment of 64 visual cortical EEGs is shown from a chronically implanted 64-channel electrode array (*left*). After the traces were digitally band-pass filtered, 64 root mean square (RMS) values were calculated and plotted as a contour map (*right*).

Table 1. Comparison of spatial patterns of the dominant components from the two EEG decomposition methods with the RMS patterns

Correlation	Number Analyzed	Significant Matches	Percentage Matches
A. Comparison of EEG decomposition techniques with visual neocortical patterns			
RMS vs. PCA	9,000	8,918	99.1
RMS vs. FFT[1]	7,377	7,377	100
PCA vs. FFT[1]	7,377	7,300	99.0
RMS vs. FFT[2]	9,000	8,764	97.4
B. Comparison of EEG decomposition techniques with PPC patterns			
RMS vs. PCA	6,750	6,348	94
RMS vs. FFT[1]	5,737	5,737	100
PCA vs. FFT[1]	5,737	5,123	89.3

Comparisons are made by nonparametric correlation of 9,000 neocortical and 6,750 PPC spatial patterns (120 ms). If the probability that the nonparametric correlation coefficient indicated a match between two spatial patterns was less than 0.001 ($p < 0.001$), then the criterion that those two spatial patterns were identical was satisfied. FFT[1] included only those records where the fast Fourier transform (FFT) decomposition was acceptable according to published criteria (Freeman 1987). FFT[2] included all FFT decompositions. EEG, electroencephalogram; RMS, root mean square; PCA principal component analysis.

comprised most of the variance, closely approximating the raw EEG (Figure 8, *lower*), an RMS calculation was more desirable in terms of computational time. Table 1A compares the modified FFT and PCA methods of neocortical EEG decomposition with the RMS algorithm by calculating the average, normalized nonparametric correlation coefficient between 9000 neocortical spatial patterns. Table 1B shows the same comparison for a PPC EEG experiment. If the nonparametric correlation coefficient indicated a match between two spatial patterns and if the probability of that match was less than 0.001, then the criterion that those two spatial patterns were identical was satisfied. Since the RMS method yielded spatial patterns significantly correlated with both the modified FFT and

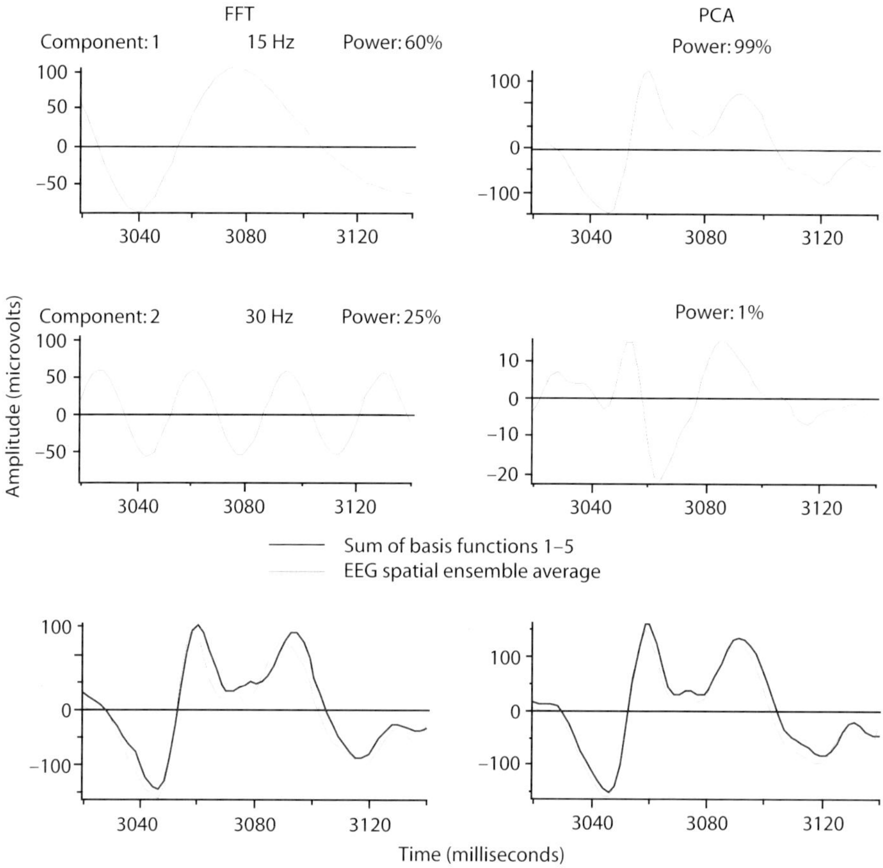

Fig. 8. Examples of the first and second modified FFT basis functions (*left*) and PCA temporal components (*right*) from the EEG traces in Fig. 7. The fraction of total variance of the EEG incorporated by the first component ranged from 60–70% for the FFT method and from 90–99% for the PCA method. The sum of all five FFT and PCA basis functions are compared (*below*) to the spatial ensemble average of the EEG traces in Fig. 7.

PCA spatial modes, the RMS method was employed throughout the remainder of this study of spatial patterns.

Conformance of neocortical spectra to Eq. [2] implied that the 1×64 FFT PSD spatiotemporal pattern vector was defined at every frequency bin by the 64 α'-coefficients of the regression curve onto each channels PSD and that the same pattern was only rescaled across frequency bins by f^{β}. Therefore, each pattern could not significantly vary given a constant PSD β'-coefficient for all 64 channels. Curves independently regressed onto the gamma frequency (20–80 Hz) domain of the PSD, from each of the 64 channels, had the same β' within 95% confidence limits but differed in α'. This meant that each spatiotemporal pattern was equivalent within the 20–80 Hz range, but that there might be other spatial patterns in the low (2–20 Hz) or high (80–100 Hz) frequency range. In order to test every pattern at every frequency below 100 Hz within a 128 ms pre- and post-stimulus time frame, the Euclidean distance from a normalized RMS reference pattern to every reconstructed, normalized FFT PSD bin spatiotemporal pattern was calculated and averaged across 20 CS– and 20 CS+ records (Figure 6 reviews the dynamic changes in α' for pre- and post-stimulus segments). Average Euclidean distances from RMS to FFT PSD patterns ±SE (99% confidence limits) are shown for the visual and somatic cortices in Figure 9. There were found to be two domains of spatiotemporal patterns: the low frequency range (2–20 Hz), and the $1/f$ range. The spatiotemporal patterns within the $1/f$ range did not statistically differ, but those within the low frequency range differed progressively with decreasing temporal frequency. No distinctive spatial patterns emerged within the 2–20 Hz domain. The centroid of the points was near the centroid for the $1/f$ cluster, but the points were widely scattered, mostly outside of the cluster. These results were replicated for the auditory cortex, and for all three neocortices with FFT decomposition windows 256 ms in length. Those FFT PSD spatiotemporal patterns within the $1/f$ domain conformed to the EEG spatial patterns resolved by the modified FFT, PCA, and RMS algorithms.

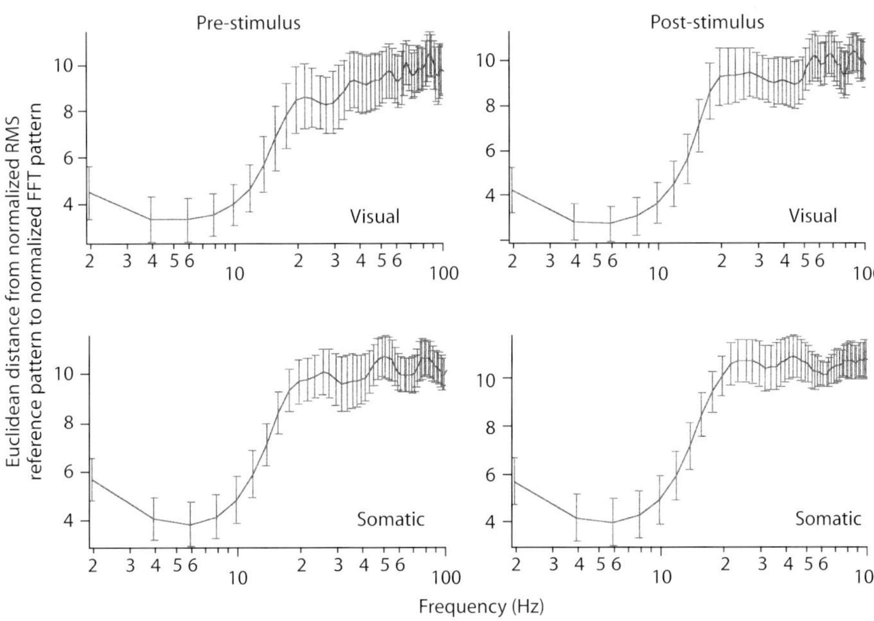

Fig. 9. Averages of Euclidean distances are shown between points in 64-space from normalized RMS reference patterns to normalized FFT PSD bin patterns from 40 CS–/CS+ pre-stimulus (*left*) traces and post-stimulus (*right*) traces. Power with an FFT spectral bin had nothing to do with the spatiotemporal pattern resolved for that bin because every pattern was normalized prior to the calculation of Eq. [4]. The graph shows the dependence of the spatial AM patterns (expressed as points) on the frequency components of the carrier, which were derived from the FFT PSD at 2 Hz intervals. Standard errors at each frequency were used to determine the 99% confidence interval. Analyses of PSD patterns from visual (*upper traces*) and somatic (*lower traces*) experiments illustrated that all FFT PSD patterns reconstructed from the 20–100 Hz (1/*f*) portion of the spectra were equivalent, and that patterns within the low frequency (2–20 Hz) range differed progressively with decreasing temporal frequency into the theta range. These results indicate that the aperiodic oscillation in each segment is a unitary event that cannot manifest a collection of limit cycle oscillators.

Spatial Pattern Classification

Segmentation of EEGs from the olfactory system was facilitated by the presence of EEG gamma bursts recurring at the respiratory rhythm. EEGs from the neocortices showed no alpha, theta, or gamma bursts and no correlation between the EEG and the respiratory rhythm. The AEP marked the arrival of the afferent volleys but failed to demarcate significant segments of post-stimulus EEGs. Therefore, successive EEG patterns were extracted by moving a fixed-length analysis window across each set of 40 records (comprising one experiment) in overlapping steps. Given the overlapped set of spatial patterns, a method of cross-classification was used to classify patterns as closer to a CS− centroid or closer to a CS+ centroid by a Euclidean distance metric. Each 40 record experiment was transformed into 40×225 (the number of analysis windows) spatial patterns. A testing of this method for false pattern classification was accomplished by generating 40×225 random (flat distribution) patterns, normalizing each pattern to zero mean and unit SD, and performing a classification analysis across the entire set. Among the 225 time-frames of random patterns, only one frame had a number of patterns classifying below the one-percent significance level.

A salient feature by which two similar groups of spatial patterns can be classified with a Euclidean distance metric was the pattern of *global* amplitude activity. Since this study concerns differences in *pattern* of neural activity and not in average *amplitude*, every spatial frame was normalized to zero mean and unit standard deviation (Viana Di Prisco and Freeman 1985). Removal of amplitude artifacts from every spatial pattern tended to improve the resolution and separation of CS− patterns from CS+ patterns (Table 2, B).

Spatiotemporal patterns from past work involved analyzing rabbit olfactory EEG bursts varying in temporal length from 75–200 ms (Freeman *et al.* 1978, 1982, 1986), or monkey visual cortical EEGs (Freeman and Van Dijk 1987). Recent results from human cortical EEG studies have demonstrated that there may be a diminishing return in terms of finding perceptual activity with analysis windows greater than 175 ms in length (Menon *et al.* 1996). The relationship between experimental spatial pattern classification efficacy and the temporal length of the RMS analysis window was explored by varying the window length, while classifying a well characterized rabbit auditory cortical EEG data set (Freeman and Barrie 1993). Increasing the RMS window length from 10 ms to 500 ms yielded a tuning curve for the number of spatial patterns classifying correctly ($p < 0.01$) versus window length (Figure 10, *upper solid trace*). Classification values were calculated by stepping an n-ms window across 40 EEG records at intervals of 10 ms and subtracting the number of pre-stimulus patterns falsely classified from the number of post-stimulus patterns correctly classified ($p < 0.01$). Classification values were plotted alongside the amount of computational time needed (on an Apple PowerPC 8100/80) to transform EEGs to spatial patterns for each RMS analysis iteration (Figure 10, *upper dashed trace*). The number of spatial patterns classifying below the one-percent significance level reached a plateau for a 150 ms RMS window and failed to improve thereafter, while the computational time steadily increased. Temporal resolution of

Table 2. Classification efficacy as a function of normalization and digital filtering

	Prepyriform $n=11$ experiments		Visual $n=6$ experiments		Somatic $n=9$ experiments		Auditory $n=14$ experiments	
Classification of averages	Pre-stim	Post-stim	Pre-stim	Post-stim	Pre-stim	Post-stim	Pre-stim	Post-stim
Raw data	0.91	3.45	1.83	24.5	2.22	30.0	2.36	6.43
Time-frame normalized data	2.09	4.73	0.83	53.2	5.22	63.2	2.64	7.71
Percent difference from raw	+130	+36.8%	−54.5%	+117%	+135%	+111%	+12.1%	+20.0%
Temporally filtered data	2.00	14.7	2.50	81.3	1.56	77.0	0.71	17.8
Percent difference from raw	+120%	+326%	+36.3%	+232%	−30.0%	+157%	−69.7%	+176%
Spatially filtered data	2.09	15.6	2.67	93.3	1.44	87.2	0.93	20.6
Percent difference from raw	+130%	+353%	+45.4%	+281%	−35.0%	+191%	−60.6%	+220%
Spatially filtered, time frame normalized data	2.00	17.7	2.83	92.2	1.11	88.7	0.71	20.7
Percent difference from raw	+120%	+413%	+54.5%	+276%	−50.0%	+196%	−69.7%	+222%

Improvements in levels of CS−/CS+ classification of RMS patterns were obtained by time-frame normalization, temporal filtering (100 Hz high-cut constant), spatial filtering, and re-normalization. A. The measure of classification efficacy was the average number of patterns classifying below the one-percent significance level across the total number of experiments. The level of classification of the pre-stimulus segments was expected to decrease, while the level of post-stimulus segment classification increased B. The RMS frames were normalized to zero mean and unit SD. This removed the frame-to-frame variation in amplitude, leaving only the pattern for classification. C. Band-pass temporal digital filters were optimized for each data set by stepping the low-cut every 5 Hz and fixing the high-cut at 100 Hz. The average number is shown of patterns classifying below one-percent for the optimal temporal filter setting for each experiment, after normalization. D. Each experimental data set was filtered at its optimal temporal filter setting before spatial filters were applied. After temporal filtering and time frame normalization, band-pass spatial digital filters were optimized for each data set by incrementing the low-cut and fixing the high-cut at 0.624 cycles/mm or 0.880 cycles/mm (just below the spatial spectral artifact) for the neocortical and PPC experiments, respectively. E. After optimal temporal and spatial filter settings were found, each set of AM patterns was re-classified following re-normalization after spatial filtering. These results reflect the finding that each segment has a center frequency that continually varies. The differing ranges of filters for optimal classification reflect the inhomogeneities of that variation across trials and sessions. For abbreviations see Table 1.

periods of spatial pattern classification was concurrently studied in order to determine whether or not a certain analysis window length incorporated neural activity from neighboring discrete neural events, in effect, blurring two or more separate events. Classification resolution was determined by calculating the average change in probability values using Eq. [10]. The hypothesis was that short temporal windows would yield spurious pattern classification resulting in a large $\overline{\Delta p}$ and that as the window length increased beyond 150 ms, individual neural events would become blurred thereby yielding a low $\overline{\Delta p}$. Calculation of Eq. [10] for a series of window lengths yielded a tuning curve for the resolution of discrete neural events (Figure 10, *lower trace*). The goal of this study was to maximize the number of patterns correctly classifying ($p < 0.01$), minimize computational time, and find some compromise between spurious and blurred neural event resolution. On the basis of the results from Figure 10, a window length of 120 ms was adopted for the remainder of this study. This length served to capture short events without compromising the resolution of longer events. The time interval between analysis windows was chosen to be 20 ms. Shorter intervals increased computational time and provided no additional information regarding the temporal location of CS−/CS+ pattern separation. These findings generalized across all neocortical areas.

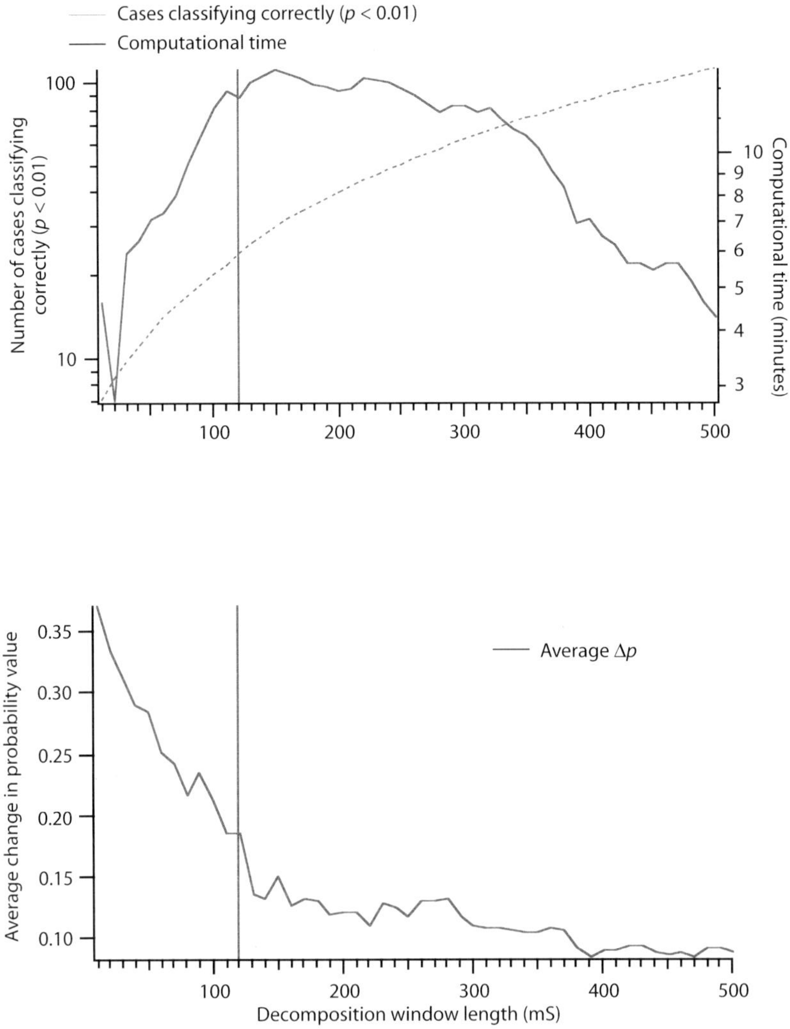

Fig. 10. The relationship is shown between the duration of the segment for RMS calculation and the spatial pattern classification efficacy. The number of pre-stimulus patterns classifying correctly ($p < 0.01$) was subtracted from the number of post-stimulus patterns classifying correctly. The difference was plotted against segment duration (*upper frame, solid trace*) and against the computational time needed (with an Apple Power*PC* 8100/80) with increasing RMS duration (*upper frame, dashed trace*). The solid vertical bar indicates a 120 ms window. The changes in probability of CS–/CS+ pattern difference were summed, as the window was stepped along the 6-second traces, on the premise that short windows would capture sudden changes in pattern difference more readily than longer windows that spanned intervals of pattern shift. Spatial pattern classification levels were maximized with a 150 ms RMS window, but a shorter length of 120 ms gave better temporal resolution and lower computational time.

EEG fluctuations over 100 Hz or below 0.1 Hz were attenuated by analog RC filters. Within this range, digital temporal low-pass, high-pass, and band-pass filters were used to optimize spatial pattern resolution. Previous studies regarding the temporal digital filtering of raw olfactory bulb (*OB*) EEGs prior to RMS analysis had indicated the facilitated extraction of perceptually-related spatial patterns of activity (Freeman and Viana Di Prisco 1986) by the removal of the low frequency respiratory wave and the high frequency fluctuations from the EEG time series. For the neocortical data sets no one set of band-pass filter settings could be applied *a priori* because the neocortical $1/f$ spectra gave no indication of the appropriate band onto which to lock filter settings. A low-pass temporal filter alone did not significantly alter the pattern classification levels by increasing the number of post-stimulus CS– patterns which were distinguished from CS+ patterns. A high-pass filter with a low-cut component of 5–20 Hz, on average, increased the number of post-stimulus patterns that were correctly classified, indicating that low frequency EEG activity tended to mask spatial AM pattern stabilizing events. Therefore, an optimal band-pass filter setting had to be derived on a per experiment basis by incrementing the low-cut parameter while keeping the high-cut setting constant at 100 Hz. After every new band-pass setting the data set was normalized and re-classified to evaluate CS–/CS+ pattern separation. Table 2(C) illustrates the degree of improvement in the number of patterns correctly classified above the one-percent significance level for the pre- and post-stimulus periods with a 100 Hz constant high-cut setting and a variable low-cut setting. Applying temporal filters to the raw EEG time series did not qualitatively affect the nature of the spatial pattern subsequently resolved, but only tended to improve the resolution of the individual CS– and CS+ patterns. This observation was supported by the $1/f$ form of the EEG time series, and the finding that each spatiotemporal pattern did not differ within the constant $1/f$ PSD domain (Figure 9).

After a set of spatial patterns had been optimized with temporal filtering, the spatial spectra of those patterns was explored. Calculation of the average 2D-PSD from PPC patterns (Figure 11) closely matched previous results from the *OB* (Freeman and Baird 1987). The lowest resolvable spatial frequency for the PPC array (0.56 mm interelectrode spacing) was 0.18 cycles/mm. The small peak near 0.9 cycles/mm was an artifact dependent on the electrode spacing of the array; since the spatial sampling frequency of the PPC array was 1.79 cycles/mm, the artifact located at 0.9 cycles/mm was equivalent to the spatial Nyquist frequency. Average spatial spectra from neocortical data sets (Figure 11) resembled the PPC spectra in the upward convexity of the curve over the low spatial frequencies but differed in resolution due to the 0.79 mm interelectrode spacing of the neocortical array. The lowest resolvable frequency for the neocortical data sets was 0.13 cycles/mm and the spectral peak near 0.62 cycles/mm was also an artifact reflecting the electrode spacing or the spatial Nyquist frequency for the neocortical arrays.

Log spatial PSDs calculated from the prepyriform, visual, somatic, and auditory cortices, when plotted against the log spatial frequency, were tested for conformity to a $1/f$ spatial spectra above 0.1 cycles/mm. Curves described by Eq. [2] were regressed onto the high frequency segment of the 2D-PSD data (defined as that between 0.1–0.9 cycles/mm for the PPC data and 0.1–0.62 cycles/mm for the neocortical data) and plotted against the average PSD ±SE expressed to the 95% confidence level (Figure 11). The criterion was adopted that the spatial spectra were $1/f$, if the regression curves lay within the standard error curves. According to the results illustrated in Figure 11, the majority of the PPC spatial spectra deviated from $1/f$, although the neocortical spectra conformed to $1/f$ for those spatial frequencies

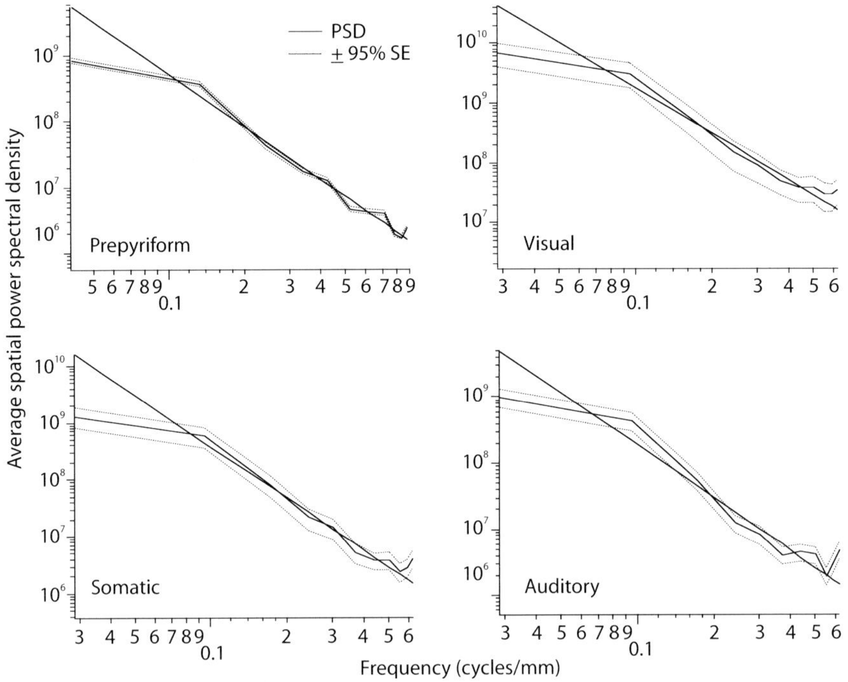

Fig. 11. Average spatial power spectral densities (*solid curves*) and ±95% confidence intervals (*dashed curves*) were calculated from 40 post-stimulus windows of EEGs from the four cortical areas. Curves were regressed onto the domain of the spatial spectra from 0.1–0.9 cycles/mm for the PPC and from 0.1–0.62 cycles/mm for the neocortices. The peaks near 0.9 cycles/mm in the PPC spectra and 0.62 cycles/mm in the neocortical spectra lay at the spatial sampling (Nyquist) frequency determined by the inter-electrode distances of the recording arrays.

between 0.1 cycles/mm and the spatial frequency artifact at 0.62 cycles/mm. Spatial frequencies above 0.62 cycles/mm formed a spectral tail significantly below the regression curve.

Previous studies had indicated that spatial digital filters applied to EEG spatial patterns increased the resolution of classifiable patterns from the olfactory bulb (Freeman and Baird 1987). The spatial digital filter parameter space was searched to determine if the prepyriform and neocortical patterns could be better resolved. Band-pass spatial filters facilitated the resolution of spatial patterns by increasing the number of post-stimulus patterns which correctly classified while concurrently decreasing the number of pre-stimulus patterns falsely classifying (Table 2, D). Pre- and post-stimulus classification figures represent the number of patterns classifying below the one-percent level for each cortical area. On average, the classification levels increased about 50% over the numbers attained with temporal filters alone. In order to test whether or not the process of spatially filtering the spatial patterns had induced any amount of artificial pattern classification, the spatially filtered data were re-normalized and re-classified. There was no significant difference between the spatially filtered data and the re-normalization of the spatially filtered data (Table 2, E).

For each experiment, across animals and sensory modalities, a statistically significant ($p < 0.01$) number of post-stimulus patterns was correctly classified when compared to the pre-stimulus section of the same experiment (Table 2, A). This finding conformed to previous results in the olfactory (Freeman and Schneider 1982; Viana Di Prisco and Freeman 1985; Freeman and Grajski 1987) and visual systems (Freeman and Van Dijk 1987). The post-stimulus temporal location, within trials, of spatial pattern separation (CS– from CS+) was determined by plotting the probability values calculated with Eq.[9]. Based on the number of CS– patterns correctly discriminated from CS+ patterns at any given time, a series of probability values was generated for each experiment (Figure 12).

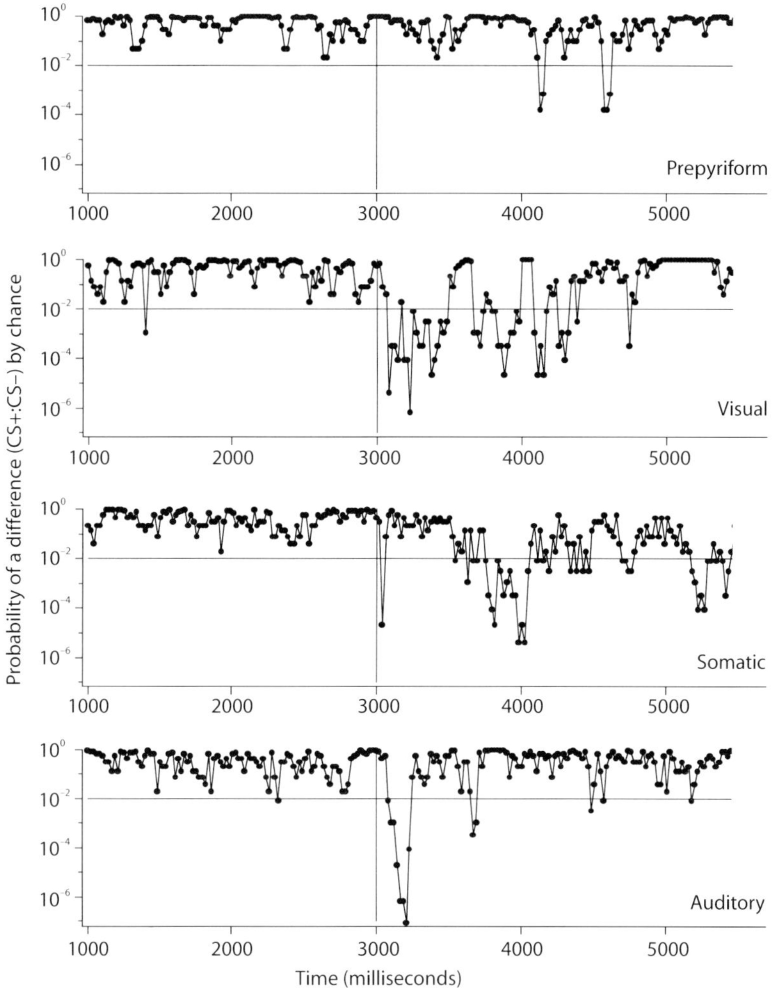

Fig. 12. The CS–/CS+ pattern differences were expressed as the probability that correct classification under cross-validation was due to chance. The pre-stimulus differences were at chance levels. The post-stimulus differences in each cortex revealed a sequence of brief episodes of declining significance after stimulus arrival, with a late increase in somatic experiments a half-second prior to the onset of the US. Each point represents one, 120 ms time frame. The onset of the CR (an increase in respiration for all experiments) occurred at ~700 ms ±500 ms post-stimulus arrival.

There were multiple temporal locations where the CS– and CS+ spatial patterns were separately classified ($p < 0.01$) within the post-stimulus section of each experiment. These took the form of a rapid sequence of multiple episodes of separation between 30–1000 ms post-stimulus, with the later peaks corresponding to periods where the animal's respiration increased in response to the CS+ stimulus (behavioral marker, CR). The time onset of the CR was ~700 ms ±500 ms post-stimulus arrival. The temporal location of PPC pattern classification was found to be lagged when compared to the neocortical data sets, due to a 500 (SE) ms delay inherent in the olfactometer's air stream. The successive spatial patterns within each episode of pattern separation differed, as measured by a nonparametric correlation coefficient, from all patterns which preceded the period of classification and all patterns which were subsequently formed. For example, in Figure 12 (*prepyriform*) there were two episodes where the CS– spatial patterns were significantly different from the CS+ patterns. The initial episode occurred at ~4100 ms while the second episode was located at ~4600 ms. Between these two episodes the CS– and CS+ patterns at 4100 ms were not repeated (i.e. identical) at 4600 ms.

Determination of whether or not peaks of significant pattern classification were due to a small sample size (40) was conducted by computing a cross-classification analysis of two 100-record experimental data sets from the visual cortex. Multiple locations of significant ($p < 0.01$) pattern separation were observed in the 100-record experiments with no increase in spatial pattern resolution, demonstrating that earlier findings (Figure 12, *visual*) were not a function of small sample size.

Temporal sections of EEG data in which spatial patterns were significantly separated ($p < 0.01$) showed no distinctive features in the raw, normalized data. Figure 12 is representative of the locations of different CS–/CS+ spatial pattern separation after time-frame normalization, temporal digital filtering, and spatial digital filtering, but as Table 2 demonstrates, the number of patterns correctly classifying as CS– or CS+ was dramatically lower when only the raw data was analyzed. Some experiments had no patterns classifying above the one-percent level prior to normalization, thereby highlighting the need to remove global EEG amplitude biasing associated with a particular stimulus. Within the visual cortical experiments, there was a 117% average increase in the number of post-stimulus patterns correctly classifying and a 54.5% average decrease in the number of pre-stimulus patterns falsely classifying (Table 2, B) after removal of all global amplitude biasing by means of time-frame normalization of each spatial pattern. Spatial pattern resolution was further increased via the elimination of low and high frequency EEG fluctuations within the temporal and spatial spectral domains. Post-stimulus pattern resolution for the visual cortical experiments improved 270% after temporal and spatial digital filtering (Table 2, C/D).

During the search of the temporal digital band-pass filter parameter space, the criteria for filter component optimization included only maximizing the number of spatial patterns correctly separated within the post-stimulus experimental period and minimizing the number of patterns falsely separated within the control period. After the band-pass temporal filters had been optimized, each experiment was only filtered at one, optimal filter setting. Since the elimination of the 2–20 Hz EEG activity facilitated the resolution of the $1/f$ domain of patterns, and the $1/f$ domain

(20–100 Hz) of patterns comprised the dominant number of patterns able to be separated, the 2–20 Hz domain of EEG activity was normally removed. The possibility still remained that patterns within the 2–20 Hz domain were able to be separated within some post-stimulus periods, but the exclusion of parts of the 2–20 Hz temporal frequency range hampered the ability to resolve those spatiotemporal patterns for this study.

Spatial patterns were compared between experiments taken from week one and week two by calculating the centroids ($\overline{C}^{(m,j)}$), using Eq.[6], from the first week's experiment and subsequently using them to classify the spatial patterns ($\overline{V}^{(i,j)}$) reconstructed, using Eq.[5], from the second week's experiment. By applying this method, we demonstrated that the patterns from week one served to classify patterns relating to identical stimuli from week two. Use of centroids calculated from week one or from week two to classify patterns from week three similarly did not succeed. The distinctive pattern of rapid episodes of pattern classification (Figure 12) was absent during the inter-week cross-classification of the spatial patterns reconstructed from the experiment where the CS–/CS+ contingency was reversed. Though the stimuli were unchanged, the spatial AM patterns were altered by the shift in context associated with each stimulus. There were still significant CS–/CS+ pattern differences within the experiment from week three regardless of contingency reversal. These findings were similar to stable spatiotemporal patterns of EEG activity previously found in the *OB* until response contingencies were changed (Freeman and Grajski 1987), and they also support recent findings that neocortical spatial patterns are slowly and measurably altered as a function of time (Barrie *et al.* 1994).

Spatial Distribution of the EEG Signal

Previous work had demonstrated that the contribution to the classification of EEG spatiotemporal patterns from the olfactory bulb was not enhanced or diminished by deletion of any one channel as a subset of the 64 recordings used to compute those spatiotemporal patterns (Freeman and Baird 1987). This finding was tested with the PPC and neocortical data by deleting groups of randomly selected channels in sets of 0, 2, 4, 8,... 63 across 50 iterations for each set from a well characterized section of an experiment, and then recomputing the cross-classification analysis. A running average was logged of the number of patterns classifying correctly after a certain channel had been deleted in respect to the number of times it was used. If any one deleted channel contained more stimulus-related EEG signal than average, the number of patterns classifying correctly would decrease on repeated classification analysis. Similarly, if a deleted channel contributed more noise than signal then the number of patterns classifying correctly would increase after its removal. The average number of patterns classifying correctly after any channel was removed remained within the 99% confidence limits. This indicated that every channel contained an equal amount of the spatial signal used to classify the data set.

The question was also asked how deletion of increasing numbers of the 64 channels impaired pattern classification. After deletion of randomly selected groups of

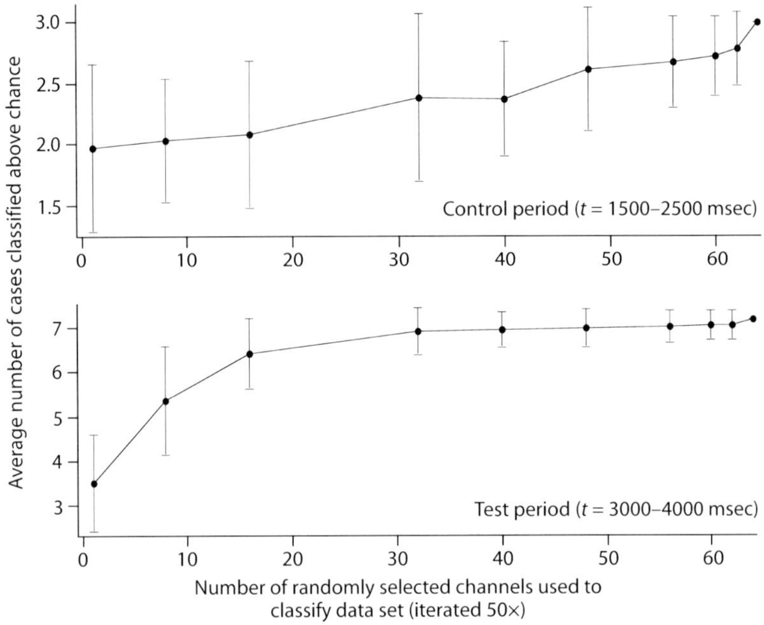

Fig. 13. These two graphs illustrate the effects of randomly deleting channels in groups of 1, 2, 4, 8, 16,... 63 and then calculating the average number of patterns classifying better than chance levels. These graphs demonstrate that spatial patterns may be resolved almost as well as with a 16- or a 64-channel array and that the EEG is homogeneously distributed across the spatial aspect of the cortex.

channels, with 50 iterations of classification per number deleted, an average number of patterns classifying above chance levels (50%) was calculated across one second of pre- and post-stimulus data. Correctly classified patterns ranged from 2–3 patterns above the chance level for the control period (Figure 13, *upper trace*) and from 7–3.5 patterns above chance for the test period (Figure 13, *lower trace*). The error bars reflect the 99% confidence levels. These results showed that spatial patterns from primary sensory cortices were significantly resolved by as few as 16 channels, but with a monotonic decrease in classification with diminishing number of channels. The spatial location of the selected or deleted channels was arbitrary.

Discussion

The hypothesis guiding this study was developed for the olfactory system. It holds that the module of perception in each sensory modality is an area of cortex that is comparable in size to that of the olfactory bulb and cortex, on the order of 1 cm^2 and containing several thousand cortical columns. The microscopic activity of the cortical neurons coursing by action potentials through the neuropil gives rise to a macroscopic cortical state, which is expressed in a spatially coherent, broad-spectrum waveform that is modulated in amplitude (AM) with respect to the surface coordinates. It is also modulated in time, owing to repeated state transitions at irregular intervals lasting on the order of 0.1 s, which manifest rapid changes from one spatial pattern to the next, owing to the activity-dependent gain inherent in the sigmoid I/O curve of local populations. The patterns are constructed by the nonlinear synaptic interactions of neurons distributed over the entire area, and they are subject to changes by modification of the synaptic web and modulation of the excitability of neurons by neuromodulators. Patterns tend to recur in respect to prior learning, and they are accessed by the cortex through the microscopic sensory input from receptors, relayed through the thalamus for neocortices. This input is re-expressed in the macroscopic state as it guides the neuropil in the construction of a pattern that depends in part on the cortical neurons selected by the stimulus and in part on the synaptic web pre-formed by learning (Freeman and Viana Di Prisco 1986).

The distinction between the microscopic and macroscopic levels of activity is crucial to this study. The number of neurons that suffices to form a local neighborhood has often been estimated to be on the order of 10,000 neurons, which is appreciably larger than the number that can be accessed by recording pulse trains simultaneously from multiple electrodes with present techniques. The fraction of variance in the pulse train of each neuron that is covariant with the population waveform has been estimated to be less than 1 in 1,000 (Freeman 1992b), so the present limit of about 100 neurons is 10 times too small. The local EEG potentials arise from thousands of neurons by their the synaptic currents that flow across the tissue resistance between the neurons and sum as scalar values (Freeman 1975, 1992a). That sum is an epiphenomenon like a noise, for it does not bind the cells together. The binding is done by the innumerable synaptic interactions between the neurons to give the local mean field of the ensemble. EEGs are signs of the macroscopic activity of populations, while the pulse densities in the populations carry the activity from one place to another. When multiple recordings of pulse trains are made simultaneously, the participation of neurons in populations is seen as covariances called "phase locking" (Gray *et al.* 1992), "binding" (Eckhorn 1991), "synfiring" (Abeles 1991) and "reverberation" (Amit 1989). Population "coding" is commonly conceived (Georgeopolis *et al.* 1986; Rolls *et al.* 1989; Kruger 1990; McNaughton 1993; Aertsen *et al.* 1994) as the scalar or vector sum of the pulse trains of a collection of neurons, which serves to represent a sensory or motor "feature" in some pattern. Macroscopic "coding" is a small fraction of the variances in relative pulse frequencies that is covariant among all neurons in a local neighborhood; it is an element in an activity density function (Freeman 1975) extending over a cooperative domain of cortical neuropil.

Cortical output consists of the simultaneous pulse trains of all of the corticofugal axons during the time window of a given state or frame. Owing to putative spatial divergence and temporal dispersion in the tracts carrying the output to cortical targets, the output undergoes spatial integration, by which all activity not near the common instantaneous frequency and phase is smoothed and attenuated. By this mechanism the cooperative activity pattern constructed by the cortex is selected, and the microscopic sensory-evoked activity is attenuated. Where cortical output is transmitted to multiple targets, all targets receive the same input, but may select differing aspects depending on their own learned spatial patterns of excitability. If it can be shown that this model for perceptual dynamics holds for all modalities, then a solution may be facilitated for the problem of multimodal convergence and integration of perceptual activity in the thalamus, entorhinal cortex, frontal lobe, and elsewhere.

This conclusion, regarding the selection by the target of the cooperative activity over the microscopic residuals transmitted to it, has been confirmed for the olfactory cortex (Bressler 1984, 1987), including allowances made for the axonal transmission delays in the output pathway, but it remains to be demonstrated for neocortical targets. Descriptions of synchronous firing of units separated by substantial distances, even interhemispheric, remain anecdotal, and in any event cannot be explained as being based on axonal and synaptic transmission without an appeal to interactive population neurodynamics.

According to the present hypothesis the optimal site for electrode placement is at the pial surface, owing to the spatial integration over neighboring columns that, to some extent, mimics the smoothing in the output transmission. Narrow-band oscillations in EEGs and in unit firing probabilities have been reported from depth recordings in the somatomotor cortex (Sanes and Donoghue 1993; Murthy *et al.* 1994) and visual cortex (Eckhorn *et al.* 1988; Gray *et al.* 1992). Other studies have failed to find oscillations or have found them uncommonly and without stable relationships to sensory or motor events (Ghose and Freeman 1992; Young *et al.* 1992). The concept of binding by phase-locking of oscillations or spike firing has been criticized on the grounds that the onset of synchrony is too slow, the number of periodically firing neurons is too few, and the number of spikes in each segment from each neuron are too few (Young *et al.* 1992; Tovee *et al.* 1993). Narrow band oscillations, by the present hypothesis, constitute transitory coherences or intermediate products of integration, which are not given as output, if their phase and frequency are inconsistent with the whole. Our failure to find narrow spectral peaks in surface EEGs is evidence not that such activity is absent in the depth, but that it is unlikely to form a significant part of the perceptual output. Its absence does not imply that cortex could not work with narrow band oscillations, but that it does not, perhaps because switching from each state to the next may be quicker and more reliable, if persistent "ringing" and "resonance" are disenhanced by the use of a broad spectrum for perceptually significant transmissions.

The result that the classificatory information in EEGs is homogeneously spatially distributed is important for the hypothesis for several reasons. The transmitted pattern is not only independent of the topography of the input axons that guided its formation; it is also independent of the topography of the output axons to multiple targets, which in the large is determined by the requirements of embryological development and not by ontological experience. The interfacing of an extremely varied set of input patterns with the varying requirements of the set of cortical targets offers a combinatorial problem of immense complexity for any attempt at modeling with switching networks. The modeling is elementary, when an AM spatial pattern is elicited by any of a class of equivalent inputs, and when the same AM pattern is sent to all targets.

Further, the precise location of experimental electrodes is no longer important, with respect either to input or output topographies. As with distributed representation in a hologram, the same information is present on every channel, and the resolution depends on the number available. Having no specificity in respect to the topography of the sensory arrays, the AM patterns are well adapted for integration into multisensory gestalts. Finally, the result offers a solution to the "binding problem" (Hardcastle 1994) in broader terms than its original conception (Milner 1974; von der Malsburg 1983). Sensory information is indeed refined by preprocessing and is manifested in the activity of "feature detector" neurons, but its effects are directed into the activity of all other neurons in the cortical area, not a select few, and the AM pattern output is shaped by the entire history comprising the context of that area, not by a hierarchical schema of elementary forms. Such schemata appear to be unlikely, owing to the lack of AM pattern invariance with respect to stimuli, and their uniqueness in spatial form as they evolve within each subject. The larger question of whether any invariants for sensory inputs exist in memory stores within brains is addressed elsewhere (Freeman 1995).

Another conclusion from this study is that 16 channels can suffice for EEG pattern classification almost as well as 64. This outcome was not predicted from previous olfactory results, so it was necessary to use the larger number. Despite wider spacing of the electrodes and therefore a larger "window" onto the pial surface in search of boundaries for cooperative domains, no "edges" were detected for the spatial patterns, in the form of channels near the margins of the arrays having less information for the classification. Preliminary recordings in other rabbits with electrodes placed in multiple sensory areas have shown that the concomitant activities of the three neocortices, as expected, differ substantially in detail, indicating that boundaries for each area of cooperation must exist in some form, but the boundaries may fluctuate from each frame to the next in an unpredictable manner.

The structures and properties of the several dynamic systems generating these EEG patterns are not known in detail, though it is likely that, as in the case of the olfactory system, the interactions of excitatory, inhibitory and modulatory neurons in large numbers and over diverse distances can be modeled with nonlinear ordinary differential equations. Large sets of such equations, when coupled with many forms of feedback, are typically governed by stable solution sets known as "attractors". The formal definition and description of an attractor require that the system be stationary and autonomous, and that it be watched long enough for it to settle into a basin of attraction. These conditions do not hold for cortex during perception, so that it is impossible to prove whether a given state, which is manifested by an AM pattern of an aperiodic carrier, manifests a strange attractor or an asymptotic approach to a point, periodic, or quasiperiodic attractor perturbed by noise from the Poisson-like continuous firing of individual cortical neurons. The question whether the cortical dynamics is chaotic (Elbert *et al.*, 1994) is not only unresolvable; at present it is unimportant. The issue is whether a mass of neurons forming an area of neuropil is capable of establishing spatial patterns of cooperative neural activity over areas far greater than the mean length of dendritic and axonal arbors, which have a characteristically broad-spectrum carrier, and which are formed and dissolved in time periods much shorter than their mean 0.1 s duration. The results from the present study show that these are reliable properties of laminated neuropil, and that the dynamics can take microscopic sensory input and transform it to a macroscopic pattern in preparation for multimodal perceptual integration in thalamocortical and limbic systems.

Acknowledgments

Reprinted by permission from the *Journal of Neurophysiology*, from: Spatiotemporal Analysis of Prepyriform, Visual, Auditory, and Somesthetic Surface EEGs in Trained Rabbits; Barrie JM, Freeman WJ, Lenhart M, *Journal of Neurophysiology* **76**: 520–539, 1996

The authors wish to thank Bret Andrews, S. Baker, Janine Chin, Joanne Leung, David Graham-Squire, and Steve Truong for their invaluable assistance with the experimental preparation and Brian Burke and Hung-Jen Chang for their review of the data analysis.

More information on the background and research interests of the authors may be obtained on the Freeman Lab WWW site by accessing: `http://sulcus.berkeley.edu/` on the Internet.

This work was supported by the National Institute of Mental Health – MH06686.

References

Abeles, M. *Corticonics: Neural Circuits of the Cerebral Cortex*. Cambridge, UK: Cambridge University Press, 1991.

Adrian, E.D. Sensory discrimination, with some recent evidence from the olfactory organ. *Br. Med. Bull.* **6**: 330–332, 1950.

Aertsen, A., Erb, M. and Palm, G. Dynamics of functional coupling in the cerebral cortex. *Physica D* **75**: 103–128, 1994.

Amit, D.J. *Modeling Brain Function: The World of Attractor Neural Networks*. Cambridge, UK: Cambridge University Press, 1989.

Barlow, J.S. *The Electroencephalogram: Its Patterns and Origins*. Cambridge, MA: MIT Press, 1993.

Barrie, J.M. and Freeman, W.J. Perceptual topography: spatiotemporal analysis of prepyriform, visual, auditory, and somesthetic EEGs in perception by trained rabbits. In: *Analysis and Modeling of Neural Systems*, edited by J. Bower. Boston: Kluwer Academic Publishers, 1994, p. 173–178.

Barrie, J.M., Freeman, W.J. and Lenhart, M.D. Cross modality cortical processing: spatiotemporal patterns of olfactory, visual, auditory, and somatic EEGs in perception by trained rabbits. *Soc. Neurosci. Abstr.* **20**: 1011, 1994.

Boudreau, J.C. and Freeman, W.J. Spectral analysis of electrical activity in the prepyriform cortex of the cat. *Exp. Neurol.* 8: 423–430, 1963.

Bressler, S.L. Spatial organization of EEGs from olfactory bulb and cortex. *Electroencephalogr. Clin. Neurophysiol.* **57**: 270–276, 1984.

Bressler, S.L. Relation of olfactory bulb and cortex. I. spatial variation of bulbocortical interdependence. *Brain Res.* **409**: 285–293, 1987.

Bressler, S.L., Coppola, R. and Nakamura, R. Episodic multiregional cortical coherence at multiple frequencies during visual task performance. *Nature* **366**: 153–156, 1993.

Bressler, S.L. and Freeman, W.J. Frequency analysis of olfactory system EEG in cat, rabbit, and rat. *Electroencephalogr. Clin. Neurophysiol.* **50**: 19–24, 1980.

Chang, H. The evoked potentials. In: *Neurophysiology*, edited by J. Field, *et al.* Washington: American Physiological Society, 1959, p. 299–314.

Davis, G.W. and Freeman, W.J. On-line detection of respiratory events applied to behavioral conditioning in rabbits. *IEEE Trans. Biomed. Eng.* **29**: 453–456, 1982.

DeMott, D.M. Cortical micro-toposcopy. *Med. Res. Eng* **5**: 23–29, 1966.

Eastman, C. Construction of miniature electrode arrays for recording cortical surface potentials. *J. Electrophysiol. Tech.* **5**: 28–30, 1975.

Eckhorn, R. Principles of global visual processing of local features can be investigated with parallel single-cell and group-recordings from the visual cortex. In: *Information Processing in the Cortex*, edited by A. Aertsen and V. Braitenberg. Berlin: Springer-Verlag, 1991, p. 385–420.

Eckhorn, R., Bauer, R., Jordan, W., Brosch, M., Kruse, W., Munk, M. and Reitboeck, H.J. Coherent oscillations: a mechanism of feature linking in the visual cortex? *Biol. Cybern.* **60**: 121–130, 1988.

Eckhorn, R., Schanze, T., Brosch, M., Salem, W. and Bauer, R. Stimulus-specific synchronizations in cat visual cortex: multiple microelectrode and correlational studies from several cortical areas. In: *Induced Rhythms in the Brain*, edited by Boston: BirhŠuser, 1992, p. 47–80.

Eeckman, F.H. and Freeman, W.J. Correlations between unit firing and EEG in the rat olfactory system. *Brain Res.* **528**: 238–244, 1990.

Eeckman, F.H. and Freeman, W.J. Asymmetric sigmoid nonlinearity in the rat olfactory system. *Brain Res.* **557**: 13–21, 1991.

Elbert, T., Ray, W.J., Kowalik, Z.J., Skinner, J.E., Graf, K.E. and Birbaumer, N. Chaos and physiology: deterministic chaos in excitable cell assemblies. *Physiol. Rev.* **74**: 1–47, 1994.

Engel, A.K., Konig, P., Kreiter, A.K., Schillen, T.B. and Singer, W. Temporal coding in the visual cortex: new vistas on integration in the nervous system. *Trends Neurosci.* **15**: 218–226, 1992.

Freeman, W.J. Topographic organization of primary olfactory nerve in cat and rabbit as shown by evoked potentials. *Electroencephalogr. Clin. Neurophysiol.* **36**: 33–45, 1974.

Freeman, W.J. *Mass Action in the Nervous System.* New York: Academic Press, 1975.

Freeman, W.J. Spatial properties of an EEG event in the olfactory bulb and cortex. *Electroencephalogr. Clin. Neurophysiol.* **44**: 586–605, 1978.

Freeman, W.J. Analytic techniques used in the search for the physiological basis of the EEG. In: *Handbook of Electroencephalography and Clinical Neurophysiology*, edited by A.S. Gevins and A. Remond. Amsterdam: Elsevier, 1987, p. 583–664.

Freeman, W.J. Development of a new science of brain dynamics with guidance from the theory of nonlinear dynamics and chaos. In: *8th International Conference on Biomagnetism*, Munster, 1991a.

Freeman, W.J. Insights into processes of visual perception from studies in the olfactory system. In: *Memory: Organization and Locus of Change*, edited by L.R. Squire, *et al.* New York: Oxford University Press, 1991b, p. 35–48.

Freeman, W.J. Nonlinear dynamics in olfactory information processing. In: *Olfaction: A Model System for Computational Neuroscience*, edited by J.L. Davis and H. Eichenbaum. Cambridge, MA: Bradford MIT Press, 1991c, p. 225–249.

Freeman, W.J. Predictions on neocortical dynamics derived from studies in paleocortex. In: *Induced Rhythms in the Brain*, edited by E. Basar and T.H. Bullock. Boston: Birkhäuser, 1992a, p. 183–199.

Freeman, W.J. Tutorial on neurobiology: from single neurons to brain chaos. *Int. J. Bifurc. Chaos Appl. Sci. Eng.* **2**: 451–482, 1992b.

Freeman, W.J. *Societies of Brains. A Study in the Neuroscience of Love and Hate.* Hillsdale, NJ: Lawrence Erlbaum Associates, 1995.

Freeman, W.J. and Baird, B. Relation of olfactory EEG to behavior: spatial analysis. *Behav. Neurosci.* **101**: 393–408, 1987.

Freeman, W.J. and Barrie, J.M. Spatiotemporal patterns of visual, auditory, and somesthetic EEGs in perception by trained rabbits. *Soc. Neurosci. Abstr.* **19**: 1229, 1993.

Freeman, W.J. and Barrie, J.M. Chaotic oscillations and the genesis of meaning in cerebral cortex. In: *Temporal Coding in the Brain*, edited by G. Buzsaki, *et al.* Berlin: Springer-Verlag, 1994, p. 13–37.

Freeman, W.J. and Grajski, K.A. Relation of olfactory EEG to behavior: factor analysis. *Behav. Neurosci.* **101**: 766–777, 1987.

Freeman, W.J. and Schneider, W. Changes in spatial patterns of rabbit olfactory EEG with conditioning to odors. *Psychophysiology* **19**: 44–56, 1982.

Freeman, W.J. and Van Dijk, B.W. Spatial patterns of visual cortical fast EEG during conditioned reflex in a rhesus monkey. *Brain Res.* **422**: 267–276, 1987.

Freeman, W.J. and Viana Di Prisco, G. Relation of olfactory EEG to behavior: time series analysis. *Behav. Neurosci.* **100**: 753–763, 1986.

Freeman, W.J., Viana Di Prisco, G., Davis, G.W. and Whitney, T.M. Conditioning of relative frequency of sniffing by rabbits to odors. *J. Comp. Psychol.* **97**: 12–23, 1983.

Galli, F., Lifschitz, W. and Adrian, H. Studies on the auditory cortex of rabbit. *Exp. Neurol.* **30**: 324–335, 1971.

Georgeopolis, A.P., Schwartz, A.P. and Kettner, R.E. Neural population coding of movement direction. *Science* **233**: 1416–1419, 1986.

Ghose, G.M. and Freeman, R.D. Oscillatory discharge in the visual system: does it have a functional role? *J. Neurophysiol.* **68**: 1558–1574, 1992.

Gould, H.J. Body surface maps in the somatosensory cortex of rabbit. *J. Comp. Neurol.* **243**: 207–233, 1986.

Gray, C.M. Synchronous oscillations in neuronal systems: mechanisms and functions. *J. Comp. Neurosci.* **1**: 11–38, 1994.

Gray, C.M., Engel, A.K., Konig, P. and Singer, W. Synchronization of oscillatory neuronal responses in cat striate cortex: temporal properties. *Vis. Neurosci.* **8**: 337–47, 1992.

Gray, C.M., Konig, P., Engel, A.K. and Singer, W. Oscillatory responses in cat visual cortex exhibit inter-columnar synchronization which reflects global stimulus properties. *Nature* **338**: 334–337, 1989.

Hardcastle, V.G. Psychology's binding problem and possible neurobiological solutions. *J. Conscious. Stud.* **1**: 66–90, 1994.

Higuchi, T. Relationship between the fractal dimension and the power law index for a time series: a numerical investigation. *Physica D* **46**: 254–264, 1990.

Hollander, H. and Halbig, H. Topography of retinal representation in the rabbit cortex: an experimental study using transneuronal and retrograde tracing techniques. *J. Comp. Neurol.* **193**: 701–710, 1980.

Hughes, A. and Vaney, D.I. The organization of binocular cortex in the primary visual area of the rabbit. *J. Comp. Neurol.* **204**: 151–164, 1982.

Kauer, J.S. Coding in the olfactory system. In: *Neurobiology of Taste and Smell*, edited by T.E. Finger and W.L. Silver. New York: John Wiley and Sons, 1987, p. 205–231.

Koch, C. Computational approaches to cognition: the bottom-up view. *Curr. Opin. Neurobiol.* **3**: 203–208, 1993.

Kruger, J. Multi-microelectrode investigation of monkey striate cortex: Link between correlational and neuronal properties in the infragranular layers. *Vis. Neurosci.* **5**: 135–142, 1990.

Lancet, D., Greer, C.A., Kauer, J.S. and Shepherd, G.M. Mapping of odor-related neuronal activity in the olfactory bulb by high-resolution 2-deoxyglucose autoradiography. *Proc. Natl. Acad. Sci. USA* **79**: 670–674, 1982.

Lilly, J.C. A method of recording the moving electrical potential gradients in the brain: the 25-channel Bavatron and electro-iconograms. In: *Conference on Electronic Instrumentation in Nucleonics and Medicine*, New York, 1949.

Lilly, J.C. Instantaneous relations between the activities of closely spaced zones on the cerebral cortex - electrical figures during responses and spontaneous activity. *Am. J. Physiol.* **176**: 493–504, 1954.

Livanov, M.N. *Spatial Organization of Cerebral Processes*. New York: John Wiley and Sons, 1977.

Maier, J., Dagnelie, G., Van Dijk, B.W. and Spekreijse, H. Principal components analysis for source localization of VEPs in man. *Vis. Res.* **27**: 165–177, 1987.

McMullen, N.T. and Glaser, E.M. Tonotopic organization of rabbit auditory cortex. *Exp. Neurol.* **75**: 208–220, 1982.

McNaughton, B.L. The mechanism of expression of long-term enhancement of hippocampal synapses: Current issues and theoretical implications. *Ann. Rev. Physiol.* **55**: 375–396, 1993.

Menon, V., Freeman, W.J., Cutillo, B.A., Desmond, J.E., Bressler, S.L., Laxer, K.D., Barbaro, N. and Gevins, A.S. Spatiotemporal correlations in human gamma band electrocorticograms. *Electroencephalogr. Clin. Neurophysiol.* **98**: 89–102, 1996.

Mickle, W.A. and Ades, H.W. Spread of cortical evoked potentials. *J. Neurophysiol.* **17**: 608–633, 1954.

Milner, P.M. A model for visual shape recognition. *Physiological Review* **81**: 521–535, 1974.

Moulton, D.G. Spatial patterning of responses to odors in the peripheral olfactory system. *Physiological Review* **56**: 578–593, 1976.

Murthy, V.N., Aoki, F. and Fetz, E.E. Synchronous oscillations in sensorimotor cortex of awake monkeys and humans. In: *Oscillatory Event-Related Dynamics*, edited by C. Pantev. New York: Plenum Press, 1994, p. 343–356.

Press, W.H., Flannery, B.P., Teukolsky, S.A. and Vetterling, W.T. *Numerical recipes in C*. Cambridge, UK: Cambridge University Press, 1988.

Rolls, E.T., Miyashita, Y., Cahusac, P.B.M., Kesner, R.P., Niki, H., Feigenbaum, J.D. and Bach, L. Hippocampal neurons in the monkey with activity related to the place in which the stimulus is shown. *J. Neurosci.* **9**: 1835–1845, 1989.

Sanes, J.N. and Donoghue, J.P. Oscillations in local field potentials of the primate motor cortex during voluntary movements. *Proc. Natl. Acad. Sci. USA* **90**: 4470–4474, 1993.

Thommeson, G. The spatial distribution of odour induced potentials in the olfactory bulb of char and trout (Salmonidae). *Acta Physiol. Scand.* **102**: 205–217, 1978.

Thompson, J.M., Woolsey, C.N. and Talbot, S.A. Visual areas I and II of cerebral cortex of rabbit. *J. Neurophysiol.* **13**: 277–288, 1950.

Tovee, M.J., Rolls, E.J., Treves, A. and Bellis, R.P. Information encoding and the responses of single neurons in the primate temporal visual cortex. *J. Neurophysiol.* **70**: 640–654, 1993.

Viana Di Prisco, G. and Freeman, W.J. Odor-related bulbar EEG spatial pattern analysis during appetitive conditioning in rabbits. *Behav. Neurosci.* **99**: 964–978, 1985.

Von der Malsburg, C. How are nervous structures organized? In: *Synergetics of the Brain*, edited by E. Basar, *et al.* Berlin: Springer-Verlag, 1983, p. 238–249.

Walter, W.G. *The Living Brain.* New York: W.W. Norton and Company, Inc., 1953.

Wells, D.P., Scott, J.W. and Harrison, T.A. Discrimination among odorants by single neurons of the rat olfactory bulb. *J. Neurophysiol.* **61**: 1161–1177, 1989.

Yao, Y. and Freeman, W.J. Model of biological pattern recognition with spatially chaotic dynamics. *Neural Netw.* **3**: 153–170, 1990.

Young, M.P., Tanaka, K. and Yamane, S. On oscillating neuronal responses in the visual cortex of the monkey. *J. Neurophysiol.* **67**: 1464–74, 1992.

14. *Stochastic Differential Equations and Random Number Generators Minimize Numerical Instabilities in Digital Simulations*

The dynamics of the olfactory system are now fairly clear (Freeman 1999b). The receptors high in the nose transduce odorant stimuli to spatial patterns of action potentials as point processes that are topographically mapped into the bulb. The bulbar KII population is the key element, providing an attractor landscape shaped by learning and attention under limbic control, the wings of which are accessed by input-dependent destabilization to generalize and classify specific inputs. The external interneurons in the glomerular layer provide logarithmic range compression, spatial coarse-graining, and excitatory bias varying with arousal (Freeman 1992a; Chang and Freeman 1999). The anterior olfactory nucleus serves as a controller of the chaotic output of the bulb (Pecora 1998; Pecora *et al.* 1997), and the prepyriform cortex provides for signal extraction and generalization over the bulbar outputs in respect to response selection.

Apparent success in simulating the aperiodic waveforms of the EEG with the lumped KIII model (Chapter 12) fulfilled one of two major criteria for accepting a model of olfactory dynamics. The other criterion was to emulate the capacity of the olfactory system to classify spatial AM patterns with an aperiodic carrier waveform after reinforcement learning to discriminate between two classes of sensory input. In pursuit of this goal the olfactory bulb was expanded into a globally coupled layer of 64 oscillators (Ahn and Freeman 1975, Yao and Freeman 1990), and a test set was obtained consisting of images of small machine parts (Buckley and Stelson 1978). Initial results of classification were encouraging but inconclusive (Yao *et al.* 1991), owing to the compromises required to engineer the software, including binarization of input, and measurement of the low frequency offset instead of the gamma oscillation of the output. The reason for these compromises was the excessive sensitivity to small changes in parameters of the KIII model in aperiodic domains. Gradually it became clear that numerical instabilities in the digital simulation were responsible for many of the properties of prior simulations, very likely including those reported in Chapter 12. The problem worsened as improvements in techniques of parameter optimization (Freeman and Shimoide 1994; Chang and Freeman 1996) brought the temporal spectrum of the output close to linear fall-off in log-log coordinates ("$1/f^{\alpha}$"), and the KIII model became ever more brittle and sensitive to exceedingly small changes in inputs as well as in parameters (Chang and Freeman 1998a).

A technical fix for the problematic lack of robustness came accidentally through a software modification, in which a floating-point division was replaced with a multiplication by the reciprocal of the divisor in a do-loop. The outputs given by the two arithmetically identical operations diverged after less than 2 seconds of simulated run time, showing that the digital embodiment of the KIII model in double-point arithmetic was sensitive to a difference in the values of the state variables by 1 part in $10^{15.4}$. The run time to divergence was modestly increased by randomizing the terminal bit (Zak 1991). Full stabilization was achieved by using additive noise (Chang and Freeman 1998b), but not multiplicative noise, giving robust simulations of the waveforms (Figure 3 in Chapter 14) and spectra (Figures 4 and 5 in Chapter 14) of the activity density functions of the main components of the olfactory system.

These results exposed some of the fundamental unsolved problems in the simulation of biological dynamics with numerical techniques on digital computers. Among these issues is the distinction between deterministic chaos and other types that can be called nondeterministic or stochastic chaos. The development of the theory of chaotic dynamics in the past two decades has been dominated by deterministic models that are low-dimensional, autonomous, stationary, and noise-free. Chaos in brains emerges from synaptic interactions of immense numbers of neurons that stimulate and yet constrain each other, creating mesoscopic order from microscopic disorder. Brain subsystems are nonstationary, unstable, constantly altered by sensory bombardment, changed in functional dimension on interactions with other brain parts, and sustained by noise, yet displaying a high degree of reliability and metastability in the appropriate environments to which they are adapted. Attempts to simulate brain dynamics with digital models encounter the limits expressed in numerical instabilities imposed by finite approximations with rational numbers, attractor crowding, collapse into quasiperiodic solutions, the lack of shadowing trajectories, and the curse of "infinite sensitivity to the initial conditions," as described in Chapter 14.

Analog embodiments (Eisenberg *et al.* 1989) may take advantage of the use of continuous variables, highly parallel integrative operations, and reliance on internally generated electronic and thermal noise that is essential as well as unavoidable in normal function. Nagging problems in analog computation of instability from bias drift, aging of components, sensitivity to temperature, and variation of parameters may be addressed by using the biological concept of homeostasis (Chang and Freeman 1999) to model the neural mechanisms that maintain local stability in the face of sensory inputs that may vary in intensity over several logarithmic units, requiring conversions of stimuli from linear to logarithmic scales (Freeman 1974d, 1992a). The connectivity problem for a large array can be solved by time multiplexing without need for analog to digital conversion (Freeman 1988a), though digital controls are required for timing and switching of adaptive gains between successive pairs of elements, so that a device for simulation of cortical dynamics and use in pattern classification (Yao *et al.* 1991; Shimoide and Freeman 1995) will optimally be a hybrid of analog and digital components. The problem is to control chaos (Schuster and Stemmler 1997), not to suppress it (Schiff *et al.* 1994), and to use it for rapid switching among alternative states in an attractor landscape, generalizing beyond stochastic resonance (Bulsara *et al.* 1993) by allowing aperiodic input, and using noise-driven transitions between "fragile" attractors in a Milnor attractor network (Kaneko 1997, 1998), which appears to more closely represent sensory cortical dynamics than any other extant formal model.

Taming Chaos: Stabilization of Aperiodic Attractors by Noise

Abstract

A model named 'KIII' of the olfactory system contains an array of 64 coupled oscillators simulating the olfactory bulb (OB), with negative and positive feedback through low-pass filter lines from single oscillators simulating the anterior olfactory nucleus (AON) and prepyriform cortex (PC). It is implemented in C to run on Macintosh, IBM or UNIX platforms. The output can be set by parameter optimization to point, limit cycle, quasiperiodic, or aperiodic (presumably chaotic) attractors. The first three classes of solutions are stable under variations of parameters and perturbations by input, but they are biologically unrealistic. Chaotic solutions simulate the properties of time-dependent densities of olfactory action potentials and EEGs, but they transit into the basins of point, limit cycle, or quasiperiodic attractors after only a few seconds of simulated run time. Despite use of double precision arithmetic giving 64-bit words, the KIII model is exquisitely sensitive to changes in the terminal bit of parameters and inputs. The global stability decreases as the number of coupled oscillators in the OB is increased, indicating that attractor crowding reduces the size of basins in the model to the size of the digitizing step ($\sim 10^{-16}$). Chaotic solutions having biological verisimilitude are robustly stabilized by introducing low-level, additive noise from a random number generator at two biologically determined points: rectified, spatially incoherent noise on each receptor input line, and spatially coherent noise to the AON, a global control point receiving centrifugal inputs from various parts of the forebrain. Methods are presented for evaluating global stability in the high dimensional system from measurements of multiple chaotic outputs. Ranges of stability are shown for variations of connection weights (gains) in the KIII model. The system is devised for pattern classification.

1. Introduction

Experiments in animals reveal that the electroencephalograms (EEGs) of the primary sensory cortices for vision, audition, olfaction and touch have spatiotemporal patterns relating to external stimuli [1, 2]. The patterns have the form of spatial amplitude modulation of a broad spectrum, aperiodic waveform [3], that is common to all the 8×8 traces simultaneously recorded at the cortical surfaces. Following training of the experimental subjects to respond to conditioned stimuli, the spatial patterns, expressed as 64×1 column vectors, can be classified in respect to the discriminated conditioned stimuli. In the language of dynamics, each sensory cortex maintains a low dimensional global aperiodic attractor with multiple "wings", one for each class of learned stimuli [4]. During an act of perception the activity of the cortex is confined into an appropriate "wing" by which the appropriate spatial pattern occurs. The process can be modeled by a stimulus-induced state transition in a complex dynamical system, which is composed of an array of coupled nonlinear oscillators based in excitatory and inhibitory neuron ensembles, and which maintains a stable chaotic state before and after step inputs are given. The long-range purpose is to use nonlinear dynamics in fast and accurate pattern recognition in simulation of the capabilities of biological intelligence [5, 6]. The present aim is to evaluate the stability of the extant model.

The KIII model of the olfactory system consists of a set of parametric ordinary differential equations [7], which become difference equations (ODEs) in a digital embodiment [8, 9, 10]. They incorporate the synaptic connections and express the physiological time delays and distributed interactions among the cell ensembles. The model has an unspecified number of point, limit cycle, quasiperiodic, and chaotic attractors [11]. Examination of solutions in the first three classes has shown them to be stable under perturbation by inputs and variation of parameter values on the order of 5% to 50% of center values, optimized by methods previously described [12].

However, the high-dimensional ODE set parameterized to give chaotic behavior reveals an exquisite sensitivity to any small change in input or in a parameter value, which is revealed by loss of a previously optimized aperiodic trajectory. Further, when free running in a chaotic mode, sooner or later the model undergoes a state transition from an assigned chaotic attractor to a new stable orbit or point. A measure of this instability is the time from initialization to the onset of the transition [13]. Stability decreases with increasing size of the array of coupled oscillators. An exceedingly small numerical perturbation in parameters or variables can change the state of the whole system from one attractor to another, indicating that it is due to attractor crowding [14, 15, 16], by which the size of the basins shrinks to the scale of the word size in double precision arithmetic that gives a digital increment of 1 in 64 bits.

Prior studies of chaotic systems have emphasized the suppression of chaos or the entrainment of two or more elements into the same aperiodic time series. In the present study the effects were studied of additive noise on the stability of chaotic attractors in an extension of preliminary work [13]. Prior work to control chaos has been focused on suppressing chaos or phase locking chaotic generators [17, 18]. Most prior uses of noise have been to achieve destabilization, as in simulated annealing and stochastic resonance [19]. Liljenström and Wu [20] found that activity expressed as temporal noise in an associative neural network reduced recall time in associative memory tasks. Billah and Shinozuka [21] reported stabilization of nonlinear networks by multiplicative noise. Wackerbauer [22] used noise to stabilize the Lorenz attractor. Fryska and Zohdy [23] put a Markovian process on every ordinary differential equation (ODE) of a three dimensional piecewise linear system to randomize the rounding-off of floating point arithmetic in numerical simulation. They used the shadowing lemma to assure the existence of a numerically shadowed trajectory. Since each point on a digital trajectory is merely close to a corresponding point of the true trajectory, the computed trajectory is shadowed by the true trajectory [24, 25]. Because the olfactory system and its KIII model both have Lyapunov exponents for which the real parts fluctuate above and below zero [7], the KIII model cannot solve the shadowing problem, owing to the resulting numerical instabilities [12, 13, 16, 25].

In Section 2 the KIII model is briefly reviewed, and the types of noise are described along with the nodes of injection and the reasons for choosing those nodes. The state variables and the equations have already been described in detail [5, 6, 8, 12]. In Section 3 the criteria are listed and described, according to which the aperiodic outputs of nodes in the KIII model are optimized to conform to biological measurements, and some representative traces, power spectra, and amplitude histograms are shown. In Section 4 algorithms are given to evaluate the global stability of the KIII model in aperiodic states. The ranges are shown for the variations of input and of the important parameters over which stability obtains. Results are discussed and evaluated in Section 5.

2. The Hybrid KIII Model with Additive Noise

2.1 KIII Topology

The central olfactory system consists of the bulb (OB), anterior olfactory nucleus (AON), and prepyriform cortex (PC). Input from receptors (R) goes to periglomerular (P) and mitral cells (M). The mitral cells transmit to granule cells (G) and to AON and PC, from which the final output is sent to the external capsule (EC) from deep pyramidal cells (C), as well as back to the OB and AON. The nodes in the topological diagram (Figure 1) represent neural aggregates. When the neurons in an aggregate have no interactions with others, they are represented by a K0 set, as for example R and C. An aggregate of interactive neurons is represented by a KI set, consisting of a transmitting K0 set and a receiving K0 set undergoing continuous renewal. Mutually excitatory neurons form a KI_e set (P, M, E and A), and mutually inhibitory neurons form a KI_i set (G, I and B), both comprising positive feedback and having only zero and nonzero point attractors. The interaction of excitatory and inhibitory neuron populations forms a KII set (OB, AON and PC), which has point, limit cycle, and quasiperiodic attractors. The interaction of the KI_p set and the three KII sets is modeled with the KIII set. Feedforward transmission through the lateral olfactory tract is fast, so delays are not explicitly introduced. Feedback through the medial olfactory tract (MOT) is slow and dispersed, owing to the

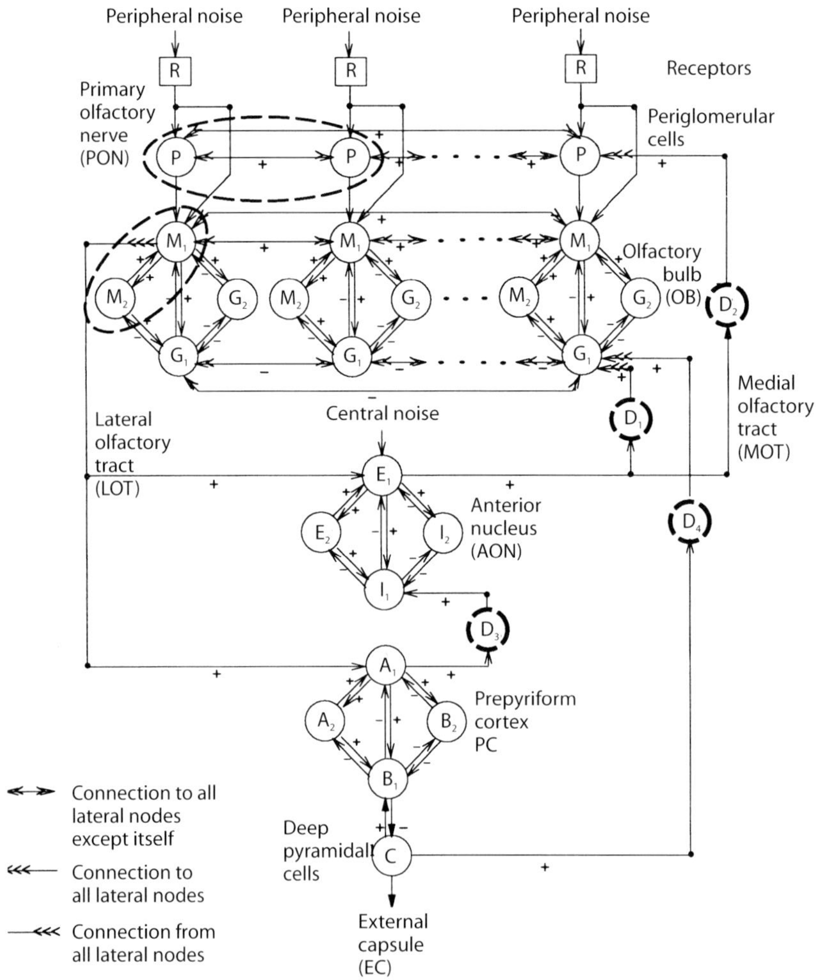

Fig. 1. The topologic diagram of the olfactory system is based in the anatomical and physiological properties. Noise from external sources is given at the receptors. Central noise is focused into the anterior olfactory nucleus (KII_{aon}).

distributions of axonal conduction velocities and distances [7]. Delays are modeled by 2nd order low-pass filters [12].

The excitatory and inhibitory neurons in the OB, AON and PC form sheets of neuropil without discrete internal boundaries in directions parallel to the surface of the brain. *R* input axons are spatially coarse-grained by glia forming glomeruli, which are nests of axons, dendrites and their synapses, The KII_{ob} set is represented by an 8×8 array of KII subsets, that are fully interconnected at the M_1 and G_1 nodes, conforming to a toroidal boundary condition. The KII_{aon} and KII_{pc} sets are in lumped form in the present stage of development of the KIII model.

In digital embodiments the dynamics is re-formulated from ODEs into difference equations. The continuous tissue is represented in the model by as many spatial compartments for the KI and KII subsets as computational resources can support. Continuous time is replaced with discrete time steps. Numerical solutions to the *ODE*s were obtained with the Livermore Solver (LSODE) [12]. The basic time step was set at 1 ms and was flexibly adjusted to shorter lengths, depending on the local rate of change of the state variable under integration.

2.2 State Variables and Reciprocal Conversions

The dynamics of each node is represented by a linear second order ODE with two real rate constants, which are evaluated from the rates of rise and decay of OB, AON and PC impulse responses in the open loop state established under deep anesthesia. State variables are continuous in time in the wave mode, v, corresponding to EEG measurements of dendritic current density, and in the pulse mode, p, corresponding to extracellular observations of action potentials, and to the outputs of the OB, AON and PC to each other and other parts of the brain. The synapses convert axonal pulse density to dendritic wave density. The synaptic weights at the inputs of axons to dendrites are represented by gain coefficients, k_{ij} from the j-th to the i-th node. Trigger zones at cell bodies convert dendritic current density to pulse density. In neural populations the wave-pulse conversion is given by a static nonlinear function, the asymmetric sigmoid curve (Figure 2), with a single parameter, Q_m, that determines the maximal pulse density, the mean firing density, and the steepness of the curve. The derivative gives the nonlinear gain, $\mathrm{d}p/\mathrm{d}v$ [4, 8], that has its maximal value at wave amplitude, $v_{\max} = \ln(Q_m)$.

The range of wave-pulse conversion holds the pulse-wave conversion at synapses within a small-signal, near-linear range, which makes it possible to represent it by a

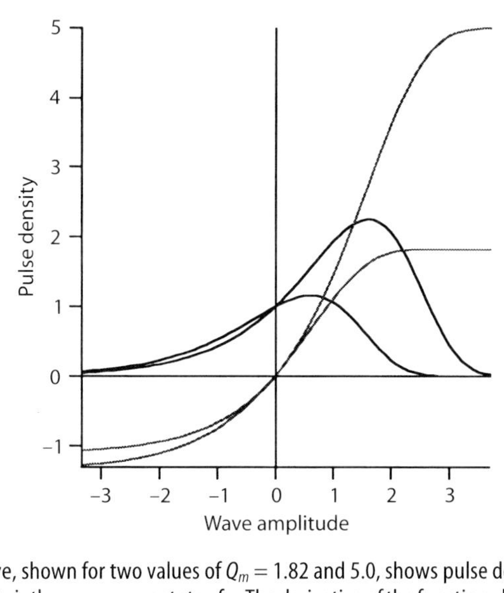

Fig. 2. The sigmoid curve, shown for two values of $Q_m = 1.82$ and 5.0, shows pulse density, q, as a function of wave density, v, where Q_m is the upper asymptote of q. The derivative of the function, dq/dv, gives the nonlinear gain. The peak gain is at the wave value, $v_{\max}$. $q = \exp[-(e^v-1)/Q_m]$, $dp/dv = \exp[v-(e^v-1)/Q_m]$, $v_{\max} = \ln(Q_m)$.

gain coefficient. It is the nonlinearity at the trigger zones that mediates the stability of the KIII set and its components. The KI_p set has a non-zero point attractor that maintains its operation above the wave value of maximal slope of the sigmoid, so that excitatory input decreases its loop gain, and inhibitory input increases it [8, 27]. The KI_m set is stabilized at a point or limit cycle attractor that keeps its operation below the value of maximal slope. Inhibitory input stabilizes it by reducing its loop gain, and excitatory input increases its loop gains. For high values of $Q_m \geq 5$, which occur in animals that are motivated to search for odors, R input destabilizes the basal state. The entire OB transits to near-periodic oscillation in the gamma range (g, $40 < f < 100$ Hz) on inhalation, and it returns to the basal state during exhalation, during which the R are not activated. The asymmetry of the nonlinear conversion of wave to pulse density at trigger zones is essential for the mechanism of input-dependent state transitions from the background, $1/f$-type state to the narrow-band oscillation in the burst state. The transition is facilitated by increase in the KI_m gains between selected subsets of excitatory neurons, during learning in accordance with the Hebb rule, which shapes a matrix of coefficients resembling those in Amari, Hopfield, Anderson and Kohonen associational networks. The state transition is blocked by reduction of the gains of selected K_{m_1} nodes to other KII_{ob} nodes. This occurs during habituation to insignificant or ambiguous or background odors during learning. The reduction enhances oscillatory responses to repetitive input in the delta range (d, $1 < f < 10$ Hz), which includes the respiratory driving frequencies, so that habituation to background inputs facilitates the destabilization to foreground odor input [4, 8], because the habituated input augments the nonspecific excitatory bias of background input. The gamma/delta (g/d) ratio in the power spectrum of KIII aperiodic time series offers a useful criterion of system performance.

2.3 Inputs of Stimuli and Noise

A small impulse to one or more of the R nodes serves to initiate activity, starting from an unstable zero point and transiting to a basal aperiodic attractor of the KIII model. Simulated sensory input given to multiple R nodes is normalized to give a fixed total amplitude, that is optimized and compressed on the scale of a power function. This operation is derived from and comparable to the normalization and dynamic range compression that is performed on olfactory receptor input into the OB [7] by the periglomerular cells (KI_p). Input from R nodes in the pulse density mode is discretized in time at 1 ms intervals and in space to the KI_{p_1} and KI_{m_1} subsets. The waveform of the simulated sensory input is a 200 ms step function with the leading and trailing edges rounded by a half cycle of a 50 Hz cosine with its mean amplitude set at half the amplitude of the step. This form models the respiratory cycle, and it reduces the ringing in the model that is observed with impulse inputs [8].

Noise is made with a random number generator giving a normal density function with zero mean and unit standard deviation (SD). The noise is added at two locations, corresponding to the two sources of external perturbation of the olfactory system [27]. Input by way of the receptor axons is purely excitatory, and their firings are locally uncorrelated. The Gaussian noise is full-wave rectified, and an independent time series is added to each R line to the KI_{p_1} and KI_{m_1} nodes. Centrifugal inputs from various parts of the brain to the olfactory system are concentrated into the AON, which serves as a subsidiary control center. Its output is broadly distributed to the OB by its divergent feedback projection. The inputs are predominantly excitatory, so random number time series is off-set by +1 SD. It is added to the LOT input to KI_{e_1} node, from which it is disseminated as spatially coherent noise. It is smoothed by the low pass filters in the MOT delay lines. The amplitudes of the two noise Standard Deviations (SDs) are optimized as described below. The input and output of the KI_{e_1} node are not significantly correlated, showing that the noise is not simply amplified and fed back into the system.

3. Simulation of EEGs and Pulse Densities

3.1 Criteria for Optimization of Performance

Parameter optimization algorithms are used to determine the parameter values required to simulate the experimental data from measurements of the EEG and action potentials from the neural populations of the olfactory system. The criteria for optimization are that the KIII outputs have:

1. stable and robust aperiodic oscillations having near $1/f$-type power spectra;
2. nearly periodic oscillations with peaks in the gamma g range during excitatory inputs corresponding to EEG bursts on inhalation;
3. spatially coherent waveforms over all subsets of the KI_{m_1} nodes and also of the KI_{g_1} nodes in the $\mathrm{KII}_{\mathrm{ob}}$ both in and between bursts;
4. transition from background to burst in < 10 ms with the stated input;
5. amplitude histograms in both wave and pulse modes that are nearly Gaussian;
6. means and distributions of periglomerular cell KI_p outputs above the point of maximal slope of the sigmoid and below the maximal (excitatory) asymptote;
7. means and distributions of mitral KI_m and granule KI_g outputs below the point of maximal slope of the sigmoid and above the minimal asymptote;
8. nearly equal means for the KI_{g_1} and KI_{g_2} nodes;
9. and nearly zero offset of the means for the mitral cell KI_{m} nodes, irrespective of the mean amplitudes of the KI_{g_1} and KI_{g_2} nodes.

Representative parameter values keyed to Figure 1 are listed in Table 1. These are useful starting guesses for optimization of the KIII model on any other platform, but

Table 1. Given the specified input waveform, the linear rates and delays, and the nonlinear sigmoid functions, there were 24 gain parameters to be optimized [12, 13].

Parameter:	*initial value*	*minimal*	*maximal*
For *n*-channel KI*p*:			
k_{pr}	0.500	0.008	1.100
k_{pp}	0.900	0.900	1.300
For *n*-channel KII_{ob}:			
k_{mm}	1.600	0.350	2.400
k_{gm}	1.723	1.400	2.400
k_{mg}	2.363	2.000	2.900
k_{gg}	2.245	1.900	2.400
k_{m1r}	1.000	0.008	3.000
k_{m1p}	0.050	0.002	0.760
For KII_{aon}:			
k_{ee}	1.200	0.390	1.202
k_{ie}	1.372	0.930	2.100
k_{ei}	1.426	0.790	2.300
k_{ii}	1.571	0.760	1.700
For KII_{pc}:			
k_{aa}	0.823	0.006	1.200
k_{ba}	1.947	0.790	2.300
k_{ab}	1.938	0.900	2.500
k_{bb}	2.354	2.300	3.200
k_{bc}	0.698	0.498	2.098
k_{cb1}	1.543	1.044	4.544
For MOT gains:			
k_{g1d1}	2.349	1.649	3.349
k_{pd2}	1.087	0.287	3.587
k_{i1d3}	2.553	0.003	9.553
k_{g1d4}	2.305	0.005	8.305

cannot be expected to reproduce on other platforms precisely the chaotic time series illustrated here.

3.2 An Assay of Sensitivity for Destabilization

As optimization of parameters to achieve these criteria proceeded, the KIII system without noise became progressively less stable and more likely to transit to a point or limit cycle attractor spontaneously. An assay of sensitivity was discovered following a routine change in programming. A line of code calling for repeated division by a constant was changed to repeated multiplication by the reciprocal. Though mathematically equivalent, one version transited to sustained aperiodic oscillation, whereas the other gave a quasi-periodic oscillation after < 1 s of simulated run time. This change implicated the round-off error in the terminal bit. Zak [28] used randomization of the terminal bits in digital state variables to simulate

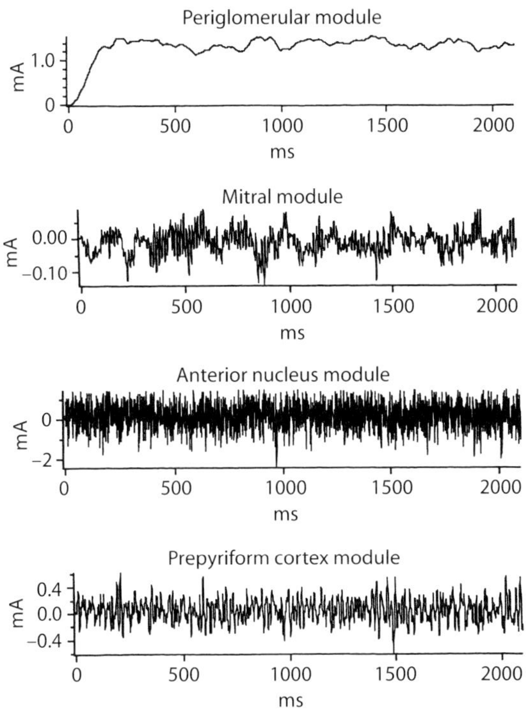

Fig. 3. Time series are from representative PG, OB, AON, PC nodes.

neurodynamics with "terminal chaos." In the absence of a shadow trajectory, one version was used to track the other one, and the sensitivity was measured by the run time to detect divergence of the two versions (Figures 12, 13 in [13]). By this test the randomization of the terminal bit in all of the state variables was found to delay but not prevent divergence of the two time series or eventual transition from an aperiodic orbit to a point or limit cycle attractor. Furthermore, any minute change in the input amplitude or in any parameter so changed the global performance of the KIII system, that the parameter set had to be re-optimized to bring the KIII output within the above criteria.

The addition of the designated noise stabilized the system, both in respect to changes of input and of parameters. An example in Figure 3 shows the time series of representative KI_p, KI_{m_1}, KI_{e_1} and KI_{a_1} nodes, with the initial transit from an unstable fixed point to its stable performance under noise. The spectrum and amplitude histogram superimposed on its nonlinear gain, $\mathrm{d}p/\mathrm{d}v$, the derivative of its sigmoid, show that the distribution of the wave density values is wholly above the wave value, $v_{\max} = \ln(Q_m)$, at peak gain for KI_p (Figure 4), and wholly to the left of $v_{\max}$ for KI_{m_1} (Figure 5), KI_{e_1} and KI_{a_1}. During bursts the distributions of wave amplitudes for both the excitatory and inhibitory nodes in the OB, AON and PC are wider but stay left of $v_{\max}$.

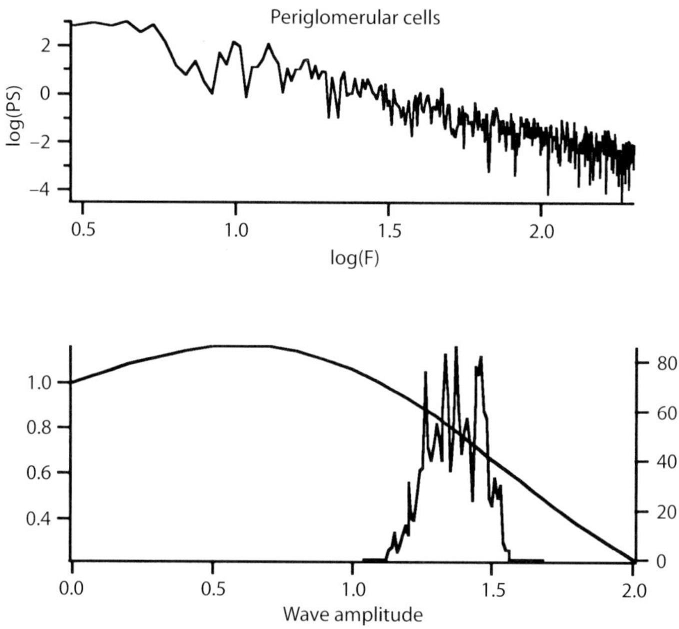

Fig. 4. The power spectrum and amplitude histogram are from a PG (KI_p) node.

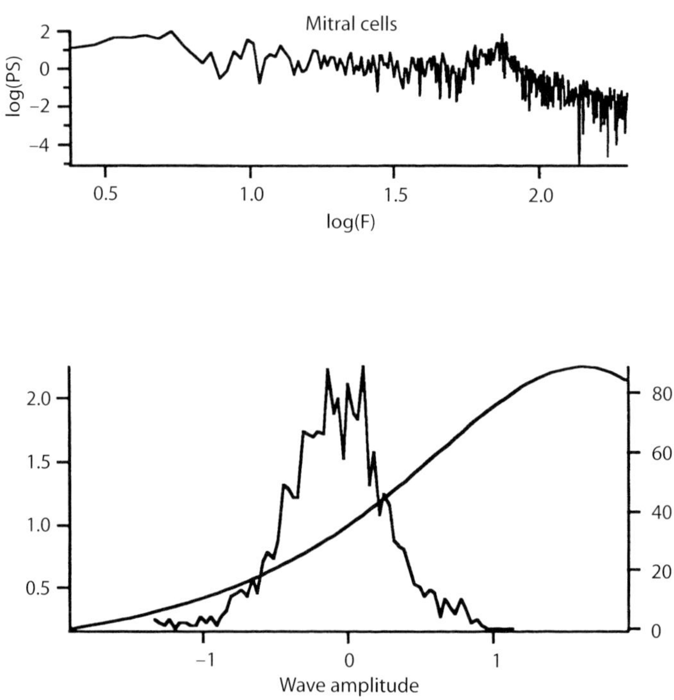

Fig. 5. The power spectrum and amplitude histogram are from the OB (KI_{m_1} mitral cell) time series.

4. Evaluation of Stability

4.1 Stability with Fixed Parameters

Thus far the state of the KIII model was evaluated by visual inspection of the ensemble of time series, by which conformance to the established patterns of the EEGs could be judged. The need was recognized for numerical descriptors of the global state, that could be used to determine whether the model persisted in a designated initial state of arbitrary duration, and whether it returned to the same state reliably after repeated induction of gamma bursts. The test would characterize the present state of the KIII model automatically without requiring detailed inspection of the time series, and it would assign a probability value to the likelihood of return of the KIII model to the same attractor basin as before an induced state transition. A method was adapted from classification of EEG spatial patterns [1, 2, 29], by which the amplitude of the common waveform was taken from each of 64 simultaneously recorded traces, formed into a 64×1 column vector, and plotted as a point in 64 space. Recurrent patterns formed a normally distributed cloud of points with a center of gravity ("centroid") that defined the class of pattern by its spatial ensemble average. The radius of the cluster measured by the standard deviation (SD) defined the probability of membership in the class. The distance of separation of two or more classes was evaluated by the distance between their centroids divided by the geometric mean of the two SDs.

For the KIII model a set of 9 traces was taken from representative nodes (P, M_1, M_2, G_1, E_1, I_1, A_1, B_1, and C). The traces were cut into 200 ms segments. Each segment gave 7 measures (the four moments: mean, SD, skewness and kurtosis; and the slope, intercept and SD from the fitted $1/f$ line of the power spectrum in log-log coordinates, $9\times7 = 63$), plus the mean correlation by z-transform between pairs of KI_m subsets to evaluate the commonality of waveform. Evaluation of the distribution of points along each of the axes in 64-space combined with visual inspection of the traces showed that the skewness, kurtosis, and spectral estimates varied too greatly, so they were dropped. The g/d ratio was added [30], giving a point in 28-space ($9\times3 + 1$) for each measured time segment of the model.

A continuous run was initiated, and after the ~700 ms required for initialization of the model, a sample of 10 segments of 200 ms at intervals of 1000 ms was evaluated to give a normally distributed cluster of 10 points with its centroid and SD. If the points showed no drift, and no points were more than 2 SD from the centroid, the system was judged to be stable. The value of the threshold SD was validated by visual inspection to confirm that all of the time series in every segment conformed to the nine criteria listed. The test was repeated with durations up to 50 s, and then with 200 ms step segments given at 1 s intervals to induce gamma bursts. Test segments taken prior to the inputs showed that, despite the unpredictable variation in the waveforms of the multiple time series, the KIII system returned to the same basal state, within 200 ms after each input step segment was ended. The same result followed delivery of step inputs with varying amplitudes and spatial patterns, after input normalization and range compression in the manner described above.

4.2 Effects of Variation of Parameters

The set of 24 gains to be optimized was too large to be evaluated in one step. Each parameter was varied from its optimized value over a range sufficient to reveal instability in both directions, with the intent to rank the parameters in order of sensitivity, and then jointly to change those few simultaneously in accordance with optimization procedures already reported [12]. Each stepwise change in a parameter changed measurements in 28-space giving apparent drift of the centroid by its mean, SD, peak frequency, and amplitude distribution of the traces. A new criterion for instability was adopted. The time ensemble average A and SD were calculated for the set of statistics from the 10 segments (mean, SD, and g/d ratio), and they were plotted as a function of the parameter value. The new criterion that emerged was the SD of the cluster, which increased markedly between stable domains outside both ends of the range of acceptable function. The simulated neural activity patterns in those upper and lower flanking domains did not conform to experimental observations from the olfactory system. By this new criterion, each parameter tested was shown to have a substantial range in which the model was stable (Table 1).

5. Discussion and Conclusions

The analysis and evaluation of the performance of the KIII model are still strongly dependent on visual inspection, so that further developments of stability assays are in progress. The criteria for conformance of the KIII model to EEG data are being expressed as fuzzy membership functions [30], either trapezoidal for average values (P-functions) or half-trapezoidal for variances (G-functions). Classification of the simulated EEG segment as acceptable or not are by the continued product (fuzzy AND operation) of the set of membership values. Membership functions are also being constructed for the stability criteria. A global stability rating on a graded scale is to be calculated from their continued product. Further options may be the use of weights to emphasize or attenuate subsets of criteria. The interactions of multiple parameters varied simultaneously are yet to be evaluated,

The robustness and stability of attractor states in the olfactory system are attested by its capacity to support the identification of familiar odors recurring over long periods of time, and by the reproducibility of the measurements of its spatiotemporal neural activity patterns. Its capacity to transit rapidly and repeatedly through its attractors underlies its service in olfactory pattern classification [31]. The global robustness of this highly unstable neural mechanism, in the face of continuing processes of growth, learning, and unlearning [32] must be maintained by chemical and neurohumoral controls exercised both locally and from the brain stem centrifugally, though little is known about them. At present the global stability of the KIII model is maintained and assayed by reference to criteria extracted by experimental studies, which appear to reflect system invariants that may or may not be used by the brain and the olfactory system, but are useful here.

Networks of nonlinear ODEs serve to simulate the olfactory EEGs and pulse densities when their parameters are adjusted to give aperiodic time series indicative of chaotic attractors. However, this approach brings three difficult problems. First, the sensitivity to initial conditions may be desirable for classification of very weak patterned inputs, but it opens a door nonselectively. Second, the aperiodic time series for successive real inputs are never identical, so the identification of an attractor for a recurring real input cannot be done by curve fitting. The problem is compounded in the high dimensional arrays of the KIII model with multiple uncorrelated time series. Third, in digital representations of the model the ODEs become difference equations. These are satisfactory for getting point and limit cycle solutions, but the truncation of the numbers in numerical approximations leads inevitably to numerical instabilities that may be difficult or impossible to distinguish from chaotic time series.

The solution for these problems by use of additive noise was suggested by analyses of the olfactory system, in which noise is always present from its receptors and its centrifugal controls. Endogenous noise is also present [27], both in additive form as a manifestation of action potentials, and in multiplicative form reflecting variation of neuronal parameters. However, neither the addition of random numbers to internal state variables, nor the random variation of parameters to give multiplicative noise, sufficed alone for stabilization, nor did they improve performance in the presence of the specified noise sources. It now appears that noise is not only unavoidable; it is also essential. This means that the attractor states manifested both in the model and in the olfactory system are not chaotic. They are hybrid. Chaotic attractors, like straight lines and perfect spheres, do not exist in nature, but are mathematical fictions to help in system design. The question, "Is brain activity chaos or colored noise?", is answered, "neither", by this hypothesis. The ratio of amplitudes by which they coexist can be calculated in models, as in the present case, but whether the simulations are accurate for the biological systems has not been determined.

Future testing of this hypothesis may be done in analog computation, in which numerical instabilities are not at issue. Earlier attempts to develop analog embodiments [33, 34] were abandoned owing to component imprecision, drift with temperature and aging, and especially the $1/f$-type noise in the operational amplifiers, which foretold that emergence of the desired property in KIII outputs could not be distinguished from capture of the global system by any one of its numerous elements. The present finding indicates that an analog approach should be re-explored to determine whether the intrinsic noise of components, particularly when they are reduced in size for LSI and VLSI, can be used to simulate action potentials in place of random number generators. Such an approach may solve another major problem, which is the amount of computational resource that is required for digital solution of ODEs. It is unlikely that models deriving from chaotic neurodynamics can be brought into the commercial and industrial arenas, until they are embodied in fully parallel analog or hybrid digital–analog devices. The taming of chaos by noise may open a path to that development.

Particular note should be taken that the use of noise that is shown in this report does not serve either to induce or suppress chaotic attractors, as described by others [34–37], but to maintain existing or desired chaotic attractors.

Acknowledgment

Reprinted with permission from: Freeman WJ, Chang H-J, Burke BC, Rose PA, Badler J (1997) Taming chaos: Stabilization of aperiodic attractors by noise. *IEEE Transactions on Circuits and Systems* **44**: 989–996.

Supported by the Office of Naval Research Grant No.: N00014-90-J-4054.

References

[1] Barrie JM, Freeman WJ, Lenhart M (1996) Spatiotemporal Analysis of Prepyriform, Visual, Auditory, and Somesthetic Surface EEGs in Trained Rabbits. *Journal of Neurophysiology* **76**: 520–539.

[2] Freeman WJ, Grajski KA (1987) Relation of olfactory EEG to behavior: Factor analysis: *Behavioral Neuroscience* **101**: 766–777.

[3] Kay LM, Shimoide K, Freeman WJ (1995) Comparison of EEG time series from rat olfactory system with model composed of nonlinear coupled oscillators. *International Journal of Bifurcation & Chaos* **5**: 849–858.

[4] Freeman, WJ (1992) Tutorial in Neurobiology: From Single Neurons to Brain Chaos. *International Journal of Bifurcation & Chaos* **2**: 451–482.

[5] Yao Y, Freeman WJ, Burke B, Yang Q (1991) Pattern recognition by a distributed neural network: An industrial application. *Neural Networks* **4**: 103–121.

[6] Shimoide K, Freeman WJ (1995) Dynamic neural network derived from the olfactory system with examples of applications. *IEICE Transaction Fundamentals* **E-78A**: 869–884.

[7] Freeman WJ (1975) *Mass Action in the Nervous System*. New York: Academic Press.

[8] Freeman WJ (1987) Simulation of chaotic EEG patterns with a dynamic model of the olfactory system. *Biological Cybernetics* **56**: 139–150.

[9] Freeman WJ (1988) Strange attractors govern mammalian brain dynamics, shown by trajectories of electroencephalographic (EEG) potential. *IEEE Transactions on Circuits & Systems* **35**: 781–783.

[10] Yao Y, Freeman WJ (1990) Model of biological pattern recognition with spatially chaotic dynamics. *Neural Networks* **3**: 153–170.

[11] Freeman WJ, Jakubith S (1993) Bifurcation analysis of continuous time dynamics of oscillatory neural networks. In: Aertsen A, von Seelen W (eds.) *Brain Theory. Spatio-temporal Aspects of Brain Function*. Amsterdam, Elsevier. pp. 183–208.

[12] Chang H-J, Freeman WJ (1996) Parameter optimization in models of the olfactory system. *Neural Networks* **9**: 1–14.

[13] Chang H-J, Freeman WJ, Burke BC (1998). Optimization of structure and parameters in a KIII model to obtain aperiodic oscillations having "1/*f*" power spectra. *Neural Networks* **11**: 449–466.

[14] Wiesenfeld K, Hadley P (1989) Attractor crowding in oscillator arrays. *Physical Review Letters* **62**: 1335–1338.

[15] Tsang KY, Wiesenfeld K (1990) Attractor crowding in Josephson junction arrays. *Applied Physics Letters* **56**: 495–496.

[16] Freeman WJ, Chang H-J, Burke BC (1995). Limitations on numerical simulations of chaotic dynamics observed in brain systems. *Proceedings, International Symposium on Nonlinear Theory and Its Applications* **I**: 10–14.

[17] Ditto WL, Rauseo SN, Spano ML (1990) Experimental control of chaos. *Physical Review Letters* **26**: 3211–3214.
[18] Pecora LM, Carroll TL (1991) Pseudoperiodic drifting: Eliminating multipole domains of attraction using chaos. *Physical Review Letters* **67**: 945–948.
[19] Carroll TL, Pecora LM (1993) Stochastic resonance and chaos. *Physical Review Letters* **70**: 576–579.
[20] Liljenström H, Wu X-B (1995). Noise-enhanced performance in a cortical associative memory model. *International Journal of Neural Systems* **6**: 19–29.
[21] Billah KYR, Shinozuka M (1996) Stabilization of a nonlinear system by multiplicative noise. *Physical Review A* **44**: 4779–4781.
[22] Wackerbauer R (1995) Noise-induced stabilization of the Lorenz system. *Physical Review* **52**: 1335–1338.
[23] Fryska ST, Zohdy MA (1992) Computer dynamics and shadowing of chaotic orbits. *Physics Letters A* **166**: 340–346.
[24] Hammel SM, Yorke JA, Grebogyi C [1987] Do numerical orbits of chaotic dynamical processes represent true orbits? *J. Complexity* **3**: 136–145.
[25] Grebogyi C, Hammel SM, Yorke JA, Sauer T (1990) Shadowing of physical trajectories in chaotic dynamics: Containment and refinement. *Physical Review Letters* **65**: 1527–1530.
[26] Dawson S, Grebogyi C, Sauer T, Yorke JA (1994) Obstructions to shadowing when a Lyapunov exponent fluctuates about zero. *Physical Review Letters* **73**: 1927–1930.
[27] Freeman WJ (1996) Random activity at the microscopic neural level in cortex ("noise") sustains and is regulated by low-dimensional dynamics of macroscopic cortical activity ("chaos"). *International Journal of Neural Systems* **7**: 473–480.
[28] Zak M (1991) Terminal chaos for information processing in neurodynamics. *Biological Cybernetics* **64**: 343–351.
[29] Freeman WJ (1987) Techniques used in the search for the physiological basis of the EEG. In: Gevins A, Remond A (eds.). *Handbook of Electroencephalography & clinical Neurophysiology* Volume 3A, Chapter 18. Amsterdam, Elsevier. pp. 583–664.
[30] Badler J (1997) A Fuzzy Logic algorithm for determining stability limits of KIII system parameters. To be posted on the Home Page of the Freeman Laboratory: `http://sulcus.berkeley.edu/`.
[31] Skarda CA, Freeman WJ [1987] How brains make chaos in order to make sense of the world. *Behavioral and Brain Sciences* **10**: 161–195.
[32] Freeman WJ (1995) *Societies of Brains. A Study in the Neuroscience of Love and Hate.* Mahwah NJ: Lawrence Erlbaum Associates.
[33] Freeman WJ (1968) Analog simulation of prepyriform cortex in the cat. *Mathematical BioScience* **2**: 181–190.
[34] Eisenberg J, Freeman WJ, Burke B (1989) Hardware architecture of a neural network model simulating pattern recognition by the olfactory bulb. *Neural Networks* **2**: 315–325.
[35] Herzel H (1988) Stabilization of chaotic orbits by random noise. *Zeitschrift angewandte Mechanik* **11**: 582–583.
[36] Matsumoto K, Tsuda I (1983) Noise-induced order. *Journal of Statistical Physics* **31**: 87–106.
[37] Kapitianek T (1988) *Chaos in Systems with Noise.* Singapore: World Scientific.

Epilog: Problems for Further Development in Mesoscopic Brain Dynamics

Further progress in mesoscopic brain dynamics can be expected in both directions, downwardly into cell biology and upwardly into macroscopic brain theories and their cognitive applications.

Links Between Microscopic Neurons and Mesoscopic Neural Assemblies

While great data banks have been built concerning the properties and functions of neurons, the concepts needed for synthesis of the data with respect to behavior have not kept pace. An understanding of the processes by which mesoscopic domains are established by neurons and in turn influence their individual actions cannot fail to change perspectives in cellular neurobiology, leading to new insights, priorities, and experimental goals (Freeman 1992b). Several examples have been raised in the preceding chapters. One of these is the need to go beyond microscopically oriented studies of topographic mapping in pathways and of point-to-point transmission in neural networks in the forebrain, in order to search for mesoscopic alternatives offered by divergent–convergent cortical projections, such as the lateral olfactory tract from the bulb to the prepyriform cortex. The need here is to derive estimates of the kernels of spatial integration by which cortical signals are laundered.

Another example stems from the paucity of measurements of chloride ion concentrations in and between neurons, leading to the widespread misconception in microscopic perspectives on neurons that the neurotransmitter GABA is solely inhibitory in the mammalian central nervous system. Modeling of mesoscopic dynamics has given evidence that some GABA-ergic interneurons are excitatory (Chapter 8; Freeman 1986), and it has led to the discovery of evidence for the accumulation of chloride ions in those neurons (Siklós *et al.* 1995), so that the action of GABA is to allow chloride ions to escape, thereby depolarizing the neurons. The possibility exists that the well-known histaminergic projections from the hypothalamus to the cerebral cortex, including the olfactory bulb, might control the uptake of chloride ions by periglomerular cells, thus regulating their excitatory bias, which may implement the well established role of histamine in arousal (Rhoades *et al.* 1988).

Histamine may also enable interneurons to switch between excitatory and inhibitory actions by controlling the uptake and intracellular concentration of chloride ions, thereby accounting for the emergence of gamma waves in the EEG shortly after birth, and on waking from deep sleep. The reasoning is based on the fact that neurons, unlike transistors, must stay active or else shrivel and even die. Their requisite background activity would be difficult to regulate if it depended on a pacemaker potential in each neuron, whereas mutual excitation provides a field of white noise easily subject to neuromodulatory control, including shut-down. The establishment of beds of connection in neuropil during embryonic growth is facilitated by the paucity of inhibitory interneurons. Unbridled excitation is apparent in waves of ultra-low frequency fluctuations of potential observed in immature cortex and retina (Kammermans and Werblin, 1992; Penn and Shatz 1999). It appears that at or near the stage of appearance of alpha, theta and gamma oscillations, GABA-ergic interneurons in neuropil shift from excitatory action to inhibitory action (Michelson and Wong 1991; Chen *et al.* 1996; Obata 1997). Whether histamine might play a role has not been determined, perhaps because it is commonly omitted from lists of brain monoamines (usually given as acetylcholine, dopamine, norepinephrine and serotonin). It might be supposed that measurements of chloride concentrations in and between neurons in human and other mammalian brains would have top priority for cellular neurobiologists, along with the effects of histamine and antihistamines (Freeman 1993), yet reports on them are rare, in part because the techniques using ion-selective pipettes or X-ray activation are so difficult, but in part because the information has not seemed important from the microscopic perspective. Chloride concentration is commonly regarded as the free parameter (the "mendicant ion") needed to balance the charge in electrochemical equations, as in the "chloride shift" that allows red blood cells to transport bicarbonate ions from active tissues to the lungs. In accord with this example mesoscopic dynamics can provide a framework for more effective guidance of cellular experiments and the interpretation of new data.

Other examples include the need for further studies of the conditional probability of neural firing on EEG and MEG amplitudes, and analyses of the regulation of time-multiplexing among neurons in local neighborhoods that convey local mean field intensities. More measurements are needed of open loop rate constants, particularly to distinguish slow postsynaptic potentials (slow EPSPs and slow IPSPs from prolonged mutual excitation or inhibition, usually coexisting, owing to positive feedback when the cortical feedback loops have, in fact, not been fully opened at the times of measurement.

More generally, the neuropharmacology of transmitter substances and their mimetics and blockers could be substantially advanced by the use of interactive models such as the KIII set to predict the complex patterns of effects revealed by localized synaptic changes on evoked potential waveforms, direct current potentials, background densities of action potentials, and EEG power spectra and coherence functions (Seyal 1976; Rhoades 1991; Freeman 1993). The classical models of the neuromuscular junction and the monosynaptic PSPs of motor neurons are simple feedforward operators that are interpreted under linear causality, and they are inadequate to handle even single feedback loops. Studies in the genesis of the various classes of epilepsy would profit from further applications of nonlinear dynamics to define the abnormal attractors of neural masses. In particular, inadequate attention has been paid to the instability that can arise by mutual inhibition, a form of positive feedback, which at high gain can result in 3/second spike trains in partial complex seizures resembling classical petit mal with trance-like states (the French word is 'absence') in animals and children (Freeman 1986), which can be treated by reducing inhibitory action instead of enhancing it (Rhoades *et al.* 1990). This instability is predicted from the two root loci that lie on the real axis in Figure 5 in Chapter 4. The root locus that runs to the left with increasing amplitude predicts a stable pole at the origin of the complex plane (an eigenvalue with zero real and imaginary parts), which determines the values of the other roots in the background state as described by Figure 3 in Chapter 8, A. The root locus that runs to the right with increasing amplitude from the open loop pole at $s = -220$/sec predicts the onset of KII instability as it crosses the imaginary axis into the right half of the s-plane. This is confirmed by numerical integration of the differential equations as shown by Figure 11 in Chapter 12.

More depth is needed in studies of the important role of gamma activity in learning. A great deal is known about the molecular biology and cellular chemistry of synaptic changes with associative learning, but relatively little attention has been paid to the properties of the populations in which the modifiable synapses are embedded, despite the well-known fact that long-term potentiation and depression (LTP and LTD) require heterosynaptic activations on a massive scale. The significance of mesoscopic neural oscillations has already been emphasized as the basis for laundering cortical outputs and extracting the cooperative signal from the incumbent noise through intercortical divergent–convergent projections. An equally important role of gamma oscillations is to facilitate Hebbian learning by concentrating the probabilities of coincident firing of cortical neurons at the crests of the excitatory cycles (p. 34 in Freeman and Barrie 1994). The efficiency with which a novel wing of an attractor forms may be proportional to the frequency of the gamma carrier wave, owing to the sharper temporal resolution provided by higher frequencies. If this were so, then the preferred mode of analysis of gamma activity in respect to learning would be first to take the time derivative by numerical differentiation (Tang *et al.* 1995), because the approximate $1/f^2$ form of cortical power spectral densities in log–log coordinates (Figure 5 in Chapter 13) yields nearly flat spectral densities over the gamma frequency range of the derivative of the EEG.

The importance of oscillations in learning mechanisms embodied in the hippocampus has repeatedly been recognized (Leung 1992; Buzsáki *et al.* 1992; Sik *et al.* 1994), but emphasis has been placed on frequencies in the range of 200–250 Hz, those of "chattering cells" in the visual cortex (Gray and McCormick 1996), "ripples" at 200 Hz in the hippocampus (Siapas and Wilson 1998), doublet ("dicrotic") firings of hippocampal neurons (Traub *et al.* 1996), and repetitively firing CA1 pyramidal cells (Buzsáki *et al.* 1994). In my view the oscillations in this very high frequency range are manifestations of reciprocal discharges, what Amit (1995) calls “reverberations”, between neurons that are coupled at the microscopic level by mutual excitation. These data show the serial actions of neurons, not the concomitant massive interactions. The period of the oscillations in this range is determined by the transit time through each neuron, about 4 to 5 msec, not by the feedback delay in a negative feedback loop, which is more than twice as long for each cycle (Chapters 2 and 3). These brief intervals are seen in the repetitive peaks of firing probability in poststimulus time histograms such as those in Figures 1 and 2 in Chapter 8A, which are time ensemble averages of responses to impulse inputs, not self-organized patterns arising from internally evolving state transitions. In contrast, the oscillations in the gamma range (here described as 20–80 Hz) are at the mesoscopic level, so that they mediate the global interrelations among all points by which spatial patterns are to be defined and classified. Serial e·e·e·e transmission is required for the regenerative feedback that underlies explosive input-dependent state transitions to learned basins of attraction (Figure 14B in Freeman 1979b; Edelman and Freeman 1990), and the pulses in that process are tightly locked to the “ripple” field potentials (Siapas and Wilson 1998; Buzsáki *et al.* 1992, 1994; Traub *et al.* 1996), whereas the pulses in gamma AM patterns appear individually as Poisson-like point processes (for example, Figure 1 in Chapter 8B) that are only remotely locked to the carrier wave, thereby reflecting the process of mesoscopic generalization from instances to the class.

Studies are needed to determine the precise time window in which the effective Hebbian conjunction takes place, and whether the induced change is gradual or all-or-none. An excellent candidate for such studies is the reciprocal axosomatic synapse on the cell bodies of mitral and tufted cells discovered by Tommy Joe Willey (1973), which according to my evidence is modified with learning, as distinct from the dendrodendritic reciprocal synapse between the dendrites of mitral and granule cells discovered by Reese and Brightman (1965) and modeled by Rall and Shepherd (1968), which I believe is not modified with associative learning. The role of this synapse is clouded in controversy. Ramón y Cajal (1955) in mice observed mitral-to-mitral synaptic connections, which he inferred served to amplify sensory inputs by excitatory “avalanche conduction” (Freeman 1984). Contemporary neuroanatomists have failed to confirm his observation, despite physiological evidence for mutual excitation (Nicoll 1971; Freeman 1975) and the obvious investment of mitral cell bodies in the internal plexiform layer, whose axons are largely contributed by collaterals from mitral and tufted cells (Willey 1973). Mesoscopic dynamics requires that these topological connections exist and are modifiable with learning, even if anatomists can’t find them.

Links from Mesoscopic Assemblies to Macroscopic Brain Dynamics

A salient finding from brain imaging of metabolic activity and cerebral blood flow is the involvement of large swatches of cerebral cortex in conjunction with diverse cognitive tasks. The spatial resolution of brain imaging suggests that the spatial grain of these activated areas is minimally on the order of a centimeter. Phase cones and domains of high spatial coherence are under 2 centimeters in diameter (Menon *et al.* 1996), which makes them comparable to the size of Brodmann's areas. In contrast, minicolumns and macrocolumns have submillimeter diameters. These are functional groupings, usually unmarked by anatomical distinctions like those of the "barrels" in somatosensory cortices and the glomeruli in the olfactory bulb, that relate to the spatial coarse-graining of sensory inputs, not to the formation of mesoscopic order parameters. Concepts of single "pontifical" neurons (Sherrington 1941), collections of "cardinal cells" (Barlow 1972), and the "binding" of "feature detector" neurons (von der Malsburg 1983; Singer and Gray 1995) are ill-posed to explain the connection between the microscopic and macroscopic domains. The question is, how might mesoscopic domains become coordinated and form the macroscopic domains that may underlie the swatches in fMRI and PET scans?

The strongest direct evidence that mesoscopic AM patterns may lock together to form macroscopic states is provided by the correlation of gamma activity in scalp EEGs with the performance of cognitive tasks by volunteers (Sheer 1989; Tallon-Baudry *et al.* 1998; Müller *et al.* 1996; Rodriguez *et al.* 1999; Miltner *et al.* 1999). The effect of volume conduction is to smooth the fields of potential generated by neural populations, so that higher the frequency of the oscillations in EEG the more susceptible they are to attenuation by axonal delays causing phase dispersion. Gamma waves should be practically invisible in scalp EEGs. For many years the oscillations in the gamma range were thought to be solely due to electromyographic muscle potentials. The mere fact that episodes of gamma activity can be observed at all gives evidence for the existence of macroscopic order parameters that coordinate the mesoscopic AM patterns. Moreover, if the activities of widely dispersed neural populations undergo spatial integration in the process of transmission by divergent–convergent pathways to targets in the basal ganglia and the brainstem red nucleus and pons, then the most important components of the signals being integrated are those which are temporally coherent, which would make them accessible to scalp recording as well. Evidence for widespread synchronization has been derived by simultaneously recording pulse trains of pairs of neurons (Roelfsema *et al.* 1997) or EEG traces (Bressler 1995) from distant sites and showing that the maximum values on time-lagged cross-correlation occur at or near zero time lag.

These findings pose an intransigent problem in the explanation of phase relations in the gamma range. Synaptic communication between areas of cortex is by propagation of action potentials, which imposes time delays that are expected to appear as phase lags between oscillatory components. Even in difficult cases as, for example, transmission of a 40 Hz gamma wave at 10 meters/second over a distance of 1 centimeters, the time delay of 1 millisecond would appear as a 14.4° phase lag, which can be measured reliably with the techniques illustrated in Figure XI, A, in the comments introducing Chapter 9. Local cortical areas <1 square centimeter regularly sustain phase gradients that are fully consistent with the conduction velocities of axons running parallel to the pial surface (Freeman *et al.* 1995). Yet between distant areas separated by distances >2 centimeters, we and other groups find that the maximal time-lagged correlations between EEG traces, even between hemispheres, are at or near zero time delay, usually in intermittent bursts relating to stages in the performance of conditioned responses by trained animals (e.g. Bressler *et al.* 1993).

Several investigators have attempted with models of microscopic interactions to explain the challenging finding of "zero lag" correlation between the gamma oscillations of neurons separated by millimeter distances (Roelfsema *et al.* 1997). Usher *et al.* (1993) and Schillen and König (1994) have modeled this phenomenon by assuming that the feedback delay within a target matches the transmission delay between the transmitter and its target in an excitatory feedback network. Traub *et al.* (1996) have overcome some of the rigidity of that model by invoking doublet firing of single neurons, which they show is enhanced in states of high amplitude gamma. The tendency of neurons embedded in excitatory feedback networks to fire twice on impulse input is well known (Freeman 1975), and the increase in gain of the excitatory positive feedback loop with learning has been shown to increase gamma amplitude substantially (Freeman 1992a), so these models are consistent with the interpretation that gamma oscillations are mesoscopic, although they do not explain the quarter-cycle phase lag of the inhibitory neural firing with respect to the excitatory firing, which is essential for mesoscopic explanation (Freeman 1975, 1992a), nor do they explain observed local phase gradients in phase cones.

Much of the evidence for zero lag correlation is based on time ensemble averages. This evidence is suspect, because the only phase value that can emerge over multiple trials with randomly varying phase gradients is at or near zero within the standard error of measurement. A phase measurement made by time ensemble averaging over successive trials of traces taken from points located in the same AM pattern domain must approach zero as the number of traces averaged increases, because the apices of the phase cones vary randomly in location and sign. Bandpass temporal filtering and digitization of EEG amplitudes both inflate zero-lag correlation coefficients. The difference between "zero" and "near-zero" may not be appreciated. Many reports have not adequately tested the confidence intervals of phase estimates by sieve bootstrap and frequency-domain jackknife techniques. It has been common to see estimates of standard deviations of zero phase values given as "±3 msec", which at "40 Hz" equals ±43°, giving a 95% confidence range of 173° on the assumption of normal distributions of phase values. Also, the levels of correlations between pulse trains are usually very low, based on covariances on the order of 0.01–1.0% of the total variance, which are above chance levels by randomized controls (e.g. Abeles 1991; Wilson and McNaughton 1993) but can scarcely be said to demonstrate phase locking and synchronization. Since there is as yet no proven neural mechanism by which "zero lag correlation" can be maintained over the distances in question, the most conservative conclusion is that zero-lag correlation is a statistical artifact. That is difficult to prove, but even more so would be any "action-at-a-distance" brain mechanism by analogy with quantum nonlocality and with instantaneous gravitational interactions of astronomical bodies (Penrose 1994), though at scales of space and time that are intermediate between the very small and the very large.

Resolution of the problem of explaining mesoscopic neural cooperation in the face of finite neural conduction velocities of microscopic neurons may hinge on the property of anomalous dispersion, in which the group velocity of a state transition may exceed the velocity of energy transmission. An example is the fact that when a metal rod is struck at one end, the sound wave reaches the other end before the impulse does. Evidence from the olfactory bulb (Freeman 1990) shows that the state transition in the jump from one wing to another of a global attractor spreads over the entire bulb in a few milliseconds, well under a quarter cycle duration of the gamma oscillation, too rapidly to be explained by serial synaptic relays, but consistent with transmission by a small number of long axon collaterals, whose impulses can push a neural system across a separatrix when it is poised on the edge of a state transition from one chaotic basin to another. Time delays afforded by neocortical axons having conduction velocities on the order of 10 meters/second appear to be negligible in neural spatiotemporal integration, by which AM patterns form in a few milliseconds and last for a tenth of a second, but they can play havoc with methods of analysis that rely on linear correlations and mutual information from traces taken at the microscopic level and even at the mesoscopic level. Moreover, just as the neurons that are subject to mesoscopic order parameters retain a high degree of autonomy, so that their pulse trains appear to be largely random with respect to each other, it appears that mesoscopic domains need lose very little of their local autonomy in combining to form macroscopic order parameters. If this were so, brains would have no particular need for or reliance on temporal synchrony in cortical macroscopic dynamics, though a modest degree of synchrony wouldn't be an impediment either, and might persist at low levels sufficient to induce researchers to spend their research careers in time-dependent linear correlation and coherence studies.

The real work of the neocortex appears to be the formation of successive spatial patterns of activity, regardless of the temporal structures of the carrier waves. This hypothesis poses major problems for devising appropriate techniques for determining the spatiotemporal relations between macroscopic EEGs from the scalp and mesoscopic EEGs simultaneously recorded from the cortical surfaces with intracranial electrodes. The presence of gamma activity in scalp EEGs suggests that the problems can be solved, and the rapidity with which animal and human cognitive processes proceed indicates that they must be solved in order to make quantitative studies of the cerebral neurodynamics. The preferred subjects for continuing studies of gamma activity are awake humans with noninvasive recordings. A major limitation in using animals to study cognition involving intentional behavior is that they can't talk. We cannot substantively relate gamma AM patterns to animal cognition, because we can't know what the animals are thinking. The major constraint in the study of human subjects is that the AM patterns are too small and too finely textured to be visible in scalp recordings. The arrays of electrodes needed to map them must be placed directly onto the pial surface of the brains of awake neurosurgical patients (Menon *et al.* 1996) in circumstances where the welfare of the patients is paramount. Detailed studies are needed of the spectral properties of scalp and subdural EEGs (Freeman *et al.* 2000), so that proposed transfer functions (Nunez 1982, 1995) can be tested and evaluated, and the requirements for intracranial data can be minimized. Such studies should give new tools for studying the neurodynamics of normal human cognition, so that the modes of failure in abnormal conditions leading pathological functions might be better understood.

A projected early application of mesoscopic dynamics to gamma activity is to use identifiable AM domains as an assay of the states of consciousness in surgically anesthetized patients and, by implication, animals. No biological assay of consciousness presently exists (Merikle and Daneman 1998); the only method to determine whether someone is or was conscious is to ask them, and when drugs such as rohypnol have been taken or administered, even that test becomes unreliable. Another area of application is in patients with the rare but distressing "locked-in" syndrome, which can accompany brain damage from accidents, penetrating wounds, infections and vascular occlusions. Conventional recordings of EEGs that reveal near-normal sleep–wake cycles suggest that perhaps as many as half of these patients are awake and aware, but are totally paralyzed, or can move only an eyelid to indicate that they are sentient (Firsching 1998; Wijdicks and Miller 1999), owing to sparing of the oculomotor nerve connections in the upper brain stem. Some patients recover and tell what they experienced; others are forever lost. Better tests are needed for diagnosis of this condition, and for research on new methods of treatment. These efforts may possibly lead to a device enabling the locked-in individuals to communicate with us if they can learn to modulate and control their gamma activity. Beyond the medical setting, such a device with an incredibly broad bandwidth for rapid data transmission might find uses that have to date only been imagined in science fiction (Thomas 1977).

References

Abeles M (1991) *Corticonics: Neural Circuits of the Cerebral Cortex*. Cambridge UK: Cambridge University Press.

Abeles M, Bergman H, Gat I, Meilijson I *et al.* (1998) Cortical activity flips among quasi-stationary states. *Proceedings, National Academy of Sciences USA* **92**: 8616–8620.

Adrian ED (1950) The electrical activity of the mammalian olfactory bulb. *Electroencephalography and clinical Neurophysiology* **2**: 377–388.

Ahn SM, Freeman WJ (1975) Neural dynamics under noise in the olfactory system. *Biological Cybernetics* **17**: 165–168.

Aihara K, Takabe T, Toyoda M (1990) Chaotic neural networks. *Physics Letters A* **144**: 333–340.

Amari S-I. (1977) Dynamics of pattern formation in lateral-inhibition type of neural fields. *Biological Cybernetics* **27**: 77–87.

Amit DJ (1989) *Modeling brain function: The World of Attractor Neural Networks*. Cambridge UK: Cambridge University Press.

Amit DJ (1995) The Hebbian paradigm reintegrated: local reverberations as internal representations. *Behavioral and Brain Sciences* **18**: 617–618.

Amit DJ (1998) Simulation in neurobiology: theory or experiment? (Viewpoint) *Trends in Neurosciences* **21**: 231–237.

Andersen P, Andersson SA (1968) *Physiological Basis of the Alpha Rhythm*. New York: Appleton-Croft-Century.

Anderson JA, Silverstein JW, Ritz SR, Jones RS (1977) Distinctive features, categorical perception, and probability learning: Some applications of a neural model. *Psychological Review* **84**: 413–451.

Axel R (1995) The molecular logic of smell. (Genetics of the human olfactory sense) *Scientific American* **273**: 154–159.

Baars B (1988) *A Cognitive Theory of Consciousness*. Cambridge UK: Cambridge University Press.

Babloyantz A, Destexhe A (1986) Low-dimensional chaos in an instance of epilepsy. *Proceedings, National Academy of Sciences USA* **83**: 3513–3517.

Barlow HB (1972) Single units and sensation: A neuron doctrine for perceptual psychology? *Perception* **1**: 371–394.

Barlow JS (1993) *The Electroencephalogram: Its Patterns and Origins*. Cambridge MA: MIT Press.

Basar E (1980) *EEG Brain Dynamics. Relation between EEG and brain evoked potentials*. Amsterdam: Elsevier.

Basar E (1998) *Brain Function and Oscillations*. Berlin: Springer-Verlag.

Basar E, Bullock TH (eds.) (1989) *Brain Dynamics*. Berlin: Springer-Verlag.

Bear MF, Cooper LN (1998) From molecules to mental states. *Daedalus* **127**: 131–144.

Becker CJ, Freeman WJ (1968) Prepyriform electrical activity after loss of peripheral or central input or both. *Physiology and Behavior* **3**: 597–599.

Biedenbach MA (1966) Effects of anesthetics and cholinergic drugs on prepyriform electrical activity in cats. *Experimental Neurology* **16**: 464–479.

Biedenbach MA, Freeman WJ (1965) Linear domain of potential from the prepyriform cortex with respect to stimulus parameters. *Experimental Neurology* **11**: 400–417, 1965.

Boudreau JC. Freeman WJ (1963) Spectral analysis of electrical activity in the prepyriform cortex of cat. *Experimental Neurology* **8**: 423–430.

Bower JM, Beeman D (1995) *The Book of GENESIS: Exploring Realistic Neural Models with NEural SImulation System*. Santa Clara CA: TELOS.

Braitenberg V, Schüz A (1991) *Anatomy of the Cortex: Statistics and Geometry*. Berlin: Springer-Verlag.

Bressler SL (1988) Changes in electrical activity of rabbit olfactory bulb and cortex to conditioned odor stimulation. *Journal of Neurophysiology* **62**: 740–747.

Bressler SL (1995) Large-scale cortical networks and cognition. *Brain Research Reviews* **20**: 288–304.

Bressler SL, Coppola R, Nakamura R (1993) Episodic multiregional cortical coherence at multiple frequencies during visual task performance. *Nature* **366**: 153–156.

Bressler SL, Freeman WJ (1980) Frequency analysis of olfactory system EEG in cat, rabbit and rat. *Electroencephalography and clinical Neurophysiology*. **50**: 9–24.

Buckley S, Stelson S (1978) Phase monitoring for automated inspection, positioning and assembly. *Transactions, Society of Manufacturing Engineering* **6**: 56–63.

Bullock TH, McClune MC, Achimowicz JZ, Iragui-Madoz VJ, Duckrow RB, Spencer SS (1995) EEG coherence has structure in the millimeter domain: subdural and hippocampal recordings from epileptic patients. *Electroencephalography and clinical Neurophysiology* **95**: 161–177.

Bulsara AR, Maren AJ, Schmera G (1993) Single effective neuron: dendritic coupling effects and stochastic resonance. *Biological Cybernetics* **70**: 145–156.

Burns, BD (1958) *The Mammalian Cerebral Cortex*. London: Arnold.

Buzsáki G. (1996) The hippocampal-neocortical dialogue. *Cerebral Cortex* **6**: 81–92.

Buzsáki G, Horváth Z, Urioste R, Hetke J, Wise K (1992) High-frequency network oscillation in the hippocampus. *Science* **256**: 1025–1027.

Buzsáki G, Bragin A, Chrobak JJ, Nádasdy Z, Sik A, Hsu M, Ylinen A (1994) Oscillatory and intermittent synchrony in the hippocampus: Relevance to memory trace formation. In: Buzsáki G, Llinás R, Singer W, Berthoz A, Christen Y (eds.) *Temporal Coding in the Brain*. Berlin: Springer-Verlag, pp 145–172.

Calvin WH (1996) *The Cerebral Code. Thinking a Thought in the Mosaics of the Mind*. Cambridge MA: MIT Press.

Chang H-T (1951) Dendritic potential of cortical neurons produced by direct electrical stimulation of the cerebral cortex. *Journal of Neurophysiology* **14**: 1–21.

Chang H-J, Freeman WJ (1996) Parameter optimization in models of the olfactory system. *Neural Networks* **9**: 1–14.

Chang H-J, Freeman WJ (1998a) Optimization of olfactory model in software to give 1/*f* power spectra reveals numerical instabilities in solutions governed by aperiodic (chaotic) attractors. *Neural Networks* **11**: 449–466.

Chang H-J, Freeman WJ (1998b) Biologically modeled noise stabilizing neurodynamics for pattern recognition. *International Journal of Bifurcation and Chaos* **8**: 321–345.

Chang H-J, Freeman WJ (1999) Local homeostasis stabilizes a model of the olfactory system globally in respect to perturbations by input during pattern classification. *International Journal of Bifurcation and Chaos* **8**: 2107–2123.

Chen G, Trombley PQ, Vandenpol AN (1996) Excitatory actions of GABA in developing rat hypothalamic neurones. *Journal of Physiology-London* **15**: 451–464.

Churchland PS, Sejnowski TJ (1992) *The Computational Brain*. Cambridge MA: MIT Press.

Cooper LN (1995) *How We Learn, How We Remember: Toward an Understanding of Brain and Neural Systems*. Singapore: World Scientific.

Davis GW, Freeman WJ (1982) On-line detection of respiratory events applied to behavioral conditioning in rabbits. *IEEE Transactions on Biomedical Engineering* **29**: 453–456.

DeMott DW (1966) Cortical micro-toposcopy. *Medical Research Engineering* **5**: 23–29.

DeMott DW (1970) *Toposcopic Studies of Learning*. Springfield IL: CC Thomas.

Dogaru R, Chua LO (1998) Edge of chaos and local activity domains of Fitzhugh–Nagumo equation. *International Journal of Bifurcation and Chaos* **8**: 211–257.

Eastman C. (1975) Construction of miniature electrode arrays for recording from cortical surface. *Journal of Electrophysiological Techniques* **10**: 28–30.

Eeckman FH, Freeman WJ (1990) Correlations between unit firing and EEG in the rat olfactory system. *Brain Research* **528**: 238–244.

Eeckman FH, Freeman WJ (1991) Asymmetric sigmoid nonlinearity in the rat olfactory system. *Brain Research* **557**: 13–21.

Eckhorn R (1991) Principles of global visual processing of local features can be investigated with parallel single-cell- and group-recordings from the visual cortex. In Aertsen A, Braitenberg V (Eds.), *Information Processing in the Cortex* (pp. 385–420). Berlin: Springer-Verlag.

Eckhorn R, Bauer R, Jordan W, Brosch M, Kruse W, Munk M, Reitböck HJ (1988) Coherent oscillations: A mechanism of feature linking in visual cortex? *Biological Cybernetics* **60**: 121–130.

Edelman JA, Freeman WJ (1990) Simulation and analysis of a model of mitral-granule cell population interactions in the mammalian olfactory bulb. *Proceedings IJCNN* **I**: 62–65.

Eichenbaum H, Otto T, Cohen NJ (1992) The hippocampus - what does it do? *Behavioral and Neural Biology* **57**: 2–36.

Eisenberg J, Freeman WJ, Burke B (1989) Hardware architecture of a neural network model simulating pattern recognition by the olfactory bulb. *Neural Networks* **2**: 315–325.

Elul R (1972) The genesis of the EEG. *International Review of Neurobiology* **15**: 227–272.

Emery JD, Freeman, WJ (1969) Pattern analysis of cortical evoked potential parameters during attention changes. *Physiology and Behavior* **4**: 67–77.

Engel AK, Kreiter AK, König P, Singer W (1991) Synchronization of oscillatory neuronal responses between striate and extrastriate visual cortical areas of the cat. *Proceedings, National Academy Sciences USA* **88**: 6048–6052.

Engel AK, König P, Kreiter AK, Singer W (1991) Interhemispheric synchronization of oscillatory neuronal responses in cat. *Science* **252**: 1177–1179.

Firsching R (1998) Moral dilemmas of tetraplegia; the 'locked-in' syndrome, the persistent vegetative state and brain death. *Spinal Cord* **36**: 741–743.

Fox RF (1997) Stochastic versions of the Hodgkin–Huxley equations. *Biophysical Journal* **72**: 2068–2074.

Freeman WJ, Hemingway A (1958) Nervous control of shivering. Further observations on brainstem unit potential during shivering. *Air Force Technical Report* 58–4.

Freeman WJ (1959) Distribution in time and space of prepyriform electrical activity. *Journal of Neurophysiology* **22**: 644–665.

Freeman WJ (1960a) Correlation of electrical activity of prepyriform cortex and behavior in cat. *Journal of Neurophysiology* **23**: 111–131.

Freeman WJ (1960b) Repetitive electrical stimulation of prepyriform cortex in cat. *Journal of Neurophysiology* **23**: 383–396.

Freeman WJ (1961a) A critique of the doctrine of "centers" in temperature regulation. In: *Symposia on Arctic Biology and Medicine, Arctic Aeromedical Lab. Ladd Air Force Base, Fairbanks, Alaska. I. Neural Factors in Temperature Regulation*. pp. 255–294.

Freeman WJ (1961b) *Q* meter for measuring frequency specificity of cortical reactivity to electrical stimulation. *Journal of Applied Physiology* **16**: 750–751.

Freeman WJ (1962) Linear approximation of prepyriform evoked potential in cats. *Experimental Neurology* **5**: 477–499.

Freeman WJ (1964a) Use of digital adaptive filters for measuring prepyriform evoked potentials from cats. *Experimental Neurology* **10**: 475–492.

Freeman WJ (1964b) A linear distributed feedback model for prepyriform cortex. *Experimental Neurology* **10**: 525–547.

Freeman WJ (1964c) Correlation of goal-directed work with sensory cortical excitability. *Recent Advances in Biological Psychiatry* **7**: 234–250.

Freeman WJ (1967a) Analysis of function of cerebral cortex by use of control systems theory. *Logistics Review* **3**: 5–40.

Freeman WJ (1967b) Statistical input–output relations of single cortical neurons. *The Physiologist* **10**: 172.

Freeman WJ (1968a) Patterns of variation in waveform of averaged evoked potentials from prepyriform cortex of cats. *Journal of Neurophysiology* **31**: 1–13.

Freeman WJ (1968b) Relations between unit activity and evoked potentials in prepyriform cortex of cats. *Journal of Neurophysiology* **31**: 337–348..

Freeman WJ (1968c) Effects of surgical isolation and tetanization on prepyriform cortex in cats. *Journal of Neurophysiology* **31**: 349–357.

Freeman WJ (1968d) Analog simulation of prepyriform cortex in the cat. *Mathematical BioScience* **2**: 181–190.

Freeman WJ (1970) Spectral analysis of prepyriform averaged evoked potentials in cats. *Journal of Biomedical Systems* **1**: 3–22.

Freeman WJ (1972a) Spatial divergence and temporal dispersion in primary olfactory nerve of cat. *Journal of Neurophysiology* **35**: 733–761.

Freeman WJ (1972b) Measurement of open-loop responses to electrical stimulation in olfactory bulb of cat. *Journal of Neurophysiology* **35**: 745–761.

Freeman WJ (1972c) Measurement of oscillatory responses to electrical stimulation in olfactory bulb of cat. *Journal of Neurophysiology* **35**: 762–779.

Freeman WJ (1972d) Depth recording of averaged evoked potential of olfactory bulb. *Journal of Neurophysiology* **35**: 780–796.

Freeman WJ (1972e) Linear analysis of the dynamics of neural masses. *Annual Review of Biophysics and Bioengineering* **1**: 225–226.

Freeman WJ (1972f) Waves, pulses and the theory of neural masses. *Progress in Theoretical Biology* **2**: 87–165.

Freeman WJ (1974a) Topographic organization of primary olfactory nerve in cat and rabbit as shown by evoked potentials. *Electroencephalography and clinical Neurophysiology* **36**: 33–45.

Freeman WJ (1974b) Average transmission distance from mitral-tufted to granule cells in olfactory bulb. *Electroencephalography and clinical Neurophysiology*. **36**: 609–618.

Freeman WJ (1974c) Attenuation of transmission through glomeruli of olfactory bulb on paired shock stimulation. *Brain Research* **65**: 77–90.

Freeman WJ (1974d) Relation of glomerular neuronal activity to glomerular transmission attenuation. *Brain Research* **65**: 91–107.

Freeman WJ (1974e) A model for mutual excitation in a neuron population in olfactory bulb. *IEEE Transactions on Biomedical Engineering* **21**: 350–358.

Freeman WJ (1974f) Stability characteristics of positive feedback in a neural population. *IEEE Transactions on Biomedical Engineering* **21**: 358–364.

Freeman WJ (1975) *Mass Action in the Nervous System*. New York NY: Academic Press.

Freeman WJ (1978a) Spatial properties of an EEG event in the olfactory bulb and cortex. *Electroencephalography and clinical Neurophysiology* **44**: 586–605.

Freeman WJ (1978b) Spatial frequency analysis of an EEG event in the olfactory bulb. In: *Multidisciplinary Perspectives in Event-Related Brain potential Research*. Otto DA (ed.) U.S. Govt. Printing Office, EPA-600/9-77-043. pp. 531–546.

Freeman WJ (1979a) Nonlinear gain mediating cortical stimulus-response relations. *Biological Cybernetics* **33**: 237–247.

Freeman WJ (1979b) Nonlinear dynamics of paleocortex manifested in the olfactory EEG. *Biological Cybernetics* **35**: 21–37.

Freeman WJ (1979c) EEG analysis gives model of neuronal template-matching mechanism for sensory search with olfactory bulb. *Biological Cybernetics* **35**: 221–234.

Freeman WJ (1980) Use of spatial deconvolution to compensate for distortion of EEG by volume conduction. *IEEE Transactions on Biomedical Engineering* **27**: 421–429.

Freeman WJ (1983) Dynamics of self-organizing neural systems in attention and perception of rabbits; A reflection on Katchalsky's contribution. In: *Biological Structures and Coupled Flows*. Oplatka A, Balaban M (eds.) New York NY: Academic Press. pp. 493–497.

Freeman WJ (1984) Premises in neurophysiological studies of learning. Chapter 13 in: *Neurobiology of Learning and Memory*. Lynch G, McGaugh JL, Weinberger NM (eds.) New York NY: Guilford Press. pp. 231–234.

Freeman WJ (1986) Petit mal seizure spikes in olfactory bulb and cortex caused by runaway inhibition after exhaustion of excitation. *Brain Research Reviews* **11**: 259–284.

Freeman WJ (1987a) Techniques used in the search for the physiological basis of the EEG. In: Gevins A, Rémond A (eds.) *Handbook of Electroencephalography and clinical Neurophysiology*. Vol 3A, Part 2, Ch. 18. Amsterdam, Elsevier. pp. 583–664.

Freeman WJ (1987b) Simulation of chaotic EEG patterns with a dynamic model of the olfactory system. *Biological Cybernetics* **56**: 139–150.

Freeman WJ (1988a) Pattern learning and recognition device. *United States Patent # 4,748,674*, May 31, 1988.

Freeman WJ (1988b) Strange attractors that govern mammalian brain dynamics as shown by trajectories of electroencephalographic (EEG) potential. *IEEE Transactions on Circuits and Systems* **35**: 781–783.

Freeman WJ (1990) On the problem of anomalous dispersion in chaoto-chaotic phase transitions of neural masses, and its significance for the management of perceptual information in brains. In: Haken H, Stadler M (eds.) *Synergetics of Cognition*. Berlin: Springer-Verlag. Volume **45**: 126–143.

Freeman WJ (1991a) The physiology of perception. *Scientific American* **264**: 78–87.

Freeman WJ (1991b) Colligation of coupled cortical oscillators by the collapse of the distributions of amplitude-dependent characteristic frequencies. Chapter 5 in: Gear CW (ed.) *First NEC Research Symposium*. Philadelphia PA: SIAM, pages 69–103.

Freeman WJ (1991c) Insights into processes of visual perception from studies in the olfactory system. Chapter 2 in: Squire L, Weinberger NM, Lynch G, McGaugh JL (eds.) *Memory: Organization and Locus of Change*. New York NY: Oxford University Press. pp. 35–48.

Freeman WJ (1992a) Tutorial in Neurobiology: From Single Neurons to Brain Chaos. *International Journal of Bifurcation and Chaos* **2**: 451–482.

Freeman WJ (1992b) Predictions in neocortical dynamics that are posed by studies in paleocortex. Chapter 9 in: Basar E, Bullock TH (eds.) *Induced Rhythms of the Brain*. Cambridge MA: Birkhaeuser Boston. pp. 183–199.

Freeman WJ (1993) Valium, histamine and neural networks. Editorial. *Biological Psychiatry* **34**: 1–2.

Freeman WJ (1995) *Societies of Brains*. Mahwah NJ: Lawrence Erlbaum Associates.

Freeman WJ (1997) Three centuries of category errors in studies of the neural basis of consciousness and intentionality. *Neural Networks* **10**: 1175–1183.

Freeman WJ (1999a) *How Brains Make Up Their Minds*. London UK: Weidenfeld and Nicolson, Orion Press.

Freeman WJ (1999b) Il sistema olfattivo: rilevamento e classificazione degli odori. Chapter in: *Frontiere della vita*, vol. 3, pp. 477–494. Roma, Italia: Istituto della Enciclopedia Italiana.

Freeman WJ (2000) Consciousness, intentionality, and causality. *Journal of Consciousness Studies* **96**(1): xxx–xxx.

Freeman WJ, Baird B (1987) Relation of olfactory EEG to behavior: Spatial analysis. *Behavioral Neuroscience* **101**: 393–408.

Freeman WJ, Baird B (1989) Effects of applied electric current fields on cortical neural activity. Chapter in: Schwartz, E. (ed.) *Computational Neuroscience*. New York NY: Plenum Press. pp. 274–287.

Freeman WJ, Barrie JM (1994) Chaotic oscillations and the genesis of meaning in cerebral cortex. In: Buzsáki G, Llinás R, Singer W, Berthoz A, Christen Y (eds.) *Temporal Coding in the Brain*. Berlin: Springer-Verlag, pp 13–37.

Freeman WJ, Barrie JM, Lenhart M, Tang RX (1995) Spatial phase gradients in neocortical EEGs give modal diameter of "binding" domains in perception. *Abstracts, Society for Neuroscience* **21**: 1649 (648.13).

Freeman WJ, Davis DD (1959) Effects on cats of conductive hypothalamic heating and cooling. *American Journal of Physiology* **197**: 145–148.

Freeman WJ, Davis GW (1990) Olfactory EEG changes under serial discrimination of odorants by rabbits. Chapter in: Schild D (ed.) *Chemosensory Information Processing*. Berlin: Springer-Verlag. pp. 375–391.

Freeman WJ, Grajski KA (1987) Relation of olfactory EEG to behavior: Factor analysis. *Behavioral Neuroscience* **101**: 766–777.

Freeman WJ , Maurer K (1989) Advances in brain theory give new directions to the use of the technologies of brain mapping in behavioral studies. Chapter in: Maurer K (ed.) *Proceedings, Conference on Topographic Brain Mapping*. Berlin: Springer-Verlag. pp. 118–126.

Freeman WJ, Rogers LJ, Holmes MD, Silbergeld DL (1999) Spatial spectral analysis of human electrocorticograms including the alpha and gamma bands. *Journal of Neuroscience Methods* **7**(1): xxx–xxx.

Freeman WJ, Schneider W (1982) Changes in spatial patterns of rabbit olfactory EEG with conditioning to odors. *Psychophysiology* **19**: 44–56.

Freeman WJ, Shimoide K (1994) New approaches to nonlinear concepts in neural information processing: Parameter optimization in a large-scale biologically plausible cortical network. Chapter 7 in: Zornitzer SF, Davis JL, Lau C, and McKenna T (eds.) *An Introduction to Neural and Electronic Networks*, 2nd Edition. New York NY: Academic Press. pp. 119–137.

Freeman WJ, Skarda CA (1985) Nonlinear dynamics, perception, and the EEG; the neo-Sherringtonian view. *Brain Research Reviews* **10**: 147–175.

Freeman WJ, Skarda CA (1989) Representations: Who needs them? In: *Third Conference, Brain Organization and Memory: Cells, Systems and Circuits*. McGaugh JL, Weinberger NM, Lynch G (eds.) New York: Guilford Press pp. 375–380.

Freeman WJ, Skarda CA (1990) Mind/Brain Science: Neuroscience on philosophy of mind. Chapter 7 in: Lepore E, van Gulick R. (eds.) *John Searle and his Critics*. Oxford UK: Blackwell pp. 115–127.

Freeman WJ, Van Dijk B (1987) Spatial patterns of visual cortical fast EEG during conditioned reflex in a rhesus monkey. *Brain Research*: **422**: 267–276.

Freeman WJ, Viana Di Prisco G. (1986) Relation of olfactory EEG to behavior: Time series analysis. *Behavioral Neuroscience* **100**: 753–763.

Freeman WJ, Viana Di Prisco G (1986) EEG spatial pattern differences with discriminated odors manifest chaotic and limit cycle attractors in olfactory bulb of rabbits. In: *Brain Theory*. Palm G, Aertsen A (eds.) Berlin: Springer-Verlag. pp. 97–119.

Freeman WJ, Viana Di Prisco G, Davis GW, Whitney TM (1983) Conditioning of relative frequency of sniffing by rabbits to odors. *Journal of Comparative Psychology* **97**: 12–23.

Freeman WJ, Yao Y, Burke B (1988) Central pattern generating and recognizing in olfactory bulb: A correlation learning rule. *Neural Networks* **1**: 277–288.

Gevins A, Smith ME, McEvoy L, Yu D (1997) High resolution EEG mapping of cortical activation related to working memory: effects of task difficulty, type of processing, and practice. *Cerebral Cortex* **7**: 374–385.

Georgeopolis AP, Schwartz AB, Kettner RE (1986) Neural population coding of movement direction. *Science* **233**: 1416–1419.

Ghose GM, Freeman RD (1992) Oscillatory discharge in the visual system - Does it have a functional role? *Journal of Neurophysiology* **68**: 1558–1574.

Ghose GM, Freeman RD (1997) Intracortical connections are not required for oscillatory activity in the visual cortex. *Visual Neuroscience* **14**: 963–979.

Gonzalez RC, Wintz P(1977) *Digital Image Processing*. Reading MA: Addison-Wesley.

Gonzalez-Estrada TM, Freeman WJ (1980) Effects of carnosine on olfactory bulb EEG, evoked potentials, and D.C. potentials. *Brain Research* **202**: 373–386.

Gray CM, Freeman WJ, Skinner JE (1986) Chemical dependencies of learning in the rabbit olfactory bulb: Acquisition of the transient spatial-pattern change depends on norepinephrine. *Behavioral Neuroscience* **100**: 585–596.

Gray CM, Koenig P, Engel KA, Singer W (1989) Oscillatory responses in cat visual cortex exhibit intercolumnar synchronization which reflects global stimulus properties. *Nature* **338**: 334–337.

Gray CM, McCormick DA (1996) Chattering cells – superficial pyramidal neurons contributing to the generation of synchronous oscillations in the visual cortex. *Science* **274**: 109–113.

Gray CM, Skinner JE (1988) Centrifugal regulation of neuronal activity in the olfactory bulb of the waking rabbit as revealed by reversible cryogenic blockade. *Experimental Brain Research* **69**: 378–386.

Grossberg S (1975) A neural model of attention, reinforcement, and discrimination learning. *International Review of Neurobiology* **18**: 263–327.

Grossberg S (ed.) (1988) *Neural Networks and Natural Intelligence*. Cambridge MA: MIT Press.

Grossberg S (1990) Content-addressable memory storage by neural networks: A general model and global Liapunov method. Chapter 6 in Schwartz EL (ed.) *Computational Neuroscience*. Cambridge MA: MIT Press. pp. 56–65.

Haberly LB, Bower JM (1989) Olfactory cortex – model circuit for study of associative memory. *Trends in NeuroSciences* **12**: 258–264.

Haken H (1983) *Synergetics: An Introduction*. Berlin: Springer-Verlag.

Haken H (1999) What can synergetics contribute to the understanding of brain functioning? In: *Analysis of Neurophysiological Brain Functioning*. Uhl C (ed.) Berlin: Springer-Verlag. pp. 7–40.

Hayashi H, Ishizuka S (1995) Chaotic responses of the hippocampal CA3 region to a mossy fiber stimulation in vitro. *Brain Research* **686**: 194–206.

Hayman LA (1992) *Clinical Brain Imaging: Normal Structure and Functional Anatomy*. St. Louis MO: Mosby Year Book.

Hebb DO (1949) *The Organization of Behavior: A Neuropsychological Theory*. New York NY: Wiley.

Herrick CJ (1948) *The Brain of the Tiger Salamander*. Chicago IL: University of Chicago Press.

Hodgkin AL (1964) *The Conduction of the Nervous Impulse*. Springfield IL: CC Thomas.

Hopfield JJ (1982) Neuronal networks and physical systems with emergent collective computational abilities. *Proceedings, National Academy of Sciences USA* **81**: 3058–3092.

Horowitz J, Freeman WJ, Stoll PJ (1973) A neural network with a background level of excitation in the cat hippocampus. *International Journal of Neuroscience* **5**: 113–123.

Imry Y(1997) *Introduction to Mesoscopic Physics*. New York NY: Oxford University Press.

Ingber L (1992) Generic mesoscopic neural networks based on statistical mechanics of Neocortical interactions. *Physical Review A* **45**: R2183–2186.

Ingber L (1995a) Statistical mechanics of neocortical interactions – constraints on 40 Hz models of short-term memory. *Physical Review E* **52**: 4561–4563.

Ingber L (1995b) Statistical mechanics of multiple scales of neocortical interactions. In: Nunez PL, *Neocortical Dynamics and Human EEG Rhythms*. New York NY: Oxford University Press. pp. 628–681.

Ingber L (1997) Statistical mechanics of neocortical interactions: Canonical moments indicators of electroencephalography. *Physical Review E* **55**: 4578–4593.

Jackson JH (1884/1958) Evolution and dissolution of the nervous system: Lecture III. (Taylor J, ed.) *Selected Writings*. New York NY: Basic Books.

Johnston D, Wu SM (1995) *Foundations of Cellular Neurophysiology*. Cambridge MA: MIT Press.

Joliot M, Ribary U, Llinás R (1994) Human oscillatory brain activity near 40 Hz coexists with cognitive temporal binding. *Proceedings, National Academy of Sciences USA* **91**: 11748–11751.

Kammermans M, Werblin F (1992) GABA-mediated positive autofeedback loop controls horizontal cell kinetics in tiger salamander retina. *Journal of Neuroscience* **12**: 2451–2463.

Kaneko K (1997) Dominance of Milnor attractors and noise-induced selection in a multiattractor system. *Physical Review Letters* **78**: 2736–2739.

Kaneko K (1998) On the strength of attractors in a high-dimensional system: Milnor attractor network, robust global attraction, and noise-induced selection. *Physica D* **124**: 308–330.

Kay LM, Freeman WJ (1998) Bidirectional processing in the olfactory-limbic axis during olfactory behavior. *Behavioral Neuroscience* **112**: 541–553.

Kelso JAS (1995) *Dynamic Patterns: The Self-organization of Brain and Behavior*. Cambridge MA: MIT Press.

Koch C, Laurent G (1999) Complexity and the nervous system. *Science* **284**: 96–98.

Koffka K (1935) *Principles of Gestalt Psychology*. New York NY: Harcourt Brace.

Köhler W (1940) *Dynamics in Psychology*. New York NY: Grove Press.

Kohonen T (1984) *Self-organization and Associative Memory*. Berlin: Springer-Verlag.

König P, Schillen TB (1991) Stimulus-dependent assembly formation of oscillatory responses. 1. Synchronization. *Neural Computation* **3**: 155–166.

Kreiter AK, Aertsen AMHJ, Gerstein GL (1989) A low-cost single-board solution for real-time unsupervised waveform classification of multineuron recordings. *Journal of Neuroscience Methods* **30**: 59–69.

Kruse W, Eckhorn R (1996) Inhibition of sustained gamma oscillations (35–80 Hz) by fast transient responses in cat visual cortex. *Proceedings, National Academy of Sciences USA* **93**: 6112–6117.

Lancet D, Greer CA, Kauer JS, Shepherd GM (1982) Mapping of odor-related neuronal activity in the olfactory bulb by high-resolution 2-deoxyglucose autoradiography. *Proceedings, National Academy of Sciences USA* **79**: 670–674.

Lesse H (1957) Amygdaloid electrical activity during a conditioned response. *Proceedings, 4th International Congress of Electroencephalography and clinical Neurophysiology*, Brussels: 99–100.

Leung LS (1992) Fast (beta) rhythms in the hippocampus: A review. *Hippocampus* **2**: 93–98.

Liley DTJ. Alexander DM, Wright JJ, Aldous MD (1999) Alpha rhythm emerges from large-scale networks of realistically coupled multicompartmental model cortical neurons. *Network-Computation in Neural Systems* **10**: 79–92.

Lilly JC (1954) Instantaneous relations between the activities of closely spaced zones on the cerebral cortex - electrical figures during responses and spontaneous activity. *American Journal of Physiology* **176**: 493–504.

Livanov MN (1977) *Spatial Organization of Cerebral Processes. Israel Program for Scientific Translations Staff* (trans.) New York NY: Wiley.

Llinás R (1990) Intrinsic electrical properties of nerve cells and their role in network oscillation. *Cold Spring Harbor Symposia on Quantitative Biology* **55**: 933–938.

Llinás R, Ribary U (1993) Coherent 40-Hz oscillation characterizes dream state in humans. *Proceedings, National Academy of Sciences USA* **90**: 2078–2081.

Llinás R, Ribary U, Contreras D, Pedroarena C (1998) The neuronal basis for consciousness. *Philosophical Transactions, Royal Society London B* **353**: 1841–1849.

Lopes da Silva F (1991) Neural mechanisms underlying brain waves: from neural membranes to networks. *Electroencephalography and clinical Neurophysiology* **79**: 81–93.

Lorente de Nó R. (1934) Studies in the structure of the cerebral cortex. *Journal von Psychologie und Neurologie* **45**: 381–438.

Luchins DJ (1990) Task force on quantitative electrophysiological assessment. *Psychiatric Research Report* **5**: 3–13.

Lumer ED, Friston KJ, Rees G (1998) Neural correlates of perceptual rivalry in the human brain. *Science* **280**: 1930–1934.

Luskin MB, Price JL (1983) The topographic organization of associated fibers of the olfactory system of the rat, including centrifugal fibers to the olfactory bulb. *Journal of Comparative Neurology* **216**: 264–291.

Maclean PD (1969) *The Triune Brain*. New York NY: Plenum Press.

Malnic B, Hirono J, Sato T, and Buck LB (1999) Combinatorial receptor codes for odors. *Cell* **96**: 713–23

Margolis FL (1974) Carnosine in the primary olfactory pathway. *Science* **184**: 909–911.

Martignon L, Von Hasseln H, Grun S, Aertsen A (1995) Detecting higher-order interactions among the spiking events in a group of neurons. *Biological Cybernetics* **73**: 69–81.

Martinez DM, Freeman WJ (1984) Periglomerular cell action on mitral cell in olfactory bulb shown by current source density analysis. *Brain Research*: 223–233.

Matsunami H and Buck LB (1997) A multigene family encoding a diverse array of putative pheromone receptors in mammals. *Cell* **90**: 775–784.

McCulloch WS, Pitts WH (1943) A logical calculus of ideas immanent in nervous activity. *Bulletin of Mathematical Biophysics* **5**: 115–133.

Menon V, Freeman WJ, Cutillo BA, Desmond JE, Ward MF, Bressler SL, Laxer KD, Barbaro NM, Gevins AS (1996) Spatio-temporal correlations in human gamma band electrocorticograms. *Electroencephalography and clinical Neurophysiology* **98**: 89–102.

Merikle PM, Daneman M (1998) Psychological investigations of unconscious perception. *Journal of Consciousness Studies* **5**: 5–18.

Michelson HB, Wong RKS (1991) Excitatory synaptic responses mediated by GABA(A) receptors in the hippocampus. *Science* **5026**: 1420–1423.

Milner PM (1974) A model for visual shape recognition. *Psychological Review* **81**: 521–535.

Miltner WHR, Barun C, Arnold M, Witte H, Taub E (1999) Coherence of gamma-band EEG activity as a basis for associative learning. *Nature* **397**: 434–436.

Miyashita Y (1993) Inferior temporal cortex: Where visual perception meets memory. *Annual Review of Neuroscience* **16**: 245–263.

Miyashita Y (1995) How the brain creates imagery: projection to primary visual cortex. *Science* **268**: 1719–1720.

Müller MM, Bosch J, Elbert T, Kreiter A, Valdes Sosa M, Valdes Sosa P, Rockstroh B (1996) Visually induced gamma band responses in human EEG - A link to animal studies. *Experimental Brain Research* **112**: 96–112.

Namba S, Hamaguchi C, Ando T (eds.) (1992) *Science and Technology of Mesoscopic Structures*. New York NY: Springer-Verlag.

Nicoll RA (1971) Recurrent excitation of secondary olfactory neurons: A possible mechanism for signal amplification. *Science* **171**: 824–825.

Nunez PL (1981) *Electric Fields of the Brain: The Neurophysics of EEG*. New York NY: Oxford University Press.

Nunez PL (ed.) (1995) *Neocortical Dynamics and Human EEG Rhythms*. New York NY: Oxford University Press.

Obata K (1997) Excitatory and trophic action of GABA and related substances in newborn mice and organotypic cerebellar culture. *Developmental Neuroscience* **19**: 117–119.

Ohl FW, Scheich H, Freeman WJ (1998) Topography of averaged auditory evoked potentials (AEPs) in gerbil auditory cortex. *Proceedings, 26th Göttingen Neurobiology Conference II*, 332.

Pecora LM (1998) Synchronization conditions and desynchronizing patterns in coupled limit-cycle and chaotic systems. *Physical Reviews E* **58**: 347–360.

Pecora LM, Carroll TL, Johnson GA, Mar DJ *et al.* (1997) Fundamentals of synchronization in chaotic systems, concepts, and applications. *CHAOS* **7**: 520–543.

Pedroarena C, Llinás R (1997) Dendritic calcium conductances generate high-frequency oscillation in thalamocortical neurons. *Proceedings, National Academy of Sciences USA* **94**: 724–728.

Penn AA, Shatz CJ (1999) Brain waves and brain wiring: The role of endogenous and sensory-driven neural activity in development. *Pediatric Research* **45**:447–458.

Penrose R (1994) *Shadows of the Mind*. Oxford UK: Oxford University Press.

Pettigrew JD, Miller SM (1998) A 'sticky' interhemispheric switch in bipolar disorder? *Proceedings, Royal Society London B* **265**: 2141–2148.

Pfurtscheller G, Lopes da Silva FH (eds.) (1988) *Functional Brain Imaging*. Lewiston NY: Hans Huber Publishers.

Pribram K (1971) *Languages of the Brain; Experimental Paradoxes and Principles in Neuropsychology*. Englewood Cliffs NJ: Prentice Hall.

Price JL, Powell TPS (1970) The mitral and short-axon cells of the olfactory bulb. *Journal of Cell Science* **7**: 631–651.

Prigogine I (1980) *From Being to Becoming: Time and Complexity in the Physical Sciences*. San Francisco: W H Freeman.

Rabin MD, Cain WS (1984) Odor recognition: familiarity, identifiability and coding consistency. *Journal of Experimental Psychology: Learning, Memory and Cognition* **10**: 316–325.

Rall W (1962) Electrophysiology of a dendritic neuron model. *Biophysics Journal* **2**: 145–167.

Rall W (1995) *The Theoretical Foundation of Dendritic Function: Selected Papers of Wilfrid Rall with Commentaries*. Segev I, Rinzel J, Shepherd GM (eds.). Cambridge MA: MIT Press.

Rall W, Shepherd GM (1968) Theoretical reconstruction of field potentials and dendrodendritic synaptic interactions in olfactory bulb. *Journal of Neurophysiology* **31**: 884–915.

Ramón y Cajal S (1955) *Studies on the Cerebral Cortex (Limbic Structures)*. Kraft LM (translator). Chicago IL: Yearbook Publishers.

Reese TS, Brightman MW (1965) Electron microscopic studies on the rat olfactory bulb. *Anatomical Record* **151**: 492.

Ressler KJ, Sullivan SL and Buck LB (1994) Information coding in the olfactory system: Evidence for a stereotyped and highly organized epitope map in the olfactory bulb. *Cell* **79**: 1245–1255.

Rhoades BK (1991) Excitatory actions of GABA in the olfactory bulb. *PhD Thesis in Physiology*, University of California at Berkeley.

Rhoades BK, Cook PB, Freeman WJ (1988) Histamine effects on EEG and evoked potentials in the rat olfactory bulb. *Neuroscience Society Abstracts* **14**: 474.2 p 1187.

Rhoades BK, Leo EW, Freeman WJ (1990) Modulation of olfactory seizures by valproate and pentylenetetrazol. *(Abstract) American Chemosensory Society (AChemS)* 263.

Robinson PA, Rennie CJ, Wright JJ, Bourke PD (1998) Steady states and global dynamics of electrical activity in the cerebral cortex. *Physical Review E* **58**: 3557–3571.

Rodriguez E, George N, Lachaux J-P, Martinerie J, Renault B, Varela F (1999) Perception's shadow: long-distance synchronization of human brain activity. *Nature* **397**: 430–433.

Roelfsema PR, Engel AK, König P, Singer W (1997) Visuomotor integration is associated with zero time-lag synchronization among cortical areas. *Nature* **385**: 157–161.

Roland PE (1993) *Brain Activation*. New York NY: Wiley-Liss.

Rolls ET, Tovée MJ (1995) Sparseness of the neuronal representation of stimuli in the primate temporal visual cortex. *Journal of Neurophysiology* **73**: 713–726.

Rosenblatt F (1962) *Principles of Neurodynamics: Perceptrons and the Theory of Brain Mechanisms*. Washington DC: Spartan Books.

Rumelhart DE and McClelland JL (eds.) (1990) *Parallel Distributed Processing (Volume 1. Foundations)*. Cambridge MA: MIT Press.

Schiff SJ, Jerger K, Duong DH, Chang T, Spano ML, Ditto WL (1994) Controlling chaos in the brain. *Nature* **230**: 615–620.

Schillen TB, König P (1994) Binding by temporal structure in multiple feature domains on an oscillatory neuronal network. *Biological Cybernetics* **70**: 397–405.

Schippers B (1990) Spatial patterns of high-frequency visual cortical activity during conditioned reflex in man. *Masters Thesis*, Laboratory for Medical Physics, University of Amsterdam, Netherlands.

Schuster HG, Stemmler MB (1997) Control of chaos by oscillating feedback. *Physical Review E* **56**: 6410–6417.

Schuster H-G, Wagner P (1990) A model for neuronal oscillations in the visual cortex. *Biological Cybernetics* **64**: 77–85.

Serafetinides EA (1993) Is there a cerebral dominance for consciousness? *Perspectives in Biology and Medicine* **36**: 420–428.

Seyal M (1976) A neuropharmacological study of evoked potentials in the olfactory bulb. *PhD Thesis in Physiology*, University of California at Berkeley.

Sheer DE (1989) Sensory and cognitive 40-Hz event-related potentials: Behavioral correlates, brain function, and clinical application. In *Brain Dynamics* (Basar E, and Bullock TH eds.). Berlin: Springer-Verlag.

Sherrington CS (1929) Some functional problems attaching to convergence. *Proceedings of the Royal Society (London)* **105B**: 332–362.

Sherrington CS (1941) *Man on his Nature*. New York: Macmillan.

Shibata T, Kaneko K (1998) Collective chaos. *Physical Review Letters* **81**: 4116–4119.

Shimoide K, Freeman WJ (1995) Dynamic neural network derived from the olfactory system. *IEICE Transaction Fundamentals* **E-78A**: 869–884.

Sholl DA (1956) *The Organization of the Cerebral Cortex*. London: Methuen.

Siapas AG, Wilson MA (1998) Coordinated interactions between hippocampal ripples and cortical spindles during slow-wave sleep. *Neuron* **21**: 1123–1128.

Sik A, Ylinen A, Penttonen M, Buzsáki G (1994) Inhibitory CA1-CA3-hilar region feedback in the hippocampus. *Science* **265**: 1722–1724.

Siklós L, Rickmann M, Joó F, Freeman WJ, Wolff JR (1995) Chloride is preferentially accumulated in a subpopulation of dendrites and periglomerular cells of the main olfactory bulb in adult rats. *Neuroscience* **64**: 165–172.

Silva LR, Amitai Y, Connors BW (1991) Intrinsic oscillations of neocortex generated by layer 5 pyramidal neurons. *Science* **251**: 432–435.

Singer W, Gray CM (1995) Visual feature integration and the temporal correlation hypothesis. *Annual Review of Neuroscience* **18**: 555–586.

Skarda CA, Freeman WJ (1987) How brains make chaos in order to make sense of the world. *Brain and Behavioral Science* **10**: 161–195.

Sperry RW (1950) Neural basis of the spontaneous optokinetic response. *Journal of Comparative Physiology* **43**: 482–489.

Tallon-Baudry C, Bertrand O, Peronnet F, Pernier J (1998) Induced gamma-band activity during the delay of a visual short-term memory task in humans. *Journal of Neuroscience* **18**: 4244–4254.

Tang RX, Sugita K, Freeman WJ, Jakubith S (1995) Using fictive powers as event markers in EEG data processing. Chapter 53 in: Bower JM (ed.) *The Neurobiology of Computation*. Boston MA: Kluwer Academic, pp. 324–328.

Taylor JG (1997) The emergence of mind. *Communication and Cognition* **39**: 301–344.

Thomas C (1977) *Firefox*. New York NY: Holt, Rinehart, Winston.

Tononi G, Edelman GM (1998) Consciousness and complexity. *Science* **282**: 1846–1851.

Tovée MJ, Rolls EJ (1992) The functional nature of neuronal oscillations. *Trends in Neurosciences* **15**: 387.

Traub RD, Miles R (1991) *Neuronal Networks of the Hippocampus*. Cambridge UK: Cambridge University Press.

Traub RD, Whittington MA, Stanford IM, Jefferys JGR (1996) A mechanism for generation of long-range synchronous fast oscillations in the cortex. *Nature* **383**: 421–424.

Uhl C, Friedrich R (1999) Spatio-temporal modeling based on dynamical systems theory, In: *Analysis of Neurophysiological Brain Functioning*. Uhl C (ed.) Berlin: Springer-Verlag. pp. 274–305.

Usher M, Schuster H-G, Niebur E (1993) Dynamics of populations of integrate-and-fire neurons, partial synchronization and memory. *Neural Computation* **5**: 570–586.

Vaadia E, Aertsen A, Nelken I (1995) Dynamics of neuronal interactions cannot be explained by neuronal transients. *Proceedings, Royal Society London B* **261**: 407–410.

Vaadia E, Haalman I, Abeles M, Bergman H *et al.* (1995) Dynamics of neuronal interactions in monkey cortex in relation to behavioral events. *Nature* **373**: 515–518.

Viana Di Prisco G, Freeman WJ (1985) Odor-related bulbar EEG spatial pattern analysis during appetitive conditioning in rabbits. *Behavioral Neuroscience* **99**: 962–978.

Villringer UD, Dirnagl U (eds.) (1997) *Optical Imaging of Brain Function and Metabolism 2: Physiological Basis and Comparison to Other Functional Neuroimaging Methods*. New York NY: Plenum Press.

von der Malsburg C (1983) How are nervous structures organized? In Basar E, Flohr H, Haken H, Mandell AJ (eds.) *Synergetics of the Brain* (pp. 238–249). Berlin: Springer-Verlag.

von Holst E, Mittelstädt H (1950) Das Reafferenzprinzip (Wechselwirkung zwischen Zentralnervensystem und Peripherie). *Naturwissenschaften* **37**: 464–476.

Vonkrosigk M, Bal T, McCormick DA (1993) Cellular mechanisms of a synchronized oscillation in the thalamus. *Science* **261**: 361–364.

von Stein A, Rappelsberger P, Sarnthein J, Petsche H (1999) Synchronization between temporal and parietal cortex during multimodal object processing in man. *Cerebral Cortex* **9**: 137–150.

Wall PD. (1962) The origin of a spinal-cord slow potential. *Journal of Physiology* **164**: 508–526.

Walter WG (1953) *The Living Brain*. New York NY: WW Norton.

Wang F; Nemes A, Mendelsohn M, Axel R (1998) Odorant receptors govern the formation of a precise topographic map (vertebrate olfactory system). *Cell* **93**: 47–60.

Wijdicks EFM, Miller GM (1999) Transient locked-in syndrome after uncal herniation. *Neurology* **52**:1296–1297.

Willey TJ (1973) The ultrastructure of the cat olfactory bulb. *Journal of Comparative Neurology* **152**: 211–232.

Wilson HR, Cowan JD (1973) Excitatory and inhibitory interactions in localized populations of model neurons. *Biophysics Journal* **12**: 1–24.

Wilson MA, McNaughton BL (1993) Dynamics of the hippocampal ensemble code for space. *Science* **261**: 1055–1058.

Wright JJ, Liley DTJ (1994) A millimetric-scale simulation of electrocortical wave dynamics based on anatomical estimates of cortical synaptic density. *Network-Computation in Neural Systems* **5**: 191–202.

Wright JJ, Liley DTJ (1996) Dynamics of the brain at global and microscopic scales: Neural networks and the EEG. *Behavioral and Brain Sciences* **19**: 285–295.

Wright JJ, Sergejew AA, Liley DTJ (1994) Computer simulation of electrocortical activity at millimeter scale. *Electroencephalography and clinical Neurophysiology* **90**: 365–375.

Yao Y, Freeman WJ (1990) Models of biological pattern recognition with spatially chaotic dynamics. *Neural Networks* **3**: 153–170.

Yao Y, Freeman WJ, Burke B, Yang Q (1991) Pattern recognition by a distributed neural network: an industrial application. *Neural Networks* **4**: 103–121.

Zak M (1991) Terminal chaos for information processing in neurodynamics. *Biological Cybernetics* **64**: 343–351.

Zhang J (1991) Dynamics and formation of self-organizing maps. *Neural Computation* **3**: 54–66.

Author Index

Subject Index